BASIC ENVIRONMENTAL TOXICOLOGY

Edited by
Lorris G. Cockerham
Barbara S. Shane

CRC Press
Boca Raton   Ann Arbor   London   Tokyo

Library of Congress Cataloging-in-Publication Data

Basic environmental toxicology / editors, Lorris G. Cockerham, Barbara S. Shane
  p. cm.
Includes bibliographical references and index.
ISBN 0-8493-8851-1
1. Pollution—Environmental aspects. I. Cockerham, L. G. II. Shane, Barbara S.
QH545.A1B385 1994
574.5′222—dc20
DNLM/DLC
for Library of Congress   93-7790
                                                 CIP

Developed by Telford Press

Editors

Lorris G. Cockerham, Ph.D., is President of Phenix Services Corporation in Little Rock, Arkansas.

Dr. Cockerham received his B.A. degree in Biology and Chemistry from Louisiana College, Pineville, Louisiana, in 1957 and his M.S. and Ph.D. in Physiology and Toxicology from Colorado State University, Fort Collins, Colorado, in 1973 and 1979, respectively. He served as Assistant Professor at the United States Air Force Academy (Colorado), Adjunct Assistant Professor at the Uniformed Services School of Health Sciences (Bethesda, MD), and Associate Professor at the School of Medicine and College of Pharmacy, University of Arkansas for Medical Sciences (Little Rock, AR). He also was a Guest Lecturer in Neurophysiology at Georgetown University School of Medicine (Washington, DC) and an Instructor in the Medical Effects of Nuclear Weapons Course, Armed Forces Radiobiology Research Institute (Bethesda, MD). As Senior Scientist, Biochemical Services, Inc. (North Little Rock, AR), he served as Principal Investigator and Senior Science Writer on a contract with the National Institute of Environmental Health Sciences, National Toxicology Program for the preparation of Technical Reports and Toxicology Reports.

Dr. Cockerham, a retired Air Force Lt. Colonel, served as Chief of the General Physiology Division, Physiology Department, Armed Forces Radiobiology Research Institute (Bethesda, MD) and as Program Manager, Advanced Bioenvironmental Research, Air Force Office of Scientific Research, Bolling AFB (Washington, DC). His military awards and decorations include 2 Distinguished Flying Crosses, the Airman's Medal, Defense Meritorious Service Medal, Air Force Meritorious Service Medal, 12 Air Medals, and 9 Vietnam Service Medals.

Dr. Cockerham is a member of the American Association for the Advancement of Science, American College of Toxicology, American Physiological Society, Council of Biology Editors, International Brain Research Organization, Japan Radiation Research Society, Radiation Research Society, Society for Neuroscience, Society of Environmental Toxicology and Chemistry, Society of Toxicology, and World Federation of Neuroscientists. Honors include membership in the honor societies of Alpha Epsilon Delta, Phi Kappa Phi and Sigma Xi, and the International Platform Association. He is listed in *American Men and Women of Science, Who's Who in Frontiers of Science and Technology, Who's Who in the East, Who's Who in the South and Southwest*, and *Who's Who in the World*. In 1989 he was selected as a Distinguished Alumnus of Louisiana College.

Dr. Cockerham is the author of more than 50 national and international publications and has been the author or co-author of 7 books. His current major research interests relate to the effects of radiation and selected chemical agents as perinatal insults to the developing nervous system.

Barbara S. Shane, Ph.D., is an Associate Professor of Toxicology at the Institute for Environmental Studies at Louisiana State University, Baton Rouge. Dr. Shane graduated in 1963 with a B.Sc. (cum laude) in Biochemistry from the University of the Witswatersrand, Johannesburg, Republic of South Africa, and in 1967, she received a Ph.D. in Biochemistry from the same institution. In 1990, Dr. Shane became a Diplomate of the American Board of Toxicology. Dr. Shane is a member of the Society of Toxicology, the Society of Environmental Toxicology and Chemistry, the Environmental Mutagen Society, and the American Chemical Society.

Dr. Shane has published more than 40 peer-reviewed articles on various toxicological subjects. Her major emphasis has been on the mutagenicity of air pollutants including nitroarenes, food constituents, and components of waste materials such as incinerator ashes. She has also published articles on the toxic effects of pollutants in aquatic ecosystems. More recently, she has turned her attention to the application of *in vivo* assays, such as transgenic animals, to assess the risks posed by environmental pollutants to man.

Preface

This volume grew originally from the need for a comprehensive textbook for introductory courses in environmental toxicology. It provides a thorough, systematic introduction to environmental toxicology and addresses many of the effects of pollutants on man, animals, and the environment. The first section (Chapters 1 through 3) of this volume introduces the reader to toxicology and ecotoxicology. The second section (Chapters 4 through 10) presents the effects of different types of toxicants. The third section (Chapters 11 through 15) discusses the effects of toxicants in different compartments of the environment. The final section (Chapters 16 through 22) addresses environmental health, occupational health, detection of pollutants, and risk assessment.

Although this text provides a comprehensive, systematic introduction to environmental toxicology and every effort has been made to keep the book suitable for use as a textbook in a beginning graduate course, it should also be of interest to those not directly involved in academia. Individuals with risk assessment or risk management responsibility either in the human health area or in the area of environmental contamination will find this volume of particular interest. This group includes state and local public health officials, environmental engineers, representatives of industrial management, and representatives of independent consulting and testing laboratories.

Basic Environmental Toxicology will be useful as a text for introductory courses in environmental toxicology, a review for the practicing toxicologist, and a reference for those interested in the state of our environment.

Lorris G. Cockerham
Barbara S. Shane

Contributors

Jason S. Albertson, M.S., Ph.D..
Toxicology Resident
Comparative Toxicology
  Laboratories
Kansas State University
Manhattan, Kansas

Cathy Anderson, M.A.
Research Assistant
Soil & Crop Science Department
Texas A & M University
College Station, Texas

Gary C. Barbee, M.S.
Research Associate
Soil & Crop Science Department
Texas A & M University
College Station, Texas

Janice E. Chambers, Ph.D.
Professor
College of Veterinary Medicine
Mississippi State University
Mississippi State, Mississippi

Louis W. Chang, Ph.D.
Professor
Departments of Pathology,
  Pharmacology, and Toxicology
University of Arkansas for
  Medical Sciences
Little Rock, Arkansas

Lorris G. Cockerham, Ph.D.
President
Phenix Consulting and Services
  Ltd. Co.
Little Rock, Arkansas

**Michael B. Cockerham, R.PH.,
M.S.**
Clinical Pharmacy Specialist
Department of Pharmacy
Overton Brooks Medical Center
Shreveport, Louisiana

K. C. Donnelly, Ph.D.
Assistant Professor
Soil & Crop Science Department
Texas A & M University
College Station, Texas

Donald E. Gardner, M.S., Ph.D.
Vice President/Chief Scientist
Department of Toxicology
ManTech Environmental
  Technology, Inc.
Research Triangle Park,
  North Carolina

Susan C. M. Gardner, M.S.
Research Assistant
School of Fisheries
University of Washington
Seattle, Washington

Camille J. George, M.D.
Research Fellow
Department of Pharmacology
Tulane University School of
  Medicine
New Orleans, Louisiana

William J. George, Ph.D.
Professor, Director of Toxicology
Department of Pharmacology
Tulane University School of
  Medicine
New Orleans, Louisiana

Mark S. Goodrich, B.S.
Manager
Aquatic Toxicology Laboratories
Woodward Clyde Consultants
Franklin, Tennessee

Russell J. Hall, Ph.D.
Senior Staff Biologist
Department of Research and
  Development
U.S. Fish and Wildlife Service
Washington, D.C.

Larry G. Hansen, Ph.D.
Professor
Department of Veterinary
  Biosciences and Institute for
  Environmental Studies
University of Illinois
Urbana, Illinois

Ing K. Ho, Ph.D.
Professor and Chairman
Department of Pharmacology
  and Toxicology
University of Mississippi
  Medical Center
Jackson, Mississippi

Arthur S. Hume, M.S., Ph.D.
Professor
Department of Pharmacology
  and Toxicology
University of Mississippi
  Medical Center
Jackson, Mississippi

Kevin M. Kleinow, D.V.M., Ph.D.
Associate Professor
Department of Pharmacology
  and Toxicology
School of Veterinary Medicine
Louisiana State University
Baton Rouge, Louisiana

Donat J. Manek, B.S.
Chemist
Nukem Development
Houston, Texas

Frederick W. Oehme
Professor and Director
Comparative Toxicology
  Laboratories
Kansas State University
Manhattan, Kansas

Edward B. Overton, P.hD.
Director
Institute of Environmental
  Studies
Louisiana State University
Baton Rouge, Louisiana

Norbert P. Page, M.S., D.V.M.
President
Page Associates
Gaithersburg, Maryland

Christine A. Purser, B.S.
GC/MS Specialist
Department of Pharmacology
  and Toxicology
University of Mississippi
  Medical Center
Jackson, Mississippi

Jerry D. Rench, M.S., Ph.D.
Director
Environmental Health Sciences
SRA Technologies
Alexandria, Virginia

Paulene Roberts, B.S.
Research Associate
Institute of Environmental
  Studies
Louisiana State University
Baton Rouge, Louisiana

Donald J. Rodier, M.S.
Biologist
Environmental Effects Branch
Chemical Screening and Risk
  Assessment Division
Office of Pollution Prevention
  and Toxics
U.S. Environmental Protection
  Agency
Washington, DC

David J. Schaeffer, Ph.D.
Associate Professor
Department of Veterinary
  Sciences
University of Illinois
Urbana, Illinois

Barbara S. Shane, Ph.D.
Associate Professor
Institute of Environmental Studies
Louisiana State University
Baton Rouge, Louisiana

Waynette D. Sharp, B.A., B.S.
Technical Writer/Editor
Biotechnical Services, Inc.
North Little Rock, Arkansas

Gregory J. Smith, Ph.D.
Director of Laboratory Programs
Wildlife International, Ltd.
Easton, Maryland

Judith S. Weis, Ph.D.
Professor
Department of Biological Sciences
Rutgers University
Newark, New Jersey

Peddrick Weis, D.D.S.
Professor
Department of Anatomy,
  Cell Biology, and Injury Sciences
New Jersey Medical School
University of Medicine and
  Dentistry of New Jersey
Newark, New Jersey

Maurice G. Zeeman, Ph.D.
Chief
Environmental Effects Branch
Office of Pollution Prevention and
  Toxics
U.S. Environmental Protection
  Agency
Washington, DC

Contents

Section I

Chapter

1

Introduction to Ecotoxicology

Barbara S. Shane

Ecology has been described as scientific natural history, or as the scientific study of the interactions between abiotic factors and biota that determine the distribution and abundance of organisms (Moriarity, 1985). Linnaeus first introduced the classification of living organisms into species and laid the foundation for the system used today in which each organism is given a generic and specific name. The environment has been defined as all the inanimate compartments such as air, soil, and water and all the animate components such as plants and animals that surround a particular individual organism or population of organisms.

Toxicology is the study of the adverse effects of chemicals and physical agents on living organisms. The term ecotoxicology which was first coined by Truhaut in 1969 has come to mean the study of the fate and effect of a toxic compound on an ecosystem (Moriarity, 1985). In some cases the major difference between classical toxicology and ecotoxicology is the species selected for the toxicological tests; in ecotoxicology acute toxicity is measured on the water flea (*Daphnia* spp.), while in classical toxicology the test animal is the laboratory rat. However, ecotoxicology attempts to relate the effect of toxic compounds on populations and communities in a particular ecosystem rather than to the effects on a single organism, as is done in classical toxicology. This approach has turned out to be extremely difficult, as the effect of a toxicant on one species in an ecosystem may be different from that on another species in the same ecosystem and may differ again from the effect of the same chemical on the same species in another ecosystem. To complicate the issue, not only

0-8493-8851-1/94/$0.00+$.50
© 1994 by CRC Press, Inc.

3

can the toxicant directly affect the organism being evaluated, but the toxicant can adversely affect both biotic and abiotic parameters such as the prey or source of food of the species being studied, thus having an indirect effect on the survival of the organism. As a result of the complex simultaneous interaction between the toxicant in question on many species and the variation in the outcome of these effects on different species, many ecotoxicological studies have resorted to studying the effect of a toxicant on only one species at a time. In a broader sense, the focus of ecotoxicology has been directed not only to the possible effects of pollutants in the environment on nonmammalian species, but also to mammals and man. An appreciation of the important position of man in the scheme of the biosphere is beginning to be acknowledged by scientists and laymen alike. Thus, in ecotoxicology, studies to evaluate the effect of pollutants on man are often included. This overlap sometimes makes it difficult to separate ecotoxicology from classical toxicology.

Classical toxicology is based on theories which were originally developed by pharmacologists who studied the effects of drugs on man. Ecotoxicology is, in many cases, following the path taken originally by classical toxicology in that several acute toxicity tests are being devised and used to understand the comparative toxicity of a particular compound on many species. The question that has been asked is, how useful are these acute toxicity tests? Many ecotoxicologists suggest that these tests are useful from the standpoint that environmental parameters can be altered in the laboratory to simulate those in a particular environment, and then the subsequent acute toxicity measurements can be made. A related concern is whether these acute tests are informative about the effect of a particular compound on populations and communities.

In evaluating the toxicity of a compound on man using classical toxicological approaches, surrogate species such as rodents are chosen. In many respects surrogate species are often physiologically different from the species on which data is actually required. Therefore, there is always some question about their validity or the applicability of extrapolating from one species to another. Even though a mammalian toxicity test may be designed to determine the effects of a compound on more than one species, extrapolations are necessary for only a small number of species. In contrast, in ecotoxicology where only a few species are approved of for use as testing organisms, extrapolations are made from these few species to an enormous number of species spanning a wide range of phyla ranging from relatively simple groups of animals such as planaria to highly complex organisms such as fish. Variations in size, physiology, life history, and geographic distribution are a few of the many important parameters that are not considered. Scientists must also be cognizant that extrapolation from data in the laboratory to that in the field must be carefully evaluated. Considering all these stumbling blocks, some of the simplest toxicity tests have proven to be enormously helpful in our understanding of the effects of pollutants in the field.

For many years the prime criterion for selecting a specific organism for ecotoxicity testing was its high degree of sensitivity to chemical compounds. Emphasis is shifting to the use of those organisms that are the most reliable predictors of the responses of other organisms or even of communities or ecosystems. Also, many species that are extremely sensitive or have high correspondence with the responses of other organisms are exceedingly difficult to maintain in the laboratory. Thus, it is more reasonable to use test organisms that have been used widely for toxicity testing and whose strengths and weaknesses for this purpose are well known.

As many chemicals are persistent, their exposure times in natural systems are likely to be much longer than in normally used routine laboratory tests. In ecotoxicology, therefore, an organism such as *Ceriodaphnia* whose life stages are completed in a relatively short time is often used. To justify this choice, some scientists have suggested that the interval from the starting point of one generation to the starting point of the next is a more important time frame to determine the toxicological response than the actual period of exposure (Cairns and Mount, 1990). End points of tests can include lethality, reproductive impairment, alterations in growth and development, behavioral aberrations, or changes in biochemical indicators (biomarkers) that can serve as early warning indicators of adverse effects on reproduction and survival (Hoffman et al., 1990).

As ecosystems receiving anthropogenic wastes from both point and nonpoint sources are difficult and expensive to study in predictive models, it is obvious that toxicity tests for hazard evaluation will need to be done either in the laboratory or in artificial systems such as microcosms or mesocosms where abiotic conditions can be manipulated. The advantage of these systems is that they are more environmentally complex than the standard test systems and therefore more amenable to the simultaneous study of toxicity, chemical transformations, and partitioning processes. Microcosms can also be used to monitor responses that are of concern in the natural environment, and thus data obtained in these systems are more reliable for extrapolation to responses in natural systems.

The field of ecotoxicology is developing in two directions. In the first, studies to understand the effects of pollutants on nonmammalian species are being undertaken. In the second, extrapolation of the effects observed in one organism are being modeled so predictions of the effects in a second organism can be made. Ecotoxicology can be subdivided into several areas of study which are often related to the effects of pollutants in the three major abiotic compartments, air, water, and soil or sediments. The largest amount of data has been collected on water pollution, because more than 50% of the population in the U.S. obtains its drinking water from surface waters which can easily become contaminated both by industrial discharges (point sources) and runoff from fields (nonpoint sources). The Clean Water Act declares the minimal goal for national waters as "fishable and swimmable". These criteria refer to the number of microorganisms permitted in the water which will not harm human health. This act has also proposed concentration limits of chemicals for potable water. Hazard assessments for these chemicals are being conducted under EPA Risk Assessment Guidelines to translate chemical concentrations into acceptable ambient levels using conservative exposure models.

Less emphasis has been placed on pollutant discharges into the atmosphere although with the recent Clean Air Act passed by Congress more emphasis on these sources are likely to be made soon. The Clean Air Act is largely concerned with the effects of discharges on human health, but regulations on ecological effects are likely to follow. Unfortunately one industry that will not be regulated by this act is the electrical power industry which is the largest emitter of sulfur oxides, which are now known to have far reaching effects on the environment. The burning of coal (85% of all sulfur emitted), smelting of ores, and refining of petroleum (15%) in the past 100 years have doubled the rate at which sulfur oxides enter the atmosphere and hydrosphere (Moriarity, 1985). It is important that these emissions are distributed unevenly around the globe, with the major deposition in northwest and central Europe, northeastern U.S., and southeastern Canada. Most of this sulfur returns to the earth surface within 3000 km of the source, so the effects of pollution are regional.

A major thrust at present is the evaluation of chemicals found at hazardous waste sites and on human and ecological health. Sites are scored for their impact on health by the Hazard Ranking System (HRS). Some scientists feel that too much effort is being diverted to the study of pollutants at hazardous waste sites that have been assigned to the National Priority List which are targeted for rapid cleanup. Many professionals believe that natural systems are resilient and can transform toxic materials into nontoxic compounds or sequester some pollutants into specific ecological compartments such as sediments. For example, breakdown of the stable highly chlorinated PCBs has been documented in sediments from the highly contaminated Hudson River (Brown et al., 1987). However, the rate of this natural degradation has proven extremely slow, and the deleterious subtle effects that these compounds may be having on the environment and biota in the river is unknown. Our inability to measure this damage at the cellular level, except in costly studies, gives us a false sense of security. However, no matter how many toxicity tests are carried out, some uncertainty about the exact response of a complex system will always exist because all the variables cannot be evaluated.

Modern insecticides are extremely potent, and not only do they poison the target species but often persist in the environment and are accumulated in nontarget organisms, including fish and wildlife. The bioaccumulation potential of lipophilic insecticides tends to be inversely related to their rates of metabolism. Metals are also bioaccumulated by plants and animals. For example, certain aquatic plants bioconcentrate selenium 500 times from the surrounding water. Subsequent consumption of these plants by invertebrates and forage fish can result in additional biomagnification of 2 to 6 times the plant concentration (Hoffman et al., 1990). Further biomagnification of selenium up the food chain has been documented in aquatic birds in California. Fledglings of these birds have displayed teratogenicity manifested as bill defects, eye malformations, reduced mandibles, and foot deformities (Hoffman et al., 1988).

About 63,000 chemicals are in common use worldwide, with 3000 compounds accounting for almost 90% of the total production (Moriarity, 1985). About 200 to 1000 new synthetic chemicals are marketed each year. Few ecological assessments on the bulk of the materials on the market or those introduced each year have been undertaken. This need was perceived as very important, thus many countries have promulgated legislation that requires both toxicological and ecotoxicological testing of chemicals. Two major acts have been advanced in the U.S. to address the effects of pollutants on ecosystems and man. The first, the Toxic Substances Control Act (TSCA), was enacted in the U.S. in 1976. This act originally concentrated on the effects of toxic substances on man but has recently expanded to incorporate effects on ecosystems. The second major legislative act was the Federal Insecticide and Rodenticide Act (FIFRA) which is mainly concerned with the promulgation of laws on the sale and use of pesticides. FIFRA is a registration law that gives the EPA legal authority to require upfront testing. TSCA and FIFRA are similar in their approach to assessing ecological risk, but TSCA assessments collect minimal amounts of data while FIFRA assessments are much more comprehensive.

Analytical instruments cannot measure toxicity, only the concentration of a chemical in the environment or in an organism (Cairns and Mount, 1990). The importance of using living organisms to assess toxicity, rather than attempting to assess the toxicity of a compound from its chemical concentration in the environment or in an animal, is being realized. However, both chemical and physical measurements are still extremely important in assessing toxicity, because the concentration of the toxicant available to the test organisms must be known. The concentration of the toxicant actually available for uptake is dependent on environmental parameters. For example, it is known that conventional test organisms such as fish and *Daphnia* spp. are more sensitive to heavy metal contamination in soft water than in hard water containing higher concentrations of $CaCO_3$.

Several problems exist for the analyst who is attempting to determine the concentration of a pollutant in an environmental sample. Difficulties arise when we must distinguish between chemical species, ions, molecules, or complexes and the fraction in which the species is found such as the soluble fraction. Heavy metals in water exemplify this particularly well as they are commonly measured as the total concentration, but in fact occur in a variety of forms. With new advances such as inductively coupled argon spectroscopy/mass spectrometry (ICP/MS), some of the difficulties in interpretation of the concentration of metals in the environment may be overcome. The chemical form and species of a metal can greatly affect all aspects of the metal's behavior and its toxicological effects. For example, the different forms of selenium have different toxicological potencies, with selenite being more toxic to rats in drinking water than selenate. Adding to this complexity, particularly in an aquatic environment, is the presence of natural components such as humic acids and particulates in suspension to which metals can adsorb. For instance, humic acids in river water may form complexes with most of the dissolved species, thus limiting its availability to biota in the river.

The polychlorinated biphenyls (PCBs) have presented a major problem to both analysts and toxicologists. When the first analytical techniques became available for measuring the concentration of these compounds, the values obtained represented all the congeners of PCB in a particular formulation such as Aroclor 1254. In concert with the analytical techniques, the toxicity of the complex mixture was determined. Due to our inability to separate and detect by gas chromatography each of the 210 congeners, we were unaware of the differential toxicity of each congener. It is now known that each congener differs in the number and position of chlorine atoms on the biphenyl ring, as well as in its toxicity and persistence in animals.

Chemicals not toxic to humans and animals can alter abiotic factors in such a way that enormous repercussions could result either at the local level or worldwide. These deleterious effects could also be vital to man's survival. For example, from a worldwide perspective the increase in the concentration of both carbon dioxide and chlorofluorocarbons (CFCs) in the atmosphere may have far reaching consequences. Carbon dioxide had a mean natural concentration at the beginning of the century of about 0.028%. Since then its emission through the combustion of fossil fuels has been increasing by about 0.002% each year and had risen to 0.034% by 1984 (Waggoner, 1984). This continual increase in carbon dioxide is predicted to elevate global temperatures, which will affect the distribution of certain crops. Such an increase in temperature could enhance the melting of the ice caps which could raise the sea level, resulting in submersion of many low lying cities of the world. This increase in CO_2 will also enhance the survival of certain pests, resulting in decreased crop production and may modify the adaptation of C3 and C4 plants (Patterson and Flint, 1980). The net results will be radical alterations in the distribution of plant species in the world. CFCs, developed due to their stability and low toxicity, have contributed to the depletion of stratospheric ozone, which is increasing the risk of skin cancer and damage to vegetation. As a result of these findings and concern for the survival of the earth, an Ecotoxicity Subcommittee was formed by the EPA in 1987 to develop a framework for ecological assessments. This framework has been designed for "top-down" assessments based on field studies and "bottom-up" assessments based on laboratory studies and bioassays (Bascietto et al., 1990). Exposure assessments of both individual chemicals and whole effluents are possible; however, extensive knowledge of an ecosystem and the effects of the pollutant(s) will be needed before comprehensive and realistic assessments can be undertaken.

At the local level, other pollutants such as DDT and acid rain, which are both relatively harmless or only slightly toxic to mammals and man, have direct and indirect effects on an ecosystem. DDT jeopardizes the survival of eagles and other birds at high trophic levels by interfering with the normal function of estrogen, the hormone involved in the deposition of calcium in the egg shell. Acid deposition is thought to be the major cause of the destruction of populations of fish and other aquatic organisms in poorly buffered lakes, particularly in the northeastern U.S. and Scandinavia (Schindler, 1988), and might be contributing to the die back of forests (Johnson and Siccama, 1983). Fish die from failure of their salt-regulating mechanisms, although tests have shown that the pH of the water has to drop below 4.6 before a physiological effect is noted. Nevertheless, some sensitive fish species have disappeared from lakes with pH readings of 5.0. The increased toxicity may be due in a large part to the leaching of aluminum from certain soils by acid rain, although other physical and chemical factors may also be important (Schindler, 1988).

In attempting to determine the impact of pollutants in the environment, we must be cognizant of other changes that are continually occurring around us. For example, agriculture, forestry, recreation, control and supply of water, and other related activities all have significant and often major effects on pollutant concentration. Despite the large amount of data available on some pollutants, we often lack quantitative information on sources. Thus, changes caused by pollutants to abiotic and biotic compartments occur against a background of changes caused by other human activities. For example, an estimated 19 different stresses are acting simultaneously on Green Bay, on Lake Michigan. Besides the above mentioned factors, the discharge of many pollutants, dredging, landfill runoff, water levels, and manipulation of fish populations are a few of the other stressors to populations in this area (Harris et al., 1990).

This text has been compiled to address many of the above mentioned effects of pollutants on man, animals, and the environment, and has been divided into four sections. The first two chapters introduce the reader to toxicology and ecotoxicology, and Chapters 3 through 10 present the effects of different types of toxicants. The next four chapters discuss the effect of toxicants discharged into different compartments, and the final seven chapters address occupational health, detection of compounds, and risk assessment.

REFERENCES

Bascietto, J., Hinckley, D., Plafkin, J., and Slimak, M. Ecotoxicity and ecological risk assessment, *Environ. Sci. Technol.* 24:10–15, 1990.

Brown, J.F., Bredgard, D.L., Brennan, M.J., Carnahan, J.C., Feng, H., and Wagner, R.E. Polychlorinated biphenyl dechlorination in aquatic sediments, *Science* 236:709–712, 1987.

Cairns, J., Jr. and Mount, D.I. Aquatic toxicology, *Environ. Sci. Technol.* 24:154–161, 1990.

Harris, H.J., Sager, P.E., Regier, H.A., and Francis, G.R. Ecotoxicology and ecosystem integrity: the Great Lakes examined, *Environ. Sci. Technol.* 24:598–603, 1990.

Hoffman, D.J., Ohlendorf, H.M., and Aldrich, T.W. Selenium teratogenesis in natural populations of aquatic birds in Central California, *Arch. Environ. Contam. Toxicol.* 17:519–525, 1988.

Hoffman, D.J., Rattner, B.A., and Hall, R.J. Wildlife toxicology, *Environ. Sci. Technol.* 24:276–283, 1990.

Johnson, A.H. and Siccama, T.G. Acid deposition and forest decline, *Environ. Sci. Technol.* 17:294A–305A, 1983.

Moriarity, F. *Ecotoxicology. The Study of Pollutants in Ecosystems*, 3rd ed. Academic Press, San Diego, California, 1985.

Patterson, D.T. and Flint, E.P. Potential effects of global atmospheric CO_2 enrichment on the growth and competitiveness of C3 and C4 weed and crop plants, *Weed Science* 28:71–75, 1980.

Schindler, D.W. Effects of acid rain on freshwater ecosystems, *Science* 239:149–157, 1988.

Waggoner, P. E. Agriculture and carbon dioxide, *Am. Sci.* 72:179–184, 1984.

Chapter

2

Principles of Ecotoxicology

Barbara S. Shane

A. INTRODUCTION

A pollutant or toxicant is defined as a substance that occurs in the environment, at least in part as a result of an anthropogenic discharge from industrial activity, and has a deleterious effect on living organisms. A contaminant on the other hand is a compound released by man's activities that does not necessarily have a deleterious biological effect. Frequently, the term pollutant is used loosely to cover both definitions. A pollutant may kill more than 50% of the individuals of a species in a population but be of little or no ecological importance unless this loss has a direct effect on the survival of the ecosystem. Some pollutants, in the amounts discharged into the environment, may not have a direct effect on living organisms, but they may alter the physical and chemical environment, impairing the survival of a species. For example, runoff of inorganic nutrients, particularly phosphorus and nitrogen derived from sewage and agricultural fertilizers into water bodies, enhance the multiplication of algae, a process known as eutrophication, resulting in a massive requirement for oxygen by these rapidly growing plants. As a result, the oxygen concentration in the water body is decreased and many fish species may not survive.

To understand the effects of pollutants on organisms, one needs to understand how the compound is absorbed, how it is distributed by the blood stream (if the organism has one), its subsequent metabolism, and its excretion. During these four basic phases of the passage of a pollutant through an organism,

0-8493-8851-1/94/$0.00+$.50

known as pharmacodynamics, several metabolic processes in the animal or plant can be altered. The outcome of these processes can be exemplified by either a drastic response such as death or a less final outcome but one in which the physiological responses of the animal or plant are altered in some deleterious way. Death usually occurs when the organism is exposed for a short time to a high concentration of a compound. Exposure to lower concentrations over a longer period may result in a subtle biochemical or physiological effect which in itself does not appear to affect the health of the organism. However, extended periods of exposure could result in changes that can initiate the beginning of a disease or condition that could affect the survival of an individual or a population.

Initially, ecotoxicology concentrated on studying the acute effects of chemicals on living organisms, but with more sophisticated analytical and biochemical methods a major thrust has been the determination of the first molecular insult that occurs following chronic exposure. These include the measurement of changes in enzyme activity or mutations in the genetic material, which can be used as indicators or biomarkers of exposure. It has been suggested that these biomarkers could be used as early warning signs of subtle biochemical and pathological lesions which could affect the health of the organism. These biomarkers are analogous to clinical chemistry parameters measured by physicians to elucidate early pathological changes in a patient. Statistically significant changes in these measurements from normal values indicates that the patient is ill and must be treated. Unfortunately, the use of biomarkers for ecotoxicological effects has not progressed as far as it has in the mammalian system. Frequently, biomarkers measured in lower vertebrates and invertebrates are indicative of exposure rather than pathological changes in the organisms. In an attempt to determine the presence of biomarkers in fish following exposure, Varanasi et al. (1982) and Malins et al. (1987) undertook comprehensive studies of fish exposed to polycyclic aromatic compounds (PAHs) in the Dwamish Waterway in Seattle. These studies have been most informative and have greatly enhanced our understanding of the interrelationship between environmental contamination, concentration of pollutants in fish, biomarkers of exposure, and disease outcomes. These studies have shown that there is a relationship between the concentration of PAHs in the environment and the concentration of these compounds in fish living in a contaminated area. In contrast, fish inhabiting a pristine area had background levels of these compounds in their tissues. Alterations in the metabolism of exposed fish were exemplified by elevated levels of PAH metabolites in the bile (Krahn et al., 1986). A statistically significant number of fish from the contaminated environment had hepatic tumors compared to fish from the pristine area.

A major difference in evaluating the effect of a pollutant on a mammal in a laboratory setting and the effect of a pollutant on an organism in the environment is the exposure of the animal in the "real world". Physical and chemical alterations and movement of a compound in the environment can ultimately affect the dose to which an animal is exposed. To understand the movement of a chemical from one area to the next, each sphere of the world is considered as a compartment. Four compartments, air, water, soil or sediments, and biota or living organisms have been identified. The area of ecotoxicology that addresses the transport of a compound from one environmental compartment to the next is known as chemodynamics. This chapter will mainly focus on chemodynamics although some aspects of pharmacodynamics will be addressed. A more detailed discussion of pharmacodynamics can be found in detailed chapters in texts devoted to mammalian toxicology (Klaassen, 1986b; Renwick, 1989).

Ecotoxicology differs from classical toxicology in that it concentrates on the effects of pollutants on populations while classical toxicology focuses on the effects on individuals. The range of variables that affects populations is greater than the range that affects individuals, particularly as environmental conditions can influence the response of organisms to a pollutant. For example, it is known that *Daphnia magna*, the water flea, is more sensitive to heavy metals in soft water than in hard water containing higher concentrations of $CaCO_3$. In an environmental setting, sublethal effects are probably as important as lethal effects as sublethal effects can impair the structure of the ecosystem and possibly cause its demise. Due to outbreeding among organisms in nature and inbreeding among rats and mice in classical toxicology, there is a greater variation in the response of individuals in a population in an ecosystem than in animals in a laboratory setting. Similarly, it can also be assumed that different populations of one species in different environmental settings would respond differently to the same concentration of a toxicant. This difference is borne out by the response of different insects to an insecticide in different areas of the country.

B. ECOLOGICAL PRINCIPLES

To understand the effect of a pollutant on organisms in an ecosystem, an understanding of relevant ecological terms and concepts are important.

A *population* is defined as a group of organisms of the same species living together in a particular ecosystem and capable of breeding and producing offspring.

A *community* is defined as the interdependence of several populations living together in a localized area with the same physical and chemical properties such as in a field or stream.

A *food chain* is the relationship between different organisms that are dependent on each other for their food supply.

A *food web* is the interrelationship between several food chains within an ecosystem so each organism is not dependent on only one source of food but

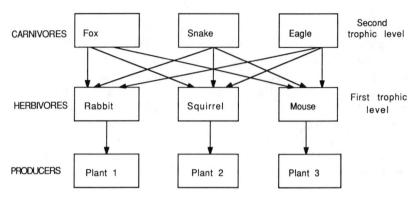

Figure 2.1. A simple food web showing the relationships with the trophic levels.

obtains its nutrients from three or four different sources of food. Those organisms which are dependent on a single food source can be in a precarious position regarding survival, if their food supply is eliminated by pollution.

A *trophic level* is the level at which an organism is placed in a food chain. The organisms in the lowest trophic level are primary consumers and depend on plants for their survival. Secondary consumers depend on primary consumers for their food and tertiary consumers depend on secondary consumers. In general, there are three to four trophic levels in each food chain (Figure 2.1).

C. PRINCIPLES OF TOXICOLOGY

1. Lethal and Sublethal Effects

Although one of the aims of ecotoxicology is to study the effect of pollutants on populations and subsequently communities, very few studies have fulfilled this aim. In most cases the effects of a pollutant on members of a species is usually studied. This information is required before more elaborate studies on complex communities can be initiated. Information on the effects of a pollutant on all the populations making up a community is the ideal situation, but in most cases it is impossible to achieve. Following exposure to a pollutant, some species in a community may be able to survive and maintain their numbers, but in other cases populations in a community may be adversely affected resulting in the demise of that species. Thus some species can adapt to adverse conditions while others cannot. With continuous exposure to the pollutant in question adaptation and tolerance may be affected, so that sublethal responses are manifest which could affect the long term survival of the species. Eventually the concentration of the pollutant may reach a high enough value to cause the death of the members of a population. Thus the effects of a pollutant on an organism can be divided into two phases, a lower concentration that results in a sublethal effect which may be detected or a lethal response in which the organism dies. Some sublethal effects may be visually undetectable but with physiological or biochemical means can be identified.

2. Toxicological Responses

The earliest changes in an organism following exposure to a pollutant occurs at the cellular level. Some of the more important effects are alteration in the structural components of the cell membrane, inhibition of certain enzymes such as microsomal enzymes, interference in protein, lipid, or carbohydrate biosynthesis or metabolism, and alterations in DNA fidelity resulting in mutations and interference with the regulation of cell growth. Most proteins have either structural or enzymatic functions and are required for cellular processes including catabolism, anabolism, and reproduction. The mechanism of the initial insult at the molecular level is known is some cases. Alterations in the normal functions of the cell membrane can occur as a result of alterations in the synthesis of phospholipid or glycoprotein components. Carbon tetrachloride can alter the integrity of cell membranes through the formation of chloromethyl free radicals which cause the peroxidation of unsaturated lipids. Mercury alters the conformation of proteins by interactin with sulfhydryl (–SH) groups of cysteine residues resulting in the formation of sulfhydryl bridges between two SH groups. Toxic compounds can also interfere with the synthesis of specific macromolecules. For example, lead interferes with the synthesis of hemoglobin by inhibiting two enzymes in the heme biosynthesis pathway. Toxic compounds may bind covalently with DNA, thus rendering it inactive or impairing its normal function. For example, many PAHs form epoxides (benzo(a)pyrene) or nitrenium ions (nitropolycyclic aromatic hydrocarbons such as 1,6-dinitropyrene) which bind to DNA molecules, thus altering its fidelity and resulting in a base substitution or frameshift mutation.

Certain toxicants are also known to interfere with metabolic pathways crucial to the survival of the plant or animal. These effects may occur as a result of competitive or noncompetitive inhibition. Fluoroacetate, a compound found in a South African plant, has been used as rodent bait. This compound replaces acetate in the condensation reaction between acetate and oxaloacetate, and forms fluorocitrate. This compound blocks aconitase in the Krebs cycle, thus inhibiting the catabolism of acetic acid derived from glucose and the subsequent generation of energy.

3. Interaction Between Compounds

In most ecosystems animals and plants are exposed to many pollutants simultaneously rather than to a single pollutant. Interaction between these pollutants can enhance or decrease the toxicity of the mixture. The mode of interaction between different pollutants can be due either to the chemical structure of the molecules or to alterations in the physiological processes within the organism. Thus either the uptake, distribution, metabolism, storage, or excretion of the compound could be altered. Also, alteration in the binding affinity of a toxicant for a receptor can be manifest. Exposure to two or more compounds can result in an additive, inhibitory (decreased), or synergistic (increased) response. An additive response is often observed when two organophosphate compounds are given simultaneously. An increased or synergistic response occurs when the combined effect of two compounds is greater than the sum of the individual compounds. Such a response has been observed in mammals when the hepatotoxins, ethanol, and carbon tetrachloride are administered together to rats. Potentiation occurs when one of the two compounds to which the organism is exposed is nontoxic alone but when given with the second compound enhances its toxic response. Isopropanol alone is nontoxic, but when administered with carbon tetrachloride it enhances the latter's toxicity. Antagonism occurs when two compounds are administered together or sequentially and one interferes with the action of the second.

An antagonistic interaction between two pollutants can occur when the two compounds interact at the same site in the target tissue. Five different types of antagonistic reactions have been described. An example of *chemical antagonism* occurs when an antagonist inactivates an agonist through a chemical interaction. For example, selenium is known to bind with mercury thus preventing its binding to –SH groups on protein molecules. *Competitive antagonism* occurs when an antagonist displaces an agonist from its site of action. This situation occurs when oxygen is administered to a patient intoxicated by carbon monoxide resulting in the displacement of the toxic carbon monoxide from the heme moiety of hemoglobin. In *noncompetitive antagonism* the antagonist interferes with the deleterious effect that is elicited by the agonist by binding to specific receptors of the agonist but not with the agonist itself. This effect can be demonstrated by atropine which binds with the specific receptors of acetylcholine in nerve cells, thus preventing the deleterious effects of acetylcholine that occurs when this neurotransmitter accumulates following inactivation of acetylcholinesterase by organophosphate insecticides. *Functional antagonism* occurs when two chemicals counterbalance each other by producing opposite effects. In barbiturate poisoning the blood pressure drop that is frequently encountered can be ameliorated by the administration of norepinephrine, a vasopressor agent. *Dispositional antagonism* occurs when the pharmacodynamics of a compound is altered so less toxic compounds reach the target organ. This situation is manifest by including piperonyl butoxide in organophosphate insecticide preparations. Piperonyl butoxide inhibits the formation of the active intermediate of the organophosphate by inhibiting the activity of cytochrome P_{450}, an enzyme required for the metabolism of the organophosphate to its active form.

Interaction between a pollutant and a natural chemical in the environment can result in the formation of new molecules or complexes, thus preventing the expected uptake of a compound by an organism. For example, in an aquatic environment metals are absorbed by organisms if they are in solution. However, addition of iron and manganese oxides coprecipitate cobalt, zinc, and copper. In the presence of these oxides, the absorption of soluble cobalt, zinc, or copper could thus be inhibited.

4. Effects of Exposure

Characteristics

A toxic effect is elicited when a pollutant reaches the target organ in an organism. Determination of the concentration of a compound in the target organ is frequently difficult and usually impossible unless the organism is sacrificed. To overcome this problem, the concentration of a compound in the surrounding medium can be used as a measure of the internal concentration. Another approach has been the development of computer models to predict the internal concentration in the target tissue based on the external concentration and pharmacodynamic principles of the compound.

A third approach that is frequently used to measure concentration is to determine the time taken for a specific pollutant to cause an effect at a particular concentration. In this case it is necessary to know the type of effect expected, the relevant information about the chemistry of the compound and the likelihood that it will reach a particular target tissue. The route of administration and the frequency of exposure can have an effect on the outcome, particularly if the organism of interest is a vertebrate.

Routes of Exposure

In vertebrates the most rapid response to a pollutant will be elicited if the toxicant is administered intravenously followed sequentially by inhalation, injection via the interperitoneal (i.p.), subcutaneous (s.c.), or intramuscular (i.m.) routes, ingestion, or topical application. Certain compounds are metabolized to a less toxic intermediate if given orally rather than if given by inhalation. If a compound is absorbed rapidly, its lethal dose is similar whether it is administered intravenously, orally, or topically. In general the toxicity of a compound is lower if applied dermally compared to ingestion because the skin acts as an effective barrier.

Duration and Frequency of Exposure

Experimental animals and animals in the environment can be exposed to toxicants over four different time frames: acute, subacute, subchronic, and chronic exposure. An acute exposure is exposure to a high dose of a compound within a short period, usually 24 h. In the other three exposures, repeated exposure to the compound occurs. Subacute exposure refers to repeated exposure for a month or less to doses lower than is present during acute exposure. Subchronic exposure refers to many exposures over a period of 1 to 3 months. Chronic exposure refers to exposure for more than 3 months at doses about 1/100 to 1/1000th the acute dose. Many invertebrates do not live for 3 months and thus these definitions of intervals can be appropriately shortened. The toxic effect encountered can be different following single or multiple exposure. Frequently, when the dose is administered in fractions, the combined fractions elicit a lesser response than the total dose administered as an acute dose. This is due to the continual metabolism of the compound to less toxic intermediates which are then excreted, while with a larger dose the metabolic and/or excretory pathways of the organism may become saturated thus resulting in a more pronounced effect. Adverse effects from a chronic exposure are observed when absorption is greater than biotransformation and excretion. If the rate of excretion is less than the rate of absorption, then the compound accumulates. A steady state concentration will be reached when the rates of these two processes are the same.

5. Dose Response Relationship

In determining the relationship between dose and response, it is necessary to distinguish between the dose or environmental concentration and the amount of the chemical that reaches the target tissue. In some cases the concentration of a compound may be high in the environmental medium, particularly if it is water soluble, but if the compound is not absorbed and does not reach the target organ, no effect will be observed. It is also necessary to distinguish between response and effect. Dose response is usually the relationship between a measurable physical, chemical, or biological reaction following exposure to a certain concentration of a compound. As it is extremely difficult in most experiments to determine the dose reaching the target tissue, the environmental concentration or the amount of compound administered is used as the dose. The reaction to the dose, or the response that is elicited, can be quantitatively measured either by the magnitude of the response or the time taken for a specific response to be recorded.

In measuring the dose response, it is assumed that the response observed is due to administration of the chemical and thus there is a causal relationship. It is also assumed that the response is related to dose and thus there is a particular molecular site with which the chemical reacts, that the response is proportional to the concentration of the compound at the site of action, and that the concentration of the compound is related to the dose administered. Thus, the concentration at the site of action is a function of the dose.

There must be a quantifiable measure of toxicity, and ideally it should be related to a molecular event. An example of a known interaction between a pollutant and its target is the binding of 2,3,7,8-tetrachlorodibenzo-*p*-dioxin with the aryl hydrocarbon hydroxylase receptor in the cytoplasm of cells. The bound receptor is then translocated into the nucleus where it initiates gene transcription and synthesis of certain cytochrome P_{450} enzymes. However, the adverse effect resulting from the induction of this enzyme has not been proven.

The measurements described above cannot be used when the toxicity of a compound is unknown. In this case the first measure of toxicity used is a quantal response or an "all or none" response which is often measured as lethality. To obtain a better understanding of the cause of the toxicity in these experiments in which this quantal response is being determined, observation of the organisms is essential. It is also important that necropsies be performed and tissues taken for histological examination.

6. Determination of the LD$_{50}$ Following Acute Exposure

The parameter used to measure mortality is the LD_{50}, a statistically derived dose of a substance that causes the death of 50% of the organisms under study. This concept was originally proposed by Trevan (1927). A typical sigmoidal curve, depicted in Figure 2.2a, shows the increase in the response that is obtained as increasing amounts of a compound are administered to a group of animals. A smooth curve is only obtained when the number of doses that kill members of the group exceeds three. The lowest dose causing an "all or none" response is known as the threshold dose.

In classical toxicological studies, and also in the evaluation of the toxicity of environmental samples such as effluents from an industrial plant, 10 animals are assigned to a group, each of which is exposed to a particular concentration of the suspected toxicant or environmental mixture. In the initial experiments, logarithmic doses are usually used to obtain an indication of the range that should be studied in depth. Depending on the species, fewer than 10 animals can be used in the range finding study. Once the range has been defined, a narrower range of the doses can be tested. The response measured can represent percent mortality or some other toxic outcome.

At each dose the incremental number of animals that die is plotted vs. the logarithm of the dose to give a sigmoidal curve (Figure 2.2a). At the lowest concentration most or all the animals survive while at the highest concentration all the animals die. For every chemical there is a minimally effective dose known as the threshold that evokes the response being measured. It is virtually impossible to determine this threshold experimentally. The sigmoid

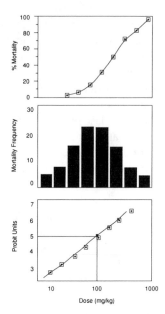

Figure 2.2. Determination of the dose response relationship using the NED and probit methods. The log dosage of the chemical is on the abscissa. In the upper panel (2.2a) the ordinate is percent mortality, in the middle panel (2.2b) the ordinate is mortality frequency, and in the bottom panel (2.2c) the mortality is plotted as probit units. (From Klaassen, C.D. and Eaton, D.L. *Casarett and Doull's Toxicology, 4th ed.* M.O. Amdur, J. Doull, and C.D. Klaassen, Eds. Pergammon Press, New York, pp. 12–49. With permission.)

dose response curve is linear between 16 and 84% of the respondents, and this range represents the limits of 1 SD from the mean in a normally distributed population. Usually a quantal dose response such as lethality exhibits a normal Gaussian distribution (Figure 2.2b). The plot of the Gaussian distribution can be made by constructing a histogram at each dose so each bar represents the percent of animals that died at each dose minus the percent that died at the immediately lower dose. The bell-shaped curve that results is known as a normal frequency distribution, and it comes from the susceptibility or biological variation between individual animals exposed to each concentration being tested. In a normally distributed population the mean ±1 SD represents 68.3% of the population, the mean ±2 SD represents 95.5% and the mean ±3 SD represents 99.7% of the population. As it is difficult to determine accurately the LD_{50} from the sigmoid curve or determine the slope of the line, a method was developed to linearize the response. Since the quantal response is normally distributed, the percent response can be converted to units of deviation from the mean or normal equivalent deviations (NED). The NED for a 50% response is 0, a NED of +1 is associated with 84.1% response, and a NED of −1 represents a 15.9% response. A further modification of this approach was suggested by Bliss (1957) when he con-

**TABLE 2.1. Relationship Between the Percent
Response of a Population, the Normal Equivalent
Deviation (NED), and Probit Units**

% Response	NED	Probit Unit
0.1	−3	2
2.3	−2	3
15.9	−1	4
50.0	0	5
84.1	+1	6
97.7	+2	7
99.9	+3	8

verted NED units to probit units by adding +5 to the value, thus avoiding the use of negative numbers. The probit value is thus NED +5 (Table 2.1). The LD_{50} is obtained by drawing a horizontal line through the probit unit of 5 to the dose response line and then a vertical line to the corresponding dose line (Figure 2.2c). Although log doses are usually used to determine the LD_{50} as they usually result in the representation of the dose response relationship as a straight line, this is not always the case. A better probit fit has been obtained using an arithmetic expression of doses when the LD_{50} of radiation is being investigated (Klaassen, 1986a).

Although the LD_{50} is determined to measure the toxic potency of different compounds, this value may not be representative of the actual toxicity of a compound as it represents the response following an acute exposure but does not represent the toxic effect following chronic exposure. The LD_{50} of the same chemical in different rodent or fish species also varies.

In ecotoxicology most LD_{50} studies have been performed on aquatic animals with fewer studies on terrestrial organisms. In these studies a LC_{50} is determined, which is a measure of the concentration in water that causes the death of 50% of the animals. In aquatic toxicity testing when the LC_{50} of an effluent is being determined, the organisms are exposed to the toxicant in question in water that simulates the water quality parameters of the receiving stream. Two approaches are used, one uses standardized or "artificial" water that is prepared by adding the appropriate salts to distilled water and the other uses the receiving stream into which the effluent is being discharged. EPA guidelines have preferred the first alternative since it is difficult, if not impossible, to replicate natural waters. The LD_{50} and LC_{50} values can be influenced by many factors including strain of animal, age, sex, as well as several environmental parameters such as temperature, water quality (hardness or softness), pH, etc.

Aquatic toxicity testing is a dynamic process in which new types of tests are continually being developed. For many years the prime criterion for the selection of an organism for toxicity testing was its high degree of sensitivity to chemical compounds. Many species that are extremely sensitive or have a high correspondence to the responses of other organisms are exceedingly difficult to keep alive in a laboratory setting. A major advantage of using test organisms that have been widely used for toxicity testing and whose strengths and weaknesses are well known is the large data base that has been collected on these species. However, emphasis on testing is now shifting to the use of those organisms that are the most reliable predictors of the response of other organisms in a community.

7. Pharmacodynamics

Uptake of Pollutants

Pollutants are transported into cells by three major mechanisms: passive transport or diffusion, filtration, and active transport. Uptake into an organism is complex since organisms consist of several tissues. To simplify the under-standing of uptake and distribution of chemicals into the body, an organism is divided into several artificial compartments into which the pollutants can be distributed. For example, in fish a pollutant is taken up by diffusion into the blood of the gills. The pollutant is then transported via the blood to a target organ such as the liver. The compound is then taken up by cells of the liver and finally partitions into the mitochondria, microsomes, or cytosol. At each of these stages the pollutant traverses a biological membrane that controls the rate and concentration of the pollutant that enters each compartment.

Cell membranes are composed of two protein backbones separated by a bilayer of phospholipid molecules. Phospholipids consist of a glycerol mol-ecule to which is attached two fatty acid moieties and one charged molecule such as ethanolamine, choline, serine or inositol (Lehninger, 1982). Some of the protein molecules are folded in such a way that they invaginate into the membrane and form a pore through which chemicals can diffuse. Uptake of molecules through these pores usually depends on the size of the molecule. Absorption of neutral molecules, particularly if they are lipophilic, is usually through passive transport, while charged molecules are absorbed either by filtration, which is dependent on molecular size, or by active transport.

Passive Transport. Lipophilic substances pass through the membrane without any hindrance according to Fick's law of diffusion. This law states that the rate of diffusion is proportional to the difference in the concentration of the compound on both sides of the membrane multiplied by the area and the tempera-ture and divided by the thickness of the membrane. When the concentration gradient has reached equilibrium, the steady-state flux (F) can be determined as follows:

$$F = \frac{DC_0}{h} \qquad (2.1)$$

where C_0 is the concentration of the compound on the donor side of the membrane, h is the membrane thickness, and D is the coefficient of diffusion.

The concentration of the compound on the receptor side of the membrane is essentially zero as it is constantly being removed by reactions within the cell. C'_0 the concentration of the compound in the membrane, so

$$Co = KC'_0 \qquad (2.2)$$

where K is the partition coefficient and C_0 is the concentration at time 0 in the donor compartment. As C_0 is usually known

$$F = \frac{KDC'_0}{h} \qquad (2.3)$$

If the diffusion of the substance follows first order kinetics the concentration in the donor compartment (C') of a specific volume V is

$$C' = C'_0{}^{eDKG/hV} \qquad (2.4)$$

D, the coefficient of diffusion, is inversely related to the radius of the molecule r which can also be expressed as $MW_{1/2}$, thus

$$D = \frac{constant}{\left(MW_{1/2}\right)} \qquad (2.5)$$

The permeability coefficient, P, can then be expressed as

$$P = \frac{K \times constant}{\left(MW_{1/2}\right)} \qquad (2.6)$$

In examining the rate at which compounds are absorbed by diffusion, it has been found that there is a direct relationship between the rate of uptake and the partition coefficient of a compound. The more lipophilic a compound is, the faster it will diffuse through biological membranes. Absorption of acids and alkalies depend on their degree of ionization which, in turn, depend on their pK_a and the pH of the surrounding medium. The pK_a of a substance is the pH at which 50% of the compound is ionized and 50% is unionized. Strong acids usually have a low pK_a while weak acids have a high pK_a. The degree of ionization of an acid and base is described by the Henderson-Hasselbach equation:

$$\text{For acids:} \qquad pK_a - pH = \log \frac{[\text{nonionized}]}{[\text{ionized}]}$$

$$\text{For bases:} \qquad pK_a - pH = \log \frac{[\text{ionized}]}{[\text{nonionized}]}$$

Since unionized compounds are more likely to be absorbed by diffusion, organic acids are more likely to be absorbed in an acidic environment and bases in a more basic environment. If the pK_a of a particular compound is known and the pH of the surrounding medium is known, uptake of the compound can be predicted. If the ratio of the nonionized to ionized fraction is greater than 10:1, the compound will be absorbed by diffusion into the tissue or animal. If this ratio is less than 10:1, this often suggests that absorption will be minimal, if at all.

Filtration. Movement of molecules through membranes by filtration depends on the size of the pores in the membrane as well as the size of the molecule entering the cell by this means. The pore sizes of membranes vary from tissue to tissue. For example, the pore size of most mammalian membranes, including capillaries, is 4Å, those of red blood cells are 40Å, while those in the glomerulus of the kidney are 400Å. In the latter case a molecule with a molecular weight of 60,000 daltons can be filtered from the blood into the glomerular filtrate.

Active Transport. Molecules that are hydrophilic and cannot be filtered due to their size or charge can enter cells by active transport. Our present understanding of this system entails the transport of compounds against an electrochemical gradient across a membrane bound to a carrier molecule. Since selective transport systems exist for different molecules, competitive inhibition can occur. Transport of the compound can also be inhibited when the system becomes saturated. Since active transport requires energy, this process can be blocked by inhibitors of oxidative phosphorylation or other pathways involving energy production in the cell.

Distribution

The distribution of a toxicant in an organism depends on the complexity of the organism. With vertebrates and some advanced invertebrates, distribution depends on the number of compartments into which an animal can be subdivided. Compartments are discussed later in this chapter. For an in-depth discussion of the distribution of toxicants in vertebrates and the storage depots of these animals, the reader should consult texts on mammalian toxicology.

D. CHEMODYNAMICS

Uptake of a chemical into a living organism from the environment depends on the concentration of the pollutant in the particular compartment in which the organism is living. This concentration, in turn, depends on the movement of the pollutant from one environmental compartment to another. Therefore, the effects of a chemical in the environment depends on its dynamics in the four compartments of the ecosphere, air, water, soil or sediments, and biota. The rate of transfer of a chemical from one compartment to another is dependent on the rate of the transfer of a compound at the interface between the two compartments. At this boundary spontaneous transfer of a compound occurs until equilibrium is reached. At equilibrium movement of the chemical in both directions is equal. The distribution of the chemical in the two phases depends on several physical parameters, including the partition coefficient of the compound between the two compartments. The partition coefficient, in turn, is dependent on the relative solubility of the compound in the two compartments.

The second factor that contributes to the concentration and resulting effect of a pollutant in the environment is its rate of chemical and biological transformation.

There are several environmental processes that are important in controlling the fate and thus the concentration of chemicals in the four environmental compartments (Table 2.2). The concentration is dependent on the kinetic and equilibrium rates of these processes (Mill, 1980). Equilibrium rates are frequently used to quantify the contribution of the individual processes to the concentration of a pollutant in a particular compartment. Intercompartmental transfers of a chemical depends on several factors: (1) the physiochemical properties of that compound, (2) the specific transport processes in that compartment, (3) the rate of transformation of the chemical in the compartment, and, (4) its rate of loss from that compartment to another (excretion in a mammalian system). Most processes in the environment that affect rates involve the interaction of a pollutant with some environmental property. To understand the movement of a compound from one compartment to another, the kinetic rate law can be applied. The rate of transfer of a compound from one compartment to another depends on different kinetic laws.

TABLE 2.2. Physical Transport in Environmental Compartments

Mechanism of Transport	Compartment	Property Controlling Transport
Diffusion and dispersion	Air	Wind velocity
Precipitation	Air	Particulate concentration, wind velocity
Adsorption	Water, soil	Organic carbon content, amount of sediment, wind velocity
Volatilization	Water, soil	Evaporation rate, turbulence, organic carbon content
Leaching	Soil	Adsorption coefficient, organic carbon content
Absorption by biota	Water, sediment	Partition coefficient, concentration of compound

Basic Kinetic Rate Law

The rate of a reaction of a specific environmental process depends on the first power of the concentration. The kinetic rate law can be expressed as follows:

$$R_n = \frac{d[A]}{dt} = k_n[A][P]_n \qquad (2.7)$$

where R_n is the rate of process n, k_n is the rate constant of process, [A] is the concentration of chemical A, and [P] is the environmental property expressed in concentration units compatible with k.

To determine the concentration of a pollutant that is available in the environment, the net rate of change in the concentration of A is determined. The net rate of change in concentration of A is equal to the sum of all the equilibrium and kinetic processes and is expressed as follows:

$$R_T = k_n[A][P]_n \qquad (2.8)$$

where R_T is the net rate of change and k_n is the rate of change for the nth process. It is usually assumed that the environmental property $[P]_n$ is constant and that a change in P is such that $[\Delta P] < [P]$, a situation which occurs with low-level pollution. Under these conditions, Equation 2.8 can be expressed as a group of first order processes:

$$R_T = k_T[A] \qquad (2.9)$$

Loss of a chemical from one compartment to another, whether it be through a physical or chemical process, occurs by either zero order or first order kinetics.

Zero and First Order Kinetics

In zero order reactions the rate of change in concentration occurs by a fixed amount per unit time. Thus,

$$k = \frac{dC}{dt}$$

where C is the concentration, t is the time, and k is a constant with units of concentration/unit time (e.g., mg/h).

In first order reactions the rate of change in the concentration of a chemical is proportional to the amount of the chemical available at a particular time for the reaction. Thus,

$$kC = \frac{dC}{dt}$$

when k represents a proportional change with time^{-1} and has time related units such as s^{-1}.

In the environment chemicals are usually present in very low concentrations, therefore most processes are first order reactions. Zero order reactions occur when a compound is present at high concentrations or when an enzyme reaction occurs at its highest rate, so an increase in the concentration of that compound will not increase the rate of the reaction.

The environmental transport phenomena that affects the environmental fate of a compound depends on the compartment in which movement occurs. The most important transport and transformation processes in each compartment are summarized in Tables 2.2 and 2.3. In the air compartment dispersion by wind, fallout from precipitation, and photolysis predominates, while in the aquatic environment sorption to suspended matter or sediments, volatilization to air, and uptake by biota are the most important. In the soil compartment adsorption, runoff, volatilization, leaching, and uptake by biota predominate.

1. Persistence of a Compound in the Environment

Single Application of a Pollutant

Following the single application of a toxicant such as a pesticide to an environmental compartment, the loss of the compound follows first order kinetics. One parameter that is important in determining the persistence of a

TABLE 2.3. Chemical Transformation in Environmental Compartments

Compartment	Mechanism of Transformation	Property Controlling Transformation
Air	Photolysis	Solar energy
	Oxidation	Catalytic sites on particulates
Water	Hydrolysis	pH of water
	Oxidation	
	Reduction	Oxygen and ferrous ion concentration
Sediment	Reduction	Oxygen and ferrous ion concentration
Biota	Oxidation/reduction	Activity of biotransformation enzymes

pesticide is its half-life, which can be calculated as follows:

$$\frac{d[A]}{dt} = -k_T[A] \tag{2.10}$$

k_T is the rate constant of the process. From this equation the following expression can be derived:

$$\frac{\ln[A_0]}{\ln[A]} = -k_T t \tag{2.11}$$

where A_0 is the concentration at zero time and A is the concentration after time t. When half of A is lost, then

$$\frac{[A_0]}{[A]} = 2 \tag{2.12}$$

By substitution of Equation 2.12 in Equation 2.11

$$t_{\frac{1}{2}} = \frac{\ln 2}{k_T} \tag{2.13}$$

$$t_{\frac{1}{2}} = \frac{0.693}{k_T} \tag{2.14}$$

Continuous Application of a Pollutant

With the continuous addition of a pollutant into the environment, such as the discharge from a point source into a water body, persistence is measured as the steady-state concentration that is reached when the inputs are equal to the losses. Under these conditions

$$\frac{d[A]}{dt} = 0 = R_I - R_L \quad \left(\text{i.e., } R_I = R_L\right) \tag{2.15}$$

where R_I is the rate of input and R_L is the rate of loss through transport and transformation of A. Since $R_I = \sum k_n [A] = k_L [A]$, where k_L equals the rate loss constant for A, [A] can be determined as:

$$[A] = \frac{R_I}{k_L} \tag{2.16}$$

2. Parameters Altering Persistence

Sorption

An important characteristic of a pollutant that affects its environmental fate is its ability to adsorb to various sorbates including sediments and biomass in aquatic systems, soils in the terrestrial compartment, and particulates in the air. The Law of Mass Action can be applied to determine the binding of the chemical to the substrate. This law states that a reaction will proceed until equilibrium is reached. Equilibrium will be maintained unless the concentration of either of the reactants is changed:

$$[A] + [S] \underset{k_{-s}}{\overset{k_s}{\rightleftharpoons}} [AS]$$

According to the Law of Mass Action

$$Ks = \frac{k_s}{k_{-s}} = \frac{[AS]}{[A\,S]}$$

$$\text{and } [A] + [S] = [A_T] \tag{2.17}$$

where [A], [S], [AS], and [A_T], respectively, represent the concentration of the unadsorbed chemical A, the concentration of the sorbent (sediment) in gm dry weight/ml water, the concentration of the adsorbed A, and the total amount of

A. k_s and k_{-s} are the rate constants for the forward and reverse reactions and K_s is the equilibrium constant in mg/l.

The loss of A can be represented by the following equation:

$$R_L = \frac{k_L [A_T]}{K_s [S] + 1} \qquad (2.18)$$

Thus, the net effect of sorption on the loss of [A] from a specific compartment at equilibrium is to reduce the loss by the factor of $1/(K_s [S] + 1)$. As a result of sorption, the compound has a longer half-life which can be determined as follows:

$$t_{1/2} = (K_s [S] + 1) \ln 2 / k_T \qquad (2.19)$$

With adsorption, a chemical has a higher steady-state concentration than when no adsorption occurs. The steady-state concentration of A, $[A_{ss}]$, can be calculated from the following formula:

$$[A_{ss}] = \frac{R_I (K_s [S] + 1)}{k_L} \qquad (2.20)$$

where R_I is the rate of input of the chemical.

The two situations in which adsorption is particularly important is the adsorption of gases or liquids to solids such as soil or sediments. Several factors influence these adsorption processes. These are the structural composition of the chemical, the organic carbon content of the soil, pH of the medium, size of soil particle, ion exchange capacity of the soil, and temperature. Although adsorption usually reaches steady-state conditions in a short period, the desorption rate is much slower.

The adsorption process can be expressed as an adsorption isotherm. Two widely used isotherms are the Langmuir and Freundlich isotherms. The former is used for the adsorption of gases to solids and the latter for adsorption of liquids to solids.

Langmuir Isotherm. The moles of gas adsorbed/gm adsorbent X is a function of the equilibrium concentration C of the gas in solution:

$$X = \frac{X_m bC}{1 + bC} \qquad (2.21)$$

where X_m is the number of moles of gas adsorbed/gm adsorbent in forming a monolayer, C is the concentration of the chemical at equilibrium, and b is a constant related to the energy of adsorption.

If the reciprocal of the amount of gas absorbed/unit mass of absorbent $1/X$ is plotted as a function of the reciprocal of the equilibrium concentration of sorbate, a straight line is obtained with an intercept of $1/X_m$ and a slope of $1(b \cdot X_m)$.

For this relationship to apply the following assumptions are made: (1) the energy of adsorption is constant and independent of the extent of surface coverage, (2) adsorption is at localized sites and there is no interaction between adsorbed molecules, and (3) the maximum adsorption that is possible, is a complete mono layer.

Freundlich Isotherm. The Freundlich isotherm is usually used to determine the adsorption of pollutants to sediments or soils but has also been used to determine the adsorption of compounds to biota, particularly microorganisms that have a very large surface area to volume ratio. Adsorption of metals and organic compounds to both living and dead bacteria (Paris et al., 1978) and algae (Roy et al., 1992) have been documented. The amount of chemical adsorbed/gm of sorbent can be determined as follows:

$$\frac{X}{m} = KC^{\frac{1}{n}} \qquad (2.22)$$

where X/m is the mass of chemical absorbed/g absorbent (m), K is the equilibrium constant indicative of strength of adsorption ($K = X/m$ when $c = 1$), C is the equilibrium concentration of the chemical, and $1/n$ is the slope of the isotherm.

The line representing this relationship is linearized in the following relationship:

$$\log \frac{X}{m} = \log K + \frac{1}{n} \log C \qquad (2.23)$$

When $\log X/m$ is plotted as a function of $\log C$, a straight line is obtained with an intercept on the ordinate of $\log K$. The slope can then be determined. The higher the intercept of $\log K$, the greater the degree of adsorption; and the larger the slope the greater the efficiency of adsorption.

The adsorption of polychlorinated biphenyl congeners to many surfaces increase as their water solubility decreases (Haque and Smedding, 1976). Adsorption to soils also depends on the organic carbon content of the soil; the higher the organic carbon content the greater the degree of adsorption. With many nonionic organic compounds, the slope of the Freundlich isotherm $1/n$ approaches 1. Under these conditions Equation 2.22 becomes

$$\frac{X}{m} = K_s C \qquad (2.24)$$

where K_s is the adsorption coefficient. A relationship between K_s, the octanol water partition coefficient K_{ow}, and organic carbon content have been found. The adsorption of a compound is closely correlated with the proportion of the organic carbon content K_{oc}. Equation 2.24 is frequently expressed as

$$\frac{X}{m} = K_{oc} C \qquad (2.25)$$

Using this equation, the concentration of a chemical adsorbed is expressed per unit organic carbon rather than per unit mass of soil. Tinsley (1979) has reported that K_{oc} is more constant among soils than K_s. As expected, adsorption of organic compounds to soils increases as the particle size of the soil decreases. Thus, the K_{oc} for an organic compound is much greater in silty soils than in sand.

In the case of ionic or neutral organic compounds which can become charged in solution, the relationships with organic carbon content do not hold. Rather the adsorption of these compounds depends on the mineral content, the functional groups of the molecule, its solubility, and its partition coefficient. With these compounds other mechanisms of binding such as van der Waal's forces, hydrogen bonding, ligand exchange, ion exchange, and chemisorption are the most important factors (Tinsley, 1979).

Volatilization

Chemicals with low solubility and low polarity evaporate more rapidly from the aquatic compartment to the atmosphere than do highly soluble compounds. Many pollutants of high molecular weight can volatilize easily due to their high activity coefficients in solution (Thibodeaux, 1979). This characteristic is important in the volatilization of many refractory compounds such as DDT, dieldrin, and polychlorinated biphenyls and may be the mechanism by which these compounds have been distributed to the polar regions. Relatively high concentrations of these chlorinated compounds have been found in the fatty tissue of many Arctic and Antarctic animals.

The partial pressure of a chemical in air can be determined from the following formula, and from this the overall mass transfer coefficient K_A of a chemical from water to air can be calculated:

$$P_A = C_A \frac{\left[P_A^0 \right]}{\left[C_A^* \right]} \qquad (2.26)$$

where P_A is the partial pressure of chemical A in air, P_A^0 is the vapor pressure of pure A, C_A^* is the equilibrium solubility of A in water, and C_A is the concentration of A in water.

The overall mass transfer coefficients of some alkanes, aromatic compounds, pesticides, and various Aroclors have been compiled by Mackay and Leinonen (1975). Alkanes and aromatic compounds such as benzene and toluene have the highest transfer coefficients (±0.12 m/h), followed by two-ringed aromatic compounds such as naphthalene and biphenyl (0.09 m/h) and Aroclors™ 1242 to 1260 with coefficients of ±0.07 m/h. Chlorinated pesticides have the lowest transfer coefficients ranging from 3.7×10^{-3} for aldrin to 5.3×10^{-5} m/h for dieldrin with DDT falling between these values.

Since the volatilization of a chemical is a first order process, the corresponding half-life of the chemical can be determined from the following formula:

$$t_{1/2} = \frac{0.693d}{K_A} \tag{2.27}$$

where d is the depth of the compound in water.

3. Bioaccumulation Processes

Many compounds absorbed by living organisms from their surroundings or from food sources are either used immediately as a source of energy or are stored in a form in which they can later be used as an energy source. However, organisms cannot use many foreign compounds (xenobiotics) as a major energy source. These compounds are either metabolized to more water soluble forms and excreted or they are stored in various tissues of the body. The xenobiotic often does not elicit any deleterious effects as long as it is sequestered in a nondividing tissue. Retention of the compound or bioaccumulation results in an elevated concentration in the organism compared to the surrounding medium. The degree of bioaccumulation depends on many factors but those of greatest importance are the lipophilicity of the compound, its rate of metabolism, and its half-life in the organism. Compounds are bioaccumulated through bioconcentration or biomagnification. The former occurs following accumulation of a compound through diffusion from an aquatic environment into the organism via the gills or skin. Biomagnification occurs when pollutants are passed from a lower trophic level through the food chain to a higher trophic level, so organisms at the higher trophic levels have higher concentrations of the pollutant.

Models of Bioconcentration

The most widely used model to explain the uptake, bioaccumulation, and elimination of a pollutant is the compartmental model which has been used in conventional pharmacokinetic approaches. A compartment is defined as an

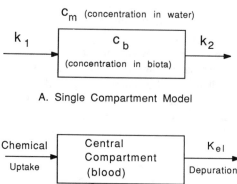

A. Single Compartment Model

B. Two Compartment Model

Figure 2.3. Single and two compartment model to determine bioaccumulation.

enclosed delineated part of an organism such as an organ in which uniform kinetics of transport and metabolism occur and whose kinetics are distinct from other compartments.

Single Compartment Model. The simplest model is the single compartment model in which only one compartment such as the blood is involved (Figure 2.3a). Bioaccumulation is the balance between two kinetic processes, uptake and excretion or depuration, which can be quantified by two first order rate constants, K_1 and K_2. The rate of change in the concentration of a pollutant in an organism can be expressed as

$$\frac{dC_b}{dt} = K_1 C_m - K_2 C_b \qquad (2.28)$$

where C_b is the concentration of the pollutant in the biota ($\mu g/g$), C_m is the concentration of the pollutant in the surrounding water, t is the time in hours, K_1 is the uptake rate constant ($ml/g \cdot h^{-1}$), and K_2 is the depuration rate constant (h^{-1}).

Integration of the equation from zero time (t = 0) when the initial concentration in the organism is 0 (i.e., $C_b = 0$) to time t is

$$C_b = \frac{K_1}{K_2} C_m \left(1 - e^{-K_2 t}\right) \tag{2.29}$$

At steady-state, uptake equals depuration:

$$\frac{dB_b}{dt} = 0 = K_1 C_m - K_2 C_b \tag{2.30}$$

$$K_1 C_m = K_2 C_b \tag{2.31}$$

The bioaccumulation factor (BCF or K_B) can be determined by

$$BCF = \frac{C_b}{C_m} = \frac{K_1}{K_2}$$

If exposure is terminated, uptake ceases and $K_1 C_m = 0$, then only depuration will occur. For depuration,

$$\frac{dC_b}{dt} = -K_2 C_b \tag{2.32}$$

Upon integration $C_b = C_{bc} e^{-K_2 t}$ where C_{b0} is the concentration of b at the beginning of depuration, and

$$\log C_b = \log C_{b0} - \frac{K_2 t}{2.303} \tag{2.33}$$

A semilog plot of C_b vs. time will be linear. The biological half-life can be calculated from

$$t_{1/2} = \frac{0.693}{K_2} \tag{2.34}$$

Two Compartment Model. In most animal systems except the simplest invertebrates which may consist of only one compartment, a pollutant does not behave uniformly and partitions into more than one compartment of the body. A certain fraction of the pollutant may be eliminated rapidly while the remainder may be excreted more slowly. For example, lipophilic compounds that partition into the fat bodies that have a poor circulation will be excreted more slowly than residues of the compound in the blood. Under these conditions a biphasic depuration

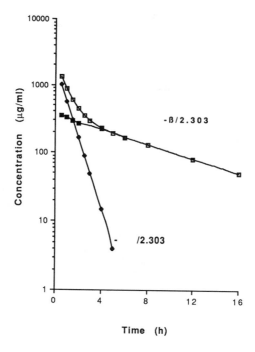

Figure 2.4. Plasma concentration of a compound vs. time in a two compartment model.

curve will be obtained with depuration occurring from "fast" and "slow" compartments. This can be visualized as shown in Figure 2.3b where uptake and depuration occur through a "fast" (central) compartment such as the blood while also depurating from a "slow" (peripheral) compartment such as a tissue.

The residual concentration in the animal can be expressed as follows:

$$C = A \cdot e^{-at} + B \cdot e^{-bt} \tag{2.35}$$

where A and B are the intercepts of the tangents drawn to the line with slopes a and b. A + B is equivalent to the initial body concentration at the start of depuration. According to Figure 2.4

$$k_{21} = \frac{Ab + Ba}{A + B} \tag{2.36}$$

$$k_{el} = \frac{ab}{k_{21}} \tag{2.37}$$

$$k_{12} = a + B - k_{21} - k_{el} \tag{2.38}$$

If the uptake constant k_{01} can be measured from the initial part of the uptake curve, the bioaccumulation factor (BCF) can be calculated as follows:

$$BCF = \frac{k_{01}}{ab}\left(a + b - k_{10}\right) \quad \text{or} \quad BCF = \frac{k_{01}}{ab}\left(k_{12} + k_{21}\right) \quad (2.39)$$

If clearance through the peripheral compartment is rapid so $k_{21} > k_{12}$, the model will reduce to a one compartment model.

Determination of Bioconcentration Factors

Many experiments to determine the BCF of a compound are performed with radiolabeled compounds. This approach is acceptable if the contribution of the radiolabeled metabolites to the elimination rate of the parent molecule is recognized (Spacie and Hamelink, 1985). If only a small percentage of metabolites are produced and excreted, the data obtained is valid. However, if a large percentage of the parent compound is metabolized and the metabolites persist for extended periods and are excreted at a much slower rate than the parent compound, then the contribution of the radioactivity of the parent compound must be determined separately. For example, with benz(a)acridine, which is rapidly metabolized by fathead minnows, the excretion rate of the parent compound was very slow as in fact, most of the parent compound had been metabolized and excreted (Southworth et al., 1981).

There is evidence relating the bioaccumulation factor K_B to the octanol/water partition coefficient P_{ow} (Mackay, 1982). P_{ow} can be measured easily and provides a means of assessing the bioaccumulation potential of a compound particularly if it is lipophilic. Two assumptions are made, namely that *n*-octanol has similar solubility properties to that of the lipids of cell walls and that uptake of a compound into an animal results in the partitioning of the compound between the organism and the surrounding water. The relationship between P_{ow} and K_B has been tested by regression analysis by Esser and Moser (1982) who showed that the following relationship held:

$$\log K_B = n \log P_{ow} + b \quad (2.40)$$

Reliable correlations between K_B and P_{ow} occur when P_{ow} is between 10^2 and 10^6 which corresponds to K_B values of 10^1 to 10^4. Mackay (1982) suggested that predictions can only be made with lipophilic compounds because compounds with a low P_{ow} are usually hydrophilic, absorbed slowly, excreted rapidly, and thus not bioaccumulated. Also, correlations for compounds with high P_{ow} values ($>10^6$) is also uncertain because steady state conditions may not be reached within short periods. Under these conditions, growth and lipid deposition must be considered. Another difficulty with compounds with $P_{ow} >$

TABLE 2.4. Key Physical and Environmental Properties in Fate Assessment

Property	Formula
Soil sorption coefficient K_d	$\dfrac{\mu g\ \text{Chemical/g of soil}}{\mu g\ \text{Chemical water/g of water}}$
Soil sorption constant K_{oc}	$\dfrac{K_d}{\mu g\ \text{Organic carbon}} \times 100$
Water/air ratio K_w	$\dfrac{\mu g\ \text{Chemical/cm}^3\ \text{of water}}{\mu g\ \text{Chemical/cm}^3\ \text{of air}}$
n-Octanol/water coefficient K_{ow}	$\dfrac{\mu g\ \text{Chemical/ml } n\text{-octanol}}{\mu g\ \text{Chemical/ml water}}$
Bioconcentration factor BCF	$\dfrac{\mu g\ \text{Chemical/g of fish}}{\mu g\ \text{Chemical/g of water}}$

Source: Swann et al., 1983 (Adapted with permission from McGrawHill, Inc.).

10^6 is their insolubility in water. Mackay (1982) estimated that the BCF of a compound is about 5% the P_{ow}.

The octanol/water partition coefficient of a compound can be estimated if its water solubility is known. The following regression equation has been proposed by Chiou et al. (1977) to calculate P_{ow}:

$$\log P_{ow} = 5.00 - 0.67 \log S \qquad (2.41)$$

where S is the solubility of the compound in $\mu mol/l$.

Transport of a pollutant between the four environmental compartments, air, soil, water, and biota, can be related to each other by a series of partition coefficients. The soil adsorption coefficient K_d can be calculated from the ratio between the concentration of the compound in the soil and its concentration in water (Table 2.4). As the organic carbon content of the soil greatly influences the K_d, the adsorption characteristics can be normalized by using K_{oc}, which is obtained by dividing K_d by percent organic carbon and multiplying this ratio by 100. Thus, it is apparent that the availability of a compound for uptake by an organism in the field is dependent on several physical partition coefficients between constituents of these compartments.

Biomagnification

When accumulation of a toxicant from a food source is being determined, a correction for growth of the organism can be included in the one compartment model according to the following formula of Thomann (1981):

$$\frac{dC_b}{dt} = K_1 C_m + \alpha R C_f - K_2 C_b \qquad (2.42)$$

where α is the assimilative efficiency of the contaminant from the food (i.e., μg absorbed/μg ingested), R is the g ingested/gm body weight/d, and C_f is the concentration of the contaminant in the food ($\mu g/g$).

The relative contribution of the amount of pollutant bioaccumulated from food and water sources and the effect of growth on this bioaccumulation can be determined. If exposure is for an extended period, the total body burden can increase as long as growth continues. Steady state is usually only reached when growth is complete. Based on the model of Thomann (1981), bioaccumulation from food is more important than bioconcentration from the water body. This is especially true when the pollutant has a long half-life. For pollutants with short half-lives, contribution from food sources is less important, particularly if depuration is rapid. This model assumes that the contribution from the two sources is additive and that the excretory rates of the residues from both sources are identical and follow first order kinetics. However, in a study on the accumulation of cadmium and zinc from crayfish by Giesy et al. (1980), it was shown that elimination of these metals from the food source was slower than the rate from water uptake. Absorption and elimination of contaminants by animals in the field are altered by the metabolic rates of the animals which, in turn, is affected by environmental conditions such as temperature. Temperature influences caloric intake, growth, and lipid content of the animal. Also the elimination rate of lipophilic compounds declines as the animal accumulates fat. Based on this finding, Bruggeman et al. (1981) suggested that chemical concentrations and rate constants be normalized to lipid content.

Kinetics of Depuration or Elimination

The rate at which a compound is eliminated from an organism has a marked effect on its toxicity and bioaccumulation. Vertebrates have several routes by which compounds can be excreted. These include urinary excretion through the kidneys, transport across respiratory surfaces such as the lung in terrestrial vertebrates and gills in aquatic animals, and secretion into the bile for subsequent fecal excretion. Arthropods can excrete toxicants through molting (Spacie and Hamelink, 1985). The gill appears to play a major role in the excretion of nonpolar compounds in both fish and invertebrates. Water soluble compounds are preferentially excreted via the kidney of vertebrates and modified organs such as malphigian bodies in insects or the green gland in crustacea. Recently, Krahn et al. (1986) showed that fish excrete polycyclic aromatic hydrocarbons into the bile.

The simplest model of excretion occurs when uptake and distribution is into a single compartment and excretion follows first order kinetics. The time

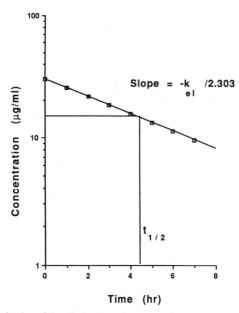

Figure 2.5. Determination of the elimination rate constant in a one compartment model using a graphic method.

required for the concentration of the chemical to decrease by one half remains constant until all the chemical is excreted. Theoretically, under these conditions a compound is never completely eliminated from the body. For most practical applications, a compound is considered completely eliminated after nine half-lives have elapsed. The half-life of a chemical that follows first order kinetics is independent of the dose. In a one compartment open model, it is assumed that a compound is eliminated from the plasma at the same rate as it is eliminated from a specific tissue. The elimination rate constant k_{el} and the half-life ($t_{\frac{1}{2}}$) of a compound can be determined by measuring the plasma concentration of a compound over time (Figure 2.5). The logarithm of the plasma concentration vs. hours or days of sampling is plotted. The line can be extrapolated to zero time.

The elimination rate constant k_{el} is calculated by first determining the slope of the line according to the following equation:

$$\text{Slope} = \frac{\Delta y}{\Delta x} = \frac{\log p_2 - \log p_1}{t_2 - t_1} \tag{2.43}$$

where p_2 is the plasma concentration at some time t_2 after measurements began and p_1 is the plasma concentration at an earlier time t_1.

$$\text{Slope} = \frac{k_{el}}{2.303} \tag{2.44}$$

Substitute to obtain k_{el}:

$$T_{\frac{1}{2}} = \frac{0.693}{k_{el}} \tag{2.45}$$

The characteristics of first order elimination kinetics include the representation of the log of the concentration of the chemical vs. time as a straight line. The rate of elimination of a chemical is directly proportional to its concentration in the organism. The $t_{\frac{1}{2}}$ is independent of dose, and the concentration of the chemical decreases by a constant fraction/unit time, known as the elimination rate constant k_{el}.

If a straight line is not obtained when plotting the logarithm of the concentration of the chemical vs. time but instead an exponential curve results, then a multicompartmental analysis will be required. If the curve can be resolved into two straight lines using the method of residuals, then a two compartment model can be used. Under these conditions the rate of removal of the chemical from the central or blood compartment and a second compartment such as the liver is different.

Elimination of most chemicals occurs by first order kinetics. However, if the concentration of a chemical is high in an animal, the initial rate of elimination may decrease. This is due to saturation of various processes such as biotransformation or active transport.

4. Degradation and Transformation of Pollutants

Several chemical and biochemical reactions are involved in the transformation and degradation of compounds. Depending on the compartment concerned, the predominant reaction will vary (Table 2.4). Chemical transformations are somewhat similar to biotransformations and include hydrolysis, oxidation, reduction, and photolytic reactions. These reactions can be catalyzed by metal ions, oxides, clay surfaces, and other organic compounds. Biotransformation reactions are performed by all living organisms, but the ability to biotransform or degrade a particular compound is species specific. A brief description of the physical parameters involved in the degradation of compounds by microorganisms is discussed below. For more information on biotransformation, the reader is referred to Chapter 3. Factors that affect the rate of degradation are pH, temperature, moisture content of soils, clay content, etc.

Photolysis

Photochemical reactions can occur in air and water compartments but are of little significance in soils. Chemicals undergo photochemical reactions after they have absorbed light in the UV range that has enough energy to break existing chemical bonds. The principal reactions are photooxidation and photoreduction, both of which proceed through the formation of free radicals which then either react with molecular oxygen or abstract hydrogen from organic compounds. Transformation of pollutants in the atmosphere is mediated by sunlight, particularly short wavelengths in the ultraviolet range, and by catalytic reactions on particulates in the atmosphere. Photolysis of pollutants depends on the energy of the incident light, the absorption spectrum of the molecule, and the presence of photosensitizers in the environment. Absorption of ultraviolet light can result in the cleavage of bonds, oxidation, hydrolysis, dimerization, and rearrangement.

Unsaturated aromatic compounds are the most likely compounds to undergo photolysis. Photochemical degradation occurs after enough radiant energy has been absorbed. Photolysis of chemicals in air, water, or soil can be represented by the following kinetic relationship (Mill, 1980):

$$I_A(\lambda) = e_\lambda I_\lambda [C] = K_a(\lambda)[C] \qquad (2.46)$$

where e_λ is the molar absorbance, I_λ is the intensity of the incident light at wavelength λ, a parameter that is available in the literature, $[C]$ is the concentration of the chemical, and K_a is the rate constant for light absorption. The rate of photolysis can be represented by a simple first order kinetic expression

$$Rp = \frac{dC}{dt} = K_a(\lambda)f_\lambda[C] \qquad (2.47)$$

$$\text{or} \quad K_p = K_a \phi \qquad (2.48)$$

where f_λ is the quantum yield, which is the efficiency of converting the absorbed light into a chemical reaction. This is measured as the ratio of the moles of substrate transformed to the einsteins of photons absorbed. K_p is the first order rate constant for direct photolysis. The half-life of a chemical undergoing photolysis can be estimated according to the equation

$$t_{1/2} = \frac{0.693}{K_a \phi} \qquad (2.49)$$

Hydrolysis

Chemical transformation by hydrolysis takes place mainly in the aquatic compartment. Similar reactions can be mediated by biological systems (see Chapter 3). Second in importance to microbial degradation is hydrolysis in water. With organic compounds a hydroxyl group is introduced into the chemical so it replaces a leaving group. Callahan et al. (1979) undertook a comprehensive study of the aquatic fate of the so-called "priority pollutants" and found that the following products could be formed:

$$RX + H_2O \rightarrow ROH + H^+ + X^-$$

$$\begin{matrix} O \\ \| \\ RC\text{–}X + H_2O \end{matrix} \quad \rightarrow \quad \begin{matrix} O \\ \| \\ RCOH + H^+ + X^- \end{matrix}$$

Salts of metals can also undergo hydrolysis according to the following equation:

$$MX + H_2O \quad \rightarrow \quad MOH + H^+ + X^-$$

As many of these reactions can be catalyzed by either hydronium or hydroxyl ions, the rate of hydrolysis can be expressed according to the following equation:

$$R_H = \frac{d[C]}{dt} k_h[A] \tag{2.50}$$

$$= k_A\left[H^+\right]\left[A\right] + k_B\left[OH^-\right] + k_N\left[H_2O\right]\left[A\right] \tag{2.51}$$

where R_H is the rate of hydrolysis, k_h is the first order rate constant at a particular pH, k_A and k_B are the second order rate constants for acid and base catalyzed reactions, and k_N is the second order rate constant for the neutral reaction of the chemical with water which can be expressed as a pseudo first order reaction; k_h is pH dependent unless k_A and $k_B = 0$.
Assuming first order kinetics, the half-life can be calculated from

$$t_{\frac{1}{2}} = \frac{0.693}{k_h \ sec^{-1}} \tag{2.52}$$

Microbial Transformations

Microorganisms including bacteria, fungi, protozoa, and algae are the major organisms involved in the biodegradation of pollutants in the terrestrial and aquatic compartments. The reactions that can be performed by these organisms include oxidation, reduction, hydrolysis, and rearrangements (see Chapter 3). The rate of the reaction is dependent on the molecular structure of the compound, its concentration, and many environmental conditions such as temperature, pH, amount of nutrients, etc. The rate of growth of a microorganism is dependent on the concentration of the growth limiting substrate.

The Monad equation is used to describe the relation between the concentration of the growth limiting compound and the specific growth rate μ (Connell and Miller, 1984); thus,

$$\mu = \frac{\frac{d[X]}{dt}}{[X]} = \frac{\mu_{max}[A]}{K_s + [A]} \tag{2.53}$$

where μ is the specific growth rate, $[X]$ is the biomass per unit volume, $[A]$ is the substrate concentration, μ_{max} is the maximum specific growth rate, and K_s is the concentration of substrate supporting half-maximum specific growth rate $(0.5 \, \mu_{max})$.

The rate of loss of chemical A or the substrate can be calculated as follows:

$$\frac{-d[A]}{dt} = \frac{\mu[X]}{X} = \frac{\mu_{max}[A][X]}{YK_s + [A]} \tag{2.54}$$

where Y is the biomass produced from a unit amount of chemical or substrate consumed and $\mu_{max}/Y = K_b$.

The constants μ_{max}, K_s, and Y are dependent on microorganism number, pH, temperature, concentration of nutrients, and other factors affecting the microbes.

When the chemical concentration is high then $[A] > K_s$ and Equation 2.54 becomes

$$\frac{-d[A]}{dt} = K_b[X] \tag{2.55}$$

In many cases in the environment, however, the concentration of A is much less than the concentration of the substrate supporting half-maximal growth rate, i.e., $[A] < K_s$; thus,

$$\frac{-d[A]}{dt} = \frac{K_b[A][X]}{K_s} = K_{b2}[A][X] \tag{2.56}$$

where K_{b2} is a second order rate constant.

Under many environmental conditions, when $[X]$ is relatively large and $[A]$ is comparatively smaller, the microbial biomass remains constant when the chemical is consumed. The degradation rate is then usually assumed to be pseudo first order; thus,

$$\frac{-d[A]}{dt} = K_b'[A] \tag{2.57}$$

K_b' is the pseudo first order rate constant which is dependent on the initial cell concentration $[X_0]$. Then,

$$K_{b2}[A][X] = K_b'[A] \tag{2.58}$$

and

$$\frac{K_b'}{X_0} = K_{b2} \tag{2.59}$$

The half-life of the chemical at a given X_0 can thus be calculated as follows:

$$t_{\frac{1}{2}} = \frac{0.693}{K_{b2}X_0} \tag{2.60}$$

After a chemical is discharged into the environment, an acclimation period is required by the unacclimated microorganisms before microbial degradation can be initiated. The time required for the concentration of the chemical to be reduced by half is the sum of the time required for acclimation and the half-life of transformation. Under these conditions, Monad kinetics as described above cannot be used to calculate the half-life as this equation only applies to degradation by acclimated microorganisms.

E. CONCLUSIONS

A few of the important basic concepts about toxicology and ecotoxicology have been briefly discussed in this chapter. Many more kinetic equilibria are important in determining the transport of chemicals from one compartment to another, but space prohibits the discussion of all these concepts. Several important texts have been written in which an in-depth discussion of these concepts can be obtained. Some of these texts have been referenced in this chapter.

The understanding of the movement of chemicals in the environment has received much attention in recent years, but still more information needs to be acquired to understand the role of kinetic processes in ecotoxicology.

REFERENCES

Bliss, C.L. Some principles of bioassay, *Am. Sci.* 45:449–466, 1957.

Bruggeman, W.A., Matron, L.B.J.M., Kooiman, D., and Hutzinger, O. Accumulation and elimination kinetics of di- tri- and tetrachlorobiphenyls by goldfish after dietary and aqueous exposure, *Chemosphere* 10:811–832, 1981.

Chiou, C.T., Freed, V.H., Schmedding, D.W., and Kohnert, R.L. Partition coefficient and bioaccumulation of selected organic chemicals, *Environ. Sci. Technol.* 11:475–478, 1977.

Connell, D.W. and Miller, G.J. Chemodynamics of pollutants, in *Chemistry and Ecotoxicology of Pollution.* D.W. Connell and G.J. Miller, Eds. Wiley Intersciences, New York, 1984, 7–42.

Esser, H.O. and Moser, P. An appraisal of problems related to the measurement and evaluation of bioaccumulation, *Ectoxicol. Environ. Safety* 6:131–138, 1982.

Giesy, J.P., Jr., Bowling, J.W., and Kania, H.J. Cadmium and zinc accumulation and elimination by fresh-water crayfish, *Arch. Environ. Contam. Toxicol.* 9:685–699, 1980.

Haque, R. and Smedding, D. A method of measuring the water solubility of very hydrophobic chemicals, *Bull. Environ. Contam. Toxicol.* 14:13–18, 1976.

Klaassen, C.D. Principles of toxicology, in *Casarett and Doull's Toxicology. The Basic Science of Poisons,* 3rd ed. C.D. Klaassen, M.O. Amdur, and J. Doull, Eds. Macmillan Publishing Company, New York, 1986a, 11–32.

Klaassen, C.D. Distribution, excretion and absorption of toxicants, in *Casarett and Doull's Toxicology. The Basic Science of Poisons,* 3rd ed. C.D. Klaassen, M.O. Amdur, and J. Doull, Eds. Macmillan Publishing Company, New York, 1986b, 33–63.

Krahn, M.M., Rhodes, L.D., Myers, M.S., Moore, L.K., MacLeod, W.D., Jr., and Malins, D.C. Associations between metabolites of aromatic compounds in bile and the occurrence of hepatic lesions in English sole *(Parophrys vetulus)* from Puget Sound, Washington. *Arch. Environ. Contam. Toxicol.* 15:61–67, 1986.

Lehninger, A.L. *Principles of Biochemistry.* Worth Publisher's, New York, 1982, 1011.

Mackay, D. Correlation of bioconcentration factors, *Environ. Sci. Technol.* 16:274–278, 1982.

Mackay, D. and Leinonen, P.J. Rate of evaporation of low solubility contaminants from water bodies to the atmosphere, *Environ. Sci. Technol.* 9:1178–1180, 1975.

Malins, D.C., McCain, B.B., Myers, M.S., Brown, D.W., Krahn, M.M., Roubal, W.T., Schiewe, M.H., Landahl, J.T., and Chan, S.-L. Field and laboratory studies of the etiology of liver neoplasms in marine fish from Puget Sound, *Environ. Health Perspect.* 71:5–16, 1987.

Mill, T. Data needed to predict the environmental fate of organic chemicals, in *Dynamics, Exposure and Hazard Assessment of Toxic Chemicals.* R. Haque, Ed. Ann Arbor Science, Ann Arbor, MI, 1980, 297.

Paris, D.F., Steen, W.C., and Baughman, G.L. Rate of physico-chemical properties of Aroclor 1016 and 1242 in determining their fate and transport in aquatic environments, *Bull. Environ. Contam. Toxicol.* 7:319–325, 1978.

Renwick, A.G. Pharmacokinetics of toxicology, in *Principles and Methods of Toxicology,* 2nd ed. A.W. Hayes, Ed. Raven Press, New York, 1989, 835–878.

Roy, D., Greenlaw, P.N., and Shane, B.S. Adsorption of heavy metals by green algae and ground rice hulls, *J. Environ. Sci. Health* A28:37–50, 1992.

Southworth, G.R., Keffer, C.C., and Beauchamp, J.J. The accumulation and disposition of benz(a)acridine in the fathead minnow, *Pimephales promelas, Arch. Environ. Contam. Toxicol.* 10:561–570, 1981.

Spacie, A. and Hamelink, J.L. Bioaccumulation, in *Fundamentals of Aquatic Toxicology.* G. Rand and S.R. Petrocelli, Eds. Hemisphere Publishing Corporation, Washington, DC, 1985, 495–525.

Swann, R.L., Laskowski, D.A., McCall, P.J., Vander Kuy, K., and Dishburger, J., II. A rapid method for the estimation of the environmental parameters octanol/water partition coefficient, soil sorption constant, water to air ratio, and water solubility, *Residue Rev.* 85:17–28, 1983.

Thibodeaux, L.J. *Chemodynamics; Environmental Movement of Chemicals in Air, Water and Soil.* John Wiley & Sons, New York, 1979, 501.

Thomann, R.V. Equilibrium model of fate of microcontaminants in diverse aquatic food chains, *Can. J. Fish Aquat. Sci.* 39:280–296, 1981.

Tinsley, J. *Chemical Concepts in Pollutant Behavior.* John Wiley & Sons, New York, 1979, 256.

Trevan, J.W. The error of determination of toxicity, *Proc. R. Soc. Lond. (Biol.)* 101:483–514, 1927.

Varanasi, U., Nishimoto, M., Reichert, W.L., and Stein, J.E. Metabolism and subsequent covalent binding of benzo(a)pyrene to macromolecules in gonads and liver of ripe English sole (*Parophrys vetulus*), *Xenobiotica* 12:417–425, 1982.

Chapter

3

Xenobiotic Metabolism

Larry G. Hansen and Barbara S. Shane

A. INTRODUCTION

Biotransformation refers to the specific phase of "metabolism" during which one chemical is transformed to another by a biotic system, most often an enzyme. Metabolism includes various processes including absorption, distribution, excretion, and biotransformation. Several factors influence the site of action of a xenobiotic in a single cell or in an organism, namely, the dose to the cell, the rate of transport to the cell, and the duration of its effect. Biotransformation influences all of these processes by altering distribution characteristics as well as activity, and it, in turn, is influenced by the other processes.

Xenobiotics (or foreign chemicals) are not necessarily foreign to all living things. Caffeine, curare, heroin, nicotine, penicillin, physostigmine, and rotenone are well-known plant substances which may be foreign to other organisms. Ethanol, acetaldehyde, formaldehyde, uric acid, and several other endogenous products which are usually produced in small amounts can have devastating effects if their concentrations become too high. The levels of steroid hormones are controlled by a variety of mechanisms, so large doses of exogenously administered hormones such as estradiol or testosterone can be considered xenobiotics. Thus, a xenobiotic need not be a synthetic chemical but may be a natural substance introduced into the wrong species, in the wrong tissue, in the wrong amount, or in the wrong proportion. Under these conditions, the compound may be toxic.

0-8493-8851-1/94/$0.00+$.50

Living systems have several mechanisms for metabolizing and detoxifying xenobiotics. Generally, the enzymes responsible are part of the normal biochemical pathways of the organism, but they may be specifically involved in the detoxification of internally generated toxicants. Many systems have been previously altered through exposure so they can deal more efficiently with xenobiotics. Such exposure may alter populations and/or individuals.

The number and variety of xenobiotics with which organisms are now challenged enhances the numbers of basic enzyme systems that may be involved. The systems and pathways involved in the metabolism of xenobiotics are categorized in various ways for orderly presentation, systematic study, or relating biotransformation and physiological and/or toxicological responses to xenobiotics. These classifications, although necessary for comprehension of the complexity of enzymes involved, are artificial and force an unnatural order onto a natural complex continuum of both subtle influences and dramatic interdependence.

B. PATHWAYS AND ENZYMES

The usual approach for presenting the pathways of xenobiotic biotransformation is to divide them into phase I and phase II pathways, thereby preserving both an interdependence and a continuum. Phase I reactions expose (e.g., by hydrolysis) or introduce (e.g., by oxidation) a reactive function to the molecule, which makes it more polar and thus more water soluble. This intermediate is also more susceptible to phase II pathways.

Phase II reactions involve the conjugation of the phase I product with an endogenous substance which usually renders the product less bioactive and more readily excreted. There are some very important exceptions, especially the conjugation of glutathione with both alkyl and aryl halides, which do not decrease the lipophilicity of the final product and actually result in greatly enhanced toxicity. These exceptions are more common than originally believed.

A third phase can also be considered during which (1) part or all of the endogenous conjugating moiety is recovered, reintroducing the phase I product or a minor variant, or (2) the phase II product is reintroduced to phase I pathways resulting in additional transformations.

Using these distinctions, representative and more common examples will be presented to establish a framework within which the roles of biotransformation in environmental toxicology can be recognized. To maintain important toxicological perspectives during the examination of enzyme systems and individual pathways, the following generalizations may also be helpful:

1. Biotransformation changes the biological activity of xenobiotics.
    a. Quantitative — both detoxication and activation
        1. Phosphorothionate pesticides
        2. Polynuclear aromatic hydrocarbons
    b. Qualitative — distribution and effect
        1. Acetanalid
            → phenylhyroxylamine (methemoglobinemia)
            → nitrosobenzene (cancer)
        2. Phenacetin → OH acetamide (renal necrosis)
        3. Acetaminophen → N-hydroxyacetaminophen →
            (liver necrosis)
2. Phase I products (usually oxidative or hydrolytic) are almost always more polar and less lipophilic.
3. Phase II reactions frequently, but not invariably, abolish biological activity and add to polarity.
4. Biotransformation requires energy or energy equivalents which are supplied by glucose, glycine, coenzyme A (CoA), NADH, NADPH, and 3-phosphoadenine-S-phosphosulfate (PAPS).
5. The enzyme systems are saturable, frequently via depletion of cofactors (especially sulfate and glutathione).
6. The systems are age, sex, and species dependent and can be altered by exogenous factors such as diet, nutrition, state of health, induction, inhibition, and "natural" selection.

1.   Phase I Reactions

Monooxygenases

The most common phase I reaction is carbon oxidation, and this is most frequently accomplished by enzymes that introduce a single oxygen atom from the oxygen molecule. The second oxygen atom is usually reduced to form water. Thus, many of these monooxygenases have been referred to as "mixed function" (oxidative + reductive) or "multifunction" (because of the wide range of reactions possible) oxidases or MFO. MFO is entrenched jargon in pharmacology as well as toxicology.

The microsomal monooxygenases are versatile enzyme systems that oxidize endogenous steroids, fatty acids, prostaglandins, and xenobiotics. These enzymes are found in the fractured endoplasmic reticulum of disrupted cells and are isolated by differential centrifugation. Oxidative as well as conjugating enzyme systems are located within this lipoprotein membrane, and potential substrates must be *lipophilic* to access these enzyme systems. Although cytochrome P_{450} receives the most attention of the microsomal monooxygenases, another monooxygenase, formerly referred to as an amine oxidase but now called the FAD-dependent monooxygenase, is especially important in the oxidation of nitrogen and sulfur atoms.

Cytochromes P$_{450}$

For sheer numbers of potential substrates and potential pathways, the microsomal monooxygenases are by far the most important and most studied. Microsomal MFOs are, indeed, extremely important in pharmacology and toxicology, and a working concept of these oxygenases is essential. Extensive studies of the cytochromes P$_{450}$, the versatile terminal oxidases in the system, have been undertaken. All phase I reactions must not be attributed to cytochromes P$_{450}$ because other enzymes including monoamine oxidases and alcohol dehydrogenases mediate equally important biotransformations.

P$_{450}$ monooxygenases consist of a three-component electron transport system which generates an active oxygen in the presence of substrate, oxygen, and NADPH. The three necessary components are

- Cytochrome P$_{450}$ (heme-containing terminal oxidase)
- NADPH-cytochrome P$_{450}$ reductase (FAD- and/or FMN-containing enzyme)
- A phospholipid (e.g., phosphatidyl choline, phosphatidyl ethanolamine) that couples the reductase and cytochrome P$_{450}$

The overall oxidation reaction of xenobiotics is

$$NADPH + O_2 + Xenobiotic + H^+ \rightarrow NADP^+ + H_2O + X_{ox}$$

Since cytochrome P$_{450}$ has a heme molecule at the active center, it binds CO as well as oxygen. Thus CO is an effective inhibitor. The CO binding spectrum with an absorbance maximum at 450 nm was responsible for the nomenclature of the enzymes in the mid 1950s. Careful examination of the wavelength for maximum absorbance showed some shifts between 448 and 452 nm, depending on the source of the enzyme. This was the first strong evidence that there were actually at least a few different isozymes of cytochrome P$_{450}$.

Distribution and Function of Isozymes of Cytochromes P$_{450}$

Rat microsomes are fairly characteristic and contain the following average amounts of P$_{450}$ (nmol/mg microsomal protein):

Liver	0.86 (many isozymes, metabolize xenobiotics)
Kidney	0.11–0.20 (mainly oxidize fatty acids)
Lung	0.035 (fewer isozymes than liver)
Gut	0.01 (induced by dietary factors, 1A1 is the most studied and may be dominant)
Testes	0.07 (mainly metabolizes steroids)

Cytochrome P_{450} is also found in the skin, brain, breast, pituitary, nasal epithelium, etc. The cytochrome P_{450}s in various tissues metabolizing steroid hormones (such as adrenal, testis, ovary, corpus luteum) have long been known to be distinct from those in the lung and liver. In addition to being located in the nuclear membrane, mitochondria, and cytosol of the cell, they are also in the endoplasmic reticulum. Most of these enzymes are highly specific:

P_{450} XIXA1 (aromatase metabolizes testosterone to estradiol)

P_{450} XIA1 (scc) — cholesterol $\rightarrow$ androstenedione

Most of the effort has focused on identifying, purifying, and sequencing several forms of microsomal P_{450} from liver and lung microsomes which are important in the biotransformation of xenobiotics.

Coon et al. (1977) identified three distinct proteins (P_{450}s) in uninduced rat liver microsomes. These were referred to as LM (liver microsome) 1, LM 4, and LM 7. LM 2 was induced by phenobarbital (PB). Each had a somewhat different specificity for model substrates:

	LM 1,7	LM 2	LM 4
Benzphetamine	7.7	56.80	4.9
Biphenyl (2-OH)	0.4	0.70	0.3
Biphenyl (4-OH)	0.6	5.40	0.4
Benzopyrene	0.5	0.04	T

Using PB and tetrachlorodibenzo-*p*-dioxin (TCDD) as inducers, Johnson (1979) purified and described three major P_{450}s from rabbit liver. He called them 2, 4, and 6 or a, c, and b, respectively. Form 2 was induced by phenobarbital and had a molecular weight of about 47,500. TCDD induced form 4 in the adult and form 6 in the neonate. Forms 4 and 6 had molecular weights of 54,500 and 57,000, respectively. The Michaelis-Menten constants of these enzymes and their inhibition constants also varied. The substrate turnover rates (mmole/min/mmole P_{450}) were

	P_{450} forms		
	2	4	6
Acetanilide	1.0	6.1	1.0, 2.0
Benzopyrene	T	T	4.1
Benzphetamine	51.0	3.0	1.0
Biphenyl (4-OH)	4.1	3.3	1.0
Ethoxyresorufin	1.2	60.0	40.0

TABLE 3.1. Examples of Cytochromes P_{450}

Protein Name (P_{450})	Other Names	Inducer[a]	Prototype Substrate
IA1	c (rat), P_1 (mouse, human), 6 (rabbit), IA1 (trout)	3MC	Benzpyrene, testosterone ethoxyresorufin
IA2	d (rat) P_3 (mouse, human) LM4 (rabbit)	3MC	Acetanilide
IIB1	b (rat)	PB	Pentoxyresorufin, benzphetamine
IIB2	e (rat)	PB	Benzphetamine
IIB4	LM 2, P-450₁, B1, b46 (all in rabbit)	PB	Benzphetamine
IIB6	LM 2 (human)	PB	
IIC6	PB1, PB-C (rat)	PB	Warfarin
IIE1	j (rat, human), 3a (rabbit)	Ethanol Acetone	Ethanol nitrophenol
VIA1	Housefly	Resistant (PB?)	No prototype
XIA1	scc (human, cow)		Cholesterol (cleavage)
XVIIA1	17a (human, cow, pig, rat, chicken)		Pregnenolone, progesterone
XIXA1	Arom (human, chicken)		Testosterone → estradiol

[a] Phenobarbital (PB) or 3-methyl cholanthrene (3MC) are the prototype inducers.

Source: Gonzalez, 1990; Nebert et al., 1989b.

Since 1979, naming of the cytochrome P_{450}s has been more standardized. The important thing to remember is that P_{450}s have different substrate specificities and are induced unequally. Table 3.1 lists some of the major P_{450}s for which the genes have been defined. The modern nomenclature, as well as historical alternate names, are given.

The biotransformation pathways are presented somewhat systematically as phase I pathways (Table 3.2) and phase II pathways (Table 3.3). The end of Table 3.3 also shows phase III type pathways where the glutathione conjugate can be further metabolized to a mercapturic acid or, through beta lyase, to a thiol derivative. The thiol derivative can be methylated, oxidized to a sulfoxide by the FAD monooxygenase, and then oxidized to a sulfone by a P_{450} monooxygenase.

2. Nonmicrosomal Oxidation

Not all oxidations are microsomal (See Table 3.2, I.B.). Although ethanol is metabolized by cytochrome P_{450} IIE1, it is mainly oxidized to acetaldehyde

TABLE 3.2. Phase I Biotransformation Reactions

I. Oxidations
  A. Microsomal cytochromes P_{450} (and FAD monooxygenase)
    1. Carbon oxidations
      a. Methyl groups
        1. Aromatic methyl — very rapid

$$RCH_3 \rightarrow RCH_2OH$$

          Toluene, xylenes, tolbutamide, TNT
        2. Methylene between two aromatic rings

$$DDT \rightarrow Kelthane, trimethoprim$$

        3. Aliphatic ring methyl
           delta-8 and delta-9 tetrahydrocannabinols
      b. Terminal (or omega)

$$RCH_2CH_3 \rightarrow RCH_2CH_2OH$$

          Fatty acids (mainly kidney), alkanes,
            ethanol (but not as with alcohol dehydrogenase)
      c. Penultimate (or omega-1)

$$RCH_2CH_3 \rightarrow RCH(OH)CH_3$$

          Fatty acids, *n*-heptane, secobarbital, chlorpropamide
      d. Alicyclic carbons
         Tetrahydrocannabinols (minor), hexobarbital, nicotine,
          cyclopentene, testosterone, diazepam
      e. Haloalkanes
         Chloroform, carbon tetrachloride, DDT
      f. Olefin epoxidation (NIH shift also possible)
         Vinyl chloride, trichloroethylene, aflatoxin, aldrin
      g. Aromatic carbons
        1. Hydroxylation
           Steroids, aniline, biphenyl, acetanilid, nitrobenzene,
            nitrophenol, bromobenzene, phenobarbital, warfarin,
            polynuclear aromatic hydrocarbons
        2. Epoxidation (arene oxide formation)
           NIH shift results in dislocation of substituents acetanilid,
            bromobenzene, halobiphenyls, polynuclear aromatics
            (naphthalene, styrene, benzo[a]pyrene, etc.), carbaryl

TABLE 3.2. (Continued)

h. *N-*, *O-* and *S*-dealkylations

    1. *N*-dealkylation

$$RNHCH_3 \rightarrow [RNHCH_2OH] \rightarrow RNH_2 + HCHO$$

    Carbamate insecticides (carbaryl, propoxur), morphine, methadone, caffeine, benzphetamine, methyl ephedrine, aminopyrene

    2. *O*-dealkylation of ethers

$$ROCH_3 \rightarrow [ROCH_2OH] \rightarrow ROH + HCHO$$

    Propoxur, phenacetin, codeine, methoxychlor, ethoxyresorufin, *p*-nitroanisole

    O-dealkylation of esters

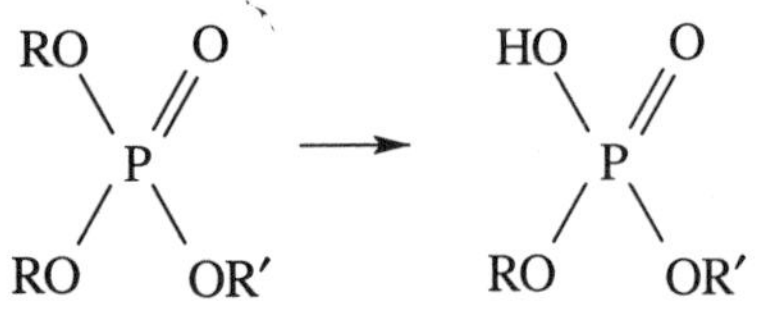

    Organophosphorous pesticides (chlorfenvinphos)

    O-dealkylation of dioxymethylenes

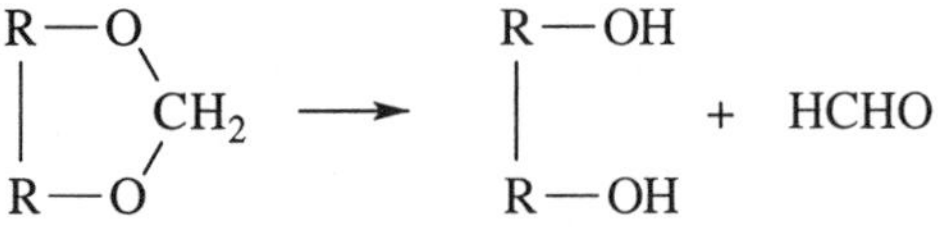

    Methylene dioxyphenyl (MDP) compounds

    Diethyl MDP acetamide (hypnotic — fast)

    Piperonyl butoxide (Insectide synergist — slow)

    3. *S*-dealkylation (may be FAD monooxygenase, not P_{450})

$$RSCH_3 \rightarrow RSH + HCHO$$

    Methitural, methylmercaptan, methylthiopurine

i. Oxidative deamination (essentially *N*-dealkylation which cleaves a relatively larger group from N)

$$RCH_2NH_2 \rightarrow RCHO + NH_3$$

TABLE 3.2. (Continued)

Amphetamine, mescaline (plasma amine oxidase),
   tyramine (monoamine oxidase)

2. Nitrogen oxidations
   a. Microsomal P_{450}

$$RNHR \rightarrow RNOHR, RN = O \text{ or } R = NOH$$

   Phenacetin >> aniline; 2-naphthalamine, 4-amino-biphenyl, urethane
      to carcinogenic products
   Amphetamine, lidocaine
   b. Microsomal FAD monooxygenase

$$R_3N \rightarrow R_3N = O$$

   N,N-demethylamphetamine, dimethylphenothiazines, nicotine,
      morphine

3. Phosphorus oxidations
   a. Microsomal P_{450}
      Oxidative desulfuration and/or dearylation

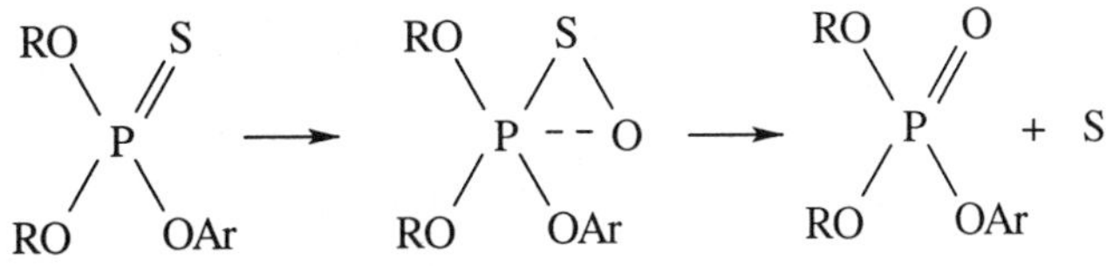

and/or

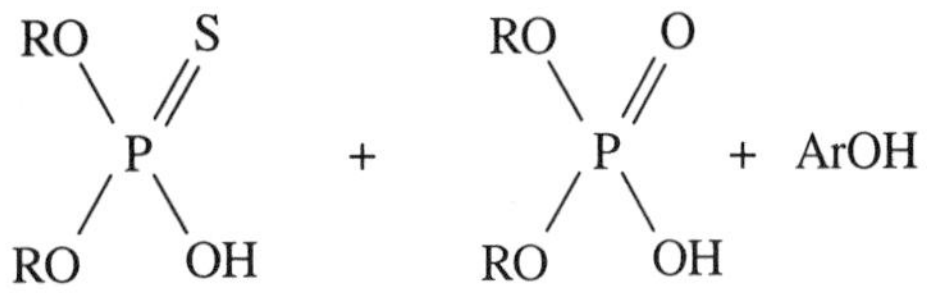

   Phosphorothionates (parathion, malathion), phosphorodithioates
      (phorate, systox) and phosphonothionates (EPN, fonophos)
   b. Microsomal FAD monooxygenase
      Oxidative desulfuration
      Phosphonothionates (fonophos, leptophos)

4. Sulfur oxidation
   a. Cytochromes P_{450}
      Sulfoxide oxidation

TABLE 3.2. (Continued)

$$RS(O)R \rightarrow RS(O)_2R \text{ (sulfone)}$$

Dimethylsulfoxide, sulfoxide metabolites generated by
   microsomal FAD monooxygenase
b. Microsomal FAD monooxygenase
Sulfoxidation

$$RSR \rightarrow RS(O)R$$

Phosphoro- and phosphonodithioates (phorate)
Carbamates (aldicarb, methiocarb)
Chlorpromazine, phenothiazines, methylthio metabolites of
   chlorinated aromatics such as DDT, PCBs and hexachlorobenzene
5. Dehalogenation (P_{450})
Halothane

$$F_3CHBrCl \rightarrow F_3C = CHBr + HCl \text{ (or } = CHCl + HBr)$$

Iodotyrosines (liver and thyroid)
B. Nonmicrosomal oxidations
   1. Dehydrohalogenation (soluble fraction, GSH required but not
       consumed)

$$RCC(Cl_3)R \rightarrow RC = C(Cl_2)R$$

$$p,p'-DDT \rightarrow DDE; \; p,p'-DDD \rightarrow TDE$$

Lindane, chlordane, chloroethanes
   2. Alcohol dehydrogenases (mostly soluble, many variants, reversible,
       NAD >> NADP as cofactor)

$$RCH_2OH \rightarrow RCHO$$

Methanol < ethanol < *n*-butanol, benzyl alcohol, chloral hydrate,
   many phase I products
   3. Aldehyde dehydrogenases (also soluble and reversible)

$$RCHO \rightarrow RCOOH$$

TABLE 3.2. (Continued)

Formaldehyde, acetaldehyde, benzaldehyde, ketamine, most alcohol
dehydrogenase products

4. Xanthine oxidase (soluble, Mo containing flavoprotein)
   Xanthine and purine analogs to uric acids, metabolic products of
   trimethylxanthines (caffeine, theophylline, theobromine), and other
   aromatics (benzaldehyde, salicyl-aldehyde) to a lesser extent

5. Monoamine oxidases (mainly mitochondrial)

$$RCH_2NH_2 \rightarrow RCHO + NH_3$$

Central nervous system (catecholamine neurotransmitters)
Liver, kidney, platelets, etc. (primary amines > secondary amines >
tertiary amines, *not* amines such as amphetamine with substituent on
carbon alpha to amine function
Dopamine, serotonin, tryptamine, tyramine (food sources)

6. Diamine oxidases (mainly soluble, Cu-containing)

$$H_2N(CH_2)_{4-8}NH_2 \rightarrow H_2N(CH_2)_{3-7}CHO$$

Primary diamines (longer chains by MAO)
Putrescine, cadavarine, histamine

7. Peroxidation, lipoxygenases
   Free radicals generated in oxidative reactions

II. Reduction

A. Reductive dehalogenation
   Halothane

$$F_3CHBrCl \rightarrow F_3CCH_3 + Cl^- + Br^-$$

$$DDT \rightarrow DDD$$

Pentachloroethane (both oxidative and reductive)

$$CCl_3-CHCl_2 \rightarrow \ldots CCl_2 = CHCl$$

B. Reverse oxidation (alcohol dehydrogenases, aldehyde reductase)

$$RCH_2OH \leftarrow \cdot \rightarrow RCHO \leftarrow \cdot \rightarrow RCOOH$$

TABLE 3.2. (Continued)

Warfarin, methadone, ketamine, acetaldehyde, benzaldehyde

C. Nitroreduction (microsomal and soluble)

$$RNO_2 \rightarrow RNO \rightarrow RNHOH \rightarrow RNH_2$$

Nitrobenzene, *p*-nitrobenzoic acid, parathion, TNT, picric acid, chloramphenicol, nitrofurazone

D. Azo reduction

$$RN = NR' \rightarrow RNH_2 + R'NH_2$$

Prontosil $\rightarrow$ Sulfanilamide
Azobenzene, aminoazotoluene, azo dyes (e.g., 4-dimethylaminoazobenzene or "butter yellow")

E. Transhydrogenases

$$RSSR' \rightarrow RSH + R'SH$$

Glutathione, insulin

F. Peroxidases

$$ROOH \rightarrow ROH$$

Catalase, glutathione dependent, ascorbate dependent

III. Hydrolysis

A. Acid catalyzed
   e.g., gastric juice
   Penicillin G

B. Epoxide hydratase

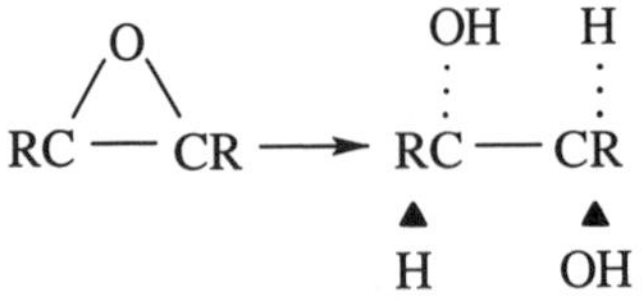

$$\underset{H}{\overset{O}{RC{-}CR}} \longrightarrow \underset{H \quad OH}{\overset{OH \quad H}{RC{-}CR}} \quad \text{(dihydrodiol)}$$

C. Amidases (also included as an esterase), lactamases

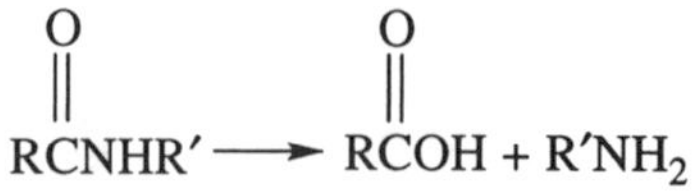

$$\overset{O}{\overset{\|}{RCNHR'}} \longrightarrow \overset{O}{\overset{\|}{RCOH}} + R'NH_2$$

TABLE 3.2. (Continued)

Dimethoate, penicillin
D. Esterases
    1. A-esterases ("arylesterases")
        Hydrolyze phosphotriesters and not inhibited
    2. B-esterases (all inhibited by paraoxon)
        a. Cholinesterases
            Acetyl cholinesterase (synapse) — acetylcholine
            Plasma cholinesterase — succinylcholine, procaine,
              organophosphates (but inhibited), atropine (in some rabbits)
        b. Carboxylesterases (and amidases)
            Deficient in insects, responsible for the selectivity of malathion
        c. Monoacylglycerol lipases
    3. C-esterases (paraoxon neither substrate nor inhibitor)
        Acetyl esters
E. Hydrolytic dehalogenation

$$RCH_2Cl \rightarrow RCH_2OH$$

Haloethanes, chlormethiazol
F. Conjugate hydrolysis
    Deacetylases, glucosidases, glucuronidases, sulfatases
    In rats, at least, glucosidases also catalyze the reverse formation of
    glucosides

TABLE 3.3. Phase II Biotransformation Reactions

I. Glycoside conjugations
    Substrates
        1. Alcohols

$$ROH \rightarrow RO\text{-glucuronic acid}$$

        Phenols, halophenols, naphthol, steroids, thyroxine, trichloroethanol,
        hydroxycoumarin
        2. Carboxylic acids

$$RCOOH \rightarrow RCOO\text{-glucuronic acid}$$

        Benzoic acids, aliphatic acids

TABLE 3.3. (Continued)

3. Amines

$$RNH_2 \rightarrow RNH\text{-glucuronic acid}$$

Anilines, naphthylamines, sulfonamides, carbamates

II. Sulfate conjugations

At least four different families of sulfotransferases have been identified in various tissues, mainly in the soluble fraction

1. Phenol sulfotransferase(s)

$$ArOH \rightarrow ArOSO_3H$$

Phenols, not steroids

2. Alcohol sulfotransferase(s): may reflect overlapping activities of other sulfotransferases

$$ROH \rightarrow ROSO_3H$$

Alcohols, polyols

3. Steroid sulfotransferases: several distinct and specific enzymes, may act in membrane transport

Estrone, androstenolone, testosterone, and deoxycorticosterone sulfotransferases

4. Arylamine sulfotransfer ($\rightarrow$ sulfamate) may be by steroid transferases

$$RNH_2 \rightarrow RNHSO_3H$$

5. Bile salt sulfotransferases

III. Phosphorylation

Very Rare

$$ROH + ATP \rightarrow ROP(O)O_2^{2-}$$

Dog: 2-naphthylamine

Cockroach: 1-naphthol and 4-nitrophenol

IV. Methylation

S-, N-, and *O*-methyl transferases

1. Catechol *O*-methyl transferases (soluble fraction of tissues in which catecholamines exert effects, highest in liver and kidney, will not methylate monophenols)

$$Ar(OH)_2 \rightarrow Ar(OH)OCH_3$$

TABLE 3.3. (Continued)

Dopamine, norepinephrine, dihydroxyamphetamine, 4-nitrocatechol
poor substrate, but good inhibitor
2. Other O-methyltransferases
   a. Hydroxyindole O-methyl transferase — soluble fraction of pineal
      gland and retina of vertebrates, diurnal variation, higher in darkness,
      melatonin synthesis, but also serotonin and hydroxyindoleacetic
      acid, not catechols
   b. Microsomal catechol O-methyl transferase,
         3-MC inducible
   c. Microsomal phenol O-methyl transferase

$$ArOH \rightarrow ArOCH_3$$

   e.g., hydroxyacetanilid
3. N-methyltransferases
   a. Soluble nonspecific N-methyltransferase
      Many tissues, but decidedly higher in lung

$$RNHR' \rightarrow RN(CH_3)R'$$

   Broad specificity, serotonin, tryptamine, benzylamine, amphetamine,
      nornicotine
   b. Soluble phenylethanolamine N-methyltransferase (PNMT)
      Many tissues, decidedly higher in adrenal

$$RCH_2(OH)NH_2 \rightarrow RCH_2(OH)NHCH_3$$

Norepinephrine, norephedrine, normetanephrine, not
phenylethylamines
   c. Histamine NMT — highly specific, soluble fraction, many tissues,
      but especially in brain
4. Thiol S-methyltransferase — microsomes may be inducible by 3-MC

$$RSH \rightarrow RSCH_3$$

   Mercaptoethanol, mercaptoacetic acid, phenylsulfide,
      2,3-dimercaptopropanol (British anti-lewisite [BAL])
   Hydrogen sulfide:

$$H_2S \rightarrow CH_3SH \text{ (toxic)} \rightarrow CH_3SCH_3$$

TABLE 3.3. (Continued)

5. Microbial biomethylation of metals and metalloids
    Increased lipophilicity very important toxicologically

$$Hg^{+2} \rightarrow CH_3Hg^+ \rightarrow CH_3HgCH_3$$

Mercury, lead, tin, thallium, arsenic, selenium, sulfur

V. Acylation: acetylation and amino acid conjugation
    Through a two step process which uses ATP, the acid is activated by
    binding with coenzyme A

$$RCOOH + ATP + CoASH \rightarrow RCOSCoA + AMP$$

A. Acetylation (soluble fraction)
    Acetyl coenzyme A: arylamine *N*-acetyl transferase

$$RNH_2 + CH_3CO\text{-}S\text{-}CoA \rightarrow RNHCOCH_3 + CoASH$$

Endogenous OH (choline) and SH (CoA) compounds but exogenous
  OH and SH acetylation not known
Aniline, benzidine, isoniazid, sulfamethazine
(high degree of genetic polymorphism, "slow" and "fast" acetylators
  show different suceptibilities to, e.g., isoniazid)
Low levels in newborns and some species (e.g., dog)
Sulfanilamide, *p*-aminobenzoic acid appear to be acetylated by a more
  genetically uniform enzyme

B. Amino acid conjugation

$$R'COSCoA + R(NH_2)COOH \rightarrow R'CONHRCOOH$$

Benzoic acids, phenylacetic acids, indolylacetic acid
Bile acids generally conjugated with taurine, aliphatic acids rarely
  conjugated with amino acids

C. Deacetylation (different enzymes: soluble, mitochondrial, and microsomal)
    Predominates in dog so that net acetylation is poor

Acetanilid → aniline

VI. Glutathione transferases
    Various glutathione *S*-alkyl, *S*-aryl, *S*-arylalkyl, *S*-epoxide, and *S*-alkene
      transferase reactions are catalyzed by several GSH transferases in insects as
      well as vertebrates

TABLE 3.3. (Continued)

$$\underset{\substack{\displaystyle | \\ \underset{\substack{\| \quad\quad\quad \| \\ O \quad\quad\quad NH_2}}{NHCCH_2CH_2COOH}}}{\overset{\displaystyle O}{\overset{\|}{HSCH_2CHCNHCH_2COOH}}} + RX \longrightarrow \underset{\substack{\displaystyle | \\ \underset{\substack{\| \quad\quad\quad \| \\ O \quad\quad\quad NH_2}}{NHCCH_2CH_2COOH}}}{\overset{\displaystyle O}{\overset{\|}{RSCH_2CHCNHCH_2COOH}}}$$

$$\longrightarrow \underset{\substack{| \\ NH_2}}{\overset{\displaystyle O}{\overset{\|}{RSCH_2CHCNHCH_2COOH}}} + \text{glutamate}$$

$$\longrightarrow \underset{\substack{| \\ NH_2}}{\overset{\displaystyle O}{\overset{\|}{RSCH_2CHCOH}}} + \text{glycine}$$

$$\longrightarrow \underset{\substack{| \\ \underset{\substack{\| \\ O}}{NHCCH_3}}}{\overset{\displaystyle O}{\overset{\|}{RSCH_2CHCOH}}} \quad \text{(mercapturic acid)} \quad (N\text{-acetylated cysteine})$$

$$\xrightarrow{\text{beta-lyase}} \text{XSH and/or XSCH}_3 \longrightarrow \overset{\displaystyle O}{\overset{\|}{\text{XSCH}_3}} \longrightarrow \underset{\substack{\| \\ O}}{\overset{\displaystyle O}{\overset{\|}{\text{XSCH}_3}}}$$

A. Epoxides (naphthalene oxide, vinyl chloride oxide, hexachlorobutadiene oxide)
B. Halo- and nitrobenzenes (PCBs, chlorodintrobenzene, TNT) and other aryl compounds (phenacetin with or without epoxide intermediate, paracetamol apparently through oxide)
C. Haloalkanes (methyl iodide is a GSH depleter)
D. Sulfate esters (bromsulfophthalein, methyl naphthyl sulfate)
E. Phosphate triesters

TABLE 3.3. (Continued)

1. Aryl transferases (paraoxon)
2. Alkyl transferases (trimethylphosphate)

F. Steroids (testosterone, cortisol, estradiol)
G. Alkenes (diethylmaleate GSH depleter)

$$\begin{matrix} CHCOOC_2CH_3 \\ \| \\ CHCOOC_2CH_3 \end{matrix} \longrightarrow \begin{matrix} CHCOOC_2CH_3 \\ | \\ GSCHCOOC_2CH_3 \end{matrix}$$

H. Some carboxylic esters

by catalase and alcohol dehydrogenases (Figure 3.1). Methanol is oxidized only by alcohol dehydrogenases and catalase. The aldehydes are very short-lived and rapidly oxidized to acetic and formic acids, respectively. Alcohol dehydrogenases can reduce the aldehydes as well as oxidize the alcohols. Ethanol, which has a greater affinity for some alcohol dehydrogenases and a higher turnover rate than does methanol, can be coadministered in methanol poisoning to slow the production of formic acid.

3. Phase II Reactions

Synthetic reactions or conjugations result in the addition of an endogenous molecule (such as acetate, glucuronic acid, sulfate, glycine, glutamine, or glutathione) to a susceptible group on the xenobiotic. The functional group to which the conjugating moiety attaches is frequently introduced during the phase I biotransformation. Substrates include many more hydrophilic chemicals which may not require phase I activation (e.g., halophenols, benzoic acids, 2,4-dichlorophenoxyacetic acid [2,4-D], 2,4,5-trichlorophenoxyacetic acid [2,4,5-T], etc.).

Conjugating groups are usually transferred by endogenous enzyme systems, and either the conjugating agent or the xenobiotic is frequently activated to a high energy intermediate. The conjugated products are almost always more water soluble and more readily excreted, and they are frequently less bioactive than the unconjugated compound. A brief discussion of the formation of five of the most important high energy conjugating agents are outlined below.

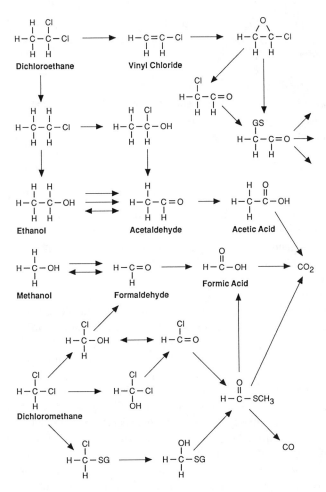

Figure 3.1. Relationships between the biotransformation of chloroalkanes and simple alcohols (Anders et al., 1977; Monks et al., 1990).

Glucuronidation

Uridine diphosphate glucose (UDPG), which is used as the conjugating moiety in invertebrates and plants, is formed from the condensation of uridine triphosphate (UTP) with glucose-1-phosphate. UDPG is dehydrogenated to form UDP-glucuronic acid (UDPGA) which is the usual conjugating moiety in vertebrates. UDPGA transferases are located in the endoplasmic reticulum and are induced by similar compounds to those which induce cytochromes P_{450}. In contrast to other conjugating enzymes which are found in the cytosol, UDPGA-transferase is found in the microsomal fraction (Hodgson and Guthrie, 1980).

At least four UDPGA transferases have been isolated from rat liver microsomes. Substrates for UDPGA transferases include alcohols, carboxylic acids, and amines (Table 3.3).

Sulfotransferases

Formation of the cofactor, 3'-phosphoadenosine-5'-phosphosulfate (PAPS), for sulfotransferase activity requires considerable energy and is represented by the following reaction:

$$SO_4^{-2} + 2ATP \rightarrow 3\text{'-phosphoadenosine-5'-phosphosulfate}$$

As the free sulfate pool is rather low in most animals, chronic exposure to high levels of xenobiotics can deplete the pool. Four different families of sulfotransferases have been identified in the soluble fraction of many tissues. Substrates that undergo conjugation by these enzymes include phenols, alcohols, steroids, arylamines, and bile salts.

Methyltransferases

The active methyl cofactor required for methyltransferases is *S*-adenosyl methionine (SAM) which is formed by the activation of methionine by ATP:

$$\text{Methionine} + ATP \rightarrow S\text{-adenosyl methionine}$$

5-Methyltetrahydrofolic acid may also be a cofactor in some cases. In contrast to other phase II reactions, methylation *generally decreases* water solubility and may merely change rather than decrease biological activity. This is especially true for endogenous substrates such as *N*-acetyl serotonin which is methylated to melatonin or norepinephrine that is methylated to epinephrine. Besides these *O*-methyltransferases, *N*-methyltransferases and thiol *S*-methyltransferases are important. Microbial biomethylation of metals including mercury, lead, tin, thallium, arsenic, and selenium play a role in the toxicity of these metals (Chapter 4).

Acetylation

Acylation in xenobiotic biotransformation involves either activation of acetate that is transferred to an amine, or direct activation of a xenobiotic to an acid so an endogenous amino acid can be conjugated to it. Through a two-step process that uses ATP, the acid is activated by coenzyme A (CoASH):

$$RCOOH + ATP + CoASH \rightarrow RCOSCoA + AMP$$

Acetylation occurs in the soluble fraction and involves the interaction of acetyl coenzyme A with an amine such as aniline, benzidine, isoniazid, and sulfamethazine

$$RNH_2 + CH_3COSCoA \rightarrow RNHCOCH_3 + CoASH$$

A high degree of genetic polymorphism in acetylation exists in man and some other vertebrates. Men who are "slow" or "fast" acetylators show different susceptibilities to certain drugs such as isoniazid. Low levels of acetylase activity are found in newborns as well as other vertebrates including dogs. Sulfanilamide and *p*-aminobenzoic acid may be acetylated by a more genetically uniform enzyme.

An activated carboxylic acid can be conjugated to an amino acid:

$$R'COSCoA + R(NH_2)COOH \rightarrow R'CONHRCOOH$$

The amino acid used varies with the species; in humans conjugation is mostly with glutamine, while other primates use glutamine and glycine and most other mammals use glycine. Ornithine is the predominant amino acid used by reptiles and birds, taurine by fish (also, cats, dogs, and ferrets), and arginine by arthropods. Some pathways use dipeptides such as glycylglycine. Benzoic acids, phenylacetic acids, and indolylacetic acid undergo this reaction. Bile acids conjugate with taurine; but aliphatic acids rarely conjugate with amino acids (Hodgson and Levi, 1987).

Glutathione *S*-Transferases

Various glutathione *S*-alkyl, *S*-aryl, *S*-arylalkyl, *S*-epoxide, and *S*-alkene transferase reactions are catalyzed by several glutathione-*S*-transferases in vertebrates as well as in insects. There may be a great deal of overlap in substrate specificity. Most of the enzymes are found in the soluble fraction of the cell, although some isozymes are microsomal. Some of the isozymes are induced by both phenobarbital and 3-methylcholanthrene, presumably via mRNA for different subunits. Almost any parent compound or metabolite with an electrophilic function can be bound to the sulfur atom of glutathione, thus protecting cellular nucleophiles from binding. However, vicinal dihaloethanes are activated to potent mutagens (episulfonium ion) (Figure 3.1) and haloalkanes are activated to nephrotoxicants. Reversed deconjugation, which usually occurs in the bladder, may release active electrophiles. In addition, microbial (gut) and kidney cysteine beta-lyases may release toxic thio- or methylthio-compounds which may be reabsorbed and/or additionally activated.

Although conjugation by phase II enzymes usually results in detoxification, in certain circumstances this reaction enhances the toxicity of an intermediate. Such a situation occurs with many chloroethanes, shown in Figure 3.1,

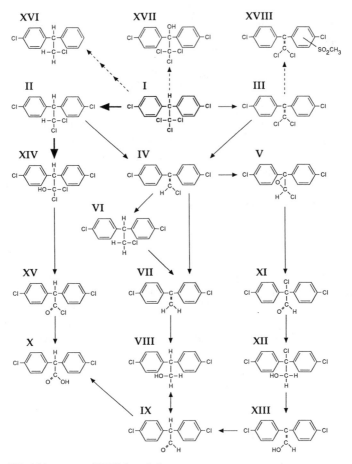

Figure 3.2. Major routes of DDT degradation.

in which dehydrohalogenation results in the formation of a reactive epoxide. For example, dehalogenation of dichloroethane results in the formation of vinyl chloride, which undergoes oxidation by cytochrome P_{450} to form a reactive epoxide. The epoxide rearranges to the chloroethanol which is conjugated to glutathione (GS). This intermediate can bind to macromolecules such as DNA (Anders et al., 1977). The GS conjugate may eventually form a mercapturic acid, but a large proportion of the cysteine conjugate is converted to reactive thiols via the beta lyase pathway. For example, trichloroethylene-extracted soybean oil meal contains the toxic dichlorovinyl cysteine conjugate (Monks et al., 1990). Dichloromethane is metabolized to toxic thiols which can break down to carbon monoxide. This degradation pathway probably requires GS-dependent routes in vertebrates, but not in bacteria.

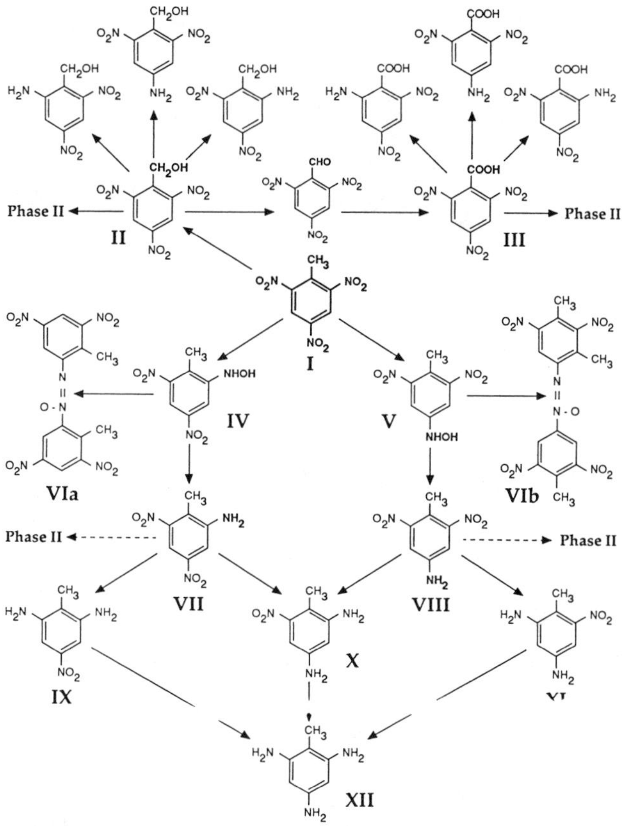

Figure 3.3. Common metabolites of 2,4,6-trinitrotoluene (TNT).

4. Complexity of Xenobiotic Metabolism

Although the metabolism of a xenobiotic is often depicted as being mediated by only one phase I and/or phase II enzyme reaction, in reality a xenobiotic can be metabolized by many cytochrome P_{450} enzymes and nonmicrosomal enzymes simultaneously. Each of the products in turn can be further metabolized by cytochrome P_{450} or other enzymes. Also the products formed after each of these steps can be rendered more hydrophilic through phase II pathways. To represent the complexity of these enzymatic pathways, schemes showing the metabolism of dichlorodiphenyltrichloroethane (DDT, Figure 3.2), trinitrotoluene (TNT, Figure 3.3) and a representative polychlorinated biphenyl (PCB 52, Figure 3.4) are described and shown as follows.

Metabolism of DDT

DDT (1,1,1-trichloro-2,2-bis[*p*-chlorophenyl]ethane, Figure 3.2, I), also

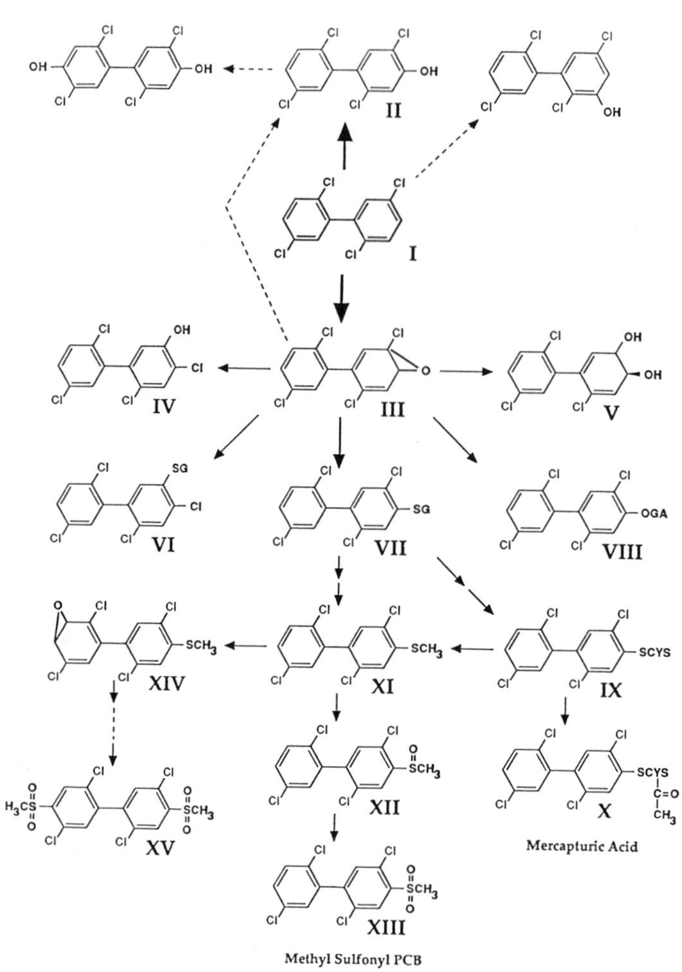

Figure 3.4. A representative pathway for PCB biotransformation.

known as dichlorodiphenyltrichloroethane, is dechlorinated to DDD (1,1-dechloro-2,2-bis[*p*-chlorophenyl] ethane, Figure 3.2, II), also known as dichlorodiphenyldichloroethane or tetrachlorodiphenylethane (TDE). In addition to being an impurity in DDT as well as being a metabolite, TDE was once marketed as a less broad spectrum insecticide under the name of Rhothane. The *o,p*-isomer has also been used in chemotherapy of adrenal tumors under the brand name of Mitotane. DDT is also dechlorinated to DDE (1,1-dechloro-2,2-bis[*p*-chlorophenyl]ethylene, Figure 3.2, III) which can also be an impurity in DDT preparations. The oxidized metabolite of DDT, known as Difocol or Kelthane (2,2,2-trichloro-1,1-bis[*p*-chlorophenyl]ethanol, Figure 3.2, XVII) which was sold as an acaracide, can also be produced from DDT by some insects and microorganisms. The stable methyl sulfonyl metabolite (Figure 3.2, XVIII) of DDE has been found in the adipose tissue of seals (Jensen and Jansson, 1976). Although most biotransformation reactions occur at the ethane bridge, evidence for biotransformation of the chlorophenyl rings has been found with the isolation of a dechlorinated compound (Figure 3.2, XVI).

DDD is rapidly formed from DDT and, in turn, is rapidly oxidized to DDMU (1-chloro-2,2-bis[*p*-chlorophenyl]ethene, Figure 3.2, IV) which is then metabolized to V, VI, or VII depending on conditions. The epoxide (Figure 3.2, V), a potent electrophile, can also rearrange to form XI and eventually, via several steps, yields DDA (2,2-bis[*p*-chlorophenyl]acetic acid, Figure 3.2, X), the major urinary metabolite of DDT (Gold and Brunk, 1984). An alternative route for the metabolism of DDD to DDA (Figure 3.2, X) is via the potent electrophile XV which may be formed in greater amounts in the kidney of some species than is the epoxide V. Most of the DDA formed is conjugated to amino acids for urinary excretion (Morgan and Roan, 1971). Some microbes can decarboxylate DDA and further oxidize the resulting methane bridge. This results in the splitting of the biphenyl moiety to chlorobenzoic acid and other products. Dehydrochlorination at the ethane bridge is considerably slower for *o,p*-chlorinated isomers.

Although dechlorination of DDT to DDD (II) is more rapid than dehydrochlorination to DDE (III), further metabolism of DDE to IV is slower; thus, DDE is more stable in the environment and occurs at higher levels than either DDT or DDD. Some DDT resistant organisms have greater DDT dehydrochlorinase activity, so conversion to DDE may be more rapid in these cases.

Metabolism of Trinitrotoluene

Oxidation of unsubstituted toluene by cytochrome P_{450} (Figure 3.3, I to II) is very rapid and is followed sequentially by further oxidation by alcohol and aldehyde dehydrogenases to benzoic acid. Benzoate is conjugated to an amino acid, most commonly glycine, to yield hippuric acid (Figure 3.3).

With the substitution of toluene at three positions by nitro groups to form trinitrotoluene (TNT), reduction reactions are favored in both mammals and microorganisms in contaminated environments. The nitro groups of TNT (Figure 3.3, I) are reduced to form 2-amino-4,6-dinitrotoluene (Figure 3.3, VII) and 4-amino-2,6-dinitrotoluene (Figure 3.3, VIII). The two compounds and their conjugates comprise significant proportions of environmental residues and mammalian excretory products (Yinon and Hwang, 1984). Both 2-amino-4,6-dinitrotoluene and 4-amino-2,6-dinitrotoluene can undergo further reduction of the nitro groups to form 2,4,6-triaminotoluene (Figure 3.3, XII).

TNT is manufactured by progressive nitration of toluene, and the product is purified by washing with massive amounts of water. The waste waters contain many of the metabolites as well as the products from partial nitration.

These TNT wastes and/or biodegradation mixtures are more toxic to rats and invertebrates than the individual products alone. Toxicity is also enhanced by exposure to near-UV light (Johnson, 1992). Glutathione appears to mediate the enhancement of toxicity of some metabolites and/or photodegradation products while participating in the detoxification of others.

The nonenzymatic condensation products (Figure 3.3, VIa and VIb) derived from the hydroxylamino intermediates, IV and V, are interesting in that a similar product (azoxy-4-ethoxybenzene) is found in the urine of rats following administration of the analgesic, phenacetin (N-[4-ethoxyphenyl]acetamide).

Metabolism of PCB 52

PCB 52 (2,2′,5,5′-tetrachlorobiphenyl)+ (Figure 3.4, I), is a moderately chlorinated symmetrical congener. The number of metabolites formed from an asymmetric and more lightly chlorinated congener would be greater. Arene oxide formation (e.g., XI → XIV) is favored by the presence of vicinal hydrogens. However, oxidation of PCB 52 (I → III) is also possible, and the NIH shift products (Figure 3.6, IV, VI) are common. Other mechanisms of hydroxylation, preferably at the 4 position are also possible, yielding II.

Glucuronyl conjugates (Figure 3.4, VIII) are readily excreted, but glutathione conjugates (Figure 3.4, VI, VII) may be further metabolized to mercapturic acids (Figure 3.4, X) or enter phase III metabolism via the beta lyase pathway. The thiomethyl (Figure 3.4, XI) intermediates can be reabsorbed, then oxidized by the FAD dependent monooxygenase to the sulfoxide (Figure 3.4, XII) which is further oxidized by P_{450} to the methyl sulfone (Figure 3.4, XIII) (Sipes and Schnellmann, 1987). Methyl sulfonyl PCBs are lipophilic and persistent and have been found in animal as well as human tissues (Jensen and Jansson, 1976).

C. FACTORS MODIFYING BIOTRANSFORMATION

1. Diet

Mineral, vitamin, and protein deficiencies can affect the rate at which biotransformation takes place. Deficiencies of calcium, copper, iron, magnesium, and zinc decrease the rate of oxidation and reduction of xenobiotics by cytochrome P_{450}. When the mineral intake returns to normal, the rates of the enzyme activities also return to their normal background values. Deficiencies in vitamins are directly or indirectly involved in the regulation of cytochrome P_{450} enzymes and can alter the redox state of cells, resulting in a deficit of high energy cofactors which are required for phase II biotransformation. Low protein diets reduce the toxicity of compounds that require bioactivation but increase the toxicity of compounds that are direct acting compounds. For example, animals maintained on a diet low in protein are less vulnerable to the toxicity of the indirect acting compound dimethylnitrosamine (DMN) (Sipes and Gandolfi, 1991). This is probably due to a reduction in the rate of the *N*-demethylation of DMN which is the first step required for the activation of this particular compound. As P_{450} enzymes are membrane bound, dietary lipids can also alter the activity of these enzymes. For example, rats fed diets high in polyunsaturated fats have a decreased concentration of cytochrome P_{450}. This is thought to be due to the increased susceptibility of unsaturated fatty acids to undergo peroxidation, resulting in the degradation of the endoplasmic reticulum and concomitant loss of cytochrome P_{450}.

In many toxicology studies, animals are deprived of food overnight to decrease the concentration of food in the gut and to enhance the absorption of orally administered compounds. However, rodents are nocturnal feeders and will actually have been deprived of food for longer than 12 h. Food deprivation increases the levels of several biotransformation enzymes which could enhance liver injury. In addition, food deprivation also reduces the concentration of cofactors and conjugating enzymes. For example, the concentration of glutathione is reduced by 50% if animals are deprived of food overnight. This results in a potentiation of the hepatotoxicity of acetaminophen, bromobenzene, and other compounds that are detoxified through the glutathione pathway. Some natural dietary components can inhibit or enhance certain biotransformation reactions. Ingestion of quinoline-type compounds, produced by charbroiling and frying certain meats, can induce certain P_{450} isozymes. All of the above mentioned factors may play a role in individual variation observed in the biological response to various xenobiotics.

2. Hepatic Injury

As the liver is the principal site of biotransformation in an animal, any adverse effect on the liver will interfere with normal cytochrome P_{450} metabolic pathways. Diseases which have been shown to reduce hepatic biotransformation capabilities of the liver include certain viral infections, carcinoma, obstructive jaundice, hepatitis, and cirrhosis (Sipes and Gandolfi, 1991). However, in the regenerating liver, biotransformation is frequently enhanced.

3. Circadian Rhythms

The time of day at which cytochrome P_{450} enzymes are isolated from animals can influence the outcome of *in vitro* studies of xenobiotic biotransformation. This may be related to variations in hormone levels which affect diurnal rhythm. A study has shown that the *in vitro* biotransformation of hexobarbital depends on the time of day at which the animals are sacrificed.

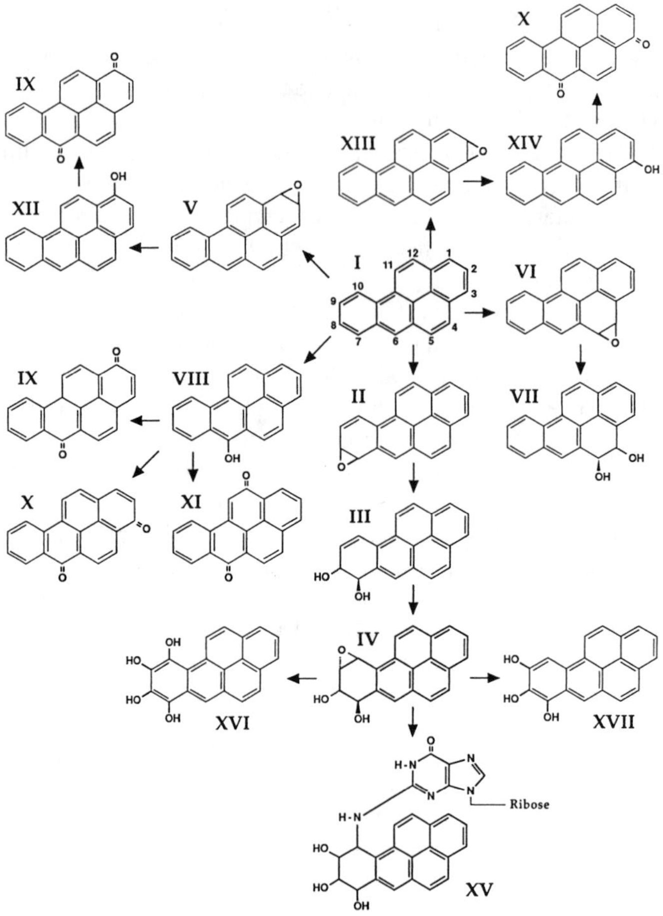

Figure 3.5. Oxidative biotransformation of benzo[a]pyrene (Sims and Grover, 1974; Levin et al., 1982).

Besides cytochrome P_{450}, glutathione is also known to be altered by circadian rhythm.

4. Bioactivation

Many xenobiotic compounds undergo biotransformation to active inter-mediates, a process known as bioactivation. The insertion of an oxygen atom, acceptance of an inducing equivalent, or conjugation with a phase II enzyme can result in the formation of a reactive intermediate that may rearrange to form an unstable compound. Reactive intermediates interact with nucleo-philic sites such as the sulfhydryl group of cysteine and methionine in

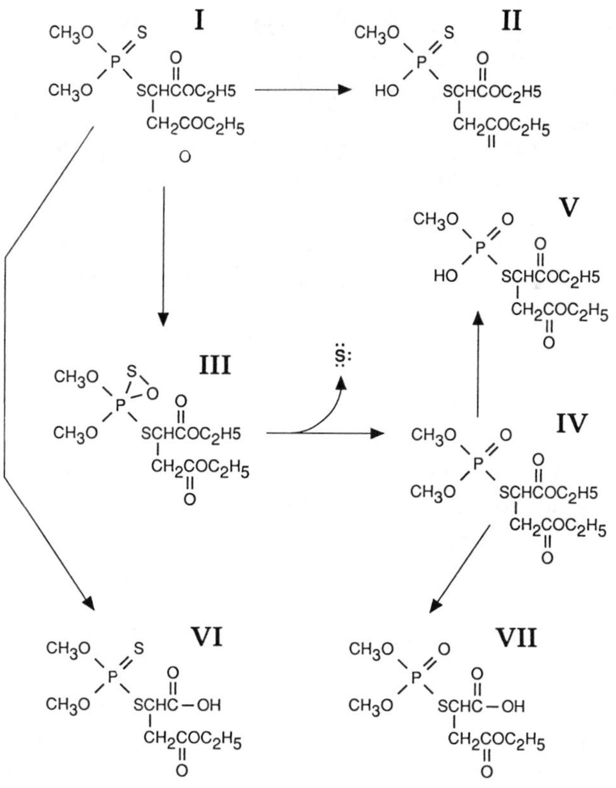

Figure 3.6. Biotransformation of the insecticide, Malathion.

proteins, or the amino or hydroxyl groups of DNA, RNA, or protein. This covalent interaction between a macromolecule and a xenobiotic is thought to be important in the elicitation of certain toxic effects. In some cases, these reactive intermediates can be detoxified. No adverse cellular effect is observed when there is a balance between the formation rate of a reactive intermediate and its detoxification. However, when this balance is disturbed either by enhanced production of reactive intermediates or a diminished capacity for detoxification reactions, injury may occur. Several factors can interact or interfere with the balance between the rate of formation and the rate of detoxification of reactive intermediates. For example, enzyme induction can increase the overall rate of biotransformation of a chemical leading to the excess production of a reactive intermediate. Examples of compounds that are bioactivated to reactive intermediates are aflatoxin B1, benzene, benzo(a)pyrene (Figure 3.5), bromobenzene, carbon tetrachloride, chloroform, dimethylnitrosamine, parathion (Figure 3.6), trichloroethylene, and vinyl chloride.

Benzo(a)pyrene (BP, Figure 3.5, I) is subject to P_{450}-mediated activation reactions at several sites and often by successive oxidations. BP can be simultaneously activated to toxic intermediates (Figure 3.5, IV, VI) and deactivated to less toxic metabolites (Figure 3.5, V, XIII) by mixed function oxidases. These latter epoxides undergo hydrolysis by epoxide hydrolase to 1 and 3-hydroxybenzo(a)pyrene (Figure 3.5, XII, XVI) which are then conjugated by phase II enzymes to form excretory products. Although the 4,5-oxide (Figure 3.5, VI) is mutagenic, the pathway of major concern is I → II → III → IV. The diol epoxide (Figure 3.5, IV) can exist as (+) or (–) isomers, both with different potencies, but both can arylate nucleosides such as guanosine to form an adduct (Figure 3.5, XV). The 7,8-*trans*-dihydrodiol-9,10-epoxide (Figure 3.5, IV) is the most potent carcinogenic metabolite of BP. Many metabolites including the 3,6-quinone (Figure 3.5, X) and 1,6-quinone (Figure 3.5, IX) can be produced by more than one pathway. Although not as toxic as the 7,8-*trans*-dihydrodiol-9,10-epoxide, the quinones can cause tissue damage through redox cycling. The scheme represents most of the known metabolites of BP, although the number of actual and potential metabolites are more extensive.

5. Induction

Induction is the process whereby the activity of an enzyme or the amount of a specific protein is increased following exposure of an animal to a specific xenobiotic. Many chemicals including drugs, pesticides, environmental chemicals, natural products, and alcohols have been shown to cause induction. Frequently, the enhanced activity of the enzyme results from an increase in the synthesis of the biotransformation enzyme, that is, there is *de novo* protein synthesis. Originally it was thought that only phase I enzymes, particularly cytochrome P_{450} enzymes, could be induced by these agents, but now it is known that phase II conjugating enzymes including glutathione-*S*-transferase and UDP-glucuronyltransferase can also be induced. The dynamics of the increase in these monooxygenase enzymes depends on many factors including the dose of the inducing agent, the substrate used to assay the enzyme activity, the species, sex, and duration of exposure of the animal to the inducing agent, and the tissue in which the activity is measured.

In an environmental setting, the compounds that have been shown to be the most widely distributed inducing agents are the polycyclic aromatic hydrocarbons (PAH), polychlorinated biphenyls (PCB), and tetrachlorodibenzo-*p*-dioxins (TCDD). A second group of inducing agents with different inducing characteristics has also been identified. Members of this group include phenobarbital, DDT, and other chlorinated compounds. The anabolic steroid testosterone is also responsible for elevated levels of certain biotransformation enzymes.

3-Methylcholanthrene (3-MC) is a classic compound representing the first group which is often used in laboratory experiments. Compounds in the first group are usually associated with the induction of cytochrome P_{450} isozymes which activate compounds to more toxic intermediates while compounds in the second group are more frequently associated with detoxification enzymes.

Mechanism of Induction

An understanding of the mechanism of induction by 3-MC-type inducers has been obtained using the very potent inducing agent TCDD. TCDD is a potent inducing agent of aryl hydrocarbon hydroxylase activity (AHH), the cytochrome P_{450} isozyme that metabolizes many polycyclic aromatic hydrocarbons. Poland et al. (1976) identified a protein in the cytoplasm of cells which bind covalently to TCDD. It is now known that this protein is a receptor for TCDD. The resulting receptor ligand complex that forms between TCDD and the receptor protein is transported into the nucleus. This complex then interacts with specific genomic recognition sites and stimulates transcription and subsequent translation of the genes that code for specific cytochrome P_{450} isozymes. It has been postulated that the mechanism of induction by benzo(a)pyrene, 3-MC, and some PCB isomers is similar to TCDD as these compounds can displace TCDD from the receptor. The major requirement for interaction with the receptor appears to depend on the lipophilicity and planar configuration of the molecule. For example, certain PCB congeners such as 2,2', 3,3', and 6,6' do not displace TCDD from its receptor and are not good inducing agents of cytochrome P_{450}, while the planar PCB congener 3,3', 4,4', 5 is a good inducing agent.

The mechanism of induction by phenobarbital and related compounds is less well known, although it is believed that synthesis of the isozymes induced by phenobarbital is regulated at the transcriptional level and results in a rapid and dramatic increase in messenger RNA (mRNA) which encodes these enzymes.

Induction by the various inducing agents follows two different time courses. With 3-MC, induction is rapid and reaches a peak 48 h after administration of the compound while that of phenobarbital requires about 5 d. Withdrawal of the inducing agent results in a return to baseline levels within 48 h with 3-MC and 5 to 7 d with phenobarbital. If a compound is highly lipophilic and poorly biotransformed, then its period of induction capacity can be over a much longer time frame.

In mammals, phenobarbital is a poor inducer in extrahepatic tissues while 3-MC and related compounds are good inducers of extrahepatic P_{450} enzymes. A cytosolic binding protein has been found in many extrahepatic tissues and is thought to participate in the induction of isozymes in these tissues in a similar manner to that in which it induces cytochrome P_{450} in the liver.

6. Inhibition

A compound is considered to be an inhibitor of an enzyme if it decreases the ability of an enzyme to metabolize a substrate relative to its activity in a control animal (Testa and Jenner, 1981). Several mechanisms may be involved in inhibition, including competition for the active sites or cofactors of the enzyme, inhibition of the transport of an enzyme or a component of an enzyme, and decreased biosynthesis or increased breakdown of the enzyme itself (Sipes and Gandolfi, 1991). Competitive inhibition and/or saturation of enzymes can be important toxicologically if many similar xenobiotics or very high concentrations are present simultaneously.

Compounds that alter protein synthesis will obviously inhibit the synthesis of biotransformation enzymes. Some inhibitors are more specific than others in that they inhibit the synthesis of cytochrome P_{450} but not the synthesis of proteins in general. Two chemicals that are known to inhibit the synthesis of heme, a coenzyme of cytochrome P_{450}, and its precursor porphyrin are, respectively, cobalt and 3-amino-1,2,3-triazole. Some compounds are known to affect the tissue levels of certain cofactors. Both L-methionine-*S*-sulfoximine and buthionine sulfoximine inhibit the synthesis of glutathione, while diethyl maleate rapidly reduces the tissue stores of glutathione. The synthesis of UDP-glucuronic acid is inhibited by galactosamine. As a result of this depletion or reduction in concentration of these cofactors, the formation of glutathione and glucuronide conjugates may be inhibited.

7. Sex Differences

Very little information has been gathered on sex differences in the biotransformation of xenobiotics in lower vertebrates (except a few fish species) and invertebrates, although this is a common finding among mammals. In general, males have a higher activity of cytochrome P_{450} enzymes and therefore metabolize many compounds more rapidly than do females. The levels of both cytochrome P_{450} and NADPH-cytochrome P_{450} reductase activity is 20 to 30% greater in male compared to female rats. Organophosphate insecticides are twice as toxic in female rats as a result of the extended biological half-life of the activated intermediate. Also, male detoxification enzymes are relatively more active than their activation enzymes (desulfurases).

In general, if a reactive intermediate is formed from a particular xenobiotic, male rats are more susceptible. Thus male rats are more likely to show signs of hepatic injury following exposure to carbon tetrachloride. Similarly, chloroform, which is metabolized to phosgene, is far more toxic to male mice than to females. It has been postulated that the levels of the hormone testosterone are important in modulating these responses. This is borne out by the finding that administration of testosterone to female rats increases their ability to biotransform several chemicals. In fact, the activity in these animals approaches that found in males. Similarly, castration of males results in a reduced capacity to metabolize certain xenobiotics.

8. Age

Xenobiotics are more toxic to very young and very old mammals due to the absence or low levels of cytochrome P_{450} and the associated enzymes in these animals. By the age of 30 d, however, the activity of cytochrome P_{450} in young rats has reached adult levels. When the animal is 20 months old, the activity of these enzymes begins to decline to 50 to 60% of the level found in the adult animal. As an organism reaches the last quarter of its life span, its ability to biotransform xenobiotics also decreases due to the reduction in the level of both cytochrome P_{450} and the corresponding reductase.

The differences in levels of cytochrome P_{450} enzymes in young animals to that of adults may be related to the amount of smooth endoplasmic reticulum present in the hepatocytes or other cells. At birth, the amount of these membranes is extremely low but increases rapidly after birth. Whether this increase is due to *de novo* synthesis or loss of ribosomes by preformed endoplasmic reticulum to form smooth endoplasmic reticulum is debatable. The low level of biotransformation activity of young animals is related to enzyme deficiency. As expected, administration of inducing agents to neonates or developing embryos results in the induction of cytochrome P_{450}. The formation of phase II enzymes in developing animals closely follows the time sequence for the formation of cytochrome P_{450} isozymes. Besides the changes in the activities of these enzymes, other factors are certainly involved in the decreased ability of older animals to detoxify xenobiotics.

9. Inherent and Acquired Resistance

A key determinant to species survival is successful competition, including both the ability to consume and the ability to deter consumption. Species that acquire the genetic capacity to produce a defensive chemical have an immediate advantage and are usually successful. These animals are a potential food source only for those species able to tolerate the defensive chemical. Although many mechanisms for tolerance coevolved with defensive chemicals, biotransformation is a major factor.

Once the acquired biotransformation potential becomes part of the species' genome, it becomes an inherent trait of the species. Such is the case for many phytophagous insects that have developed midgut monooxygenases that enable them to feed on potentially toxic plants (Krieger et al., 1971).

A genetic predisposition toward biodegradation has enhanced the development of pesticide resistance in several species of both plants and animals exposed to synthetic pesticides. Acquired resistance became inherent in certain strains, and enhanced biotransformation became a major factor to contend with in pest control. The previous history of pesticide use in an area also influences the rates and pathways by which xenobiotics will be degraded because of the enhanced biotransformation potential of some of the resident species.

D. VARIATION IN BIOTRANSFORMATION ENZYMES

Variations in biotransformation in different species can be divided into two categories, qualitative and quantitative differences. Qualitative differences are due to differences in metabolic routes, activation and deactivation due to defects in the metabolic pathway, and specific peculiar reactions in a particular species. Quantitative differences result from variations in enzyme levels, presence of particular inhibitors, activity of reverse enzyme reactions, or the extent of competing reactions. Among mammals, qualitative differences occur mainly in phase II enzyme systems and differences in phase I reactions are related to quantitative differences. Variations in phase II reactions are associated with evolutionary development. These variations result either from differences in the ability of an animal to synthesize the necessary cofactor, the concentration of a particular transferase, or the nature of the xenobiotic involved. For example, glucuronic acid conjugation is one of the most important phase II reactions, yet it is absent in the cat family. Acetylation of aromatic amines is not generally detected in the dog. Primates use glycine and glutamine as amino acid conjugates while birds use ornithine.

Species difference in biotransformation capability may be the reason for susceptibility or resistance to certain toxicants, particularly neurotoxicants. For example, the three enzymes involved in the metabolism of organophosphates are mixed function oxidases, glutathione *S*-transferase, and esterases. Domestic chickens are unable to hydrolyze *p*-nitrophenol from ethyl paraoxon by either plasma or microsomal A-esterases, although they have some ability towards methyl paraoxon. In contrast, rats and rabbits have greater A-esterase activity towards the ethyl analog than they do towards the methyl analog, although mammalian plasma methyl paraoxon esterase activity is higher than in the chicken. Also, glutathione *S*-alkyl transferases are more active towards methyl esters of phosphorothionates, which could explain the attenuated toxicity of methyl esters. Domestic chicken glutathione *S*-alkyl transferase activity towards methyl paraoxon is about 1/10 that found in the rat (Hollingworth, 1976). With phenylphosphonothionates, which are delayed-type neurotoxicants, the reverse relationship has been found (Hansen, 1983). Thus, not only are cytochrome P_{450} isozymes important in the activation and deactivation of organophosphates, but other enzyme systems such as esterases must also be considered. One problem encountered in comparing these activities across species is the biochemical optima for the various enzyme systems that are being studied. There are definite differences between species in their metabolic capabilities, but if optimum conditions are not used to measure these activities, comparison of the results may be meaningless.

1.   Nonmammalian Vertebrates

Monooxygenases of Birds

The concentration of hepatic cytochrome P_{450} in birds is lower than that found in the hepatic microsomes of mammals. The lowest hepatic microsomal monooxygenase activity has been reported for fish eating birds. These low levels of P_{450} in certain species of birds may make them more susceptible to toxic environmental chemicals that are detoxified by the system.

The range and concentration of cytochrome P_{450} in 16 species of birds is 0.15 to 0.51 nmol/mg protein with an overall mean of 0.30 (Ronis and Walker, 1989). In contrast, the concentration of cytochrome P_{450} in mammals ranges from 0.8 to 1.9 nmol/mg protein. Much higher levels of cytochrome P_{450} were found in Passeriformes and Galliformes compared to fish eating species. The relative activities of aldrin hydroxylase and aminopyrine O-demethylase were about 1/10 that found in mammals (Ronis and Walker, 1989). In contrast, aniline hydroxylase had a similar activity in birds compared to the rat, while activity of ethoxyresorufin O-deethylase was twice as high in birds as in mammals. This higher level of ethoxyresorufin O-deethylase may be related to a relatively high concentration of a particular cytochrome P_{450} involved in its oxidation. Little information on sex variations and age differences in the activity of cytochrome P_{450} in birds is available (Walker and Ronis, 1989).

Induction of cytochrome P_{450} by 3-MC, TCDD, and PCB gives a similar response in birds, reptiles, and amphibians. Cytochrome P_{450} induced by these agents had absorption maxima of 448 nm when complexed with carbon monoxide. Differences in the response among various species of birds have been noted. Of all the species studied, aminopyrine N-demethylase activity was only increased in the domestic fowl. In Japanese quail, similar induction patterns to that observed in the rat was noted except that aflatoxin metabolism was not increased (Walker and Ronis, 1989). Exposure of the pigeon to Aroclor 1254 resulted in the induction of a cytochrome P_{450} that was recognized by monoclonal antibodies raised to scup $P_{450}E$, which is similar to the induced rat $P_{450}I1A1$ (Walker and Ronis, 1989).

With phenobarbital, variable responses have been reported in birds. In contrast to the findings with the rat in which four cytochrome P_{450} forms were induced and in the rabbit in which five forms were found, with the domestic fowl only two cytochrome P_{450} isozymes were induced (Darby et al., 1986). One study showed that DDT and dieldrin caused a phenobarbital-type induction in the duck but not in the fowl (Davidson and Sell, 1972). No induction has been recorded with phenobarbital given at extremely high doses to cormorants. Prochloraz, an ergosterol biosynthetic inhibitor, that induced cytochrome P_{450} 10-fold in the red legged partridge (Ronis and Walker, 1989), is a poor inducer in the rat. Induction of this isozyme led to an increased activity of ethoxyresorufin-*O*-deethylase (EROD) and aldrin epoxidase. The absorption maximum of the prochloraz-induced cytochrome P_{450} was 453 nm, suggesting that a different and perhaps unique cytochrome P_{450} isozyme had been induced.

There appears to be evidence for the presence of isozymes of cytochrome P_{450} in birds and other lower vertebrates as reported for mammals, but as less is known of the monooxygenases in these phyla, assignment of these isozymes to families identified in mammals must be approached with caution. A proposal by Nebert and Gonzalez (1987) suggests that a primitive form of cytochrome P_{450} (P4501A1) exists in all phyla of the animal kingdom. There seems to be an under representation of isozymes belonging to the cytochrome $P_{450}II$ family in birds.

Monooxygenases of Reptiles

Alligators. The concentration of cytochrome P_{450}, cytochrome b_5, NADH, and NADPH cytochrome C reductase in *Alligator mississippiensis* liver is about 33, 16, 10, and 18%, respectively, of that found in the rat (Jewell et al., 1989). The low level of MFO activity in alligators may make them more vulnerable to toxic compounds in the environment (Schwen and Mannering, 1982a). In fact reptiles are more susceptible to the toxicity of the pesticide endrin than are birds or mammals (Hall, 1980). Alligator cytochrome P_{450} hydroxylated benzo(a)pyrene and dealkylated ethoxycoumarin and aminopyrine but did not hydroxylate aniline. The activity (pmol/min/mg protein) of these enzymes were 9, 11, and 4% of those recorded for the uninduced rat (Jewell et al., 1989). Induction by 3-MC of cytochrome P_{450} in the rat is about 2.4-fold while in the alligator it is 1.6-fold. Although the activities of both ethoxycoumarin *O*-deethylase and aminopyrine *N*-demethylase were increased with 3-MC treatment, the turnover numbers (nmol product formed/min/nmol P_{450}) were not altered. Induction of benzo(a)pyrene hydroxylase by 3-MC in the alligator was more than 10-fold that of control animals. 3-MC was found to induce two cytochrome P_{450} isozymes which cross-reacted with antibodies raised to rat $P_{450}1A1$ and $P_{450}1A2$ (Jewell et al., 1989).

Snakes. Treatment of the garter snake *Thamnophis* with 3-MC resulted in increased levels of cytochrome P_{450} and benzo(a)pyrene hydroxylase but no shift in the absorption maximum from 450 to 448 nm was noted. In contrast to the insensitivity of the 3-MC inducible benzo(a)pyrene hydroxylase by *a*-naphthaflavone in the snake, alligator enzyme was found to be sensitive (Jewell et al., 1989).

Monooxygenases of Amphibians

Little information on the metabolism of xenobiotics by amphibians has been published. The concentration of cytochrome P_{450} in the leopard frog was reported to be 0.4 nmol/mg protein (Ronis and Walker, 1989). The activity of aminopyrine O-demethylase and benzo(a)pyrene hydroxylase is much lower in the leopard frog than that recorded for birds. The only activities in the frog comparable to that in the rat are benzo(a)pyrene hydroxylase and aldrin epoxidase in the frog. Administration of phenobarbital to the leopard frog did not significantly induce the activity of aminopyrine N-demethylase and benzo(a)pyrene hydroxylase (Schwen and Mannering, 1982b).

Monooxygenases of Fish

Second only to mammals is the extremely large database that has been collected on the metabolism of xenobiotics by fish. Not only are the activities of these enzymes altered by sex and hormonal differences, but environmental parameters such as temperature and diet can also modulate the response in fish (Stegeman and Kloepper-Sams, 1987). Differences exist in phase II metabolism of aflatoxin in fish and rats, the glucuronide conjugate being the major conjugate in the rainbow trout and glutathione in the rat (Degen and Neumann, 1981).

Sex differences have been noted with rainbow and brook trout. For example, the level of cytochrome P_{450} and the activity of aminopyrine O-demethylase in the liver and kidneys of mature rainbow and brook trout were greater in males than females (Stegeman and Chevion, 1980). Sex differences were not noted with NADPH cytochrome c-reductase activity or cytochrome b_5 content but were observed with hepatic microsomal NADH cytochrome c-reductase. It is likely that these differences are due to the level of testosterone, since the level of cytochrome P_{450} in juvenile brook trout was elevated following testosterone administration but decreased with estradiol 17β (Stegeman et al., 1982).

It has been proposed that the measurement of certain parameters (biomarkers) in fish could be used as early warning signals of toxicity in water bodies. Studies have indicated that a good correlation exists between the concentration of certain environmental pollutants including PAH and the ensuing diseases including papillomas and cancer in fish living in these environments. For instance, fish exposed to PAH or PCB have higher levels of cytochrome P_{450} and associated enzymes than fish living in a pristine area (Winston et al., 1988).

Differences have been noted in the metabolism of PAH by fish and rats. Most of the studies have been performed with the benchmark PAH, benzo(a)pyrene (BP) (Figure 3.5). A greater proportion of BP was converted to the 7,8-diol by microsomes from sole and flounder than by rat microsomes (Varanasi et al., 1986). Within fish species, wide variations in their ability to metabolize BP and other xenobiotics have been documented. In general, microsomes from teleost fish produce more 7,8-diols from BP than do cartilaginous fish that metabolize BP to more phenolic metabolites (Stegeman and Woodin, 1980). However, differences among cartilaginous fish have also been reported. The metabolism of BP was higher in mirror carp *Cyprinus carpio* microsomes than in brown bullhead *Ictalurus nebulosus*, two bottom dwelling species. Although the proportion of BP metabolites formed by the two fish species was similar and included 4,5- and 7,8-diol, 3- and 9-hydroxy, and 1,6-, 3,6-, and 6,12-quinones, the relative amounts of these intermediates were different in the two fish species. The carp produced a much greater amount of 7,8-diol, which is the proximate carcinogen of BP, than did the bullhead microsomes, yet the bullhead was more susceptible to tumor formation than the carp (Sikka et al., 1990). Other factors that can influence cancer causation in these two fish species are the rate of phase II metabolism and DNA repair, two parameters that were not measured in this study. The relative amount of diols formed by the carp compared to the bullhead suggested that the carp has a higher level of epoxide hydratase than does the brown bullhead. The major metabolites formed by the brown bullhead were BP quinones which are usually the predominant metabolites of BP formed under peroxidatic conditions as a result of one electron oxidation (Cavalieri and Rogan, 1985). The method by which BP metabolites are measured can affect the results that are obtained. The fluorescent assay (Nebert and Gelboin, 1968) which has been used in many studies, but not in the one described above, preferentially measures phenolic metabolites, and thus an underestimation of quinones and other metabolites may result.

Inducing agents of cytochromes of the $P_{450}1$ family including BP and 3-MC have been shown to be good inducers in many species of fish. In contrast to mammals, no analogous cytosolic aryl hydrocarbon hydroxylase receptor has been detected in rainbow trout, salmon, or lake trout (Denison et al., 1985). There is much controversy about phenobarbital-type induction in fish, with most studies showing no induction. Two documented exceptions are the demonstration that DDT induced benzphetamine-N-demethylase activity in the mummichog (Pohl et al., 1974) and aminopyrine N-demethylase and aniline hydroxylase in rainbow trout (DeWaide, 1971).

Environmental factors are very important in the induction process in fish. For example, temperature can alter the levels of cytochrome P_{450} enzymes. Fish, in general, have lower temperature optima than do mammals. Also, fish respond to acclimation temperature in a compensatory manner. Rainbow trout acclimated to colder temperatures have greater enzymatic activity than those acclimated to warmer temperatures. Aryl hydrocarbon hydroxylase activity in rainbow trout was found to be identical when measured at environmental temperatures of 20°C in August and 5°C in November, but when these same activities were measured at 18°C from animals acclimated to the above mentioned temperatures, microsomes from trout acclimated to the lower temperature had higher activity. In addition, microsomes from trout acclimated to 7°C metabolized BP to a greater extent than microsomes from trout acclimated to 16°C when the activities were measured at 29°C (Kleinow et al., 1987).

Although several fish cytochrome P_{450} isozymes have been isolated and characterized, nomenclature has been confusing and assignment with the equivalent mammalian isozyme has been difficult. Scup cytochrome $P_{450}E$ which is related to trout $P_{450}LM_{4b}$ are the primary forms of P_{450} involved in metabolizing polycyclic aromatic hydrocarbons in these species (Williams and Buhler, 1984). Both isozymes metabolize BP to the corresponding 7,8-dihydrodiol and are induced by PAH, PCB, and β-naphthaflavone (Kloepper-Sams et al., 1987; Williams and Buhler, 1984). One of the isozymes induced by β-naphthaflavone, shares epitopes with both rat 1A1 and 1A2 isozymes (Kloepper-Sams et al., 1987).

$P_{450}E$ has also been identified in the endothelial cells of the gill lamellae of scup and rainbow trout, but increased enzyme activity was only detected in scup induced with β-naphthaflavone (Miller et al., 1989). Gill microsomes have been shown to metabolize BP to the 7,8-dihydrodiol (Stegeman et al., 1984).

2. Biotransformation by Invertebrates

Mollusks

When first cytochrome P_{450} enzyme activity was investigated in mollusks it was not found. This was because these enzymes are located in the hepatopancreas, which contains several proteases and other hydrolytic enzymes that deactivated the mixed function oxidases. Later, when trypsin inhibitor and sodium dithionite and other protective agents were added during preparation of the mixed function oxidases, activity was restored. The molluscan mixed function oxidase system is similar to that of the mammalian system in that it is primarily membrane-bound and found in the microsomal fraction of the hepatopancreas (Livingstone, 1985) and gills and leukocytes (Moore et al., 1980; Livingstone and Farrar, 1984). Cytochrome P_{450}, which has been identified in the digestive glands of 23 species of mollusks including mussels, clams, cockles, periwinkle, chiton, and snails, ranged from 3 to 134 pmol/mg protein (Livingstone et al., 1989). Multiple forms of P_{450} belonging to the IVA family have been detected.

Controversy exists about whether cytochrome P_{450} is induced in mollusks by benchmark inducing agents and/or polycyclic aromatic hydrocarbons in the field. In the laboratory, induction of mixed function oxidases has been documented in *Crossostrea virginica* exposed to BP, 3-MC, and PCB, in *Mytilus edulis* following exposure to PAH in crude oil and to PCB (Moore et al., 1980), and in *M. galloprovincialis* exposed to 3-MC and perylene (Gilewicz et al., 1984). In contrast Stegeman (1985) did not find an increase in PAH concentration in *M. edulis* sampled from sites heavily contaminated with creosote waste, nor did he find a concomitant induction of cytochrome P_{450}. Besides one report of phenobarbital induction of microsomal NAPH-cytochrome c-reductase in *M. galloprovincialis* (Galli et al., 1988), no other evidence exists in this order or other lower orders (Schwen and Mannering, 1982b) that phenobarbital is an inducer.

Stegeman (1985) found that BP is metabolized by *M. edulis* microsomes both in the presence and absence of NADPH. As a different metabolite profile resulted under these conditions it was concluded that more than one pathway is involved in the metabolism of BP (Figure 3.5). In the presence of NADPH, 4,5-, 7,8-, and 9,10-dihydrodiols and phenols comprised only 35% of the metabolites with the remainder being 1,3-, 3,6-, and 6,12-quinones (Stegeman, 1985). Higher levels of quinone compared to phenols were found with NADH with both *M. edulis* (Livingstone et al., 1988) and *Cryptochiton stelleri* microsomes (Schlenk and Buhler, 1988). Other NADPH independent P_{450} activities including ethoxycoumarin *O*-deethylase and benzphetamine demethylase have been reported (Kirchin et al., 1988). Interestingly, NADPH and other reduced compounds are inhibitory to the metabolism of many substrates by molluscan enzymes *in vitro*. This may be due to the presence of endogenous reducing equivalents, activated oxygen, one electron oxidation as opposed to two electron oxidation, or peroxidation (Cavalieri and Rogan, 1984). The latter hypothesis is consistent with the formation of a higher percentage of quinone metabolites following BP metabolism (Cavalieri and Rogan, 1985).

The metabolism of other xenobiotics has been investigated in mollusks. The hard shell clam *Mercenaria mercenaria,* the mussel, *M. edulis,* and the oyster, *Crossostrea gigas,* have been shown to activate aromatic amines to mutagenic intermediates (Knezovich and Crosby, 1985; Kurelec and Britvic, 1985). Pulmonate snails *Physa elliptica* and *Lymnea pallustris* can dealkylate methoxyclor and *p*-nitrophenol ethers *in vivo* (Hansen et al., 1972).

As has been recorded in fish species, levels of cytochrome P_{450} in *M. edulis* varies seasonally and declines with the approach of spawning (Livingstone, 1987). During certain times of the year, benzo(a)pyrene hydroxylase activity is higher in the female mussel than in the male in contrast to findings with cytochrome P_{450} activity in vertebrates.

Crustacea

The hepatopancreas is the major site of cytochrome P_{450} dependent monooxygenase activity in crustacea with minor roles being assumed by the green (antennal) glands, which have similar functions to the kidneys in vertebrates, and the stomach. As reported for mollusks, cytochrome P_{450} in the hepatopancreas was found to be extremely labile (James et al., 1977), thus, data in the early literature should be interpreted which caution. Inclusion of protease inhibitors in the isolation buffer followed by cholate digestion resulted in the recovery of an active preparation (James, 1984). Early studies of the monooxygenase activities of aldrin (Khan et al., 1972), parathion (Elmamlouk and Gessner, 1976), and ethoxycoumarin (James et al., 1979) indicated that the activity was extremely low in many crustacea. These are all substrates metabolized by the cytochrome $P_{450}II$ family. More recently when precautions were taken in the isolation of P_{450}, ethoxycoumarin *O*-deethylase activity was observed with hepatopancreas microsomes from the crayfish *Procambarus clarkii* (Jewell and Winston, 1989a). Cytochrome P_{450} from the hepatopancreas of the spiny lobster *Panulirus argus* has been shown to metabolize vertebrate sex steroids such as progesterone and testosterone, as well as several xenobiotics including benzphetamine, aminopyrine, benzo(a)pyrene, phenoxazones, and ethoxycoumarin. Aminopyrine and benzphetamine *N*-demethylases have the highest activity (James, 1989a). The rate of metabolism of benzo(a)pyrene in lobsters varies; it is metabolized rapidly in the spiny lobster *P. argus* (Little et al., 1985) but very slowly by *Homarus americanus* (James et al., 1989).

Although relatively high levels of cytochrome P_{450} have been recorded in hepatopancreas microsomes, the concentration of NADPH cytochrome c-reductase in most of the crustaceans examined including lobsters, crabs, and crayfish is very low while NADPH cytochrome P_{450} reductase has only been detected in hepatopancreas microsomes of only a few species (James, 1989b). To overcome the lack of this enzyme in spiny lobster microsomes, James (1990) purified cytochrome P_{450} from this organism and added porcine or murine NADPH cytochrome P_{450} reductase to it in the assay system. Under these conditions, benzo(a)pyrene, 7-ethoxycoumarin, and phenoxazones were metabolized with the first compound being the best substrate (James, 1990).

Treatment of the spiny lobster (James and Little, 1984), crayfish (Lindstrom-Seppa and Hanninen, 1986), and spiny crab (Batel et al., 1988) with 3-MC, Aroclor™ 1254, β-naphthaflavone, or phenobarbital did not result in an increase in cytochrome P_{450} content. In general, it does not appear that crustacean cytochrome P_{450} is induced by polycyclic aromatic hydrocarbons found in the environment, as has been documented with fish (James and Bend, 1980). For this reason, it seems unlikely that crustaceans can be used as indicators of environmental pollution. Besides cytochrome P_{450}, there is evidence to suggest that other pathways including the production of oxyradicals and hydrogen peroxide may be important in the metabolism of xenobiotics by crustacea (Jewell and Winston, 1989b).

Insects

In insects as in mammals, cytochrome P_{450} is involved in the metabolism of endogenous substrates such as hormones, in the oxidation of xenobiotics such as pesticides, and in the metabolism of plant secondary metabolites (Hodgson, 1983; Wilkinson, 1983). Cytochrome P_{450} is found in the endoplasmic reticulum of cells of the midgut, fat body, and malpighian tubules (equivalent to the kidney in mammals). The system in insects is similar to the mammalian system and consists of the flavoprotein NADPH cytochrome P_{450} reductase, the hemeprotein cytochrome P_{450} and cytochrome b_5. Cytochrome P_{450} is involved in the synthesis and degradation of various terpenoid juvenile hormones (Hammock, 1975; Yu and Terriere, 1978) and the oxidation of fatty acids such as lauric acid in the housefly (Ronis et al., 1988). Insect P_{450} enzymes also mediate several different reactions such as desulfuration, ester cleavage, epoxidation, and aliphatic hydroxylation of xenobiotics (Kulkarni and Hodgson, 1980).

There are dramatic changes in the levels of cytochrome P_{450} during the developmental cycle. For example, cytochrome P_{450} is undetectable in the egg and pupae of the southern armyworm and the housefly (Krieger and Wilkinson, 1969) but rises and falls during each larval instar reaching its highest level at the intermoult period. Cytochrome P_{450} reaches its maximum concentration in the adult 7 to 9 days post emergence depending on sex and strain (Schonbrod et al., 1965; Hodgson, 1983).

Differences in insect diets are an important factor in monooxygenase activity. With several lepidopterous larvae there is a 10-fold difference in the activity of aldrin epoxidase in the midgut of polyphagous compared to monophagous insects, with oligophagous showing immediate values (Krieger et al., 1971). A similar relationship between monooxygenase activity and diet has been shown with fish eating sea birds and raptores which have restricted diets and lower relative monooxygenase activity than do herbivorous and omnivorous bird species (Walker et al., 1986).

Although there are pronounced differences in monooxygenase activity between resistant and susceptible insect species, no greater than a twofold increase in cytochrome P_{450} has been observed between resistant and susceptible strains of the housefly. In recent years, evidence has accumulated that there are qualitatively different cytochrome P_{450} isozymes in microsomes derived from resistant compared to susceptible strains. Two or more dominant genes for high oxidase activity associated with pesticide resistant houseflies have been mapped to chromosomes II and V (Plapp, 1984). These genes apparently control the synthesis of different forms of cytochrome P_{450} rather than coding for differences in the physical characteristics or catalytic properties of a single form.

Both the housefly (Ronis and Hodgson, 1989) and fruit fly (Hallstrom, 1987) monooxygenase systems are quite complex, consisting of at least six and five different forms of cytochrome P_{450}, respectively. Two isozymes known as isozyme A and B have been characterized in the fruit fly. The $P_{450}A$ isozyme appears to be common to all insect species whereas the $P_{450}B$ isozyme appears to be associated with resistance to DDT and malathion (Waters and Nix, 1988). As noted with some crustacea, insect NADPH cytochrome P_{450} reductase is extremely labile. Reconstitution experiments with purified insect cytochrome P_{450}, rat cytochrome P_{450} reductase, and phosphatidylethanolamine have been used to measure the activity of aldrin epoxidase, ethoxycoumarin, and *p*-nitroanisole-*O*-deethylases (Agosin, 1982; Ronis et al., 1988). A polyclonal antibody has been raised against a purified housefly cytochrome P_{450} to obtain a cDNA clone. The sequence of this fly cytochrome P_{450} is quite different from any known mammalian isozyme and has been assigned to the P_{450} VI family (Feyereisen et al., 1989; Nebert et al., 1989a).

As in other animals, insect cytochrome P_{450} is inducible. Both phenobarbital and chlorinated insecticides which induce P_{450} gene family II in mammals are good inducers of insect cytochrome P_{450} and P_{450} reductase, but polycyclic aromatic hydrocarbons are not effective inducers. Purified cytochrome P_{450} from phenobarbital-treated flies oxidizes aldrin to dieldrin (Fisher and Mayer, 1984). Isosafrole is a good inducer of benzo(a)pyrene hydroxylase and ethoxyresorufin *O*-deethylase in the armyworm *Spodoptera* (Marcus et al., 1986).

The complexity of the activation and detoxification pathway of a typical organophosphate insecticide, malathion, in insects and mammals is shown in Figure 3.6, I. Cytochrome P_{450} enzymes activate malathion through oxidative desulfuration to form the more toxic intermediate, malaoxon (Figure 3.4, IV). In the conversion of P=S to P=O, a "phosphooxythirane" intermediate (Figure 3.6, III) is formed followed by the releases of a reactive sulfide. Malaoxon (Figure 3.6, IV) is the active intermediate that phosphorylates acetylcholinesterase. With phosphorothioates containing an aromatic leaving group in place of the mercaptosuccinate diethyl ester, the phosphooxythirane intermediate may rearrange in such a way as to cause dearylation. Detoxification of malathion by cytochrome P_{450} may also occur by removal of a methoxy group (I → II or IV → V).

Most vertebrates contain high carboxyesterase activity relative to insects which results in the formation of the less toxic intermediates of VI from I and VII from IV. This deesterification is a detoxification reaction and is primarily responsible for the excellent selectivity of malathion. However, carboxyesterases can be inhibited by other phosphoro- and phosphonothioates so unexpected toxicity can occur if the organism is coexposed to parathion or EPN. Selection pressures have resulted in the survival of some strains of insects with higher carboxyesterase activities and, thus, resistance to malathion.

Planaria

As planaria can metabolize several xenobiotics to active metabolites, these ubiquitous organisms could be used as sentinel species for the monitoring of acute toxicity and tumorigenicity. Exposure of planaria to DDT and Sevin resulted in the development of abnormal, lethal growths. Interestingly, several species of planaria metabolize DDT via a similar pathway to that found in mammals forming DDE and DDD (Kouyoumjian and Uglow, 1974; Kouyoumjian and Villeneuve, 1979). The cause of the abnormalities observed with DDT could have been related to the formation of DDD by reductive dechlorination or DDE by dehydrochlorination, both of which are toxic to planaria of the genus *Phagocata* spp. (Bonner and Wells, 1987). Additional evidence that planarians can activate xenobiotics in a similar manner to mammals comes from studies with mammalian protumorigens and proteratogens which require bioactivation. When *Dugesia dorotocephala* was exposed to either 1,2-benzo(a)anthracene, 3-methyl-4-dimethylaminoazobenzene, benzo(a)pyrene, or 3-methylcholanthrene, the planaria developed lethal tumors and malformations.

3. Biotransformation by Plants

The major endogenous pathways in which cytochrome P_{450} is involved in plants is in the metabolism of fatty acids and terpenes and the production of the lignin precursor cinnamic acid. The concentration of cytochrome P_{450} in the microsomes of plants ranges from 0.015 nmol/mg protein in uninduced Jerusalem artichoke tuber to 0.320 nmol/mg protein in tulip bulbs (Karasaki et al., 1983). The highest concentration in plants is still only 30% of that found in uninduced rat liver. Cytochrome P_{450} metabolizes *trans*-cinnamic acid to 4-hydroxy-cinnamic acid, an important intermediate in the synthesis of several plant specific metabolites (Higashi, 1988). The oxidation of kaurene, the precursor of the plant growth hormones gibberelins, is mediated through cytochrome P_{450} and its associated reductases (Hasson and West, 1976).

Less is known of the metabolism of xenobiotics in plants than in any other group of living organisms. Due to the very low concentration of cytochrome P_{450}, its role in the metabolism of xenobiotics is probably small. In fact, there is some evidence that plant peroxidases play a major role in the metabolism of xenobiotics by plants. Two types of peroxidase reactions are catalyzed, one of which requires hydrogen peroxide and the second oxygen. Plant peroxidases catalyze the oxidation of phenols and aromatic amines and mediate decarboxylation, *N*-demethylation, ring hydroxylation, and many other cytochrome P_{450}-type reactions (Lamoureux and Frear, 1979). The metabolism of 2-aminofluorene and *m*-phenylenediamine to active intermediates has been shown to be mediated by peroxidases in tobacco cells (Plewa et al., 1991).

Metabolism of the benchmark PAH, benzo(a)pyrene (BP), by both terrestrial plants and algae have been demonstrated. The major pathway in terrestrial plants is thought to be mediated by peroxidases and active oxygen species and to a lesser extent by cytochrome P_{450} (Van der Trenck and Sandermann, 1980; Sandermann, 1988), while in algae such as *Selenastrum capricornutum* metabolism is via a dioxygenase pathway (Schoeny et al., 1988). Similarly, marine algae and diatoms metabolize naphthalene via a dioxygenase pathway (Cerniglia et al., 1980). The major BP metabolites produced by *S. capricornutum* are the *cis* isomers of 11,12-, 7,8-, and 4,5-diols, all of which are conjugated with either sulfate or α and β glucose, and the unconjugated 9,10-diol and 3,6-quinone (Warshawsky et al., 1990). The fate of the metabolites produced by algae and terrestrial plants is different. Algae frequently release the conjugates into the surrounding water, while terrestrial plants store them in the plant body which could result in their consumption by animals and man.

Crop tolerance to sulfonylureas is thought to be due to the metabolism of these herbicides by cytochrome P_{450} monooxygenases. A study with primisulfuron by Fonne-Pfister et al., (1990) suggested that cytochrome P_{450} enzymes hydroxylate this herbicide and that the *in vitro* hydroxylation is stimulated by pretreatment of the plant with crop safeners. In contrast, weeds such as *Sorghum* spp. metabolize primisulfuron much more slowly and are thus susceptible to the herbicide (Sweetser et al., 1982).

Microsomes from the Jerusalem artichoke metabolize BP without NADPH (Van der Trenck and Sandermann, 1980) suggesting that the PAH is metabolized through a peroxidatic mechanism. The major metabolites of benzo(a)pyrene produced by pea and soybean microsomes were 1,6-, 3,6-, and 6,12-quinones (Van der Trenck and Sandermann, 1980) all of which can be formed from 6-hydroxybenzo(a)pyrene by auto-oxidation or through other pathways not involving cytochrome P_{450} enzymes (Figure 3.5).

Other xenobiotics besides BP are also metabolized by plants. Wheat cell cultures and horse radish peroxidase metabolize pentachlorophenol to the corresponding tetrahydrocatechol (Schafer and Sandermann, 1987) which is then conjugated to form β-*D*-glucosides. The high activity of a *N*-dealkylase in cotton, buckwheat, and plantain results in the resistance of these plants to the phenylurea herbicide monuron, while corn, potatoes, celery, and soybeans which have much lower activities are susceptible to this herbicide (Frear, 1968).

Induction of cytochrome P_{450} in plants by physical agents and compounds of varying chemical structures have been documented. An enhancement in the levels of cytochrome P_{450} is mediated through wounding and exposure of cells to moist atmospheres (Higashi, 1988) by phenobarbital, herbicides (Reichhart et al., 1980), 2,4-dichlorophenoxyacetic acid, manganese, and aminopyrine (Werck-Reichhart et al., 1990). Wounding of Jerusalem artichoke cells resulted in an increase in the rough endoplasmic reticulum with an increased capacity for protein synthesis resulting, in turn, in an increase in cytochrome P_{450} and enhanced activity of cinnamate-4-monooxygenase (Benveniste et al., 1977). Although 2,4-dichlorophenoxyacetic acid induced cytochrome P_{450} and cinnamate-4-monooxygenase in Jerusalem artichoke tubers, activities of other cytochrome P_{450} related enzymes were not enhanced, suggesting heterogeneity of cytochrome P_{450} in plants. Induction by manganese and aminopyrine of two different cytochrome P_{450} enzymes from Jerusalem artichoke tubers have been described (Werck-Reichhart et al., 1990). One enzyme mediated the *O*-

deethylation of 7-ethoxycoumarin and the second the *O*-deethylation of 7-ethoxyresorufin. The artichoke microsomes also metabolized 7-ethoxycoumarin but not 7-ethoxyresorufin in the presence of cumene hydroperoxide.

4. Biotransformation by Microorganisms

Bacteria constitute a large biomass which actively degrade xenobiotics in the environment. Selection of a dominant species that metabolizes a specific xenobiotic frequently involves parameters such as pH, oxygen, energy sources, and light, but resistance to antimicrobials is also a significant determinant. Among gram negative bacteria, multiple drug resistance involves plasmid transmissible genes, resulting in direct increases in biotransformation potential in lactamases (penicillin) and acetyl transferases (aminoglycosides and chloramphenicol) (Sande and Mandell, 1980). Bacteria resistant to Hg in the environment also have an enhanced ability to bioalkylate inorganic Hg, rendering it volatile so it escapes from the medium.

Due to the enormous number of genera of microorganisms, numerous enzymatic pathways have been developed to metabolize xenobiotics. Microorganisms metabolize organic compounds by dehalogenation, hydrolysis, oxidation, reduction, conjugation, and methylation. In this chapter a snapshot of the degradation of the simple and complex organic compounds will be outlined.

Oxidation of simple *n*-alkanes to alkan-1-ol by cytochrome P_{450} enzymes in *Candida* spp. has been documented. Two alkane inducible cytochrome P_{450}s have been identified in *Candida* spp. Sequence analysis of one of these cytochromes from *C. tropicalis* has shown that it is distinctly different from any previously reported cytochrome P_{450} enzyme. These two fungal enzymes, named alk I and alk II, that metabolize alkanes have been assigned to a new family designated as P_{450}LII (Sanglard et al., 1987; Sanglard and Loper, 1989). Less common, but of importance, is the oxidation of *n*-alkanes by dioxygenases to the corresponding hydroperoxides followed by reduction to the corresponding alkan-1-ol (Watkinson and Morgan, 1990). Subsequent metabolism of the alcohol results in the formation of the aldehyde and acid through pyridine-linked dehydrogenases. Alkenes can be metabolized through four major pathways: (1) through oxygenase attack on the terminal methyl group to form the corresponding aldehyde and acid, (2) through oxidation across the double bond to form the epoxide, (3) through subterminal oxygenase attack to produce the corresponding alcohol and acid, and (4) through epoxidation across the double bond to form the diol (Britton, 1984).

The metabolism of naturally occurring halogenated compounds by bacteria occurs rapidly, while that of highly chlorinated compounds produced by anthropogenic processes is much slower. Although methane, ethylene, benzene, and biphenyl are metabolized rapidly by bacteria, their corresponding perchlorinated compounds, carbon tetrachloride, tetrachloroethylene, hexachlorobenzene, and decachlorobiphenyl, are metabolized and degraded slowly (Wackett, 1991). In fact, tetrachloroethylene and hexachlorobenzene, which are commonly used as a solvent and fungicide, respectively, are thought to be biodegraded only in an anaerobic environment (Vogel et al., 1987). Cleavage of the carbon-halogen bond by bacteria occurs through four different mechanisms. Hydrolysis results in the replacement of the halogen by a hydroxyl group derived from water while oxidation results in the formation of the corresponding aldehyde. Bacteria producing these oxygenases cannot use the halogenated compound as the sole source of carbon, a situation known as cometabolism, but must have a simpler carbohydrate source for energy production. The third metabolic pathway of reductive dehalogenation may be the most important in the metabolism of highly chlorinated compounds. Finally, these compounds can be dehalogenated by two types of elimination reactions, the first involves the elimination of a proton and halide atom from two adjacent carbon atoms to yield an olefin and hydrogen halide and the second involves the removal of two adjacent chlorine atoms to form an olefin and a chlorine molecule.

Methylotrophic bacteria that usually use methanol or dimethylsulfoxide as their sole source of carbon also contain dehalogenases which metabolize chlorinated alkanes and alkenes. For example, dichloromethane is dechlorinated to formaldehyde which is then assimilated into more complex molecules (Scholtz et al., 1988). Trichloroethylene is metabolized to the corresponding epoxide by soluble methane monooxygenases found in several *Pseudomonas* species. The epoxide then undergoes hydration followed by carbon-carbon bond scission to form carbon dioxide and formic acid. The oxygenases that metabolize trichloroethylene are compound specific and do not metabolize tetrachloroethylene or PCBs. Reductive dehalogenation of these latter mentioned compounds usually takes place in highly anaerobic environments (Vogel and McCarty, 1985). Until recently reductive dechlorination could only be mediated by mixed cultures, but in the past two years a unique bacterium was isolated that synthesized adenosine triphosphate by coupling the oxidation of formate and acetate to the reduction of chlorobenzoate to form benzoate (Dolfing, 1990). Nonspecific dehalogenation by titanium (III) citrate, vitamin B_{12}, coenzyme F_{430}, and transition metals that can mediate electron transfer and catalytic reactions (Wackett, 1991) have been implicated in the degradation of highly chlorinated organics, but the rates are extremely slow. Degradation of tetrachloroethylene under anaerobic conditions in a laboratory setting resulted in the formation *cis*-1,2-dichloroethylene which was further reduced to the toxic vinyl chloride (Freedman and Gossett, 1989).

Genera of bacteria that degrade PCBs include *Pseudomonas*, *Alcaligenes*, *Arthrobacter,* and *Acinetobacter*. Mineralization to carbon dioxide and water are enhanced by sunlight. With many congeners, complete mineralization does not take place and chlorinated benzoates accumulate. Mixed consortia are more effective in the degradation of PCB than a single species. The first step in the degradation is the 2,3-dioxygenation of the least substituted ring followed by cleavage and further degradation to the chlorobenzoate. Nonchlorinated vicinal *ortho* and *meta* positions favor dioxygenation. Congeners with five or more chlorine atoms, or with halogen substituents on both rings or substituted in the *ortho* positions, are the most recalcitrant (Commandeur and Parsons, 1990). Thus, 2,2′,3,3′-tetrachlorobiphenyl was more rapidly degraded than 3,3′,4,4′-TCB by an *Alcaligenes* strain (Parsons et al., 1988). Anaerobic dehalogenation of PCB with more than seven chlorine atoms does not occur. *Ortho* substituents are not cleaved by anaerobic dehalogenation but if all the other halogen atoms have been removed, the *ortho* substituted PCB can be aerobically degraded. Co-metabolism of mono- , and di-substituted dioxins by the biphenyl degrading organism *Beijerinckia* spp. results in the formation of *cis*-1,2-dihydrodiols, but subsequent ring cleavage does not occur (Klecka and Gibson, 1980).

The metabolism of polycyclic aromatic hydrocarbons by bacteria is different from that reported for the animal kingdom in that cytochrome P_{450} is not involved. Bacteria can metabolize both simple one-ringed benzene to the complex five-membered ring PAH benzo(a)pyrene. In contrast to the metabolism of these compounds to *trans* dihydrodiols by cytochrome P_{450} and epoxide hydrolase in mammalian systems (Figure 3.5), bacteria metabolize PAH to *cis*-dihydrodiols through the incorporation of both atoms of molecular oxygen into the PAH nucleus. The initial reaction is catalyzed by a dioxygenase consisting of several enzymes, while the terminal oxygenase is an iron-sulfur protein (Cerniglia, 1984). The *cis*-dihydrodiols are then oxidized by a *cis*-dihydrodiol dehydrogenase to yield catechols. Cleavage of catechol can occur at two positions, (1) at the *ortho* position between the two carbon atoms to which the hydroxyl groups are attached to yield *cis,cis*-muconic acid or (2) at the *meta* position to yield 2-hydroxymuconic semialdehyde. These intermediates are then rapidly metabolized, thus supplying energy to the organism.

Fungi metabolize PAH in a similar manner to that reported for mammals. Cytochrome P_{450} enzymes and epoxide hydrolase oxidize PAH to *trans*-dihydrodiols (Cerniglia, 1981). Fungi can use certain PAH (anthracene and phenanthrene) as their sole source of carbon and produce both toxic and nontoxic intermediates. The stereochemistry of the cytochrome P_{450} enzyme in fungi differs from that in mammals, as the major enantiomer of the fungal enzyme is (+)- (1S, 2S) while that of the mammalian enzyme is (−)- (1R, 2R) (Jerina et al., 1970). Fungi conjugate phenols derived from PAH to glucuronides and sulfates (Cerniglia et al., 1982). Cell extracts of *Cunninghamella elegans* contain cytosolic UDP-glucuronyltransferase activity (Wackett and Gibson, 1988), while in mammalian tissues this enzyme is membrane-bound.

E. CONCLUSION

The study of the metabolism of xenobiotics by living organisms encompasses several aspects, of which only a few have been documented here. For more in-depth information, the review articles and books cited in this chapter should be consulted. In summary, cytochrome P_{450} is an extremely important enzyme having many functions in the metabolism of the cell. In some higher invertebrates and vertebrates the enzyme is also involved in the metabolism of anthropogenic compounds, while in lower vertebrates and plants, although cytochrome P_{450} is present, its role in the detoxification of xenobiotics appears to be of lesser importance. Peroxidases and dioxygenases probably play a significant role in the metabolism of xenobiotics in plants and lower vertebrates.

REFERENCES

Agosin, M. Multiple forms of insect cytochrome P-450: role in insecticide resistance, in *Cytochrome P-450 Biochemistry, Biophysics and Environmental Implications.* R. Hietanen, M. Latinen, and O. Hanninen, Eds. Elsevier Publishing, New York, 1982, 661–669.

Anders, M.W., Kubic, V.L., and Ahmed, A.E. Metabolism of halogenated methanes and macromolecular binding, *J. Environ. Pathol. Toxicol.* 1:17–124, 1977.

Batel, R., Bihari, N., and Zahn, R.K. 3-Methylcholanthrene does induce mixed functions oxidase activity in hepatopancreas of spiny crab *Maja crispata, Comp. Biochem. Physiol.* 90C:435–438, 1988.

Benveniste, I., Salaun, J.P., and Durst, F. Wounding-induced cinnamic acid hydroxylase in Jerusalem artichoke tuber, *Phytochemistry* 16:69–73, 1977.

Bonner, J.C. and Wells, M.R. Comparative acute toxicity of DDT metabolites among American and European species of planarians, *Comp. Biochem. Physiol.* 17C:437–438, 1987.

Britton, L.N. Microbial degradation of aliphatic hydrocarbons, in *Microbial Degradation of Organic Compounds.* D.T. Gibson, Ed. Marcel Dekker, New York, 1984, 89–129.

Cavalieri, E.L. and Rogan, E.G. One-electron and two-electron oxidation in aromatic hydrocarbon carcinogenesis, in *Free Radicals in Biology, Vol VI.* W.A. Pryor, Ed. Academic Press, New York, 1984, 323–369.

Cavalieri, E.L. and Rogan, E.G. One-electron oxidation in aromatic hydrocarbon carcinogenesis, in *Polycyclic Aromatic Hydrocarbons and Carcinogenesis*, ACS Symposium Series 283. R.G. Harvey, Ed. The American Chemical Society, Washington, DC, 1985, 289–305.

Cerniglia, C.E. Aromatic hydrocarbons: metabolism by bacteria, fungi and algae, *Rev. Biochem. Toxicol.* 3:321–361, 1981.

Cerniglia, C.E. Microbial metabolism of polycyclic aromatic hydrocarbons, *Adv. Appl. Microbiol.* 30:31–71, 1984.

Cerniglia, C.E., Gibson, D.T., and Van Baalen, C. Oxidation of naphthalene by cyanobacteria and microalgae, *J. Gen. Microbiol.* 116:495–500, 1980.

Cerniglia, C.E., Dodge, R.H., and Gibson, D.T. Fungal oxidation of 3-methyl cholanthrene: formation of proximate carcinogenic metabolites of 3-methyl cholanthrene, *Chem. Biol. Interact.* 38:161–173, 1982.

Commandeur, L.C.M. and Parsons, J.R. Degradation of halogenated aromatic compounds, *Biodegradation* 1:207–220, 1990.

Coon, M.J., Vermilion, J.L., Vatsis, K.P., French, J.S., Dean, W.L., and Haugen, D.A. Biochemical studies on drug metabolism: isolation of multiple forms of liver microsomal cytochrome P-450, in *Drug Metabolism Concepts.* D.M. Jerina, Ed. American Chemical Society, Washington, DC, 1977, 46–71.

Darby, J.J., Lodola, A., and Burnet, F.R. Testosterone metabolite profiles reveal differences in the spectrum of cytochrome P-450 isozymes induced by phenobarbitone, 2-acetylaminofluorene and 3-methylcholanthrene in the chick embryo liver, *Biochem. Pharmacol.* 35:4073–4076, 1986.

Davidson, K.L. and Sell, J.L. Dieldrin and pp'-DDT. Effects on some microsomal enzymes of chickens and mallard ducks, *J. Agric. Food Chem.* 20:1198–1205, 1972.

Degen, G.H. and Neumann, H.B. Differences in aflatoxin B_1 susceptibility of rat and mouse are correlated with the capability *in vitro* to inactivate aflatoxin B_1-epoxide, *Carcinogenesis* 2:299–306, 1981.

Denison, M.S., Hamilton, J.W., and Wilkinson, C.F. Comparative studies of aryl hydrocarbon hydroxylase and the Ah receptor in nonmammalian species, *Comp. Biochem. Physiol.* 80C:319–324, 1985.

DeWaide, J.H. Metabolism of xenobiotics. Ph.D. thesis, University of Nijmegen, The Netherlands, 1971.

Dolfing, J. Reductive dechlorination of 3-chlorobenzoate is coupled to ATP production and growth in an anaerobic bacterium, strain DCB 1, *Arch. Microbiol.* 53:264–266, 1990.

Elmamlouk, T.H. and Gessner, T. Species differences in metabolism of parathion: appearance inability of hepatopancreas fractions to produce paraoxon, *Comp. Biochem. Physiol.* 53C:19–24, 1976.

Fisher, C.W. and Mayer, R.T. Partial purification and characterization of phenobarbital-induced housefly cytochrome P-450, *Arch. Insect Biochem. Physiol.* 1:127–138, 1984.

Fonne-Pfister, R., Gaudin, J., Kreuz, K. Ramsteiner, K., and Ebert, E. Hydroxylation of primisulfuron by an inducible cytochrome P450-dependent monooxygenase system from maize, *Pestic. Biochem. Physiol.* 37:165–173, 1990.

Frear, D.S. Microsomal *N*-demethylation by a cotton leaf oxidase system of 3-(4'-chlorophenyl)-1,1-dimethylurea (monuron), *Science* 162:674–675, 1968.

Freedman, D.L. and Gossett, J.M. Biological reductive dechlorination of tetrachloroethylene and trichloroethylene to ethylene under methanogenic conditions, *Appl. Environ. Microbiol.* 55:2144–2155, 1989.

Galli, A., Del Chiero, D., Nieri, R., and Bronzetti, G. Studies on cytochrome P-450 in *Mytilus galloprovincialis*: induction by Na-phenobarbital and ability to biotransform xenobiotics, *Mar. Biol.* 100:69–73, 1988.

Gilewicz, M., Guillaume, J.R., Carles, D., Leveau, M., and Bertrand, J.C. Effects of petroleum hydrocarbons on the cytochrome P450 content of the mollusc bivalve *Mytilus galloprovincialis*, *Mar. Biol.* 80:155–159, 1984.

Gold, B. and Brunk, G. A mechanistic study of the metabolism of 1,1-dichlor-2,2-bis(*p*-chlorophenyl) ethane (DDD) to 2,2-bis(*p*-chlorophenyl) acetic acid (DDA), *Biochem. Pharmacol.* 33:979–982, 1984.

Gonzalez F.J. Molecular genetics of the P-450 superfamily, *Pharmacol. Ther.* 45:1–38, 1990.

Hall, R.J. *Effects of Environmental Contaminants on Reptiles: A Review.* Special Scientific Report-Wildlife No. 228. Fish and Wildlife Service, U.S. Department of the Interior, Washington, DC, 1980.

Hallstrom, I. Genetic variation in cytochrome P-450-dependent demethylation, in *Drosophila melanogaster, Biochem. Pharmacol.* 36:2279–2282, 1987.

Hammock, D.D. NADPH dependant epoxidation of methyl farnesoate to juvenile hormone in the cockroach, *Blaberus giganticus, Pestic. Sci.* 17:323–328, 1975.

Hansen, L.G. Biotransformation of organophosphorus compounds relative to delayed neurotoxicity, *Neurotoxicology* 4:97–112, 1983.

Hansen, L.G., Kapoor, I.P., and Metcalf, R.L. Biochemistry of selective toxicity and biodegradability: comparative *O*-dealkylation by aquatic organisms, *Comp. Gen. Pharmacol.* 3:339–344, 1972.

Hasson, D.P. and West, C.A. Properties of the system for the mixed function oxidation of kaurene derivatives in microsomes of the immature seed of *Marah macrocarpus, Plant Physiol.* 58:479–484, 1976.

Higashi, K. Metabolic activation of environmental chemicals by microsomal enzymes of higher plants, *Mutat. Res.* 197:273–288, 1988.

Hodgson, E. The significance of cytochrome P-450 in insects, *Insect Biochem.* 13:237–265, 1983.

Hodgson, E. and Guthrie, F.E. *Introduction to Biochemical Toxicology.* Elsevier Publishing Company, New York, 1980, 437.

Hodgson, E. and Levi, P.E. *A Textbook of Modern Toxicology.* Elsevier Publishing Company, New York, 1987, 336.

Hollingworth, R.M. The biochemical and physiological basis of selective toxicity, in *Insecticide Biochemistry and Physiology.* C.F. Wilkinson, Ed. Plenum Press, New York, 1976, 431–506.

James, M.O. Catalytic properties of cytochrome P-450 in hepatopancreas of the spiny lobster, *Panulirus argus, Mar. Environ. Res.* 14:1–11, 1984.

James, M.O. Cytochrome P450 monooxygenases in crustaceans, *Xenobiotica* 19:1063–1077, 1989a.

James, M.O. Biotransformation and disposition of polycyclic aromatic hydrocarbons in aquatic invertebrates, in *Metabolism of Polycyclic Aromatic Hydrocarbons in the Aquatic Environment.* U. Varanasi, Ed. CRC Press, Boca Raton, FL, 1989b, 69–91.

James, M.O. Isolation of cytochrome P450 from hepatopancreas microsomes of the spiny lobster, *Panulirus argus,* and determination of catalytic activity with NADPH cytochrome P450 reductase from vertebrate liver, *Arch. Biochem. Biophys.* 282:8–17, 1990.

James, M.O. and Bend, J.R. Polycyclic aromatic hydrocarbon induction of cytochrome P-450 dependent mixed-function oxidases in marine fish, *Toxicol. Appl. Pharmacol.* 54:123–133, 1980.

James, M.O. and Little, P.J. 3-Methylcholanthrene does not induce *in vitro* xenobiotic metabolism in spiny lobster hepatopancreas or affect *in vivo* disposition of benzo(a)pyrene, *Comp. Biochem. Physiol.* 78C:241–245, 1984.

James, M.O., Fouts, J.R., and Bend, J.R. Xenobiotic metabolizing enzymes in marine fish, in *Pesticides in the Aquatic Environment*. M.A.Q. Kahn, Ed. Plenum Press, New York, 1977, 171–189.

James, M.O., Khan, M.A.Q., and Bend, J.R. Hepatic microsomal mixed-function oxidase activities in several marine species common to coastal Florida, *Comp. Biochem. Physiol.* 62C:155–164, 1979.

James, M.O., Schell, J.D., and Magee, V. Bioavailability, biotransformation and elimination of benzo(a)pyrene and benzo(a)pyrene-7,8-dihydrodiol in the lobster, *Homarus americanus, Bull. Mount Desert Island Biol. Lab.* 28:119–121, 1989.

Jensen, S. and Jansson, B. Methylsulfone metabolites of PCB and DDE, *Ambio* 5:257–260, 1976.

Jerina, D.M. Daly, J.W., Witkop, B., Zaltzman-Nirenberg, P., and Udenfriend, S. 1,2-Naphthalene oxide as an intermediate in the microsonal hydroxlation of naphthalene, *Biochemistry* 9:147–156, 1970.

Jewell, C.S.E. and Winston, G.W. Characterization of the microsomal mixed-function oxygenase system of the hepatopancreas and green gland of the red swamp crayfish, *Procambarus clarkii, Comp. Biochem. Physiol.* 92B:329–339, 1989a.

Jewell, C.S.E. and Winston, G.W. Oxyradical production by hepatopancreas microsomes from the red swamp crayfish, *Procambarus clarkii, Aquat. Toxicol.* 14:27–46, 1989b.

Jewell, C.S.E., Cummings, L.E., Ronis, M.J.J., and Winston, G.W. The hepatic microsomal mixed-function oxygenase (MFO) system of *Alligator mississippiensis*: induction by 3-methylcholanthrene (MC), *Xenobiotica* 19:1181–1199, 1989.

Johnson, E. Multiple forms of cytochrome P-450: criteria and significance, *Rev. Biochem. Toxicol.* 1:1–26, 1979.

Johnson, L.R. Light enhanced toxic/tumorigenic potential of trinitrotoluene (TNT) and analogs. Ph.D. Dissertation, University of Illinois at Urbana-Champaign, 1992.

Karasaki, Y., Ikeuchi, K., and Higashi, K. Highly variable distribution of *trans*-cinnamic 4-monooxygenase in a variety of plant tissues, *J. Univ. Occup. Environ. Health* 5:329–335, 1983.

Khan, M.A.Q., Coello, W., Kahn, A.A., and Pinto, H. Some characteristics of the microsomal mixed-function oxidase in the freshwater crayfish, *Cambarus. Life Sci.* 11:405–415, 1972.

Kirchin, M.A., Wiseman, A., and Livingstone, D.R. Studies on the mixed function oxygenase system of the marine bivalve *Mytilus edulis, Mar. Environ. Res.* 24:117–118, 1988.

Klecka, G.M. and Gibson, D.T. Metabolism of dibenzo-*p*-dioxin and chlorinated dibenzo-*p*-dioxins by a *Beijerinckia* species, *Appl. Environ. Microbiol.* 39:288–296, 1980.

Kleinow, K.M., Melancon, M.J., and Lech, J.J. Biotransformation and induction: implications for toxicity, bioaccumulation and monitoring of environmental xenobiotics in fish, *Environ. Health Perspect.* 71:105–119, 1987.

Kloepper-Sams, P.J., Park, S.S., Gelboin, H.V., and Stegeman, J.J. Specificity and cross reactivity of monoclonal and polyclonal antibodies against cytochrome P-450E of the marine fish scup, *Arch. Biochem. Biophys.* 253:268–278, 1987.

Knezovich, J.P. and Crosby, D.G. Fate and metabolism of *o*-toluidine in the marine bivalve molluscs *Mytilus edulis* and *Crassostrea gigas, Environ. Toxicol. Chem.* 4:435–446, 1985.

Kouyoumjian, H.H. and Uglow, R.F. Some aspects of the toxicity of p,p'-DDT, p,p'-DDE, and p,p'-DDD to the freshwater planarian *Polycelis felina* (Tricladida), *Environ. Pollut.* 7:103–109, 1974.

Kouyoumjian, H.H. and Villeneuve, J.P. Further studies on the toxicity of DDT to planaria, *Bull. Environ. Contam. Toxicol.* 22:109–122, 1979.

Krieger, R.I. and Wilkinson, C.F. Microsomal mixed function oxidases in insects. I. Localization and properties of an enzyme system effecting aldrin epoxidation in larvae of the southern armyworm (*Prodenia eridana*), *Biochem. Pharmacol.* 18:1403–1415, 1969.

Krieger, R.I., Feeny, P.P., and Wilkinson, C.F. Detoxification enzymes in the guts of caterpillars: an evolutionary answer to plant defences, *Science* 172:579–581, 1971.

Kulkarni, A.P. and Hodgson, E. Multiplicity of cytochrome P-450 in microsomal membranes from the housefly, *Musca domestica, Biochem. Biophys. Acta* 632:573–588, 1980.

Kurelec, B. and Britvic, S. The activation of aromatic amines in some marine invertebrates, *Mar. Environ. Res.* 17:141–144, 1985.

Lamoureux, G.L. and Frear, D.S. Pesticide metabolism in higher plants: *in vitro* enzyme studies, in *Drug Metabolism*. D.V. Parke and R.L. Smith, Eds. Taylor and Francis, London, 1979, 1091–1217.

Levin, W., Wood, A.W., Chang, R.L., Ryan, D., Thomas, P.E., Yagi, H., Thaker, D.R., Vyas, K., Boyd, C., Chu, S-Y., Conney, A.H., and Jerina, D.M. Oxidative metabolism of polycyclic aromatic hydrocarbons to ultimate carcinogens, *Drug Metab. Rev.* 13:555–580, 1982.

Lindstrom-Seppa, P. and Hanninen, O. Induction of cytochrome P-450 mediated monooxygenase reactions and conjugation activities in freshwater crayfish *(Astacus astacus)*, *Arch. Toxicol. Suppl.* 9:374–377, 1986.

Little, P.J., James, M.O., Pritchard, J.B., and Bend, J.R. Temperature depends on disposition of [^{14}C] benzo(a)pyrene in the spring lobster, *Panulirus argas, Toxicol. Appl. Pharmacol.* 77:325–333, 1985.

Livingstone, D.R. Response of the detoxification/toxification enzyme systems of molluscs to organic pollutants and xenobiotics, *Mar. Pollut. Bull.* 16:158–164, 1985.

Livingstone, D.R. Seasonal responses to diesel oil and subsequent recovery of the cytochrome P-450 monooxygenase system in the common mussel, *Mytilus edulis* L. and the Periwinkle *Littorina littaca* L., *Sci. Total Environ.* 65:3–20, 1987.

Livingstone, D.R. and Farrar, S.V. Tissue and subcellular distribution of enzyme activities of mixed-function oxygenase and benzo[*a*]pyrene metabolism in the common mussel, *Mytilus edulis*, *Sci. Total Environ.* 39:209–235, 1984.

Livingstone, D.R., Garcia Martinez, P., Stegeman, J.J., and Winston, G.W. Benzo(a)pyrene metabolism and aspects of oxygen radical generation in the common mussel, *Mytilus edulis* L., *Biochem. Soc. Transact.* 16:779, 1988.

Livingstone, D.R., Kirshin, M.A., and Wiseman, A. cytochrome P-450 and oxidative metabolism in molluscs, *Xenobiotica* 19:1041–1062, 1989.

Marcus, C.B., Murray, M., Wang, C., and Wilkinson, C.F. Methylene-dioxphenyl compounds as inducers of cytochrome P-450 monooxygenases activity in the southern armyworm *(Spodoptera eridania)* and the rat, *Pest. Physiol. Biochem.* 26:310–322, 1986.

Miller, M.R., Hinton, D.E., and Stegeman, J.J. Cytochrome P-450 induction and localization in gill pillar (endothelial) cells of scup and rainbow trout, *Aquat. Toxicol.* 14:307–322, 1989.

Monks, T.J., Anders, M.W., Dehant, W., Lau, S.S., Stevens, J.L., and van Bladeren, P.J. Glutathione conjugate mediated toxicities, *Toxicol. Appl. Pharmacol.* 106:1–19, 1990.

Moore, M.N., Livingstone, D.R., Donkin, P., Bayne, B.L., Widdows, J., and Lowe, D.M. Mixed function oxygenases and xenobiotic detoxification/toxification systems in bivalve molluscs, *Helgol. Wiss. Meeresunters.* 33:278–291, 1980.

Morgan, D. and Roan, C. Absorption, storage and metabolic conversion of ingested DDT and DDT metabolites in man, *Arch. Environ. Health* 22:301–308, 1971.

Nebert, D.W. and Gelboin, H.V. Substrate-inducible microsomal aryl hydroylase in mammalian cell culture. I. Assay and properties of inducible enzyme, *J. Biol. Chem.* 243:6242–6249, 1968.

Nebert, D.W. and Gonzalez, F.J. P450 genes: structure, evolution, and regulation, *Annu. Rev. Biochem.* 56:945–993, 1987.

Nebert, D.W., Nelson, D.R., and Feyereisen, R. Evolution of the cytochrome P-450 genes, *Xenobiotica* 19:1149–1160, 1989a.

Nebert, D.W., Nelson, D.R., Adesnik, M., Coon, M.J., Estabrook, R.W., Gonzalez, F.J., Guengerich, F.P., Gunsalus, I.C., Johnson, E.F., Kemper, B., Levin, W., Phillips, I.R., Sato, R., and Waterman, M.R. The P-450 superfamily: updated listing of all genes and recommended nomenclature for the chromosomal loci, *DNA* 8:1–13, 1989b.

Parsons, J.R., Sijm, D.T.H.M., van Laar, A., and Hutzinger, O. Biodegradation of chlorinated biphenyls and benzoic acids by a *Pseudomonas* strain, *Appl. Microbiol. Biot.* 29:81–84, 1988.

Plapp, F.W., Jr. The genetic basis of insecticide resistance in the housefly: evidence that a single locus plays a major role in metabolic resistance to insecticides, *Pestic. Biochem. Physiol.* 22:194–201, 1984.

Plewa, M.J., Smith, S.R., and Wagner, E.D. Diethyldithiocarbamate suppresses the plant activation of aromatic amines into mutagens by inhibiting tobacco cell peroxidase, *Mutat. Res.* 247:57–64, 1991.

Pohl, R.J., Bend, G.R., Guarino, A.M., and Fouts, J.R. Hepatic microsomal mixed function oxidase activity of several marine species from coastal Maine, *Drug Metab. Dispos.* 2:545–555, 1974.

Poland, A., Glover, E., and Kende, A.S. Stereospecific high affinity binding of 2,3,7,8-tetrachlorodibenzo-*p*-dioxin by hepatic cytosol. Evidence that the binding species is the receptor for the induction of aryl hydrocarbon hydroxylase, *J. Biol. Chem.* 251:4936–4946, 1976.

Reichhart, D., Salaun, J.P., Benveniste, I., and Durst, F. Induction by manganese, ethanol, phenobarbital and herbicides of microsomal cytochrome P-450 in higher plant tissues, *Arch. Biochem. Biophys.* 196:301–303, 1980.

Ronis, M.J.J. and Hodgson, E. Cytochrome P-450 monooxygenases in insects, *Xenobiotica* 19:1077–1092, 1989.

Ronis, M.J.J. and Walker, C.H. The microsomal monooxygenases of birds, *Rev. Biochem. Toxicol.* 10:301–384, 1989.

Ronis, M.J.J., Dauterman, W.C., and Hodgson, E. Characterization of multiple of forms of cytochrome P-450 from an insecticide resistant strain of housefly, *Musca domestica*, *Pest. Physiol. Biochem.* 32:74–90, 1988.

Sande, M.A. and Mandell, G.L. Antimicrobial agents: general considerations, in *Goodman and Gilman's The Pharmacological Basis of Therapeutics*, 6th ed. A.G. Gilman, L.S. Gordon, and A. Gilman, Eds. Macmillan Publishing Company, New York, 1980, 1080–1105.

Sandermann, H. Mutagenic activation of xenobiotics by plant enzymes, *Mutat. Res.* 197:183–194, 1988.

Sanglard, D. and Loper, J.C. Characterization of the alkane-inducible cytochrome P450 (P450alk) gene from the yeast *Candida tropicalis:* identification of a new P450 gene family, *Gene* 76:121–136, 1989.

Sanglard, D., Chen, C., and Loper, J.C. Isolation of the alkane inducible cytochrome P450 (P450alk) gene from the yeast *Candida tropicalis*, *Biochem. Biophys. Res. Commun.* 144:251–257, 1987.

Schafer, W. and Sandermann, H. Metabolism of pentachlorophenol in cell suspension cultures of wheat *(Triticum aestivum L.):* tetrachlorocatechol as a primary metabolite, *J. Agric. Food Chem.* 36:370–377, 1987.

Schlenk, D. and Buhler, D.R. Cytochrome P-450 and phase II activities in the gumboot chiton *Cryptochiton stelleri*, *Aquat. Toxicol.* 13:167–182, 1988.

Schoeny, R., Cody, T., Warshawsky, D., and Radike, M. Metabolism of mutagenic polycyclic aromatic hydrocarbons by photosynthetic algal species, *Mutat. Res.* 197:289–302, 1988.

Scholtz, R., Wackett, L.P., Egli, C., Cook, A.M., and Leisinger, T. Dibromoethane dehydrogenase with improved catalytic activity isolated from a fast-growing dichloromethane-utilizing bacterium, *J. Bacteriol.* 170:5698–5704, 1988.

Schonbrod, R.D., Philleo, W.W., and Terriere, L.C. Hydroxylation as a factor in resistance in houseflies and blowflies, *J. Econ. Entomol.* 58:74–77, 1965.

Schwen, R.J. and Mannering, G.J. Hepatic cytochrome P-450-dependent monooxygenase systems of the trout, frog, snake. II. Monooxygenase activities, *Comp. Biochem. Physiol.* 71B:437–443, 1982a.

Schwen, R.J. and Mannering, G.J. Hepatic cytochrome P-450 dependent monooxygenase systems of the trout, frog and snake. III. Induction, *Comp. Biochem. Physiol.* 71B:445–453, 1982b.

Sikka, H.C., Rutkowski, J.P., and Kandaswami, C. Comparative metabolism of benzo(a)pyrene by liver microsomes from brown bullhead and carp, *Aquat. Toxicol.* 16:101–112, 1990.

Sims, P. and Grover, P.L. Epoxides in polycyclic aromatic hydrocarbon metabolism and carcinogenesis, *Adv. Cancer Res.* 20:165–274, 1974.

Sipes, I.G. and Gandolfi, A.J. Biotransformation of toxicants, in *Cassarett and Doull's Toxicology: The Basic Science of Poisons,* 4th ed. M.O. Amdur, J. Doull, and C.D. Klassen, Eds. Pergamon Press, New York, 1991, 88–126.

Sipes, I.G. and Schnellmann, R.G. Biotransformation of PCBs: metabolic pathways and mechanisms, *Environ. Toxicol. Rev.* 1:97–110, 1987.

Stegeman, J.J. Benzo(a)pyrene oxidation and microsomal enzyme activity in the mussel *(Mytilus edulis)* and other bivalve mollusc species from the Western North Atlantic, *Mar. Biol.* 89:21–30, 1985.

Stegeman, J.J. and Chevion, M. Sex differences in cytochrome P-450 and mixed function oxygenase activity in gonadally mature trout, *Biochem. Pharmacol.* 29:553–558, 1980.

Stegeman, J.J. and Kloepper-Sams, P.J. Cytochrome P-450 isozymes and monooxygenase activity in marine animals, *Environ. Health Perspect.* 71:87–95, 1987.

Stegeman, J.J. and Woodin, B.R. Patterns of benzo(a)pyrene metabolism in liver of the marine fish, *Stenotomus versicolor*, *Fed. Proc.* 39:1752, 1980.

Stegeman, J.J., Pajor, A.M., and Thomas, P. Influence of estradiol and testosterone on cytochrome P-450 and monooxygenase activity in immature brook trout, *Salvelinus fontinalis*, *Biochem. Pharmacol.* 31:3979–3989, 1982.

Stegeman, J.J., Woodin, B.R., and Binder, R.L. Patterns of benzo(a)pyrene metabolism by varied species, organs and developmental stages of fish, *Natl. Cancer Inst. Monog.* 65:371–377, 1984.

Sweetser, P.B., Schow, G.S., and Hutchison, J.M. Metabolism of chlorsulfuron by plants: biological basis for selectivity of a new herbicide for cereals, *Pestic. Biochem. Physiol.* 17:18–23, 1982.

Testa, B. and Jenner, P. Inhibitors of cytochrome P-450s and their mechanism of action, *Drug Metab. Rev.* 12:1–117, 1981.

Van der Trenck, Th. and Sandermann, H. Oxygenation of benzo(a)pyrene by plant microsomal fractions, *FEBS Lett.* 119:227–231, 1980.

Varanasi, U., Nishimoto, M., Reichert, W.L., and Eberhart, B.T. Comparative metabolism of benzo(a)pyrene and covalent binding to hepatic DNA in English sole, starry flounder and the rat, *Cancer Res.* 46, 3817–3824, 1986.

Vogel, T.M. and McCarty, P.L. Biotransformation of tetrachloroethylene to trichloroethylene, dichloroethylene, vinyl chloride and carbon dioxide under methanogenic conditions, *Appl. Environ. Microbiol.* 49:1080–1083, 1985.

Vogel, T.M., Criddle C.S., and McCarty, P.L. Transformations of halogenated aliphatic compounds, *Environ. Sci. Technol.* 21:722–736, 1987.

Wackett, L.P. Dehalogenation reactions catalyzed by bacteria, in *Biological Degradation of Wastes*. A.M. Martin, Ed. Elsevier Applied Science, Amsterdam, 1991, 187–205.

Wackett, L.P. and Gibson, D.T. Degradation of trichloroethylene by toluene dioxygenase in whole-cell studies with *Pseudomonas putida* F1, *Appl. Environ. Microbiol.* 54:1703–1708, 1988.

Walker, C.H. and Ronis, M.J.J. The monooxygenases of birds, reptiles and amphibians, *Xenobiotica* 19:1111–1121, 1989.

Walker, C.H., Newton, I., Hallam, S., and Ronis, M.J.J. Activities and toxicological significance of the hepatic microsomal enzymes of the kestrel and sparrowhawk, *Comp. Biochem. Physiol.* 86C:359–363, 1986.

Warshawsky, D., Keenan, T.H., Reilman, R., Cody, T.E., and Radike, M.J. Conjugation of benzo(a)pyrene metabolites by freshwater green alga *Selenastrum capricornutum*, *Chem. Biol. Interact.* 74:93–105, 1990.

Waters, L.C. and Nix, W.E. Regulation of insecticide resistance related cytochrome P-450 expression in *Drosophila melanogaster*, *Pestic. Biochem. Physiol.* 30:214–227, 1988.

Watkinson, R.J. and Morgan, P. Physiology of aliphatic hydrocarbon-degrading microorganisms, *Biodegradation* 1:79–92, 1990.

Werck-Reichhart, D., Gabriac, B., Teutsch, H., and Durst, F. Two cytochrome P-450 isoforms catalysing *O*-de-ethylation of ethoxycoumarin and ethoxyresorufin in higher plants, *Biochem. J.* 270:729–735, 1990.

Wilkinson, C.F. Role of mixed function oxidases in pesticide resistance, in *Pest Resistance to Pesticide: Challenge and Prospects*. G.P. Georghiou and M. Saito, Eds. Plymouth Press, New York, 1983, 175–205.

Williams, D.E. and Buhler, D.R. Benzo(a)pyrene hydroxylase catalyzed by purified isozymes of cytochrome P-450 from *b*-naphthaflavone fed rainbow trout, *Biochem. Pharmacol.* 33:3743–3753, 1984.

Winston, G.W., Shane, B.S., and Henry, C.B. Hepatic monooxygenase induction and promutagen activation in channel catfish from a contaminated river basin, *Ecotoxicol. Environ. Saf.* 6:258–271, 1988.

Yinon, J. and Hwang, D. Metabolic studies of explosives. I. El and Cl mass spectrometry of metabolites of 2,4,6-trinitrotoluene, *Biomed. Mass Spectrom.* 11:594–600, 1984.

Yu, S.J. and Terriere, L.C. Metabolism of juvenile hormone I by microsomal oxidase, and epoxide hydratase of *Musca domestica* and some comparisons with *Pharmia regina* and *Sarcophaga bullata*, *Pestic. Biochem. Physiol.* 9:237–246, 1978.

Section II

Chapter

4

Toxic Metals in the Environment

Louis W. Chang and Lorris Cockerham

A. INTRODUCTION

Over 40 elements in the environment are classified as metals. Many, such as the alkaline earth group and some trace elements, are essential for life; others have great potential for toxicity. Macronutrients such as calcium, magnesium, iron, potassium, and sodium are particularly important in sustaining life but may become toxic in excessive concentrations. Trace elements such as chromium, cobalt, copper, manganese, nickel, selenium, and zinc are structurally part of important molecules and may serve as cofactors of enzymes in metabolic processes. Excessive concentrations of these elements are also toxic (Mailman, 1980; Hayes, 1989). Some elements such as lead, cadmium, and mercury have deleterious effects on biological tissues at any concentration.

Metals are probably the oldest toxin known. Lead and silver may have been used before 2000 B.C. Theophrastus, who lived around 380 B.C., cited both arsenic and mercury in his writings, and Hippocrates may have been the first to link metal exposure with abdominal colic (Goyer, 1986). Lead is implicated in the decline and fall of the Roman empire through its use in water pipes. Lewis Carroll, in *Alice in Wonderland*, introduced a character known as the "Mad Hatter" who exhibited symptoms typically seen in milliners after prolonged exposure to inorganic mercury. However, it was the Minamata Bay incident in Japan that related the toxic effect of mercury with environmental exposure and heightened the concern about the increased environmental levels of heavy metals (Takeuchi et al., 1959; Tokuomi, 1961; Mailman, 1980; Mance, 1987).

0-8493-8851-1/94/$0.00+$.50
© 1994 by CRC Press, Inc.

In determining hazards associated with heavy metal contamination of the environment, the most important step is to determine the most sensitive environmental organism. Animals usually are more sensitive to heavy metal contamination than are plants, and humans are universally considered as the most sensitive, as well as the most important target species. However, heavy metal toxicity in humans is considered only briefly in this chapter because of its extensive coverage elsewhere (Gossel and Bricker, 1984; Goyer, 1986; Hayes, 1989).

B. SOURCE OF ENVIRONMENTAL METALS

The earth's crust is the ultimate source of all metallic elements found in the environment. Metals are neither created nor destroyed by humans but are redistributed naturally in the environment by both geological and biological cycles. The industrial and technological activities of man simply shortens the ore phase of the metal, forms new metallic compounds, and introduces metals into the atmosphere mainly through fossil fuel combustion which augments worldwide distribution.

1. Industry

Metal contamination of the environment arises not only from natural sources, but from industrial activity. Combustion of fossil fuels releases about 20 toxicologically important metals into the environment including arsenic, beryllium, cadmium, lead, and nickel (Goyer, 1986). Industrial products and used industrial material may contain high concentrations of toxic metals. For example, mercury is used by the chlor-alkali industry to produce chlorine and caustic soda in the pulp and paper industry and in the production of battery cells, fluorescent bulbs, electrical switches, paints, agricultural products, dental preparations, and pharmaceuticals (Mailman, 1980).

Although both adults and children may be exposed to lead dust from such industries as battery manufacturing through contact with workers' clothing, children are frequently exposed to lead through ingestion of peeling paint chips. This latter exposure of children may be a major health hazard; however, the environmental release of lead from the combustion of tetraethyl lead containing auto or industrial fuels remains the greatest source of exposure.

Cadmium, a by-product of zinc and lead mining, is an important environmental pollutant. It has many industrial uses in paints, pigments, batteries, and plastics. Another use is as an anticorrosive agent for steel, iron, copper, brass, and other alloys.

2. Agricultural Products

Heavy metals present in fertilizers include cadmium, chromium, copper, manganese, molybdenum, nickel, and zinc. The major environmental sources of arsenic are pesticides, herbicides, and other agricultural products. Lead arsenate, in addition to being a component of industrial effluents, has been used as an agricultural pesticide. Fungicides containing mercury contribute to environmental contamination. Eventually, many of these metals may accumulate in agricultural soils and pose a hazard to plant growth and animal nutrition.

3. Food and Food Additives

For humans, the ingestion of food and food additives may represent the largest source of exposure to metals. The ingestion of metals normally follows two primary routes (Mailman and Sidden, 1980). The first involves direct, accidental contact with metal-bearing food or material. Ingestion of foreign substances such as lead-containing paint chips by children is one of the best examples of this method. Since paint chips may contain 50 to 100 mg of lead in a 1 cm^2 paint chip, even a few chips a day may exceed the acceptable daily intake.

Another example of direct ingestion of lead occurred during the rule of the Roman empire (509 B.C. to 476 A.D.). The Romans used lead pipes to carry water to the homes of prominent families of the empire, thus lead was ingested in large quantities by these families. The use of lead-containing pottery glazes also added to their chronic intake of lead. This chronic ingestion of lead probably resulted in neurological symptoms and may well have contributed to the decline and fall of the Roman empire. More recently, instances of lead poisoning have occurred in the U.S. through the use of inexpensive pottery as containers for acid foods (Mailman, 1980).

The second route of ingestion of metals is through the bioaccumulation of metals in the food chain. The well known Minamata disease is the result of massive human exposure to methylmercury-contaminated fish. Microorganisms in sediments methylate mercury ions into methylmercury. Methylmercury can then be assimilated by fish through ingestion of contaminated food sources and accumulation in their tissues (Stirling, 1980). Uptake by fish in Minamata Bay apparently occurred as the industrial release of mercury compounds into the bay preceded the accumulation of mercury by edible fish. Other toxic metals such as cadmium are found in meat, fish, and fruit. Shellfish such as mussels, scallops, and oysters are a major source of dietary cadmium.

Sewage sludge containing nitrogen, phosphorous, and sulfur as well as the toxic metals cadmium, chromium, copper, lead, mercury, nickel, and zinc has been applied to agricultural soils to improve crop production (Dowdy and Volk, 1983). Although leaching of heavy metals through soils is limited, plant uptake may translocate the metals into the food chain, eventually being consumed by humans at the highest trophic level.

C. ENVIRONMENTAL FACTORS AFFECTING METAL TOXICITY

Various factors influence the metabolism and effects of metals. Those factors that include specific characteristics of the organisms exposed are known as host factors. Host factors include age, diet, immune status, sex, species, and interphyletic and circadian biorhythms (Nordberg et al., 1979). Metal toxicity can also be modified by simultaneous or previous exposure to certain abiotic conditions such as light, humidity, wind speed, amount of rainfall, ionizing radiation, and variations in temperature.

1. Temperature

Variations in ambient temperature affect the metabolism of xenobiotics in both homeothermic and poikilothermic animals. There are two basic types of temperature effects. An increase in toxicity may occur with either an increase or decrease in ambient temperature (Hodgson, 1987). With only a few exceptions, an increase in water temperature will increase the toxicity of metals to marine invertebrates (Mance, 1987). However, it is not possible to infer any specific relationship between metal toxicity and temperature with marine or freshwater fish. The effect varies with the metal and the species of fish. For example, when salmonid fish are exposed to silver nitrate toxicity varies directly with the temperature. However, when salmonid fish are exposed to copper the toxicity varies inversely with the temperature. Likewise, the toxicity of cadmium to some nonsalmonid fish varies inversely with increased temperature, while with others such as *Oryzias latipes* it does not. In homeothermic, terrestrial vertebrates the effect of temperature variations are mediated by hormonal interactions, while in terrestrial invertebrates and plants it is through an increase in metal assimilation at higher temperature (Mance, 1987; Hopkin, 1989).

2. Light

The diurnal pattern of metabolism depends on light cycles rather than light intensity. Enzymes associated with detoxification display a diurnal rhythm with the greatest activity occurring during specific phases of the light cycle. The greatest activity of microsomal cytochrome P_{450}, a major enzyme in detoxification reactions, occurs at the beginning of the dark phase of the cycle (Hodgson, 1987). Variations in the ratio of light to dark in the light cycle also may have a profound effect on metal metabolism in some organisms such as the isopod *Oniscus asellus* which molts according to a 16 h light to 8 h dark cycle but will stop molting if the light/dark ratio is reversed. The profound physiological changes that accompany molting may have an effect on metal metabolism (Hopkin, 1989).

3. pH

pH is the most important abiotic factor determining metal availability in soil to plants. Decay of leaf litter decreases the pH of soil through the release of organic acids (Hopkin, 1989). Decreasing pH enhances the mobilization of metal salts in soil (Nordberg et al., 1985). Acid precipitation enhances the mobilization of both toxic and nontoxic metals in water and soil compartments and may also enhance the conversion of some metals to more toxic forms.

4. Acclimation, Fluctuating Exposure, and Mixtures

Acute toxicity test data, but not chronic laboratory studies, support acclimation to the toxic effects of metals. Acute toxicity test with *Salmo gairdneri* parr showed that with a preexposure to copper for periods of up to 21 d there was a marked relationship between the preexposure copper concentration and the acute LC_{50}. However, organisms such as the *S. gairdneri* parr that are preexposed to a particular toxic metal have a transient resistance to most metals and will lose that acclimation after entering an environment that is not contaminated. Mortality may be increased by as much as 30% or a decrease in the growth of the survivors may be manifest with intermittent or fluctuating exposure. In most cases, mixtures of metals produce an additive effect on toxicity. In only a few instances will the addition of one metal reduce the toxicity of another (Mance, 1987). In one such instance silver interfered with the toxic effects of cadmium on the eggs of the flounder (Voyer et al., 1982).

D. TOXIC METALS IN THE ATMOSPHERE

The atmosphere is a dynamic system consisting of four principal zones, the troposphere, the stratosphere, the mesosphere, and the thermosphere. These four zones are separated by three regions of temperature inversions, called the tropopause, the stratopause, and the mesopause (Fergusson, 1990). The zone closest to the earth is the troposphere and is of the greatest concern for transport of metals.

Temperature and winds are a major influence on the rate and volume of movement of heavy metals, particulate matter, aerosols, or even vapor in the atmosphere. Changes in air temperature are responsible for vertical air movement. Even though the temperature of tropospheric air normally decreases inversely with an increase in altitude, temperature inversions can also occur, resulting in horizontal movement and mixing (Sheets, 1980; Fergusson, 1990). Horizontal movement of air, or wind, arises primarily from the interaction of three factors: a pressure gradient, Coriolis deflection, and friction with the earth's surface. The pressure gradient results in the movement of air from areas of high pressure to those of low pressure. Coriolis deflection results from the influence that the spin of the earth has on the surrounding air mass. As a result, air masses move in an easterly direction in the northern hemisphere. This deflection causes the air to move along isobars, or lines of equal barometric pressure, rather than directly from high pressure to low pressure areas. As a result, air moves counterclockwise around a low pressure area and clockwise around a high pressure area in the northern hemisphere (Fergusson, 1990). Air movement is in the opposite direction in the southern hemisphere.

An excellent example of the meteorologic effects on long-distance movement and deposition of atmospheric metals was seen following the Chernobyl reactor explosion in 1986. As discussed in Chapter 9 of this text, the radioactive clouds moved in two separate directions and covered most of Europe in approximately two weeks.

Although radioactive elements from nuclear accidents and explosions are widely distributed, volcanic emissions contain the highest concentrations of metals found in any natural contributor of metals into the atmosphere (Fergusson, 1990). Antimony, arsenic, cadmium, lead, and selenium are discharged into the atmosphere from volcanic sources while several heavy elements are found in effluents from other geothermal sources (Sabadell and Axtmann, 1975). Although the highest concentrations of metals are in volcanic emissions, windblown material is the largest contributor of total heavy metals to the atmosphere with forest fires, vegetation, and sea spray contributing a greater amount of lead and cadmium than volcanic emissions (Fergusson, 1990).

Anthropogenic sources of metal contamination in the atmosphere are more concentrated in highly industrialized urban areas. These sources include (1) industrial sites and engines involved in the combustion of fossil fuels such as coal, oil, and natural gas, (2) metal manufacturing plants and foundries, (3) mines and smelters, (4) refuse incinerators, and (5) cement production sites (Amasa, 1975; Comar and Nelson, 1975; Johnson et al., 1975; Lindberg et al., 1975; Hopkin, 1989; Fergusson, 1990). Atmospheric levels of heavy metals are significantly lower in rural and remote areas and are mainly a reflection of emissions from natural sources (Johnson et al., 1975; Fergusson, 1990).

Combustion of either coal or oil results in the discharge of arsenic, bismuth, cadmium, chromium, copper, lead, manganese, mercury, nickel, selenium, and zinc into the atmosphere as global aerosols. Antimony and indium are also found in significant quantities in urban area aerosols (Fergusson, 1990). A more detailed discussion of these pollutants and their biological effects can be found in Chapter 11 of this text.

Metal speciation of aerosols is difficult but more work has been done on lead than any other metal. Information is available on lead species from different sources (Fergusson, 1990). Pollutants derived from the above mentioned sources are eventually deposited on earth either in a dry state or as wet deposits of rain or snow. It is through this mechanism that deposition of toxic metals adds to the contamination of aquatic and terrestrial environments on earth.

E. TOXIC METALS IN WATER AND SEDIMENTS

The hydrosphere accounts for a greater area of the earth's surface than the lithosphere and is divided into lakes, rivers, estuaries, and oceans (Fergusson, 1990). Metals exist in the hydrosphere as dissolved material and suspended particulates and are in deposited sediments. Sediments in rivers, lakes, estuaries, and oceans account for the main sinks of heavy metals in the hydrosphere. In estuaries, heavy metals from the atmosphere and rivers accumulate, thus permitting chemical and physical reactions to occur before being washed out into the ocean.

Atmospheric deposition, soil leaching, runoff, erosion, and breakdown of mineral deposits all contribute to the concentration of metals in natural water supplies. Anthropogenic sources include mining, smelting, burning of fossil fuels, leaching from waste dumps, urban runoff, sewage effluent, waste dumping, and agricultural runoff.

The levels of trace metals are usually higher in rivers than in oceans because metals from point and nonpoint sources are discharged into the rivers. Changes in concentrations of metals in rivers are easily detected because of their rapid rate of transport. Levels of certain trace metals such as cadmium, mercury, and lead are closely correlated with variations in population density along a river and with seasonal variations in flow rate (Rand and Barthalmus, 1980; Fergusson, 1990). Concentrations vary inversely with velocity of the river. Likewise, sediment concentrations of a metal in a river varies inversely with distance from the source. Studies of the Rhine River in western Europe have shown that the cadmium concentrations fall with an increase in velocity of the river in winter and rise with a decrease in velocity in summer and fall. A rapid rise in cadmium concentration at the confluence of the Rhine and Ruhr Rivers is followed by a gradual decrease in the next 50 km (Fergusson, 1990).

Among its other functions, the Rhine River in western Europe has served for hundreds of years as an international cesspool for municipal and industrial waste (Rand and Barthalmus, 1980). From Lake Constance to the North Sea, the Rhine flows for over 1230 km through some of the most heavily industrialized areas of western Europe. Heavy metals are a serious form of aquatic pollution because, as stable compounds, they are difficult to remove by any natural process. Other metallic cations such as calcium, sodium, potassium, magnesium, and iron form ionizable salts with chlorides, sulfates, nitrates, bicarbonates, and phosphates. In the Rhine River huge quantities of these salts are found as a result of erosion, atmospheric fallout (snow and rainfall), and discharges from mines and industry.

Levels of heavy metals in estuaries, coastal waters, and sediments vary considerably, depending on inputs. Much of the input to estuaries is due to atmospheric fallout, and as much as 93% of heavy metal entering an estuary is retained (Fergusson, 1990). An excellent discussion of the heavy metal components of an estuarine environment is presented in Chapter 14.

The atmosphere is the major contributor of heavy metals to the oceans. This is especially true for lead, with 90% of the lead found in ocean waters derived from the atmosphere (Fergusson, 1990). Lead in surface waters of the North Pacific has been linked to atmospheric inputs from automobile and smelter emissions. Atmospheric inputs from the U.S. may also be responsible for seasonal variations in mercury concentrations in surface waters of the Atlantic (Leland and Kuwabara, 1985).

Concentrations of heavy metals in ocean sediment vary with geographical locations (Fergusson, 1990). Higher levels are often found in coastal waters because of nearby sources of pollution. The profile of heavy metals such as lead in ocean sediment is similar to that found in the water column, with the highest concentrations near the surface. Any disturbance of the sediment will alter the stratigraphy of the metals in the sediment. Bioturbation, or disturbance by living organisms, may occur through burrowing or the bioaccumulation of the heavy metals by the organism. Obviously, underwater volcanism or earthquakes will alter the profiles.

The pollution of aquatic environments by metals is well-documented worldwide. Metal contamination in several rivers in Wales has been documented from the early 19th century, with some rivers having only invertebrate communities and showing no sign of fish life by the early 20th century (Mance, 1987). However, despite documentation, concern over aquatic metal pollution did not intensify until the human fatalities from mercury at Minimata Bay and cadmium poisoning (Itai-itai disease) occurred in Japan (Mance, 1987).

The effect of metals on aquatic organisms is difficult to determine as many physical and chemical properties such as flow rate contribute to the outcome. Also, the size and nature of the particulates to which the metals are attached affect the toxicity of the metal. In order for a metal to enter an organism it must either be phagocytized or in a solubilized form. Chelation of a metal with an organic molecule may alter chemical reactions and membrane permeability, thereby altering adsorption by an organism. Metals may also adsorb to colloidal particles in the water and thus alter the degree of adsorption by the aquatic organisms (Leidy, 1980).

Toxicity of metals in the aquatic environment is evident over a wide range of effects, from a minimal reduction in growth to death. In general, young life stages of aquatic organisms are more sensitive to metals than adults (Leland and Kuwabara, 1985; Mance, 1987). However, there are strong exceptions to this generality. Eggs of some freshwater fish do not represent the most sensitive life stage of the species, and insect larvae are more resistant to metals than the adult forms. For most metals except lead, crustaceans, particularly *Daphnia*, are consistently the most sensitive aquatic organisms tested.

Notable interspecies and interphyletic variations exist in both freshwater and marine aquatic organisms.

F. TOXIC METALS IN THE TERRESTRIAL ENVIRONMENT

Many of the heavy metals that occur naturally in the earth's crust are released into the soil through the process of weathering, but the largest natural source of metals to the terrestrial environment is fallout from volcanoes (Hopkin, 1989; Fergusson, 1990). Anthropogenic sources of metal contamination to the terrestrial environment via the atmosphere includes fuel combustion, metal manufacture, foundries, refuse incineration, and cement production. Metals also are introduced into soil as plant nutrients, pesticides, and constituents of waste products. Plant uptake of heavy metals from sewage sludge applications to croplands has been documented (Dowdy and Volk, 1983).

The predominant metal pollutants found in soil are arsenic, cadmium, lead, mercury, and selenium, followed by antimony, bismuth, indium, tellurium, and thallium. Levels of metals higher than those shown in Table 4.1 can be found in soils contaminated by mining and agricultural activities and the addition of pesticides and sewage sludge to soil (Fergusson, 1990).

Elevated levels of antimony and cadmium may be associated with soil in the vicinity of smelters. High levels of arsenic are associated with soils around metal processing plants and with soils contaminated by arsenical pesticides, sometimes reaching 600 μg g^{-1} in the latter case. Sources of lead contamination include metal working, internal combustion engines using tetraethyl lead in gasoline, paint, and smelting. In some surface soils the concentration of lead can reach as high as 10%. Mercury contamination of soils is usually associated with mining activities, the production of dichlorine and caustic soda, and the agricultural use of mercury compounds such as fungicides.

Heavy metals may enter plants via uptake from the soil through the roots or through foliar uptake (Fergusson, 1990). The more common process of incorporation of metals into plants is mainly through the roots by active transport against a concentration gradient. Metals are absorbed by the roots by passive diffusion through cell membranes. Once inside the plant the metal is transported through the xylem to various parts of the plant. The mobility varies from one metal to another, but in general, the levels will be highest in the roots and lowest in the seeds.

The principle route for foliar uptake is through the leaf cuticle. This route of uptake is very significant, especially for the entry of aerosols containing

TABLE 4.1. Levels of Selected Metals in Surface Soils Worldwide

Metal	Range (μg g^{-1})	Mean (μg g^{-1})
Antimony (Sb)	0.05–260	0.9
Arsenic (As)	<0.1–97	11.3
Bismuth (Bi)	0.13–10	0.2
Cadmium (Cd)	0.1–1.0	0.62
Indium (In)	0.7–3.0	1.0
Lead (Pb)	1–888	29.2
Mercury (Hg)	0.005–1.11	0.098
Selenium (Se)	0.005–230	0.4
Tellurium (Te)	0.5–37	—
Thallium (Tl)	0.03–5.0	0.4

Source: Fergusson, 1990.

metals. The route of entry also changes the relative levels of metals found in various structures of the plant.

The concentrations of heavy metals vary widely in the edible parts of the plants, ranging from 0.001 μg g^{-1} with tin to 20 μg g^{-1} with lead. Plant species also vary in their uptake of heavy metals, and a comparison of the relative adsorption of some heavy metals by selected plants can be seen in Table 4.2.

Although many species of plants are unable to survive on soils containing high concentrations of heavy metals, some do survive and flourish. Studies have shown that when the pollutant exerts selection pressures, and if there is a suitable range of genetic variation, a considerable number of plant species, especially grasses, can produce genotypes resistant or tolerant to one or more metals (Moriarty, 1988). However, apple orchards in the northwestern U.S. become nearly sterile after exposure to heavy metal contamination (Mailman, 1980). If the germination of such an important plant is inhibited at levels below those effecting animal health, the levels of contamination are unacceptable.

Plants growing near metal smelters normally contain high levels of arsenic, which depends on the type of arsenical compound, plant species, soil type, and climate. For instance, vegetables accumulate significant levels of arsenic when grown in soils containing high concentrations of lead arsenate, but not when grown in soils containing high levels of arsenic trioxide (Dickerson, 1980). With increasing levels of lead arsenate, beets, lettuce, and radishes show an increased arsenic uptake while the uptake by broccoli, carrots, eggplants, and tomatoes is unaffected.

Cadmium is toxic to some edible plants at even relatively low levels and, like zinc, can be transported into plants with relative ease (Fassett, 1980). Some plant species bioconcentrate cadmium and zinc to levels greater than in the local environment, with a plant/soil concentration factor of 10 compared with 0.6 for zinc, 0.45 for copper, and 0.45 for lead. The uptake of cadmium in major

TABLE 4.2. Relative Adsorption of Selected Metals by Plants Consumed by Man

Plant	Arsenic	Cadmium	Lead	Mercury
Beans	Low	Low-medium		
Cabbage	High	Medium-Low	Low	
Lettuce		High	High	High-medium
Peas	Low	Medium-low	Low	
Radish	Medium	Low-medium	High	
Spinach	Low	High		
Tomato	High	Medium		
Turnip	Medium	High	High	
Corn	Medium	Medium	Low	
Oats	High	Low	Low	
Rice	Low	Low		
Wheat	High	Low	Low	
Apples	High			
Cherries	Medium			
Peaches	Low			
Pears	High			

Source: Fergusson, 1990.

seed crops is greater in soybeans and wheat and least in field beans, corn, and rice. While there is little evidence of bioconcentration in plants, the occurrence of cadmium in edible plants is important since a large portion of the human daily cadmium uptake is derived from this source.

Humans may be primary consumers of metal containing plant material, or metals may be transported between trophic levels of a food chain by other means. Terrestrial invertebrates such as earthworms, millipedes, and isopods stimulate decomposition of dead plant material by increasing the surface area for microbial attack. Centipedes and spiders are predators of terrestrial isopods and transfer metals from primary consumers to secondary consumers.

The number and diversity of these and other terrestrial invertebrates are reduced in areas contaminated by metals from industry and metal-containing pesticides (Hopkin, 1989). These primary consumers are particularly vulnerable since metals accumulate in the leaf litter layer and surface soil. For instance, a population of earthworms may be completely decimated in orchards or even golf courses contaminated with copper-based fungicides (Barthalmus, 1980; Hopkin, 1989). Since earthworms are the most important organisms in the mixing of soil, their absence leads to accumulation of partially decomposed, metal-containing leaf material on the soil surface. Other indirect effects, such as greater susceptibility of vegetation to defoliating insects and erosion of soil due to decrease in plant cover may result from the reduction in number and diversity of microorganisms and invertebrates (Hopkin, 1989).

Relatively high environmental levels of arsenic exist naturally, and arsenic is readily absorbed by animals via food and water. While the major occupational exposure for humans in the U.S. is in the manufacture of pesticides, herbicides, and other agricultural products, there are no examples of terrestrial wildlife being harmed by industrial arsenic pollution. However, ingestion of large doses of arsenic (70 to 180 mg) may be acutely fatal and could account for the deaths of deer and livestock after licking sodium arsenite from vegetation (Barthalmus, 1980; Dickerson, 1980; Goyer, 1986).

Many arsenic compounds are relatively low in toxicity and are highly species specific for their toxic effects. Copper acetoarsenite (Paris green) is used for mosquito control because it is nontoxic to most organisms other than mosquito larvae. Birds exposed to the larvicide through ingestion of the mosquito larvae accumulate negligible amounts of arsenic (Barthalmus, 1980).

Lead intoxication of wild birds is a very serious problem. Lead toxicosis from lead shotgun pellets either lodged in the flesh of birds or by ingestion of lead shot may cause more deaths among some birds than avian tuberculosis, coccidiosis, idiopathic impaction of the proventriculus, botulism, aspergillosis, or filariasis. Birds affected by lead include bobwhite, quail, pheasant, mourning doves, Canada geese, mallard and pintail ducks, pigeons, and the Andean condor (Barthalmus, 1980). In terrestrial animals, chronic lead poisoning results in a wide variety of symptoms that primarily involve the hematopoietic system, the gastrointestinal tract, and in more advanced stages, the nervous system, neuromuscular system, and kidney.

Until the mid 1960s methylmercury was used in Sweden, the U.S., and numerous other countries as a seed dressing to prevent grain spoilage because of its ability to counteract fungi and mold. Following the deaths of both seed-eating and predator birds in Scandinavia and a decrease in the reproduction in these species, a ban was placed on alkylmercury seed treatment. Wild bird populations in Scandinavia then began to return to normal (Barthalmus, 1980; Dickerson, 1980; Gossel and Bricker, 1984).

The importance of the food chain and the movement of methylmercury up the food chain was emphasized following the discovery in Sweden that the mortaliy in birds was linked to grain crops treated with methylmercury fungicides. The methylmercury treated seeds were eaten by seed-eating birds and rodents which were eaten by predators such as the peregrine falcon, hawk, and kestrel. As a result, there was a dramatic decrease in the number of predator birds in Sweden. Man, at the top of the food chain, came into the picture when it was learned that eggs from hens contained methylmercury. These hens had eaten methylmercury treated grain and grain grown from treated seeds (Dickerson, 1980).

The Minamata Bay incident in Japan was not the first outbreak of methylmercury poisoning, nor was it the last. There have been several mercury poisoning events throughout the centuries, but the largest single event occurred in 1971–1972. In this episode 459 persons died of the nearly 6350 persons hospitalized in Iraq after eating bread made from flour of methylmercury treated wheat and barley seeds that were destined for sowing. The 1971–1972 episode occurred in Iraq less than 10 years after the Scandinavian incident and just over 10 years after 1000 persons were hospitalized after eating methyl-mercury treated grain, also in Iraq (Fergusson, 1990).

G. TOXIC EFFECTS OF METALS

While most inorganic salts of metals (e.g., lead, mercury, arsenic, and cadmium) are nephrotoxic and produce various degrees of injury to the kidney, some metals, particularly those in their organic forms such as alkylmercury, organolead, and organotin, are highly neurotoxic. Increasing attention has also been drawn to the carcinogenic potential of certain metals, such as arsenic, chromium, and nickel.

1. Generalized Toxic Mechanisms of Selected Metals

Arsenic (As)
Arsenic may be considered a general cytotoxicant eliciting injury to most cells and organ systems. The trivalent As ion has a high affinity for sulfhydryl (–SH) groups. It chelates with alpha-lipoic acid forming a ring-like structure. Alpha-lipoic acid is an essential cofactor for pyruvate dehydrogenase, an enzyme required for energy production from the Krebs cycle.

Arsenate, the usual form in which arsenic is found in the biological system, is an oxyanion that mimics the phosphate oxyanion in cells. Such "substitution" of phosphate by arsenate would disrupt a variety of metabolic reactions that require phosphate. This is demonstrated in the replacement of phosphate by arsenic in ATP synthesis, resulting in the inhibition of ATP formation and "toxicity" to the cells (Jennette, 1981; Clarkson, 1991).

Cadmium (Cd)
Cadmium is known to injure the renal, pulmonary, skeletal, testicular, and nervous systems. The cytotoxicity of Cd is heavily dependent on the cellular production of the cadmium-binding protein, metallothionein (M). Thus, renal tubular epithelial cells, for example, can "store" high concentration of Cd (as Cd-M complexes) without showing signs of injury. The breakdown of the Cd-M complex within the cell releases Cd to sensitive sites, such as Zn-dependent enzymes, in the cell and triggers cell injury, particularly of the proximal tubule cells in the mammalian kidney. Independent toxic events, mechanisms, and the results of these toxic effects have been extensively reviewed by Chang et al. (1981) and will not be addressed further.

Lead (Pb)

One of the well known toxic effects of inorganic lead is its inhibition of hemoglobin synthesis leading to the development of anemia. Two mitochondrial enzymes, delta- or α-aminolevulinic acid dehydrogenase (ALAD) and ferrochelatase, are inhibited, which results in a decrease in the synthesis of protoporphyrin and a reduction of the insertion of Fe^{++} into heme. The reduction in heme stimulates an increased ALA synthetase (ALAS) activity. With an inhibition of ALAD and an increased activity of ALAS, a pronounced increase in the intermediate δ-amino levulinic acid (ALA) results (Hammond and Bililes, 1980).

Since Pb ions have an ionic structure similar to calcium (Ca) ions, they may be taken into the nervous system cells and mitochondria in a similar manner to calcium. Thus by "mimicking" Ca, Pb replaces or displaces Ca in its many cellular functions and metabolic pathways, leading to a reduction in ATP production and generalized cellular dysfunction. This alteration in function is particularly important because of its effects on synaptic transmission (Cooper et al., 1984; Minnema, 1989). The inhibiting effect of lead on acetylcholine action on the nictitating membrane is probably the result of an inhibitory action of Pb on Ca transport into the synaptic terminal.

Organolead (R_nPb)

The first comprehensive descriptions of the pathological changes in experimental animals due to the tetraalkyllead compounds (R_4Pb) were made by Davis et al. (1963) and Schepers (1964). In these studies, adult rats or dogs were given multiple daily exposures of the compounds by inhalation. The general signs and symptoms of organolead poisoning that were observed are summarized in Table 4.3. Levels of lead in the brain at the time of death averaged 10 $\mu g/g$. Pathological changes recorded in the brains included various degenerative changes affecting neurons in the neocortex, spinal cord, medulla oblongata, pons, cerebellar cortex, midbrain, thalamus, and basal nuclei. Diffuse neuroglial reactions were also reported. The overall neuropathology of organolead intoxication is presented in Table 4.4.

Tetraalkyllead (R_4Pb) is not very toxic until it is dealkylated to trialkyllead (R_3Pb) by hepatic microsomal enzymes. Trialkyllead is believed to disturb the chloride ion flux and transport into the neurons. Such ionic changes reverse the GABAergic neuron's hyperpolarization action, allowing excessive depolarization of the neuron with exaggerated excitation (hyperexcitation) which in turn leads to neuronal damage (Cremer, 1984). The increased neuronal excitation also stimulates increased glucose metabolism (Collins et al., 1980) with accumulation of pyruvic and lactic acids which are extremely harmful to neurons. Organolead, like inorganic lead, also disrupts cellular and mitochondrial calcium metabolism leading to cellular dysfunction and death.

TABLE 4.3.　Clinical Signs and Symptoms Following Et₄Pb or Et₃Pb Exposure[a]

Phase I	Lethargy
Phase II	Inappetance, tremor, hypermotility, hyperexcitability,aggression
Phase III	Hypothermia, convulsion, incoordination, ataxia, paralysis
Phase IV	Death

[a] Symptoms and signs resemble those observed in trimethyltin (TMT) intoxication.

TABLE 4.4.　Neuropathological Involvement Following Acute Exposure to Et₄Pb[a]

Neuronal necrosis and pyknosis primarily in pyriform/entorhinal cortex, hippocampal formation — fascia dentata and Ammon's horn, amygdaloid nuclei, neocortex

Neuronal chromatolysis, swelling, and necrosis in the brainstem and midbrain nuclei, pontine nuclei, basal nuclei, anterior cervical cord

In certain species, involvement of cerebellar Purkinji cells are also observed

[a] The neuropathological changes are quite similar to those in trimethyltin (TMT) poisoning.

Mercury (Hg)

Mercury is one of the oldest toxic metals known to man. Inorganic mercuric ions do not cross the blood brain barrier very effectively, and are primarily nephrotoxins producing necrotizing damage in the renal proximal tubules.

Mercury vapor and organomercury such as methylmercury, however, enter the central nervous system readily and are, therefore, potent neurotoxicants. Since the outbreak of methylmercury poisoning in Japan, methylmercury has become known as an extremely toxic substance in the environment. The overall pathology and toxic mechanism of mercury (Minimata disease) have been extensively reviewed by Chang in previous articles (Chang, 1979, 1980, 1982, 1985, 1987; Goering et al., 1987). The general clinical signs and symptoms seen in human cases of Minamata disease are summarized in Table 4.5. The pathology, both in humans and in animal models, focuses on the cerebellum (granule cells), calcarine cortex, and dorsal root ganglia. These lesions correlate well with the neurological signs and symptoms of the patients, namely ataxia, constriction of visual field (tunnel vision), and sensory disturbance.

The general toxic mechanism of methylmercury (CH_3Hg) may be outlined as follows. CH_3Hg has a high affinity for sulfhydryl (–SH) ligands. It complexes with the –SH rich amino acid cysteine which serves as a "carrier" for mercury in the blood and into the central nervous system (CNS). The blood brain barrier (BBB) is a complex system which regulates the uptake of metabolites from the

TABLE 4.5. Frequency of Clinical Signs and Symptoms in Minamata Disease

Symptom or Sign	Frequency (%)
Constriction of visual fields	100
Sensory disturbance	100
Ataxia	94
Impairment of speech	88
Impairment of hearing	85
Impairment of gait	82
Tremor	76
Mental disturbance	71
Exaggerated tendon reflexes	38
Hypersalivation	24
Hyperhydrosis	24
Muscular rigidity	21
Ballism	15
Chorea	15
Pathologic reflexes	12
Athetosis	9
Contractures	9

blood and the CNS. The BBB may be disrupted as CH_3Hg crosses it. Once within the nerve cells, CH_3Hg will bind to other –SH rich components of the cell, particularly the membranous structures in the cytoplasm (namely the endoplasmic reticulum, mitochondria, Golgi complex, and nuclear envelope). All these systems are important in energy production and protein synthesis for the cell.

Furthermore, the structure of the CH_3Hg-cysteine complex resembles that of methionine, an essential amino acid which is important in the initiation of the polypeptide chain in protein synthesis. Thus, the presence of CH_3Hg-cysteine complex could disrupt normal protein synthesis. Indeed, protein synthesis was found to be significantly suppressed or inhibited in cells of methylmercury exposed animals (see reviews by Chang).

Demethylation of CH_3Hg may also occur to some extent within the cells. Homolytic cleavage of the carbon-mercury bond may generate methyl free radicals which are cytotoxic to the cell via their initiation and promotion of lipid peroxidation. Thus, the toxic effects of methylmercury can be reduced by antioxidants such as selenium and vitamin E (see Chang's reviews). This protective effect has been documented by Chang in mice.

Disruption of microtubules by methylmercury has also been reported. This disruption of the neuronal cytoskeletal system will certainly contribute to cellular dysfunction. This is particularly important in developing neurons which may be in active stages of differentiation and migration, requiring the supportive function of the microtubules.

Tin (Tn)

Although inorganic tin compounds are not known to be particularly toxic, recent investigations have demonstrated that organotin compounds such as trimethyltin (TMT) and triethyltin (TET) are potent neurotoxins producing rapid and extensive damage to the CNS. While TET is primarily myelinotoxic in the CNS (Watanabe, 1977, 1980), TMT is extremely neuronotoxic (Brown et al., 1979; Bouldin et al., 1981; Chang et al., 1982a, b, c, 1983a, b, c; Chang and Dyer, 1983a, b; Chang, 1986).

Watanabe (1977) examined the effect of TET on developing brains of mice using light and electron microscopy. Animals were given various doses of TET at different postnatal ages. A definite trend toward longer survival and greater tolerated dose was observed in older animals, suggesting an increased sensitivity in young animals. However, myelinic vacuolation was identified 24 h after exposure regardless of the animal's age or the administered dose. The vacuoles were most frequently observed in the large myelinated fibers, while small myelinated fibers were only slightly involved. However, the myelinic vacuolation disappeared rapidly. Ultrastructural examination demonstrated intramyelinic edema and swelling of both oligodendroglia and astroglia. Massive neuronal necrosis such as that reported by Suzuki (1971) was not observed, suggesting that while there was increased myelinic destruction, a reduction in neuronal vulnerability also occurred in older animals.

The most distinct neuropathology for TMT is in the limbic system (entorhinal cortex, dentate fascia granule cells, hippocampal Ammon's horn). Based on biochemical data as well as the unique pattern of the development of the pathological lesion, Chang proposed that "hyperexcitation" was the mechanism of TMT toxicity in the hippocampal formation (Chang and Dyer, 1983a; Chang, 1984a, b; Chang, 1986).

A disturbance in brain glutamate metabolism and the GABAergic system by TMT has been reported by various laboratories (Table 4.6). The reduction in glutamate uptake and synthesis depletes the neuronal glutamate, which together with a reduction in brain taurine, probably produces the tremor seen in the animals. Reduced GABA synthesis was also reported. This reduction of GABA synthesis in the GABAergic neurons (inhibitory), together with an increased release of glutamate under the influence of TMT, enhances "hyperexcitation". The toxic effect of TMT on the hippocampus can either be promoted or alleviated by manipulating the level of corticosterone (Table 4.7), a known inhibitor of hippocampal activity (Pfaff et al., 1971; McEwen et al., 1975).

Chang further suggested that this "chain" of hyperexcitation activity can be propagated along the limbic circuiter pathway, from the entorhinal cortex to dentate fascia granule cells, to Ammon's horn CA3 neurons, to Ammon's horn CA1 and CA2 neurons. Investigation by Chang and his co-workers tested this hypothesis by confirming that the elimination of any given group of cells, for example, dentate fascia granule cells, will spare the neuronal groups such as Ammon's horn neurons at the lower end of the circuit.

TABLE 4.6. TMT Effects on Brain Glutamate Metabolism and GABAergic Neurons

Effects[a]	References
↓ Glutamate uptake	Naalsund et al., 1985
	Patel et al., 1990
↓ Glutamate synthesis	Patel et al., 1990
↓ GABA synthesis	Docter et al., 1982
	De Haven et al., 1984
	Mailman et al., 1985
↑ Glutamate release	Patel et al., 1990
↑ Brain tissue glutamine	Wilson et al., 1986
and serum ammonia	Hikal et al., 1988
↑ Damage to GABAergic neurons	Chang and Dyer, 1985
(basket cells)	
↑ Cl-flux, reverse GABA's	
inhibitory (hyperpolarization) effect	

[a] All these events would lead to neuronal hyperexcitation.

TABLE 4.7. Hippocampal Corticosterone

Binding	CA1, CA2 > CA3 > dentate fascia granule neurons (McEwen et al., 1975)
Function	Inhibition and modulation of neuronal firing rate in the hippocampus (Pfaff et al., 1971)
General vulnerability to TMT toxicity	Dentate fascia granule neurons > CA3 > CA1, CA2 (Chang, 1986)
Effect of adrenalectomy on TMT toxicity	Adrenalectomized animals show more lesions than intact animals
	Corticosterone supplementation blocks TMT-induced lesion development (Chang et al., 1989)

While the foci of lesion development in methylmercury intoxication correlates well with the metal distribution in the nervous system, there is no correlation between metal distribution and foci of lesion development in TMT poisoning. That is, while the lesions are fairly selective in the limbic system, TMT distribution in the brain is very diffuse. Thus, methylmercury may be considered as a "direct" neurotoxicant, exerting its toxic effects directly on the area and cells in which it is deposited, while TMT is an "indirect" neurotoxicant, exerting its toxic effect at a distance from its deposition (in this case, in the hyperexcitation of the neuronal pathway of the limbic system).

Manganese (Mn) and Aluminum (Al)

While both manganese and aluminum have unique neurotoxic properties, their precise mechanisms of action are still unclear. Manganese is believed to be a dopaminergic toxicant inducing Parkinson-like syndromes and degeneration of the caudate nucleus, basal ganglia, and substantia nigra. Free radical generation, particularly hydroxyl radicals ($\cdot$OH), may be related to its mechanism of toxicity. A recent review by Donaldson and Barbeau (1985) provides extensive insight to this problem.

Aluminum is believed to alter calcium homeostasis, energy metabolism, and production of macromolecules such as RNA. It may also play a role in neuronal degeneration associated with neurofibrillary degeneration similar, though not identical, to those observed in Alzheimer's disease. A review was recently written by McLachlan and Farnell (1985) on this subject.

2. Metal Carcinogenesis

Knowledge of the carcinogenicity of metal compounds has developed slowly, and information about metals as a primary cause of cancer is limited. Although only a few metals show any evidence of direct carcinogenicity, increasing numbers of metal compounds have been found to have carcinogenic potential in man and in animal models (Table 4.8). Distinguishing between the direct and secondary mechanism is a major challenge in understanding metal-induced carcinogenesis, and little is known about the role of metals as initiators, promoters, or progressors in the carcinogenic processes (Furst, 1987).

The carcinogenic potential of a metal is influenced by its crystalline structure, particle size, and negative surface charge. All these factors affect the uptake of the metal by a cell. Other factors that will influence the carcinogenicity of a metal are its oxidative state (e.g., Cr [III] is not carcinogenic while Cr [IV] is carcinogenic), its ability to cross cell membranes (e.g., manganese dioxide compared to acetoacetomyl manganese), and its effects on DNA structure or metabolism. Thus, metals such as nickel and chromium in a chronic condition may alter DNA structure and function. Other metals, although they may not be carcinogens alone, may enhance carcinogenicity (Table 4.9).

Carcinogenesis has been demonstrated with many different metal containing compounds, but the true mechanism of action of carcinogenic metals is unknown (Furst, 1987). It is not even known which metals act solely by genetic mechanisms, either directly or indirectly or by epigenetic mechanisms. Experiments are needed to examine a series of metals and the corresponding compounds containing these metals to determine which are carcinogenic. Through such a study, information on those metals that are carcinogenic and those that have mutagenic activity can be accrued. Until then, any attempt to explain the mechanism of metal-induced carcinogenesis through a single mechanism may be misguiding, if not futile.

TABLE 4.8. Metal-Induced Carcinogenesis in Man

Metal Carcinogen	Tumor
Cd	Testicular
Ni, Pb	Lung adenoma
Ni_3S_2	Muscle
Ni, Cd, Pb	(Various)
Cu	Lung

TABLE 4.9. Metal Enhancement of Carcinogenic Potentials

Metal	Carcinogen
Ni_3S_2	Benzo(a)pyrene
Zn	4-Nitroquinoline-*N*-oxide
	Methylbenzylnitrosamine
Co	3-Methylcholanthrene
Cd	Methylnitrosourea
Mg	Cd (lung adenoma)

REFERENCES

Amasa, S.K. Arsenic pollution at obuasi goldmine, town, and surrounding countryside, *Environ. Health Perspect.* 12:131–135, 1975.

Barthalmus, G.T. Terrestrial organisms, in *Introduction to Environmental Toxicology.* F.E. Guthrie and J.J. Perry, Eds. Elsevier North Holland, Inc., New York, 1980, 114–115.

Bouldin, T.W., Goines, N.D., Bagnell, C.R., and Krigman, M.R. Pathogenesis of trimethyltin neuronal toxicity, ultrastructural and cytochemical observations, *Am. J. Pathol.* 104:237–249, 1981.

Brown, A.W., Aldridge, W.N., Street, B.W., and Verschoyle, R.D. The behavioral and neuropathologic sequelae of intoxication by trimethyltin compounds in the rat, *Am. J. Pathol.* 97:59–81, 1979.

Chang, L.W. Pathological effects of mercury poisoning, in *Biogeochemistry of Mercury.* J.O. Nriagu, Ed. Elsevier Publishing, New York, 1979, 519–580.

Chang, L.W. Neurotoxic effects of mercury, in *Experimental and Clinical Neurotoxicology.* P.S. Spencer and H.H. Schaumberg, Eds. Williams & Wilkins, Baltimore, 1980, 508–526.

Chang, L.W. Pathogenetic mechanisms of the neurotoxicity of methylmercury, in *Mechanisms of Neurotoxic Substances.* K.N. Prasad and A. Vernadakis, Eds. Raven Press, New York, 1982, 51–66.

Chang, L.W. Hippocampal lesions induced by TMT in the neonatal rat brain, *Neurotoxicology* 5:205–216, 1984a.

Chang, L.W. Trimethyltin induced hippocampal lesions at various neonatal ages, *Bull. Environ. Contam. Toxicol.* 33:295–301, 1984b.

Chang, L.W. Neuropathological effects of toxic metal ions in, *Metal Ions in Neurology and Psychiatry*. S. Gabay, J. Harris, and B.T. Ho, Eds. Alan R. Liss, Inc., New York, 1985, 207–230.

Chang, L.W. Neuropathology of trimethyltin in rats — a proposed pathogenetic mechanism, *Fund. Appl. Toxicol.* 6:217–232, 1986.

Chang, L.W. Neuropathological changes associated with accidental or experimental exposure to organic metallic compounds: CNS effects, in *Structural and Functional Effects of Neurotoxicants: Organometals*. H.A. Tilson and S.B. Sparber, Eds. John Wiley & Sons, New York, 1987, 81–116.

Chang, L.W. and Dyer, R.S. Effects of trimethyltin on sensory neurons, *Neurobehav. Toxicol. Teratol.* 5:673–696, 1983a.

Chang, L.W. and Dyer, R.S. A time-course study of trimethyltin induced neuropathology in rats, *Neurobehav. Toxicol. Teratol.* 5:443–459, 1983b.

Chang, L.W. and Dyer, R.S. Early effects of trimethyltin in the dentate gyrus basket cells: a morphological study, *J. Toxicol. Environ. Health* 16:641–653, 1985.

Chang, L.W. Reuhl, K.R., and Wade, P.R. Pathological effects of cadmium, in *Biogeochemistry of Cadmium. II. Health Effects*. J.O. Nriagu, Ed. Elsevier, Publishing, New York, 1981, 783–840.

Chang, L.W., Tiemeyer, T.M., Wenger, G.R., and McMillan, D.E. Neuropathology of mouse hippocampus in acute trimethyltin intoxication, *Neurobehav. Toxicol. Teratol.* 4:149–156, 1982a.

Chang, L.W., Tiemeyer, T.M., Wenger, G.R., McMillan, D.E., and Reuhl, K.R. Neuropathology of trimethyltin intoxication. I. Light microscopy study, *Environ. Res.* 29:435–444, 1982b.

Chang, L.W., Tiemeyer, T.M., Wenger, G.R., McMillan, D.E., and Reuhl, K.R. Neuropathology of trimethyltin intoxication. II. Electron microscopic study of the hippocampus, *Environ. Res.* 29:445–458, 1982c.

Chang, L.W., McMillan, D.E., Wenger, G.R., and Dyer, R.S. Comparative studies of the neuropathological effects of trimethyltin in mice and rats, *Toxicologist* 3:76, 1983a.

Chang, L.W., Tiemeyer, T.M., Wenger, G.R., and McMillan, D.E. Neuropathology of trimethyltin intoxication. III. Changes in the brain stem neurons, *Environ. Res.* 30:399–411, 1983b.

Chang, L.W., Wenger, G.R., McMillan, D.E., and Dyer, R.S. Species and strain comparison of acute neurotoxic effects of trimethyltin in mice and rats, *Neurobehav. Toxicol. Teratol.* 5:337–350, 1983c.

Clarkson, T.W. Metal toxicity in the central nervous system, *Environ. Health Perspect.* 75:59–64, 1987.

Clarkson, T.W. Mechanisms in metal toxicology, in *Toxicity of Agents: Metals (Continuing Education Course #3)*, Soc. of Toxicology 30th Annual Mtg., Dallas, TX, 1991, 26–44.

Collins, R.C., McLean, M., and Olney, J. Cerebral metabolic response to systemic kainic acid: ^{14}C-deoxyglucose studies, *Life Sci.* 27:855–901, 1980.

Comar, C.L. and Nelson, N. Health effects of fossil fuel combustion products: report of a workshop, *Environ. Health Perspect.* 12:149–170, 1975.

Cooper, G.P., Suszkiw, J.B., and Manalis, R.S. Heavy metals: effects on synaptic transmission, *Neurotoxicology* 5:247–266, 1984.

Cremer, J.E. Possible mechanism for the selective neurotoxicity, in *Biological Effects of Organolead Compounds.* P. Grandjean, Ed. CRC Press, New York, 1984, 145–160.

Davis, R.K., Horton, A.W., Larson, E.E., and Stemmer, K.L. The acute effects of lead alkyls, *Arch. Environ. Health* 6:467–472, 1963.

De Haven, D.L., Walsh, T.J., and Mailman, R.B. Effects of TMT on dopaminergic and serotonergic functions in the CNS, *Toxicol. Appl. Pharmacol.* 74:182–289, 1984.

Dickerson, O.B. Arsenic, in *Metals in the Environment.* H.A. Waldron, Ed. Academic Press, London, 1980, 1–24.

Doctor, S.V., Costa, L.G., Kendall, D.A., and Murphy, S.D. Trimethyltin inhibits uptake of neurotransmitters into mouse forebrain synaptocomes, *Toxicology* 25: 213–223, 1982.

Donaldson, J. and Barbeau, A. Manganese neurotoxicity: possible clues to the etiology of human brain disorders, in *Metal Ions in Neurology and Psychiatry.* S. Gabay, J. Harris, and B.T. Ho, Eds. Alan R. Liss, Inc., New York, 1985, 259–285.

Dowdy, R.H. and Volk, V.V. Movement of heavy metals in soils, in *Chemical Mobility and Reactivity in Soil Systems.* Soil Science Society of America Special Publication No. 11, Madison, Wisconsin, 1983, 229–240.

Fassett, D.W. Cadmium, in *Metals in the Environment.* H.A. Waldron, Ed. Academic Press, London, 1980, 61–110.

Fergusson, J.E. *The Heavy Elements: Chemistry, Environmental Impact and Health Effects.* Pergamon Press, New York, 1990, 614.

Furst, A. Toward mechanisms of metal carcinogenesis, in *Biological Effects of Metals.* L. Fishbein and A. Furst, Eds. Plenum Press, New York, 1987, 295–327.

Goering, P.L., Mistry, P., and Fowler, B.A. Mechanisms of metal-induced cell injury, in *Toxicology.* T.J. Haley and W.O. Berndt, Eds. Hemisphere Publishing Corporation, Washington, 1987, 384–425.

Gossel, T.A. and Bricker, J.D. Metals, in *Principles of Clinical Toxicology.* Raven Press, New York, 1984, 153–187.

Goyer, R.A. Toxic effects of metals, in *Casarett and Doull's Toxicology. The Basic Science of Poisons*, 3rd ed. C.D. Klaassen, M.O. Amdur, and J. Doull, Eds. Macmillan Publishing Company, New York, 1986, 582–635.

Hammond, P.B. and Bililes, R.P. Metals, in *Casarett and Doull's Toxicology. The Basic Science of Poisons*, 2nd ed. J. Doull, C.D. Klaassen, and M.O. Amdur, Eds. Macmillan Publishing Company, New York, 1980, 409–467.

Hayes, J.A. Metal toxicity in *A Guide to General Toxicology*, 2nd ed. J.A. Marquis, Ed. Karger, New York, 1989, 179–189.

Hikal, A.H., Light, G.W., Shikker, W., Scarlet, A., and Ali, A.F. Determination of amino acid in different regions of rat brain application to acute effects of TMT, *Life Sci.* 42:2029–2035, 1988.

Hodgson, E. Modification of metabolism, in *A Textbook of Modern Toxicology.* E. Hodgson and P.E. Levi, Eds. Elsevier Science Publishing Co., Inc., New York, 1987, 85–121.

Hopkin, S.P. *Ecophysiology of Metals in Terrestrial Invertebrates*. Elsevier Science Publishing Co., Inc., New York, 1989, 366.

Jennette, K.W. Role of metals in carcinogenesis: biochemistry and metabolism, *Environ. Health Perspect.* 40:233–352, 1981.

Johnson, D.E., Tillery, J.B., and Prevost, R.J. Levels of platinum, palladium, and lead in populations of Southern California, *Environ. Health Perspect.* 12:27–33, 1975.

Leidy, R.B. Aquatic organisms, in *Introduction to Environmental Toxicology*. F.E. Guthrie and J.J. Perry, Eds. Elsevier North Holland, Inc., New York, 1980, 128–131.

Leland, H.V. and Kuwabara, J.S. Trace metals, in *Fundamentals of Aquatic Toxicology*. G.M. Rand and S.R. Petrocelli, Eds. Hemisphere Publishing Corporation, New York, 1985, 374–415.

Lindberg, S.E., Andren, A.W., Raridon, R.J., and Fulkerson, W. Mass balance of trace elements in walker branch watershed: relation to coal-fired steam plants, *Environ. Health Perspect.* 12:9–18, 1975.

Mailman, R.B. Heavy metals, in *Introduction to Environmental Toxicology*. F.E. Guthrie and J.J. Perry, Eds. Elsevier North Holland, Inc., New York, 1980, 34–43.

Mailman, R.B., Krigman, M.R., Frye, G.D., and Hannin, Z. Effects of postnasal trimethylin or triethylin treatment of CNS catecholamines, GABA, and acetyl choline systems in the rat, *J. Neurochem.* 40:1423–1429, 1983.

Mailman, R.B. and Sidden, J.A. Food and food additives, in *Introduction to Environmental Toxicology*. F.E. Guthrie and J.J. Perry, Eds. Elsevier North Holland, Inc., New York, 1980, 321.

Mance, G. *Pollution Threat of Heavy Metals in Aquatic Environments*. Elsevier Science Publishing Co., Inc., New York, 1987, 372.

McEwen, B.S., Gerlach, J.L., and Micco, D.J. Putative glucocorticoid receptors in hippocampus and other regions of the rat brain, in *The Hippocampus*, Vol. I. R.L. Isaacson and K.H. Pribram, Eds. Plenum Press, New York, 1975, 285–322.

McLachlan, D.R.C. and Farnell, B.J. Aluminum and neuronal degeneration, in *Metal Ions in Neurology and Psychiatry*. S. Gabay, J. Harris, and B.T. Ho, Eds. Alan R. Liss, Inc., New York, 1985, 69–87.

Minnema, D.J. Neurochemical alterations in lead intoxication — an overview, *Comments on Toxicology* 3(3):207–224, 1989.

Moriarty, F. Genetics of populations, in *Ecotoxicology. The Study of Pollutants in Ecosystems*, 2nd ed. Academic Press, London, 1988, 71–102.

Naalsund, L.V., Suen, C.N., and Fonnum, F. Changes in neurobiological parameters in the hippocampus after exposure to TMT, *Neurotoxicology* 6:145–158, 1985.

Nordberg, G.F., Parizek, J., and Piscator, M. Factors influencing effects and dose-response relationships of metals, in *Handbook on the Toxicology of Metals*. L. Friberg, G.F. Nordberg, and V.B. Vouk, Eds. Elsevier North Holland Biomedical Press, Amsterdam, 1979, 143–157.

Nordberg, G.F., Goyer, R.A., and Clarkson, T.W. Impact of effects of acid precipitation on toxicity of metals, *Environ. Health Perspect.* 63:169–180, 1985.

Patel, M., Ardelt, B.K., Yim, G.K.W., and Isom, G.E. Interaction of trimethylin with hippocampal glutamate, *Neurotoxicology* 11:601–608, 1990.

Pfaff, D.W., Silva, M.T.A., and Weiss, T.M.. Telemetered recording of hormone effects on hippocampal neurons, *Science* 172:384–385, 1971.

Rand, G.M. and Barthalmus, G.T. Case history: pollution of the Rhine River, in *Introduction to Environmental Toxicology*. F.E. Guthrie and J.J. Perry, Eds. Elsevier North Holland, Inc., New York, 1980, 245–246.

Sabadell, J.E. and Axtmann, R.C. Heavy metal contamination from geothermal sources, *Environ. Health Perspect.* 12:1–7, 1975.

Schepers, G.W.H. Tetraethyllead and tetramethyllead. Comparative experimental pathology. I. Lead absorption and pathology, *Arch. Environ. Health* 8:277–283, 1964.

Sheets, T.J. Transport of pollutants, in *Introduction to Environmental Toxicology*. F.E. Guthrie and J.J. Perry, Eds. Elsevier North Holland, Inc., New York, 1980, 154–160.

Stirling, L.A. Microorganisms and environmental pollutants, in *Introduction to Environmental Toxicology*. F.E. Guthrie and J.J. Perry, Eds. Elsevier North Holland, Inc., New York, 1980, 336–337.

Suzuki, K. Some new observations in triethyl-tin intoxication of rats, *Exp. Neurol.* 31:207–213, 1971.

Takeuchi, T., Kambara, T., Morikawa, N., Matsumoto, H., Shiraishi, Y., and Ito, H. Pathologic observations of the Minamata disease, *Acta Pathol. Jpn.* 9:768–783, 1959.

Tokuomi, H. Minamata disease, an unusual neurological disorder occurring in Minamata, Japan, *Kumamoto Med. J.* 14:47–64, 1961.

Voyer, R.A., Cardin, J.A., Heltshe, J.F., and Hoffman, G.L. Viability of embryos of the winter flounder *Pseudopleuronectes americanus* exposed to mixtures of cadmium and silver in combination with selected fixed salinities, *Aquat. Toxicology* 2:223–233, 1982.

Watanabe, I. Effect of triethyltin on the developing brain of the mouse, in *Neurotoxicology*. L. Roizin, H. Shiraki, and N. Grcevic, Eds. Raven Press, New York, 1977, 317–326.

Watanabe, I. Organotins, in *Experimental and Clinical Neurotoxicology*. P.S. Spencer and H.H. Schaumburg, Eds. Williams and Wilkins, Baltimore, 1980, 545–557.

Wilson, W.E., Hudson, B.M., Kanamatsu, D., Kanamatsu, D., Walsh, T., Tilson, H., and Hong, J. TMT-induced alterations in brain amino acid, amines, and amine metabolites: relationship to hyperammoniemia, *Neurotoxicology* 7:63–74, 1986.

Chapter

5

Toxicity of Petroleum

Edward B. Overton, Waynette D. Sharp, and Paulene Roberts

A. PETROLEUM AND ENVIRONMENTAL IMPACTS

Petroleum existed long before humans developed the technological ability to retrieve it from the earth and use it as a source of energy. Natural seeps within the ocean floor have been releasing hydrocarbons for thousands of years, creating ecosystems with adaptive microorganisms that utilize petroleum effectively. However, the ecological balance in environments that are not adjusted to assimilating large amounts of spilled or released petroleum may be disrupted by oil release from large quantity transportation and extraction practices. These environmental disturbances can range from short-term destruction to long-term alteration of the ecosystem. Examples of the effects of petroleum spills have been seen throughout the world.

The Ixtoco I, an offshore exploratory well in Balvia de Campeche, Mexico, was destroyed by fire on June 3, 1979. The fire created an explosive pressure that removed all containment caps on the well. An estimated 140 billion barrels (5880 billion gal) of crude oil were released before the well was sealed in June 1980. Because of prevailing summer currents in the Gulf of Mexico, petroleum impacts were noted as far north as the Texas shoreline by August 1980. Two areas that were oiled, Padre Island and Laguna Madre, are important staging and wintering grounds for waterfowl, shorebirds, and colonial water birds. Fortunately, the weathered oil that reached the northern Texas shoreline was cleaned up before it caused a significant environmental impact.

0-8493-8851-1/94/$0.00+$.50
© 1994 by CRC Press, Inc.

The Exxon Valdez released approximately 11 million gal of North Slope (Prudhoe Bay) crude oil in Prince William Sound, Alaska after navigational errors resulted in her grounding on rocks on March 24, 1989. The oil was spread by winds and currents and eventually spread over approximately 25,000 miles of coastline and adjacent waters. Damage from the spill included 1630 oiled but surviving birds from 71 different species, with over 36,000 carcasses of birds being recovered, an estimated 3500 to 5500 sea otters killed, and substantial disruption of marine habitats in the Prince William Sound area of Alaska. The cleanup cost for the first year alone following the spill was $2.5 billion, and monitoring is still going on in selected locatiŏns of Prince William Sound.

The Persian Gulf War of 1990 resulted in one of the largest areas of petroleum impact caused by humans. More than 300 petroleum wells in Kuwait burned for approximately 10 months. Additionally, stored oil was deliberately released into the Persian Gulf. Roughly 400 miles of shoreline were oiled; the amount of desert oiled was considerably greater. The ecological damage to the resources along the shorelines and in coastal waters will affect many species, including humans.

In addition to oil spills, petroleum hydrocarbons are released into the aquatic environments from natural seeps as well as nonpoint source urban runoffs. Acute impacts from massive one-time spills are obvious and substantial. The impacts from small spills and chronic releases are the subject of much speculation and continuing research. Clearly, these inputs of petroleum hydrocarbons have the potential for significant environmental impacts, but the effects of chronic low-level discharges can be minimized by the net assimilative capacities of many ecosystems, resulting in little detectable environmental harm.

B. PETROLEUM COMPOSITION

To understand the potential environmental impacts that can occur during an oil spill or release, the molecular components of petroleum hydrocarbons and their structures must be considered. The natural composition of petroleum is very complex and, to some extent, varies between petroleum reservoirs. In general, petroleum is a complex mixture of hydrocarbons (molecules made up of only carbon and hydrogen) and nonhydrocarbons (molecules made up of carbon, hydrogen, and other elements such as sulfur, nitrogen, and oxygen) that forms from the partial decomposition of biogenic material. The slow breakdown process, known as diagenesis, produces a range of hydrocarbons and hydrocarbon complexes significantly altered from the structure found in the original biomass. The resulting breakdown products occur in a combination of three phases:

1. Gaseous form — natural gas
2. Liquid form — crude oil
3. Solid form — tar or bitumen

Each petroleum reserve is a unique combination of biomass breakdown products. Hence petroleums each have a unique compositional complexity, with variations occurring within the individual petroleum reservoir (Speight, 1991). Libraries of these compositional variations, known as molecular fingerprint patterns, have been developed using analytical methods for selective petroleum compound analysis. A gas chromatograph/mass spectrometer (GC/MS) is one such analytical system that is commonly used for analysis of complex mixtures of petroleum hydrocarbons. A GC/MS system is capable of isolating individual compounds within the complex crude mixture. These separated compounds are then identified by electron impact mass spectra and ion fragment pattern recognition. The analytical data are stored in a computer and displayed in various formats that allow researchers to identify and quantify specific molecular components of crude oils. These data displays are frequently in the form of gas chromatograms of components with the same molecular weight (i.e., mass chromatograms) or gas chromatograms composed of the sum of all ions detected during each scan of the mass spectrometer (total ion chromatograms or TICs). For example, Figure 5.1 shows the TICs from GC/MS analyses of four crude oils (Prudhoe Bay Crude [MS2017B.D], Basrah Kuwait Crude [MS2017D.D], and offshore [MS2017C.D] and inland [MS2017E.D] Louisiana Crudes). The TIC data display the eluting compounds in sequence by retention times and the relative abundances of the compounds by their peak heights. As is seen from an examination of this data, the general range of components for these four different crude oils is qualitatively similar. However, detailed analyses of specific components and their ion patterns provide insight into the quantitative differences in the makeup of individual components in these oils. As a general rule, all crude oils are complex mixtures composed of the same compounds, but the quantities of the individual components differ in crude oils from different locations. This rule of thumb implies that the quantities of some compounds can be zero in a given mixture of components that comprise a crude oil from a specific location.

Since there are compositional differences in petroleum, no specific definition or composition statement is valid for all crude oils except the preceding rule of thumb. A broad functional definition of petroleum hydrocarbons is that hydrocarbons are primarily composed of many organic compounds of natural origin and low water solubility. The molecular compositional differences in crude oils result in physical and chemical properties that are unique for each type of oil. These differences are important for refinery processing and for predicting the potential impacts to the environment from oil spills and other releases.

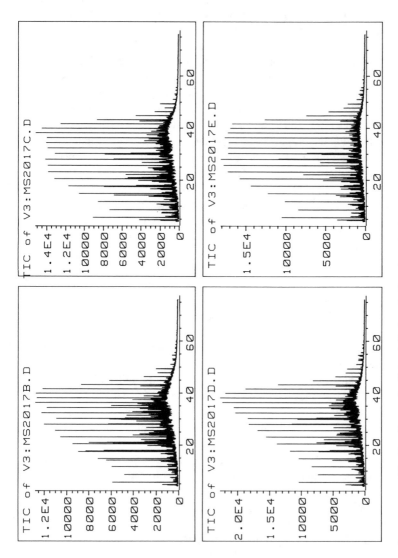

Figure 5.1. Gas chromatogram of Prudhoe Bay crude oil (top left), Basrah, Kuwait crude oil (top right), offshore Louisiana crude oil (bottom left), and inland Louisiana crude oil (bottom right).

Crude oils are used for the production of fuels and lubricants for transportation and energy applications and for feed stocks in the petrochemical industries. The refinery processes that convert crude oils to useful products include distillation techniques that separate petroleum products with different boiling point (bp) ranges and catalytic cracking processes that create usable products from the unusable components of oil. The general range for distillation products, as established by the American Standards and Testing Materials (ASTM), are listed below.

Boiling Point Range	Petroleum Distillation Ranges
<20°C	Natural gas
20–200°C	Straight-run gasoline
185–345°C	Middle distillates — kerosene, jet fuel, diesels, heating oil
345–540°C	Light to heavy lubricating oils and waxes, feed stock for catalytic cracking to gasoline
>540°C	Residual oil

C. PETROLEUM CHARACTERISTICS

1. Physical Characteristics

The physical characteristics of an oil are determined by the characteristics of the individual components within the petroleum and their relative quantities. These physical characteristics, including density, specific gravity, viscosity, pour point, and flash point, are generally derived from the total composition of petroleum and petroleum byproducts as a weighted average of the properties of individual compounds in the complex mixture.

Density
Density is the weight per unit volume of a substance. The units expressed are grams per cubic centimeter for solids and liquids and grams per liter for gases. This unit is based on the definition that 1 cm^3 water weighs 1 g.

Specific Gravity
Specific gravity (sg) relates the density of a substance to the density of a standard substance. Water is commonly used as the standard for solids and liquids and has an sg of 1.0000. For solids and liquids, the sg term is generally equivalent to density (assuming the density of water is 1.0000). In the petroleum industry, sg is generally expressed in "degrees API" in accordance with a scale established by the American Petroleum Institute (API). The relationship between sg and degrees API is defined as

$$sg = 141.5/(131.5 + \text{degrees API})$$

Viscosity

Viscosity is determined by internal friction and is measured as resistance to flow. Absolute viscosity is the force required to move a 1 cm^2 planar surface area above another planar surface at a rate of 1 cm s^{-1} when the two surfaces are separated by a layer of fluid 1 cm in thickness. The unit of absolute viscosity is the poise. The ratio of absolute viscosity to density is called the kinematic viscosity and is measured in stokes (St) or centistokes (cSt). In general, viscosity decreases with increasing API sg.

Pour Point

The lowest temperature at which an oil can be poured is the pour point. The pour point is recorded as 5°F above the solid phase temperature (temperature of no fluid movement). Pour points for crude oils generally range from 45 to 110°F (7 to 43°C). If the pour point of the oil is higher than the temperature of the environment, the oil will tend to aggregate rather than spread as a liquid.

Flash Point

The flash point is the temperature at which a liquid or volatile solid gives off vapors sufficient to form an ignitable mixture with air near the surface of the liquid or solid or within a test vessel. The flash point is an exception to the general rule that physical properties are determined by the total petroleum composition. Instead, the flash point temperature is determined by the composition of the more volatile components in petroleum. Petroleums that primarily contain volatile compounds of low molecular weight have low flash points; petroleums containing primarily compounds of high molecular weight and low volatility have high flash points.

These physical characteristics provide insight into the collective molecular composition in crude oils. For example, a "light" crude oil is one with high API sg and low viscosity and pour point temperature when compared to other oils. This generally implies a higher percentage of saturates and lower asphaltene and polar content in the petroleum mixture. Asphaltenes are substances in petroleum that are insoluble in solvents of low molecular weight such as pentane or hexane. These compounds are composed of very large cyclic and planar molecules and are solids at normal temperatures. Consequently, oils that have high asphaltene contents are very viscous, have a high pour point, and are generally nonvolatile in nature. Table 5.1 lists the physical characteristics and the chemical components of several oils and refined products. These types of data are available for most oils transported in bulk, while detailed GC/MS analytical data on oils may not be readily accessible.

TABLE 5.1. Properties of Crude and Refined Oil Products

Parameter	Crude Oils (type)				
	California	California	Basrah Kuwait	Prudhoe Bay	South Louisiana
Physical Properties					
API gravity	10.30	13.2	31.4	27.0	37.0
Density (at 20°C)	0.998	0.978	0.869	0.893	0.840
Pour point (°C)	0.0	9.0	32.0	27.0	—
Flash point (°C)	28.0	12.0	—	30.0	—
Chemical Comp. (wt %)					
Saturates	13.7	13.7	50.2	61.2	65.1
Aromatics	29.8	36.4	28.4	35.6	26.3
Polars	31.4	24.1	17.9	2.9	8.4
Asphaltenes	24.8	25.8	3.5	1.2	0.2
Sulfur (%)	3.3	5.5	2.4	0.82	0.21

Parameter	Refined Oils (type)				
	Gasoline	Kerosene	#2 Fuel Oil	10W/30°C Oil	#6 Fuel Oil
Physical Properties					
API gravity	60.0	37.0	31.6	29.0	10.0
Density (at 20°C)	0.734	0.83	0.84	0.87	0.966
Pour point (°C)	<−40.0	−18.0	−20.0	−37.0	6.0
Flash point (°C)	−40.0	38.0	55.0	188.0	80.0
Chemical Comp. (wt %)					
Saturates	39.6	85.0	61.8	73.7	24.4
Aromatics	46.2	15.0	38.2	25.4	54.6
Polars	—	—	0.0	0.9	14.9
Asphaltines	N/A	N/A	0.0	0.0	6.2
Sulfur (%)	0.07	0.5	0.32	0.37	2.0

Source: Bobra, 1989.

2. Elemental Composition

Most of the chemical components in petroleum are made up of five main elements. The most abundant element by weight is carbon, followed by hydrogen and then sulfur. Most oils contain a low percentage of sulfur and are called "sweet crudes". Crudes with a higher percentage of sulfur, or "sour crudes", are associated with a carbonate rock source that lacks inorganic iron to bind with the sulfur. The next two most common elements in oils are nitrogen and oxygen,

TABLE 5.2. Elemental Composition of Crude Oil

Element	Range (wt %)
Carbon	82–87
Hydrogen	11–15
Sulfur	0–8
Nitrogen	0–1
Oxygen	0.0–0.5

Source: Butt et al., 1986.

TABLE 5.3. Relative Trace Metal Composition of Crude Oil

Element	Concentration (ppm)
Aluminum	11
Antimony	11
Barium	3
Boron	<1
Cadmium	<1
Calcium	37
Chromium	3
Lead	2
Magnesium	8
Nickel	12
Phosphorus	98
Silicon	7
Tin	13
Vanadium	38
Zinc	7

which are also among the most common elements in the environment. A wide range of metals are found in trace amounts in crude oils. The elemental and trace metal components of petroleum are shown in Tables 5.2 and 5.3.

3. Molecular Composition

The nomenclature for petroleum-based hydrocarbons focuses primarily on the carbon bond and several of its structural formations. The following classification scheme breaks petroleum components into five major groups. The subcategories are based on the majority of the compounds in that subcategory, but many more classifications could exist.

Saturates	Normal paraffins (straight-chain hydrocarbons)
	Isoparaffins (branched-chain hydrocarbons)
	Cycloparaffins or naphthenes (cyclic saturated and partially saturated hydrocarbons)
Aromatics	Aromatic hydrocarbons (AH)
	Polynuclear aromatic hydrocarbons (PAH) and their C_1 to C_4 alkyl homologs
Polar	Sulfur-containing aromatic compounds
	Nitrogen-containing aromatic compounds
	Oxygen-containing aromatic compounds
Porphyrins	Complex large cyclic carbon structures derived from chlorophyll and characterized by the ability to contain a central metal atom (trace metals are commonly found within these compounds)
Asphaltenes and resins	Composition is dependent upon source (these structures have the highest individual molecular weight of all crude oil components and are basically colloidal aggregates)

The saturated carbon bond, or carbon bound to four other atoms, in hydrocarbon molecules can be arranged as either straight-chained, branched, or cyclic hydrocarbon structures. The straight-chain compounds are found in a homologous series (same molecular structure separated by only CH_2 groups) in most crude oils, with carbon numbers ranging from 5 or 6 up to 30 or more. These compounds are generally the major components of crude oils and are known as normal paraffins. The branched chain compounds, or isoparaffins, are primarily methylated straight chain hydrocarbons. From nine carbons and higher, most isoparaffins are isoprenoids originating from a side chain of the chlorophyll molecule, with a methyl group attached to every third carbon in the chain. These branched and isoprenoid hydrocarbons are generally degraded more slowly than the normal paraffins by microorganisms. Two common paraffins found in most petroleums are heptadecane, a straight 17-carbon chain molecular structure, and pristane, a branched 19-carbon chain isoprenoid structure. These two components have molecular weights of 240 and 268, respectively, although their bp and gas chromatographic retention times are similar. Their microbial degradation rates in the environment are very different. Heptadecane rapidly degrades in the environment, while pristane remains and is often used as a biomarker for weathering and degradation studies. Other abundant structures in most fresh crude oils are the cyclic hydrocarbons. The two basic cyclic structures are cyclohexane and cyclopentane, with five and six carbon rings, respectively. Through combinations of these two-ring structures assembled with up to five rings, many complex molecules are formed during petroleum diagenesis. Two of the larger structures that are resistant to weathering processes are the triterpanes and the steranes, with a subclassification of hopanes. These compounds, which are not degraded by microorganisms to any appreciable extent, are also used as degradation biomarkers.

Carbon atoms that are bound to three or fewer other atoms in a hydrocarbon molecule are referred to as "unsaturated". Their physical structure is a straight, branched, or aromatic form. For the purposes of petroleum toxicology, the following discussion is limited primarily to the aromatic forms, since crude oils do not contain unsaturated hydrocarbons other than the aromatic structures. However, refined petroleum products may contain unsaturated compounds with one or more double bonds in their molecular structure. As with the saturate petroleum hydrocarbons, aromatic ring structures in crude oils range from one- to five-ring combinations. Two or more five- or six-member carbon rings are fused together to form polycyclic aromatic compounds (PACs). These petroleum aromatic hydrocarbons have abundant alkyl group substitution on their ring structures. The alkyl groups generally have from one to four saturated carbon atoms and thus can produce many different structural isomers and homologs for each aromatic hydrocarbon family. The most abundant aromatic hydrocarbon families have two and three fused rings with one to four carbon atom alkyl group substituents.

The toxicological implications from petroleum occur primarily from exposure to or biological metabolism of these aromatic structures. The most acutely toxic compounds are the single aromatic ring variations, including benzene, toluene, and the xylenes. All are volatile compounds with relatively high water solubility for insoluble compounds. Therefore, both atmospheric and hydrospheric impacts must be assessed when considering toxic implications from a petroleum release containing significant quantities of these single-ring aromatic compounds. The dibenzene ring structures, or naphthalene and its homologous series, are less acutely toxic than benzene but are more prevalent for a longer period during oil spills. Table 5.4 shows a comparison of Threshold Limit Value (TLV) and Immediately Dangerous to Life and Health (IDLH) exposure levels, according to the standard National Institute for Occupational Safety and Health (NIOSH) and American Conference of Governmental Industrial Hygienists (ACGIH) manuals, to solubility and flash point data for selected aromatic compounds. The larger and heavier aromatic structures (with four to five aromatic rings), which are more persistent in the environment, have potential for chronic toxicologic effects. Since these compounds are nonvolatile and are relatively insoluble in water, their main routes of impact are through ingestion and epidermal contact. Some of the compounds in this classification are considered possible human carcinogens; these include benzo(a and e)pyrene, benzo(a)anthracene, benzo(b, j, and k)flouranthene, benzo(ghi)perylene, chrysene, dibenzo(ah)anthracene, and pyrene. These structures and their alkyl homologs are present at low

TABLE 5.4. Comparison of Aromatic Components of Petroleum

Compound	Flash Point (°F)	TLV TWA (ppm)	IDLH (ppm)	Solubility (mg/l at 25°C)
Benzene	12	10	Carcinogenic	1,800
Toluene	40	10	2,000	470[a]
Xylenes	63	100	10,000	198[b]
Naphthalene	174	10	500	31.7

[a] Solubility at 16°C.

[b] Average solubility of three constituents.

Source: Henry, 1991.

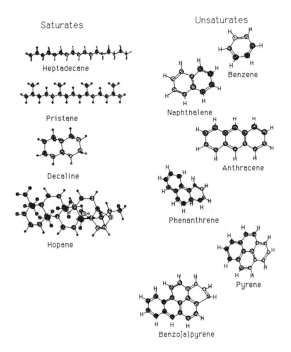

Figure 5.2. Chemical structures of hydrocarbons.

concentrations in crude oils. Their toxicological impacts to different species vary considerably. The chemical structures of various hydrocarbons are shown in Figure 5.2.

The polar compounds are also known as the nonhydrocarbons. For these compounds, the primary functional group is associated with nitrogen, sulfur, or oxygen atoms. The classification applies to sulfur compounds such as cyclic

TABLE 5.5. Elemental Composition of Crude Oil

Compound	Functional Group	Solubility (ng/µl)
Benzene	—	1,800
Phenol	–OH	82,000
Benzaldehyde	–COH	3,300
Benzoic acid	–COOH	2,900
Methylbenzene	–CH$_3$	515
Biphenyl	–C$_6$H$_6$	7.5
Benzo(a)pyrene		<0.01[a]

[a] Almost insoluble.

Source: Henry, 1991.

sulfides and thiophenes, basic and neutral nitrogen compounds, and oxygen compounds, primarily naphthenic acids.

The porphyrins, asphaltene, and resin compounds in petroleum are considered the residual oil, or residuum. During weathering processes, this fraction is the last to degrade, and its persistence over years has been noted. This residual fraction is commonly used in roofing compounds, hot melt adhesives, and heating oils.

4. Chemical Characteristics

Other properties of petroleum that are not readily observable can be extrapolated from the physical characteristics of the individual compounds that form the specific petroleum product. The most significant characteristics are water solubility and vapor pressure; these characteristics are indicators of the longevity of petroleum within the environment.

Water Solubility
One of the most important properties in determining the environmental fate of pollutants is their water solubility. Also, the water solubility of a substance determines the routes of exposure that are possible. Table 5.5 shows examples of the influence that different functional groups have on the solubility of benzene, alkylated benzene, and complex benzene-derived compounds.

Vapor Pressure
Vapor pressure is the pressure exerted by a vapor in equilibrium with the liquid from which it is derived at a given temperature (usually 20°C). This value is significant for predicting volatility of crude oils and refined products. Refined light petroleum products, under increased temperature and pressure, form explosive mixtures in air.

Relationship Between Vapor Pressure and Molecular Weight

Compound	Vapor Pressure (mm Hg at 68°F)
methane	>760
ethane	>760
propane	>760
n-butane	>760
n-pentane	426
n-hexane	124
n-heptane	40
n-octane	11
n-nonane	3
n-decane	2

Table 5.6 provides a comparison of some physical characteristics and toxicity of light-, medium-, and heavyweight petroleums.

D. WEATHERING

"Weathering" encompasses the physical and chemical changes that crude oils and refined petroleum products undergo as a result of their interaction with the environment. Of the possible environments impacted, the hydrosphere presents the most complicated and interactive zone. The hydrosphere will be the focus of the following topics of petroleum weathering and toxicology due to its complexity and dynamics; it includes the marine, estuarine, and freshwater systems.

The fate of oil in an environment is governed by the cumulative physical and chemical properties of the individual constituents in the bulk oil (Payne and McNabb, 1984). The factors affecting the fate of oil spilled or released in a water environment include spreading, evaporation, dissolution, dispersion, emulsification, absorption by sediments, and photochemical and microbial degradation.

TABLE 5.6. Selected Physical Characteristics and Toxicities of Petroleum

Weight	No. Carbon Atoms	Boiling Point (°C)	Evaporation Rate	Solubility	Toxicity
Light	<12	<150	Rapid	High	Highly acute
Medium	10–12	150–400	Moderate	Low	Moderately acute
Heavy	>20	>400	Very low	Very low	Chronic

The ultimate fate of petroleum spilled in the marine environment is degradation, primarily biological and, to a lesser extent, photolytic degradation. The rate at which these processes occur is controlled by many factors, such as the physical and chemical composition of the oil, the exposure to physical processes, abiotic environmental factors, and the preexistence or absence of hydrocarbon-degrading microorganisms (bacteria, mold, yeast, and fungi). Most chemical breakdown processes affecting whole oil occur at the oil/water interface and in the atmosphere. Chemical actions in aquatic systems include biodegradation and photolytic degradation, which are surface-active processes. Therefore, the amount of surface area affects the rate of degradation. Certain factors hinder degradation — one of these is emulsification, which slows biodegradation because the stable "mousse" formed has a smaller surface area compared to very thin oil sheens. Another factor, cool weather, often results in slowed rates of microbial activity, which results in less consumption or degradation of the oil (Malins, 1987). The microbes utilize the oil as a carbon source which is converted to energy, biomass, carbon dioxide, and water. Physical processes such as high energy storms increase the dispersion of oil, thus increasing surface areas and generally reducing the persistence of oil in the environment.

Weathered oil can be distinguished from fresh or unweathered oil by the loss of the low molecular weight constituents such as the normal hydrocarbons less than n-C_{12} (a 12-carbon molecule), the alkylbenzenes, and the naphthalene homologs. These initial changes are primarily due to evaporative losses and, to a lesser degree, dissolution. For example, less than 5% of benzene is generally lost to dissolution, while more than 95% is lost to evaporation. The constituents of low molecular weight are more volatile and more water soluble than the constituents with high molecular weight. Many refined products such as bunker oils appear as "weathered" crude oils, since they lack the compounds of lower molecular weight. The high viscosities of these refined products may make them very resistive to physical weathering in much the same way as weathered crude oils and may also make them slow to further degrade. The rate and extent of evaporation, dissolution, sedimentation, and degradation of individual components are factors that affect possible toxic responses from crude oils.

E. TOXICITY AND ROUTES OF EXPOSURE

The toxicity of complex mixtures such as crude oils and refined products is extremely difficult to assess, even when the toxicity of each individual component is known. This difficulty arises from the paucity of information on additive, synergistic, or antagonistic effects of various components of these mixtures. In addition, the chemical composition of each crude oil and petroleum product, which varies significantly, can have diverse effects on organisms within the same ecosystem. These differences in toxic effects are due not only to qualitative compositional differences of various products but also to concentration differences of various chemical constituents.

1. Acute Toxicity Components

The monoaromatic (benzene) and diaromatic (naphthalene) components in fresh crude oil are considered to be the most toxic and most abundant compounds during the initial phases of petroleum spills or releases. The naphthalene- and benzene-derived compounds, other than benzene, are considered noncarcinogenic. At the initial stages of the release, when these compounds are present at their highest concentrations, acute toxic effects are most common. These noncarcinogenic effects range from subtle changes in detoxifying enzymes to liver damage and interference with reproductive behavior (NOAA, 1992). The severity of the effects depends on the organism being exposed, the concentration of the compound, and the route of exposure.

2. Chronic Toxicity Components

As oil weathers, it becomes less acutely toxic, but it remains toxic via injection because it contains polynuclear aromatic hydrocarbons (PAHs) of high molecular weight. Many of these compounds, including at least one known human carcinogen, benzo(a)pyrene, have been identified as mutagenic; others are listed as experimental carcinogens. The remaining compounds may become enriched in the residual bulk oil due to the loss of the volatile, soluble, and easily biodegraded compounds. It is important to point out that crude oils contain primarily the alkyl homolog of aromatic compounds and relatively small quantities of the unsubstituted "parent" aromatic ring structures. It is these unsubstituted aromatic structures that are potentially responsible for the majority of the known toxic impacts of crude oils (Payne et al., 1979).

3. Routes of Exposure

The possible routes of exposure for the various classes of compounds found in petroleum depend upon their physical and chemical characteristics. If a spill or release occurs on land, containment practices are readily available; if the spill or release occurs on a body of water, containment and cleanup operations are affected by the availability of regional resources and by the weather. Petroleum is naturally weathered according to its physical and chemical properties, but during this process living species within the local environment may be affected via one or more routes of exposure, including ingestion, inhalation, dermal contact, and, to a much lesser extent, bioconcentration through the food chain. The more water-soluble petroleum compounds are dispersed in the water column, where they may be ingested or inhaled by marine species. Exposure of marine and avian populations to petroleum remaining on the water surfaces or shorelines occurs by ingestion through preening and by dermal contact. The bioconcentration process applies to the predator/prey relationship, primarily by inspections of oil-soiled animals.

Experimental biological studies and studies of past petroleum spills document the physical and toxicological effects of petroleums on various indigenous species. The common conclusion of many studies is that physical contact with weathered oil may reduce the natural protective barrier of birds and marine mammals to the elements, resulting in hypothermia (NOAA, 1992). Though most animals do not intentionally ingest petroleum, studies have shown that endangered sea turtles have ingested tar balls, apparently as a food source. Turtles that were raised in captivity were fed small brown food pellets as juveniles. After the turtles were tagged and released, the corpses of some of these turtles were found washed ashore. Autopsies indicated tar balls or globules of petroleum within the stomachs. The tar balls were apparently mistaken for food pellets (Carr, 1987).

Bioavailability, sensitivity, and bioaccumulation affect the toxic responses of a given species. Sensitivity to petroleum varies for each species, and differing degrees of sensitivity occur at different stages of life. To assist in a general understanding of the various species responses, the following categories are discussed: birds, marine mammals, pelagic species, and epifauna. Trends can be observed for each of these categories according to the route of exposure.

Birds

Bird species with water habitats are the species most commonly affected by oil spills and releases. These birds are dependent upon the marine or freshwater environment for food sources and protection and are most likely to be exposed to petroleum. The ingestion of petroleum-contaminated food sources is not common for most of these birds, though scavenging or hunting birds may feed on petroleum-impacted animals washed ashore. Exposure of water fowl occurs primarily through the fouling of plumage. Oil disrupts the fine strand structure of the feathers, resulting in loss of water repellency and in decreased body insulation. As the oiled plumage becomes matted, water penetrates the feathers and chills the body; the combined results are a loss of buoyancy and possible hypothermia. The quantity of petroleum required to produce these effects varies; 5 ml of petroleum upon the breast feathers of Cassin's auklets have been reported to cause hypothermia and death (Birkhead et al., 1973). The natural response to oil-matted plumage is preening; oiled birds often ingest petroleum while attempting to remove the petroleum from their feathers. The effects of ingested petroleum include anemia, pneumonia, kidney and liver damage, decreased growth, altered blood chemistry, and decreased egg production and viability.

Chicks may be exposed to petroleum by ingesting food regurgitated by impacted adults. For example, the growth of young black guillemots (Peakall et al., 1980) and herring gulls (Miller et al., 1978; Heath, 1983) has been shown to be slowed by ingestion of petroleum; the growth of young Leach's storm petrels (Trivelpiece et al., 1984), wedge-tailed shearwaters (Fry et al., 1986), and fork-tailed storm petrels (Boersma et al., 1988) show slight or no effects of exposure on growth, although the survival of the Leach's storm-petrel chicks that ingested petroleum was decreased. Herring gull and Atlantic puffin nestlings that received oral doses of Prudhoe Bay crude oil developed anemia (Leighton et al., 1983).

Petroleum on the external surface of eggs has also been noted to have toxic effects. Coatings of some crude and fuel oils covering less than 10% of the egg surface have been shown to cause reduction in hatching, teratogenicity, and stunted growth. GC/MS analysis of mallard eggs treated with a synthetic petroleum hydrocarbon containing benzo(a)pyrene, chrysene, or 7,12-dimethylbenz(a)anthracene indicated that the aromatic hydrocarbons, including chrysene, passed through the shell and membranes to the developing embryos (Hoffman and Gay, 1981). Mallard embryos displayed decreased body weight, crown-to-rump length, and bill length and increased teratogenicity and mortality.

Marine Mammals

Marine mammals and pelagic species have similar routes of exposure: inhalation, surficial contact, and ingestion. The marine species include cetaceans, pinnipeds, and sea otters. Very rarely are the larger mammals exposed to petroleum slicks on the water; they have been known to alter their routes to avoid petroleum. It is the sea otter that resides in a localized region that is most affected and has the most reported deaths from direct and indirect exposure (NOAA, 1992).

Inhalation of oily water occurs more frequently for those species that feed on the water surface or in shallow impacted waters. This leads to absorption by the circulation system, irritation, and possibly permanent damage to respiratory surfaces and mucous membranes. The sea otters are the most sensitive, with effects to the lungs and nervous system.

Surficial contact has little or no effect for most whales due to the insulating qualities of their blubber and skin. The exception is temporary feeding reduction for the baleen whales as petroleum coats the baleen (Geraci and St. Aubin, 1982; Engelhardt, 1983). The adult seals, sea lions, and walruses have insulating qualities similar to the whales and are not greatly affected by surficial contact. The only surficial membranes noticeably affected are the mucous eye membranes (Nelson-Smith, 1970; Engelhardt, 1983). Juvenile seals, fur seals, and sea otters, which are dependent on fur insulation, are subject to potential hypothermia (Blix et al., 1979; Engelhardt, 1983). The hypothermic response, similar to that of water fowls, occurs when petroleum-matted fur has reduced heat insulation. Affected animals must expend much more energy to maintain a normal body temperature, stressing the body systems. The majority of the sea otter deaths during petroleum spills have been associated with this hypothermic response.

Ingestion occurs through direct routes, by consumption of contaminated food, or indirectly, by grooming in attempts to remove the petroleum matting. The results are both acute and chronic and include kidney failure, destruction of intestinal lining, neural disorders, and bioaccumulation. No biomagnification instances have been reported, presumably due to the effective assimilation of petroleum compounds from food sources.

Pelagic Species

Effects of petroleum on the pelagic species are varied. Effects on plankton are short-lived — zooplankton are more sensitive than phytoplankton. The impacts to the fish population are primarily to the eggs and larvae, with limited effects on the adults. The sensitivity varies by species; pink salmon fry are affected by exposure to water-soluble fractions of crude oil (Rice et al., 1977; Thomas and Rice, 1979), while pink salmon eggs are very tolerant to benzene and water-soluble petroleum. Coho salmon smolts exposed to water-soluble fractions of Cook Inlet crude oil accumulated up to 30 times more hydrocarbon in tissues than coho salmon jacks (Thomas et al., 1989). The swimming performance of juvenile Coho salmon is also affected (Thomas et al., 1987). Effects such as increased breathing rates may be due to the accumulation of PAHs in brain tissue (Russell and Fingerman, 1984). Winter flounder larvae are more sensitive than the eggs to 100 ppb concentrations of number 2 fuel oil. Pollock embryos exposed to water-soluble fractions of Cook Inlet crude oil had reduced survival before and after hatching, slowed initial development, and increased morphological abnormalities (Carls and Rice, 1988). Liver effects have been noted in English sole in areas with high concentrations of PAHs in the sediment. The general effects are difficult to assess and quantitatively document due to the seasonal and natural variability of the species. Fish rapidly metabolize aromatic hydrocarbons due to their enzyme system, particularly the cytochrome P_{450}-dependent mixed-function oxidase, epoxide hydrolase, and glutathione-s-transferase. These systems facilitate the removal of hydrocarbons and metabolites from the tissues (Malins and Hodgins, 1981). The higher concentrations and accumulation of aromatic hydrocarbons are found in the lipid content of the viscera.

Epifauna

Epifauna includes the mussels, bivalves, crustaceans, and barnacles. The mussels and bivalves often survive oiling as adults due to their protective shells, but they have no enzymatic system for purging. Therefore, bioaccumulation occurs, resulting in reduced feeding absorption efficiency followed by growth reductions. In the crustacean *Daphnia magna*, exposure to water-soluble fractions of number 2 diesel oil resulted in lowered respiration rates and food assimilation (Ullrich and Millemann, 1983). Many crustaceans, including barnacles and crabs, are acutely sensitive to petroleum impacts and have high mortality rates and chronic effects. In immature quahog clams, cell division was affected by water-soluble fractions of number 2 fuel oil (Byrne, 1989).

F. MITIGATION METHODS

Numerous mitigation techniques have been developed during the past 20 years. Due to the complex nature of petroleum hydrocarbons, no technique can be considered a "cure all". When a petroleum release occurs, various factors must be taken into consideration in order to mitigate the impact without increasing the potential damage. Factors that must be considered are the environmental energy, the sensitivity of the pelagic and epifauna species, whether the habitat is a reproductive or nesting region, and the shoreline geology. Mitigation technologies include physical methods that are used on petroleum on the water surface before shoreline impact has occurred, chemical methods, and shoreline methods.

1. Physical Methods

The physical methods employed on the water surface are containment methods followed by removal. Containment is usually achieved by placing a system of floating booms around the spilled petroleum. The removal is generally accomplished by a skimming machine on a barge that sucks up oil on oleophilic belts. Chemical additives have been developed that aid this process by decreasing the solubility of the petroleum, thereby limiting the amount of water that must be skimmed. Some additives combine with petroleum to produce a stringy or rope-like substance that can be raked from the surface of the water. This method of recovery is limited by wave height, current speed, and oil thickness and type. The light petroleum products are difficult to boom due to their volatility and dissolution.

Another option that is available for use on open water is *in situ* burning. This process involves booming the petroleum and igniting it. The hydrocarbons remaining after burning are the extremely weather-resistant asphaltenes and resins. Preliminary tests of this method indicate the effective removal of up to 90% of the boomed petroleum or 30% to 50% of the total oil spilled (Buist, 1987). The disadvantages of this method are time and weather limitations (burning is not effective if significant evaporation has occurred) and smoke emissions, which are a potential problem with incomplete combustion (Benner et al., 1990). The incomplete combustion of the heavier petroleum products produces three- to five-ring aromatic hydrocarbons, some potentially carcinogenic, in the smoke and soot. These products are dispersed into the atmosphere, possibly impacting a larger region than the original oil spill.

2. Chemical Methods

Chemical methods may be applied to both the water surface and the shoreline mitigation techniques. To protect sensitive shorelines, chemical additives can disperse petroleum on the water surface. The chemicals used to promote this effect are fittingly called dispersants and are made from lipophilic and hydrophilic components, such as sorbitan monooleates, ethoxylated sorbitan monooleates, and polyethylene glycol esters of unsaturated fatty acids. The process of dispersion occurs as oil droplets are surrounded by a dispersant chemical, with the lipophilic (or hydrophobic) ends of the dispersant molecules surrounding the droplets and the hydrophilic ends allowing the droplets to be suspended in the water column. As with burning, there is a limited window for application of this method. Dispersion is most effective before a significant amount of evaporation has occurred and requires a concentration of 1 part dispersant to 20 parts petroleum, with proper mixing. Dispersants may react in various ways with different petroleum products, and the guiding factor for effectiveness is related to the asphaltene content, which is associated with viscosity as well as weather conditions at the scene of a spill.

A major concern with dispersants is their toxicity (NRC, 1989). The first application of a dispersant occurred in 1967 on the coast of England during the Torrey Canyon spill, where 10,000 tons of toxic dispersants were applied to 14,000 tons of crude oil. The intertidal impacts were significant and well-documented (Smith, 1968). Dispersants with low toxicity have since been developed; however, toxicity tests must be completed for the dispersal oil as well as dispersant alone. One question that must be addressed is whether bioavailability increases with the dispersion. Effects to eiders and mallards have shown that it takes less petroleum with dispersant for a reduction of plumage water repellency and an increase in water absorption and heat loss than petroleum alone (Jenssen and Ekher, 1991). Dispersants have also been proven detrimental to the eukaryotic process. Basically, dispersants increase surface area and bioavailability for microorganism degradations (Foght and Westlake, 1982). Other points to consider are the reduction of impacts to the sensitive shoreline ecosystem by use of the dispersants. Mangrove habitats are very sensitive; if petroleum covers the aerobic roots, oxygen transport is removed and the mangroves die (Mitch and Gosselink, 1986).

3. Shoreline Methods

The shoreline mitigation methods employ a variety of chemical and physical techniques. One chemical method involves application of a surfactant. When applied to oiled rocks, the surfactant lifts the petroleum from the solid surface onto the water, where the floating product is collected by a boom.

The physical methods involve manually cleaning shoreline rocks with absorbent pads and tilling the shoreline material so that the petroleum submerged in the sediments resurfaces and is exposed to weathering processes.

Another shoreline technique is bioremediation, which is an attempt to enhance microbial biodegradation. Biodegradation is a natural process in which commonly found microbial organisms, fungi, bacteria, yeast, and mold utilize oil as a carbon source for subsistence. The preferred compounds for petroleum degradation are the saturates. To accentuate the process, fertilizers containing growth-stimulating reagents, specifically nitrogen and phosphorus, are applied. The benefits of these additives and the increased degradation by foreign microbes over the indigenous species have yet to be conclusively proven (Kennicutt, 1988).

All mitigation methods affect the environment. The physical methods increase the physical abuse of an area; this may be detrimental to sensitive ecosystems. Chemical methods have been linked with increasing toxicity in some species, either by the chemical components in the compounds or by increasing the bioavailability of the toxic petroleum components. *In situ* burning has the potential to increase the carcinogenic aromatic components and spread the compounds over a wider range. Microbial degradation is a slower process that could alter the ecosystem if nutrients and foreign microbes are added. Understanding the potential environmental impacts of the various mitigation methods will lead to the use of the most effective cleanup method for a given oilspill or oil release situation.

The choice of appropriate physical or chemical mitigation technologies is influenced by a series of trade-offs between impacts of oil if no mitigation procedures are used versus impacts caused by the use of the mitigative technologies (e.g., which is least damaging to the environment — action or no action?). In essence, value decisions have to be made on which environmental resources are most valuable, most easily damaged by the spilled oil, and most resilient from impacts. Once these determinations have been made, decisions on the use of mitigating technologies, including a no response option, are based on actions that provide the least environmental damage and the maximum resource protection.

REFERENCES

Benner, B.A., Jr., Nelson, P., Bryner, S.A., Wise, G.W., Mulholland, R., Lao, C., and Fingas, M.F. Polycyclic aromatic hydrocarbon emissions from combustion of crude oil on water, *Environ. Sci. Tech.* 24:1418–1427, 1990.

Birkhead, T.R., Lloyd, C., and Corkhill, P. Oiled seabirds successfully cleaning their plumage, *Br. Birds* 66:535–537, 1973.

Blix, A.S., Grav, H.J., and Ronald, K. Some aspects of temperature regulation in newborn harp seal pups, *Am. J. Physiol.* 236:R188–R197, 1979.

Bobra, M.A. *A Catalogue of Crude Oil and Oil Product Properties.* Environment Protection Directorate, Environment Canada, Ottawa, 1989.

Boersma, P.D., Davies, E.M., and Reid, W.V. Weathered crude oil effects on chicks of fork-tailed storm-petrels (*Oceanodroma furcata*), *Arch. Environ. Contam. Toxicol.* 17:527–531, 1988.

Buist, I.A. A preliminary feasibility study of *in-situ* burning of spreading oil slicks, in *Proceedings, 1987 Oil Spill Conference.* Philadelphia, PA, 1987, 359–367.

Butt, J.A., Duckworth, D.F., and Perry, S.G. *Characterization of Spilled Oil Samples.* John Wiley & Sons, New York, 1986.

Byrne, C. Effects of the water-soluble fractions of no. 2 fuel oil on the cytokinesis of the quahog clam (*Mercenaria mercenaria*), *Bull. Environ. Contam. Toxicol.* 42:81–86, 1989.

Carls, M.G. and Rice, S.D. Abnormal development and growth reductions of pollock *Theragra chalcogramma* embryos exposed to water-soluble fractions of oil, *Fishery Bull. U.S.* 88:29–37, 1988.

Carr, A. Impacts of nondegradable marine debris on the ecology and survival outlook of sea turtles, *Mar. Pollut. Bull.* 18:352–356, 1987.

Engelhardt, F.R. Petroleum effects on marine mammals, *Aquat. Toxicol.* 4:199–217, 1983.

Foght, J.M. and Westlake, D.W.S. Effect of the dispersant corexit 9527 on the microbial degradation of Prudhoe Bay oil, *Can. J. Microbiol.* 28:117–122, 1982.

Fry, D.M., Swenson, J., Addiego, G.A., Grau, C.R., and Kang, A. Reduced reproduction of wedge-tailed shearwaters exposed to weathered Santa Barbara crude oil, *Arch. Environ. Contam. Toxicol.* 15:453–463, 1986.

Geraci, J.R. and St. Aubin, D.J. Study of the effects of oil on cetaceans, Report for U.S. Department of the Interior, Bureau of Land Management, Contract No. AA-551-CT9-29, 1982.

Heath, G.W. Toxicity of crude oils and their fractions to nesting herring gulls. I. Physiological and biochemical effects, *Mar. Environ. Res.* 8:63–71, 1983.

Henry, C.B. Petroleum training material and communications. Institute for Environmental Studies, Louisiana State University, Baton Rouge, LA. Unpublished material, 1991.

Hoffman, D.J. and Gay, M.L. Embryotoxic effects of benzo(a)pyrene, chrysene and 7,12-dimethylbenz(a)anthracene in mallard ducks (*Anas platyrhynchos*), *J. Toxicol. Environ. Health* 7:775–788, 1981.

Jenssen, B.M. and Ekher, M. Effects of plumage contamination with crude oil dispersant mixtures on thermal regulation in common eiders and mallards, *Arch. Environ. Contam. Toxicol.* 20:398–403, 1991.

Kennicutt, M.C., II. The effect of biodegradation on crude oil bulk and molecular composition, *Oil Chem. Pollut.* 4:89–112, 1988.

Leighton, F.A., Peakall, D.B., and Butler, R.G. Heinz-body hemolytic anemia from the ingestion of crude oil: a primary toxic effect in marine birds, *Science* 220:871–873, 1983.

Malins, D.C. *Effects of Petroleum on Arctic and Subarctic Marine Environments and Organisms, Vol. I, Nature and Fate of Petroleum.* Academic Press, Inc., New York, 1987.

Malins, D.C. and Hodgins, H.O. Petroleum and marine fishes: a review of uptake, disposition, and effects, *Environ. Sci. Technol.* 15:1272–1279, 1981.

Miller, D.S., Peakall, D.B., and Kinter, W.B. Ingestion of crude oil: sub-lethal effects on herring gull chicks, *Science* 199:315–317, 1978.

Mitch, W.J. and Gosselink, J.G. *Wetlands.* Van Nostrand Reinhold, New York, 1986.

National Oceanic and Atmospheric Administration (NOAA). An introduction to coastal habitats and biological resources for oil spill response, Report No. HMRAD 92-4. Hazardous Materials Response and Assessment Division, National Oceanic and Atmospheric Administration, Washington, DC, 1992.

National Research Council (NRC). *Using Oil Spill Dispersants on the Sea Marine Board.* Marine Board, Commission on Engineering and Technical Systems, National Research Council, National Academy of Science, National Academy Press, Washington, DC, 1989.

Nelson-Smith, A. The problem of oil pollution of the sea, *Adv. Mar. Biol.* 8:215–306, 1970.

Payne, J.R. and McNabb, D.G., Jr. Weathering of petroleum in the marine environment, *MTS Journal* 18:24–42, 1984.

Payne, J.F., Maloney, R., and Rahimtula, A. Are petroleum hydrocarbons an important source of mutagens in the marine environment?, in *Proc. 1979 Oil Spill Conference.* Los Angeles, 1979, 533–536.

Peakall, D.B., Hallett, D.J., Miller, D.S., Butler, R.G., and Kinter, W.B. Effect of ingested crude oil on black guillemots: a combined field and laboratory study, *Ambio* 9:28–30, 1980.

Rice, S.D., Thomas, R.E., and Short, J.W. Effect of petroleum hydrocarbons on breathing and coughing rates and hydrocarbon uptake-depuration in pink salmon fry, in *Physiological Responses of Marine Biota to Pollutants.* F.J. Vernberg, A. Calabrese, F.P. Thurberg, and W.B. Vernberg, Eds. Academic Press, New York, 1977, 259–277.

Russell, L.C. and Fingerman, M. Exposure to the water soluble fraction of crude oil or to naphthalenes alters breathing rates in gulf killifish, *Fundulus grandis. Bull. Environ. Contam. Toxicol.* 32:363–369, 1984.

Smith, J.E. *Torrey Canyon Pollution and Marine Life.* Cambridge University Press, New York, 1968.

Speight, J.G. *The Chemistry and Technology of Petroleum*, 2nd ed. Marcel Dekker, Inc., New York, 1991.

Thomas, R.E. and Rice, S.D. The effect of exposure temperatures on oxygen consumption and opercular breathing rates of pink salmon fry exposed to toluene, naphthalene, and water-soluble fractions of Cook Inlet crude oil and no. 2 fuel oil, in *Physiological Responses of Marine Biota to Pollutants.* F.J. Vernberg, A. Calabrese, F.P. Thurberg, and W.B. Vernberg, Eds. Academic Press, New York, 1979, 39–52.

Thomas, R.E., Rice, S.D., and Korn, S. Reduced swimming performance of juvenile coho salmon *(Oncorhynchus kisutch)* exposed to the water-soluble fraction of Cook Inlet crude oil, in *Pollution Physiology of Estuarine Organisms.* W.B. Vernberg, A. Calabrese, F.P. Thurberg, and F.J. Vernberg, Eds. University of South Carolina Press, Columbia, SC, 1987, 127–137.

Thomas, R.E., Rice, S.D., Babcock, M.M., and Moles, A. differences in hydrocarbon uptake and mixed function oxidase activity between juvenile and spawning adult coho salmon *(Oncorhynchus kisutch)* exposed to Cook Inlet crude oil, *Comp. Biochem. Physiol.* 93:155–159,1989.

Trivelpiece, W.Z., Butler, R.G., Miller, D.S., and Peakall, D.B. Reduced survival of chicks of oil-dosed adult leach's storm-petrels, *Condor* 86:81–82, 1984.

Ullrich, S.O., Jr. and Millemann, R.E. Effect of temperature on the toxicity of the water-soluble fractions of diesel fuel no. 2 and H-coal crude oil to *Daphnia magna,* Government Reports Announcements and Index, Issue 6. Department of Energy, Washington, DC, 1983.

Chapter

6

Toxicity of Solvents

Arthur S. Hume and Ing K. Ho

A. INTRODUCTION

Solvents are a chemically diverse group of compounds. They are usually liquids, with a wide range of specific gravities, viscosities, vapor densities, and flash points. Solvents are not all highly lipophilic, but lipophilicity is a toxicologically significant physical characteristic of some types of solvents. The distribution of the solvent to target organs in an organism is determined by cardiovascular efficiency and the physicochemical properties of the solvent, that is, partition coefficient. Bohlen et al. (1973) caution that consideration of the significance of the partition coefficient of solvents must include solvent induced lipid accumulation by the tissue component of the partition coefficient. Blood pH should not affect the transfer of most solvents across cellular membranes since solvents are usually not ionized in solution. The extent and rate of transfer will thus depend mostly upon their partition coefficient. Absorption of solvents from the lungs is unobstructed, and dissolution in the blood occurs in accordance with their solubility in blood expressed as an Ostwald coefficient.

Solvents usually have high vapor pressures which result in the formation of vapors; therefore, the primary exposure route of solvents is by inhalation. The intake rate of solvents received by inhalation depends on the rate and depth of respiration. It is appropriate, then, to express the toxicity of solvent vapors as a LC_{50}, air concentration resulting in the death of 50% of exposed animals, or the concentration in air multiplied by the time of exposure which results in death in 50% of the animals.

Due to the many different uses of solvents, exposure through the skin is frequent. As solvents are lipophilic substances, it is expected that they are

0-8493-8851-1/94/$0.00+$.50
© 1994 by CRC Press, Inc.

absorbed through the skin into the bloodstream. Systemic toxicity can occur from percutaneous absorption of some solvents. It is noted that in the use of solvents, absorption may occur by percutaneous absorption and inhalation simultaneously.

It is interesting that although heterogeneous in physical and chemical properties, most solvents exhibit a depressing effect upon the central nervous system (CNS) and are therefore classified as CNS toxicants. CNS symptoms produced by solvents are similar to those of general anesthetic agents. This suggests that the mechanism of action is that of a physical interaction of the solvent with the membranes of certain cells within the CNS that results in changes in cell membrane permeability to ions. It appears that the depressant effects of solvents are related to concentration and that the concentration that is required to produce an effect is much greater than would be necessary were a receptor mechanism involved. The fact that similar effects are elicited by solvents of such diverse chemical structure also supports a mechanism of general physical interaction. The effects of organic solvents on the CNS are related to the concentrations of the chemicals that reach the brain. Symptoms of CNS depression resulting from exposure to solvents range from nausea, dizziness, euphoria, confusion, and loss of coordination at low concentrations to convulsions, coma, and death at high concentrations. Solvents also cause cardiac sensitization. In a Swedish study, an increased susceptibility to cardiac arrhythmia was observed after exposure to solvents. Soderlund (1975) concluded that exposure to the solvents toluene, methyl chloroform, styrene, and white spirit resulted in a decreased ability to perform tasks and was even more pronounced after light work or exertion. It is interesting that the metabolites of some solvents such as *n*-hexane and methanol are responsible for more severe toxicological effects than the parent chemical. In addition to other toxic effects, some solvents or their metabolites affect specific organs that are called target organs. Listed below are examples of solvents and target organs:

Solvent	Target Organ
Methanol	Ocular nerve or retina
Chlorinated hydrocarbons	Liver
Ethanol	Liver, brain
Hexane	Nerves
Ethylene glycol ethers	Reproduction
Benzene	Bone marrow
Carbon disulfide	Nerves

The effects of chronic exposure to organic solvents on the behavioral performance of workers are of both a health and an economic interest. Issues of productivity and worker safety and health are involved. In a study of paint workers who had been exposed to a variety of solvents for more than 10 years, the workers performed less well on psychological tests than nonexposed workers (Soderlund, 1975).

The neurotoxic effect of organic solvents is of a growing concern since irreversible structural damage may be involved. Most of the research on this aspect of chronic solvent toxicity has been undertaken in Denmark, Sweden, and Finland. Schaumburg and Spencer (1976) have assigned five solvents to a category of proven neurotoxicity in humans: carbon disulfide, *n*-hexane (with or without methyl ethyl ketone), methyl *n*-butyl ketone (with or without methyl ketone), toluene (in abusive doses only) and impure trichloroethylene (TCE from dichloroacetylene). These authors suggested that other monocyclic and dicyclic aromatic hydrocarbons be studied for possible neurotoxic effects. The toxic effects of the alcohols, particularly ethanol and methanol, on the nervous system are well recognized (Sterman and Schaumburg, 1980).

Schmidt et al. (1984) attributed the neurotoxicity of solvents (e.g., *n*-hexane and methyl ethyl ketone) to a disturbance of enzymatically controlled processes and an impairment of synthesis of energy related products. Based upon their work on the intrapulmonary nerve system of rats, Spencer et al. (1979) postulated that glycolytic enzyme activity could be inhibited by a neurotoxic chemical, consequently inhibiting metabolism throughout the axon. Sabri et al. (1979) had reported earlier that glycolysis in the neurons was inhibited by hexacarbon solvents, resulting in a degeneration of the neurons.

With the enormous use of chemicals in industry, many workers are exposed to solvents, either in the ambient air or by skin contact. Therefore, it has become necessary for the sake of health and safety to attempt to establish air concentration limits for handling procedures for chemicals, including solvents. Establishment of these limits is of benefit not only to the workers and management for health and economic reasons but also provides guidelines for the general population which may be exposed to hazardous waste sites, spills, etc. The establishing of threshold limit values (TLV) has, by necessity, involved toxicological studies and reevaluations of existing data about the toxic effects of chemicals on persons in situations of prolonged exposure.

The American Conference of Governmental Industrial Hygienists (ACGIH) has listed TLVs for some chemicals, including solvents. TLV is defined as the concentration at which workers can be exposed for 8 h/d during a 40 h week without suffering harmful effects. A threshold limit value-ceiling (TLV-C) has also been proposed. This is an exposure concentration that should not be exceeded at any time during the working day. ACGIH has also published biological exposure indices (BEI) which represent the levels of chemicals or their metabolites which are likely to be detected in specimens collected from a healthy worker who has the same exposure to chemicals as the worker who is exposed by inhalation to the TLV. The National Institute for Occupational Safety and Health has proposed recommended exposure limits (REL). The Occupational Safety and Health Administration enforces exposure conditions based upon permissible exposure limits (PELS). Rather than work with a single designated limit of threshold concentration for a chemical, an average concentration limit for an 8-h day, or time weighted average (TWA), has been established for some chemicals. Also, there is a short-term exposure limit (STEL) which is the maximum concentration to which workers can be continuously exposed for 15 min. No observable effect levels (NOEL) have been listed for many chemicals that have exposure limits at which no effect will occur.

These limits of exposure are applied in a 40-h/week working situation consisting mostly of adults. They should not be applied to household situations in which children or the ill and elderly are included, who are exposed in 24-h/d situations.

Concentrations of solvents and vapors in permissible limits are expressed in parts (solvent) per million (air) or ppm, while fumes and vapors are expressed in milligrams (fumes) in a cubic meter (air) or mg/m^3. The relationship of these units is seen in the following equation: $mg/m^3 = ppm \times mol\ wt/24.45$, where mol wt is the molecular weight of the solvent and 24.45 is the volume of air at standard temperature and pressure (stp = 25°C and 760 mmHg).

B. CLASSIFICATION OF SOLVENTS

1. Alkanes

This group of solvents contains hydrocarbons, sometimes referred to as paraffins or alkanes, with the general formula C_nH_{2n+2} above C_4H_{10} and includes pentanes to hexadecanes (C_5H_{10} to $C_{19}H_{28}$).

Petroleum and petroleum products are recognized as the source of many of the chemicals of the alkane group. Petroleum products are used as motor fuel, lubricants, solvents, and in the synthesis of many different chemicals. Some hydrocarbon mixtures are listed in Table 6.1.

Most members of this group are highly lipophilic and are distributed rapidly throughout the body. They are absorbed readily by the lungs and skin, and defat the skin upon contact. Members of this group are not excreted very readily, and they are metabolized to their more hydrophilic derivatives, alcohols. These alcohols are conjugated with glucuronic acid or metabolized to carbon dioxide and water (Toftgard and Gustafsson, 1980).

TABLE 6.1. Physical Properties of Some Alkyl Hydrocarbons

Alkane (Number of Carbons)	Product	Boiling Range (C°)
C_4–C_6	Petroleum ether	20–60
C_5–C_7	Petroleum benzin	40–90
C_6–C_8	Petroleum naptha	65–120
C_5–C_{10}	Gasoline	36–210
C_7–C_9	Mineral spirits	150–210
C_9–C_{16}	Kerosine	170–300
C_5–C_{16}	Jet and turbine fuels	40–300

TABLE 6.2. Toxicological Properties of Some Alkyl Hydrocarbons

Hydrocarbon	Animal	Exposure	Toxic Effects
Pentane	Mice	90,000–120,000 ppm (5–60 min)	Narcosis
Hexane	Mice	30,000 ppm	Narcosis
Heptane	Mice	10,000–15,000 ppm (30–60 min)	Narcosis
Octane	Mice	6,600–13,700 ppm (30–90 min)	Narcosis

Alkanes are depressants to the CNS. In fact, gasoline was once evaluated as an anesthetic agent. The principal toxicity effect of alkanes in acute toxic ingestions is that of aspiration of the hydrocarbon into the lungs either at the time of ingestion or upon regurgitation. The aspiration of only 0.1 ml into the lungs can result in chemical pneumonitis. The probability of a hydrocarbon solvent being aspirated is related to the viscosity of the solvent.

The symptoms of toxicity induced by alkanes include narcosis, dizziness, headache, nausea, confusion, and in very severe cases, a loss of consciousness. The alkyl hydrocarbon solvents with a carbon number from C_1 to C_8 (octane) exhibit increasing toxic effects on the CNS. As shown in Table 6.2, octane is about 15 times more potent as a narcotic than pentane (C_5) (Stoughton and Lamson, 1936).

Sudden deaths have been reported in persons who have inhaled vapors of hydrocarbons such as gasoline. These deaths are not attributed to CNS depression but to fatal cardiac arrhythmias which could be the result of myocardial sensitization to circulating catecholamines (Krantz et al., 1948; Reinhardt et al., 1971).

Although gasoline vapors at a low concentration usually do not present significant toxicological problems, there are reports of serious, even lethal, effects from exposure of large areas of the body such as immersion (Walsh et al., 1974). Cutaneous injuries have also been reported as a result of contact with gasoline (Hansbrough et al., 1985). Gasoline has been reported to induce gross pulmonary hemorrhages (Gerarde, 1963). There are also reports of renal damage (glomerulonephritis) associated with solvent inhalation (Beirne and Brennan, 1972). In fact, renal tubular acidosis has been associated with exposure to gasoline (Taher et al., 1974).

In studies done in animals, none of the alkanes induced teratogenic, mutagenic, or carcinogenic effects (Sandmeyer, 1981).

n-Hexane

n-Hexane is considered separately from the other alkanes because of its frequent use and neurotoxicity. This solvent is widely used in industry in the synthesis of products such as plastics, rubber, paint thinners, lacquers, and glue. Paulson and Waylonis (1976) reported that an estimated 2.5 million workers are exposed to *n*-hexane. In fact, the terrestrial population is exposed on occasion to *n*-hexane while pursuing hobbies, performing home repairs, etc.

Toxicokinetics. *n*-Hexane is absorbed by ingestion of the solvent, inhalation of vapors, or percutaneous absorption. About 15% of inhaled vapors of *n*-hexane is absorbed by the lungs (Couri and Milks, 1982). About 50 to 60% of inhaled hexane is expired by the lungs.

n-Hexane, a highly lipid chemical, is distributed to tissues high in lipid content. The concentration of hexane is greatest in the brain, liver, and kidneys. Bohlen et al. (1973) reported that tissue lipid content affects the time required to saturate these tissues. Blood, brain, adrenal, kidneys, and spleen reached a saturation level at 4 to 5 h, while liver did not reach a saturation level until 10 h after exposure.

In a study by Ruff et al., (1981) a biological half-life of 2 h was determined in humans who were exposed to 100 ppm of hexane. Using a mathematical model, Perbellini et al. (1986) suggested a half-life of 64 h for hexane in the fat of humans, with a total of 10 d for the complete removal of hexane from the fat tissue.

It is of toxicological significance that hexane is metabolized in the liver by the mixed function oxidase system to 2-hexanol, which in turn is oxidized to 2,5-hexanediol (Perbellini et al., 1981). Further, 2,5-hexanediol is oxidized to 2,5-hexanedione. The metabolism of hexane is shown as follows:

n-hexane (C$_6$H$_{14}$)

↓

2-hexanol (C$_6$H$_{14}$O)

↓

2,5-hexandiol (C$_6$H$_{12}$O$_2$)

↓

5-hydroxy-2-hexanone (C$_6$H$_{12}$O$_2$)

↓

2,5-hexanedione (C$_6$H$_{10}$O$_2$)

The metabolites of n-hexane have all been shown to cause a peripheral neuropathy (functional disturbance in the peripheral nervous system) in experimental animals. Of the metabolites of n-hexane, 2,5-hexanediol is the most neurotoxic.

Toxicology. n-Hexane is a CNS depressant. At low exposure levels, headache and anoxia are noted. Confusion, stupor, and coma can result from exposure to high concentrations of n-hexane.

Chronic exposure to n-hexane produces a polyneuropathy. Incidence of polyneuropathy has been reported in workers who were exposed to ambient air containing high concentrations of n-hexane in glue vapors from recreational use (Goto et al., 1974). A hexane-induced polyneuropathy is characterized by fatigue, muscular weakness, and distal paresthesia (tingling sensation) in upper and lower extremities (Paulson and Waylonis, 1976).

The mechanism by which hexane and hexane metabolites actually damage neurons is not known. 2,5-Hexanedione has been shown to interact with several cellular molecules which would result in alteration of the neuronal functions.

n-Hexane-induced peripheral neuropathy is characterized by giant axonal swellings in the more distal sections of the axons (Ishii et al., 1972). Degeneration of the nerve fibers may occur distally to the swellings. *n*-Hexane-induced peripheral neuropathy is enhanced by the presence of other solvents (Altenkirch et al., 1977); however, some of these results are not conclusive. Methyl *n*-butyl ketone (MBK) and hexane both are metabolized to 2,5-hexanedione, which has a neurotoxic effect greater than either parent chemical (Schaumburg and Spencer, 1976). Exposure to the combination of hexane and MBK results in an almost predictable neuropathy.

An excellent review of the neurotoxicity of hexane and other solvents is found in Spencer and Schaumburg (1980).

Paulson and Waylonis (1976) reported that of a total of 50 employees who had been exposed to *n*-hexane, eight suffered from a mild neuropathy. After chronic exposure to *n*-hexane the symptoms of peripheral neuropathy were described as numbness, weakness in the lower extremities, absent or decreased patellae reflex, and myalgia (Proctor and Hughes, 1978).

2. Aromatic Hydrocarbons

This group of solvents includes benzene and derivatives of benzene. Benzene is considered separately because its toxicities are different from those of the alkyl benzenes.

Benzene

Occurrence. Benzene is used as an additive for gasoline and other fuels and as a solvent by many industries, that is, paint, plastics, and rubber. Ninety percent of the benzene produced is used in the manufacture of medicinal chemicals, dyes, and other organic compounds. It is estimated that 2 million U.S. workers are exposed to benzene in an occupational setting.

Benzene has been classified as a priority toxic pollutant, a hazardous substance (Clean Water Act), a hazardous air pollutant (Section 112 of the Clean Air Act), and a hazardous waste constituent.

Toxicokinetics. The principle route of exposure to benzene is by inhalation. About 30 to 80% of the benzene inhaled is absorbed into the circulating blood. Benzene can be absorbed through the skin, in fact, Susten et al. (1985) reported that 4 to 8 mg of benzene can be absorbed when it is applied topically. Benzene is distributed widely in the body with the greatest concentration being found in the fat. About 10 to 15% of absorbed benzene is returned to the lungs and excreted unchanged by expiration. Benzene is metabolized in the liver by the P_{450} mixed function oxidase system (Rickert et al., 1979).

Benzene oxide is the most likely toxic metabolite; however, metabolism in the bone marrow may produce a different metabolite which is toxic to hematological processes (Longacre et al., 1981).

Benzene is considered to be a human carcinogen by the EPA, OSHA, WHO, and the International Agency for Research on Cancer (IARC). The report of the Carcinogen Assessment Group (CAG) of the EPA should be consulted for further information.

Toxicology. The toxic effects of benzene are of particular concern since benzene is added to gasoline to increase its antiknock properties, and because many in the general population are exposed to gasoline fumes and to automobile exhaust to some extent. However, Parkinson (1976) reported that in a study of personnel handling gasoline at filling stations and bulk loading stations, excretion of phenol (the major metabolite of benzene) did not indicate a significant exposure to benzene. Sherwood (1972) reported that of two loaders and one weigher who loaded gasoline, only the loader who was exposed to 20 ppm benzene in air showed a urine phenol level of 83 mg/l.

Acute exposure to high concentrations (3000 to 5000 ppm) of benzene results in CNS depression characterized by headache, nausea, insomnia, agitation, stupor, coma, and convulsions. Death has resulted from large, acute exposures to benzene. Death also may be due to myocardial sensitization as the result of the production of endogenous catecholamines.

Chronic exposure to benzene can result in hematotoxicity (Vigliani and Saita, 1964; Brandt et al., 1977). Benzene is a recognized myelotoxic agent (destroys bone marrow) capable of producing fatal aplastic anemia (defective development of erythrocytes) and leukemia.

Blood dyscrasias with hemolytic effects caused by exposure to benzene are anemia, aplastic anemia, hemocytoblastia, reticulocytosis, leukopenia, thrombocytopenia, and eosinophilia. Sandmeyer (1981) divides benzene induced hematopoietic changes into three stages: (1) blood clotting defects, (2) bone marrow hypo or hyperplasia, and (3) bone marrow aplasia, which may be progressive.

Acute myeloblastic leukemia was reported by Goldstein (1977) to be the most common malignancy associated with benzene exposure. The occurrence of chromosomal aberrations in benzene-exposed workers was reported by Picciano (1979) at a rate of 21% at 1 ppm, 25% at 1 to 2.25 ppm and 33% at 2.5 to 10 ppm (only 3% of unexposed workers showed abnormalities). Also, Aksoy (1985) implied that benzene exposure may be involved in the development of bronchiogenic carcinoma.

Snyder and Kocsis (1975) have demonstrated hematotoxicity in Sprague Dawley rats and AKR mice which had been exposed to 300 ppm of benzene. Also, based on studies in animals, Kuna and Kapp (1981) have proposed the possibility of toxic effects of benzene exposure on the reproductive system of humans.

In animal studies of toxicity of benzene by inhalation, Drew and Fouts (1974) determined a LC_{50} of 13,700 ppm for a 4 h exposure while Smyth et al. (1962) reported that in rats exposed to 16,000 ppm of benzene, four of six animals died.

3. Alkyl Benzenes

Substitution on the benzene ring with alkyl groups results in the formation of compounds with different toxicological properties than benzene. The primary difference is that apparently the alkylbenzenes do not affect the hematopoietic system. Also, alkyl substitution on the benzene ring results in changes in physicochemical characteristics, that is, partition coefficients which affect the pharmacokinetics of the chemicals (e.g., absorption and metabolism).

Ethyl Benzene (Ethylbenzol, EB)
Ninety-five percent of the ethyl benzene produced is used in the manufacture of plastic (styrene). It is also used in the production of synthetic rubber and as an additive to fuels to increase the octane rating. Gasoline may contain as much as 20% ethyl benzene (Settig, 1985).

Ethyl benzene is absorbed readily by inhalation of vapors, ingestion of the liquid, and percutaneous absorption. About 64% of the ethyl benzene vapor inhaled is absorbed (Bardodej and Bardodejova, 1970). In a study involving humans in which ethyl benzene was applied topically, 118 $\mu g/m^2/h$ was absorbed through the skin (Dutkiewicz and Tyras, 1967).

Toxicokinetics. El Masry et al. (1956) reported that when ethyl benzene was given orally to rabbits, 90% of the dose was excreted as metabolites. Inhalation of ethyl benzene by humans resulted in its metabolism to mandelic acid (64%) and phenylglyoxylic acid (25%) (Bardodej and Bardodejova, 1970). Most of the inhaled dose is excreted as metabolites in the urine within 24 h after exposure (Engstrom and Bjurstrom, 1978).

Toxicology. Ethyl benzene is an irritant to the eyes and nasal passages but is not toxic to the hematopoietic system as is benzene (Fishbein, 1985b). However, it does produce a CNS depressant effect at high concentrations (Gerarde, 1960). Although the literature is scant on the embryotoxic, genetic, and teratogenic effects of ethyl benzene, ethyl benzene did not cause an increase above the spontaneous recessive lethal frequency in the Drosophila recessive lethal test (Donner et al., 1980).

Toluene (Methylbenzene, Toluol)
Toluene, xylene, and trimethylbenzene (mesitylane) are the most frequently used compounds of this group of solvents. Their toxicology is of

particular importance since they are produced, used, and disposed of in such large quantities. Production of all grades of toluene in the U.S. in 1983 was 7.15×10^9 lb — 6×10^6 tons of toluene are lost into the environment annually (Fishbein, 1985a).

Toluene is more lipid soluble and less volatile than benzene. Toluene is used in the chemical industry to produce several other chemicals such as solvents for paints, lacquers, coatings, and glues and fuels in automobiles and aircraft.

It is estimated that some 100,000 U.S. workers are exposed to toluene. With the widespread use of glues, thinners, etc., there is the additional possibility of the exposure of a greater number of the general population.

Toxicokinetics. Toluene is rapidly absorbed by inhalation with peak blood concentrations occurring 15 to 30 min after inhalation (Astrand et al., 1972). After ingestion of toluene, peak blood levels are observed after 1 to 2 h (Bergman, 1979). Topical application of toluene also results in slow but significant absorption (Cohr and Stokholm, 1979). Toluene distributes throughout the body and accumulates in the adipose tissue (Baselt, 1982).

Toluene (80%) is metabolized to benzyl alcohol and then to benzoic acid which is conjugated to produce hippuric acid. Some toluene is excreted unchanged by expiration from the lungs. Toluene has a termination half-life of 15 to 20 h (Brugnone et al., 1986).

Toxicology. CNS effects are predominant in a scenario of acute exposure to toluene. Symptoms include headache, dizziness, fatigue, muscular weakness, collapse, and coma. Extensive studies show that toluene does not cause adverse effects on the liver, kidney, lungs, and heart (Low et al., 1988).

In contrast to benzene, exposure to toluene over long periods causes general malaise and CNS effects but does not result in hematopoietic injury, cancer, or bone marrow damage (Fishbein, 1985a).

In studies on experimental animals (Fischer 344 rats) no chronic toxicity or oncogenicity was observed after exposure to 300 ppm toluene for 24 months (Gibson and Hardisty, 1983). In other studies, Hersh et al. (1985) reported that three children whose mothers had inhaled large quantities of toluene during pregnancy suffered from microencephaly, CNS dysfunction, attention deficits, and variable growth deficiencies.

Workers who had been exposed to toluene vapors over a long term did not suffer from clinically significant adverse effects on the CNS (Juntunen et al., 1985). However, peripheral nervous system changes are associated with toluene exposure (Cherry et al., 1985).

Grabski (1961) reported the first case in which cerebellar degeneration was associated with chronic toluene inhalation. Subacute reversible ataxia from occupational exposure and irreversible, severe persistent cerebellum and pyramidal syndrome after glue sniffing of toluene were reported by Boor and

Hurtig (1977). Knox and Nelson (1966) reported a case of permanent brain damage which was attributed to chronic toluene inhalation.

Carlton et al. (1989) reported on a patient who displayed persistent neurological symptoms some 8 months after exposure to toluene while painting a truck. This is the first report of neurological sequelae following acute toluene exposure.

Xylene (Dimethylbenzene)

Xylene exists in three isomeric forms: ortho (1,2-dimethylbenzene), meta (1,3-dimethylbenzene), and para (1,4-dimethylbenzene). All three isomers are found in xylenes, thus the toxicity of each isomer must be considered.

Xylene is used in the production of other chemicals including solvents for inks, resins, adhesives, and degreasers and as a fuel component. According to NIOSH about 140,000 U.S. workers are potentially exposed to xylene annually.

Studies in animals have shown that xylene is rapidly distributed to the tissues after absorption with the greatest concentration in the kidney, subcutaneous fat, nerves, liver, lungs, brain, muscle, and spleen (Carlsson, 1981).

Toxicokinetics. Xylene is absorbed rapidly by inhalation, with 60 to 65% of xylene being absorbed. Xylene is absorbed after topical exposure through the skin at a rate of 4.5 to 9.6 $mg/cm^2/h$ (Dutkiewicz and Tyras, 1968). Of the inhaled xylene, 90% is metabolized to methyl benzoic acid (toluic acid) and excreted as methyl hippuric acid. Approximately 5% of xylene absorbed through the lungs is distributed to the body fat in man. Xylene accumulates in the adipose tissue and is slowly excreted. Less than 5% of the xylene is excreted unchanged in the lungs. Xylene has a half-life of 20 to 30 h and binds tightly to proteins.

Toxicology. Xylene is very similar to toluene in its toxicological effects, in that it is a primary skin irritant, and like other organic solvents, it can produce defatting dermatitis upon prolonged exposure. Xylene is an anesthetic at concentrations above 5000 ppm (Carpenter et al., 1975). Xylene also depresses the CNS with symptoms of fatigue, nausea, headache, and ataxia. It appears that the acute toxicity of xylenes on the CNS is greater than either the toxicity of toluene or benzene. Exposure to high concentrations of xylene results in confusion, respiratory depression, and coma (Low et al., 1989).

Exposure of animals to pure xylene has not resulted in toxicity to the hematopoietic system, characteristic of benzene. Sabri et al. (1960) noted bone marrow hyperplasia in rabbits that had been exposed to xylene at a level of 1150 ppm 8 h/d 6 d/week for 55 d; however, there was no tendency toward aplasia. In studies of rats exposed to xylene at a concentration of 860 ppm, no toxic effects on the liver, kidney, or lungs were reported by Carpenter et al.

(1975). Xylene isomers have not been shown to be carcinogenic, mutagenic, or teratogenic.

Fetotoxic effects were reported in rats following maternal inhalation to mixed xylenes. Excellent summaries of studies of animal exposure to xylene can be found in Jori et al. (1986) and Low et al. (1989).

Of three workers who were exposed to xylene vapors at a concentration of 10,000 ppm, one died and two survived (Morley et al., 1970).

Human volunteers who were exposed to *m*-xylene vapors of 100 to 400 ppm for 6 h/d for 5 d showed some changes in psychological performance, that is, impairment of body balance and an increase of reaction times. These effects were considered to be much weaker than those of ethyl alcohol (Riihimaki and Savolainen, 1980).

The simultaneous occupational exposure to xylene and ingestion of ethyl alcohol has received some attention. Savolainen et al. (1978) reported that ethanol ingestion might alter the toxicological risk to xylene and also enhance some of the CNS symptoms of xylene exposure. These findings could be significant because of the frequency of ethanol ingestion by workers who are exposed to xylene in the workplace.

4.   Glycol Ethers (Glycol Alkyl Ethers, Polyalkylene Oxide Ethers)

These chemicals can be considered as a group since they are toxicologically closely related:

- Ethylene glycol-monomethyl ether — methyl cellosolve
- Ethylene glycol-monoethyl ether — cellosolve
- Ethylene glycol-monoisopropyl ether — isopropylcellosolve, IPE
- Ethylene glycol-monobutyl ether — butyl cellosolve

Ethylene glycol ethers are used in hydraulic fluid as solvents in dye resins, in cleaning compounds, and in liquid soaps. They are metabolized to the corresponding alkoxy acetic acid derivative, e.g., ethyl—> ethyoxyacetic acid.

The effects of the ethylene glycol ethers on the CNS are more pronounced in acute exposure. Savolainen (1980) reported hind limb paresis (partial or incomplete paralysis), glial cell damage, and demyelination in animals following severe exposure to ethylene glycol ethers. Chronic exposure causes blurred vision and personality changes in humans resulting in headaches, dysarthria, somnolence, and lethargy (Zavon, 1963). Cohen (1984) reported on the occurrence of macrocytic anemia following exposure to ethylene glycol monomethyl ether (EGME). Aplastic anemia following exposure to ethylene glycol monomethyl ether was reported by Parsons and Parsons (1938).

However, the system most affected by exposure to ethylene glycol monomethyl ether is the reproductive system. Studies in animals have shown dose-related toxic effects on the reproductive system in males and females. It is interesting that these effects are observed at concentrations lower than the permissible limits. Hanley et al. (1984) reported that offspring from pregnant rabbits exposed to 50 ppm had an increased incidence of malformations while young from rats and mice exposed to the same concentration were not affected. However, Toraason et al. (1986) reported that the administration of 100 mg/kg of ethylene glycol monomethyl ether by oral gavage to pregnant rats resulted in resorption of fetuses in all animals tested.

5. Ketones

Ketones are a group of chemicals characterized by the formula

$$R-CO-R$$

Ketones are used as solvents in plastics, paints, and dyes and in the chemical industry as chemical reactants in the synthesis of other chemicals.

Three of these ketones are of particular interest: dimethyl ketone (acetone), methyl ethyl ketone (MEK), and methyl-*n*-butyl ketone (MBK). These ketones are rapidly absorbed from the gastrointestinal tract, lungs, and skin. Acetone is excreted in the urine unchanged. MBK is metabolized to carbon dioxide and 2,5-hexanedione. In guinea pigs, 2-butanol, 3-hydroxy-2-butanone, and 2,3-butanediol have been identified by DiVincenzo et al. (1976) as metabolites of methyl-*n*-butyl ketone.

Methyl Ethyl Ketone (MEK 2-Butanone)

Perbellini et al. (1984) studied the ambient air concentration of methyl ethyl ketone and the excretion of urinary acetylmethylcarbinol, a metabolite of MEK, in industrial workers. Correlations between air MEK concentrations and urinary concentrations of MEK were statistically significant.

The ketones, as a group, cause depression of the CNS when given in high concentrations and can be irritating to the eyes and mucous membrane. MBK is exceptional toxicologically because it is the only ketone used singly that has been definitely associated with peripheral neuropathies (Mendell et al., 1974). Moreover, polyneuropathy was reported in three women who were exposed to MEK and toluene at concentrations below the threshold limit values of these chemicals (Dyro, 1978). The interaction of the chemicals must be considered. This "hexacarbon" neuropathy is attributed to the primary metabolite of MBK which is 2,5-hexanedione, a recognized neurotoxic agent and also a metabolite of *n*-hexane. The effects in peripheral nerves induced by MBK are potentiated by MEK (Saida et al., 1976; Takeuchi et al., 1983). Both MEK and MBK potentiate the neurotoxicity of other hexacarbon containing solvents (Altenkirch et al., 1982; Yang, 1986).

Lethargy, ataxia, headache, nausea, and incoherent speech are observed with mild exposure while stupor and coma are seen with severe exposures to acetone. The toxicity of peroxides of ketones are sometimes confused with that of the parent ketones. The ketone peroxides are irritating and corrosive; therefore they are potentially more toxic than the ketones themselves.

Accidental ingestion of an indeterminate amount of MEK resulted in coma, hyperventilation, metabolic acidosis, and tachycardia. No hepatorenal or long-term after effects were reported (Kopelman and Kalfayan, 1983).

6. Alcohols

Alcohols are rapidly absorbed through the lungs, gastrointestinal tract, and skin and distributed throughout the body tissues and fluids. Evidence exists, however, that ethyl alcohol is absorbed percutaneously only in limited amounts in short-term or single applications. However, symptoms have been observed in children after repeated applications to wide areas of the body. Perhaps the skin is defatted before significant absorption can occur (Browning, 1965).

The enzymes, alcohol and aldehyde dehydrogenases, are involved in the metabolism of ethyl and methyl alcohols. Metabolism occurs primarily in the liver. The rate of metabolism of ethyl alcohol is constant with time and follows zero order kinetics. Data on the involvement of the microsomal mixed function oxidative systems (MFO) are somewhat conflicting. There is some evidence that MFO are activated in the metabolism of ethanol when concentrations are high (Lieber, 1976). It is interesting that induction of the liver MFO activity does not result in an increase in the rate of metabolism of ethanol (Tephly et al., 1969).

Alcohols at sufficiently high doses are recognized depressants of the CNS. Ethyl alcohol is not a significant CNS depressant until a blood level of 30 to 50 mgm% is attained. Inhalation of sufficiently concentrated vapors of ethyl or methyl alcohol can result in significant effects on the CNS. These effects are seen at concentrations as high as 5000 to 10,000 ppm of ethyl alcohol.

Methyl Alcohol

Methyl alcohol is used widely as a solvent in lacquers, paints, varnishes, and in the manufacture of other chemicals. It is absorbed by all routes of administration (Haley, 1987) and is distributed throughout the body proportional to the distribution of body water (Yant and Schrenk, 1937).

Since methyl alcohol has been proposed as a motor fuel, interest in its toxicology, particularly of long-term exposure at low concentrations, has increased. Effects upon the immune system and the possibility of carcinogenicity are of particular interest. Very little data are available on this aspect of exposure to methyl alcohol.

Acute exposure to methyl alcohol results in depression of the CNS. The symptoms include headache, vertigo, severe upper abdominal pain, blurred vision, bradycardia, slow shallow breathing, coma, and death (Harger and Forney, 1967). Visual disturbances, for example, blurring and loss of acuity, are characteristics of methanol poisoning. The effect upon the eyes has been attributed to optic neuritis which regresses. However, if the optic nerve atrophies, permanent blindness will result.

Methyl alcohol is metabolized in the liver by alcohol dehydrogenase to formaldehyde and then to formic acid (Tephly et al., 1979; McMartin et al., 1980). Either formaldehyde or formic acid, more likely formaldehyde, is responsible for damage to the optic nerve which may result in blindness following ingestion of large amounts of methanol (Benton and Calhoun, 1952; Cooper and Kini, 1962).

Ethyl Alcohol

Ethyl alcohol is absorbed through all routes of exposure. It is distributed throughout the body according to body water content and is readily transferred across the blood brain barrier.

Most ethyl alcohol used as a solvent is denatured by the addition of other chemicals,i.e., methyl alcohol to make it unsuitable for human consumption. Ethyl alcohol is used as a solvent in shellacs and varnishes and as a solvent for perfumes, mouthwashes, and liniments.

Ninety percent of the absorbed ethyl alcohol is metabolized to acetaldehyde, then to acetic acid, and then via the glycolytic pathway to carbon dioxide and water. The remainder is eliminated by exhalation or in the urine, saliva, and perspiration.

Its major toxicological effect is on the developing fetus. The placenta is permeable to ethyl alcohol, and it diffuses into the fetal circulation resulting in fetal alcohol syndrome (FAS) (Pratt, 1982). Repeated ingestion of large quantities of ethyl alcohol can result in numerous toxic effects including hepatic toxicity (cirrhosis) and cardiac toxicity.

Headache and irritation of the eyes, nose, and throat occur at exposures above 1000 ppm. Prolonged exposure to low concentrations results in drowsiness, lassitude, and inability to concentrate. Sufficient exposure to ethyl alcohol fumes results in inebriation: talkativeness, alterations in behavior, delayed reflexes and reaction time, loss of muscular coordination, and diplopia. Although industrial exposure to vaporized ethanol is of no practical importance according to Hamilton and Hardy (1983), fatal intoxication has resulted from the effects of inhalation of ethyl alcohol vapors (Ritchie, 1975).

The toxicological interaction of ethyl alcohol with other solvents is of importance. When ethyl alcohol is ingested before exposure to chlorinated hydrocarbons, the hepatotoxicity of carbon tetrachloride, chloroform, trichloroethane, and trichloroethylene is increased (Zimmerman, 1978; Strubelt, 1980). However, ethyl alcohol can be used as a competitive inhibitor to lessen the toxicological effect of methyl alcohol (blindness) and ethylene glycol (kidney damage).

A review of the general toxicology of alcohols is presented in *Alcohols Toxicology* (Wimer et al., 1983).

7. Chlorinated Hydrocarbons

Aromatic chlorinated hydrocarbons are discussed in Chapter 9, but aliphatic chlorinated hydrocarbon solvents will be presented in this section. Although there are many aliphatic chlorinated hydrocarbon chemicals used as solvents, the seven most frequently used are discussed below (Hamilton and Hardy, 1983).

Chloroform ($CHCl_3$, Trichloromethane)

Perhaps the most studied chemical of this group is chloroform which became known not for its properties as a solvent but as an anesthetic agent. Although used widely in medicine, its toxic effects on the liver and heart resulted in disuse. Chloroform is classified by IARC as a suspected carcinogen in animals and by the EPA as a hazardous substance, a hazardous waste, and a priority pollutant.

Chloroform is widely used as a solvent, particularly in lacquers, and in the manufacture of plastics and fluorocarbons (refrigerants). It is estimated by OSHA that some 360,000 workers are potentially exposed to chloroform. Exposure to low levels of chloroform in drinking water is common, due to its formation during the chlorination of organic chemicals in water. This is particularly important to the public as water is consumed and used continuously on a daily basis.

Chloroform is absorbed by inhalation, ingestion, and the dermal route. The first products of chloroform metabolism are carbon dioxide and chlorine. It is of interest that carbon tetrachloride is metabolized to a free radical (CCl_3^-) but chloroform does not form this free radical. Thus, as expected, it is not as hepatotoxic as carbon tetrachloride. However, exposure to chloroform does induce centrizonal hepatic necrosis and steatosis. Renal and myocardial damage has also been noted in other studies (von Oettingen, 1964). Inhalation of chloroform also has resulted in sensitization of the heart to catecholamines which can result in cardiac arrhythmia (von Oettingen, 1937).

Although chloroform is irritating to the skin and eyes, inhalation in very high concentrations (389 ppm) can be tolerated without complaint (Lehmann and Flury, 1943).

Chloroform is somewhat unique in that it is the only halogenated hydrocarbon of low carbon number that is considered as a mildly teratogenic but highly embryotoxic agent (Schwetz et al., 1974; NIOSH, 1985–86).

Chloroform has been studied in the National Cancer Institute (NCI) bioassay program (Weisburger, 1977). In these gavage studies, renal epithelial cell tumors developed in male rats, while in white mice hepatocellular carcinomas were observed. A major difficulty is the extrapolation of these results to inhalation doses of chloroform.

Carbon Tetrachloride (CCl_4, Tetrachloromethane)

Carbon tetrachloride has been used also as an anesthetic and as an anthelmintic agent. It also was used widely as a dry cleaning and degreasing agent and as a solvent for oil, grease, fats, and waxes. Use of carbon tetrachloride has recently been discouraged since its classification as a carcinogen (NIOSH, 1985–86) and as a hazardous substance, hazardous waste, and priority pollutant by the EPA (Settig, 1985). OSHA estimates that 3.4 million workers are exposed to carbon tetrachloride in the workplace.

In recent years, increasing analyses of water supplies have revealed significant concentrations of carbon tetrachloride. It is estimated that 20 million people ingest carbon tetrachloride in contaminated water (Settig, 1985). Emissions of carbon tetrachloride into the atmosphere have resulted in the exposure of another segment of the population.

Acute exposure to high concentrations of carbon tetrachloride results in CNS depression which is exhibited as dizziness, vertigo, headache, mental confusion, and loss of consciousness. Also, some exposed persons show gastrointestinal effects such as nausea, vomiting, abdominal pain, and diarrhea. Liver and kidney damage may also occur with acute exposures, but these effects are usually seen in chronic exposure (Torkelson and Rowe, 1981).

In studies on rats, guinea pigs, rabbits, and monkeys exposed to 400 ppm of carbon tetrachloride for 7 h/d for 5 d/week, histological evaluation revealed central fatty degeneration with cirrhosis of the liver and degeneration of the tubular epithelium of the kidneys. Depression of the CNS was exhibited following severe intoxication. More than half of the animals (rats and guinea pigs) had died after 127 exposures over 173 d (Adams et al., 1952).

Methylene Chloride (CH_2Cl_2, Dichloromethane)

Methylene chloride is used as a solvent for oils, fats, and waxes and is also widely used as an aerosol propellant, paint remover, and degreaser. NIOSH estimates that 70,000 workers are exposed to methylene chloride annually, and it is classified by the EPA as a potential carcinogen, hazardous waste, and priority pollutant (Settig, 1985).

Methylene chloride is considered to be the least toxic of the methene-derived chlorinated compounds (Hamilton and Hardy, 1983). After inhalation, it is absorbed into the circulating blood (rats 55%, humans 35%), but it is not absorbed through the skin in sufficient quantities to result in systemic effects (Stewart and Dodd, 1964).

About 5% of methylene chloride absorbed by inhalation is exhaled unchanged while 25 to 34% is metabolized to carbon monoxide (DiVincenzo and Kaplan, 1981). In fact, carboxyhemoglobin blood levels can reach significant levels of saturation after a severe exposure to methylene chloride.

As with many solvents, methylene chloride is irritating to the skin, eyes, and upper respiratory tract, particularly upon repeated exposure. It produces depression of the CNS with the symptoms produced being directly proportional to its concentration and exposure time. Early symptoms include lightheadedness, nausea, headache, and loss of coordination (Ellenhorn and Barceleaux, 1988). Winneke and Fodor (1976) reported that exposure to methylene chloride in concentrations of 500 ppm resulted in decreased performance and lapses of attention.

Toxic encephalopathy, pulmonary edema, coma, and death resulted following severe exposure to methylene chloride. Although pulmonary edema is a direct effect of methylene chloride, phosgene (CH_3OCl), an extremely toxic gas, is produced upon combustion of methylene chloride. Production of this pulmonary toxicant can be very important in fires in which solvents are involved.

Methylene chloride has not been associated conclusively with liver disease among workers (Ellenhorn and Barceleaux, 1988). As with other solvents, sensitization of the myocardium by methylene chloride can result in cardiac arrhythmias. The degradation of methylene chloride to carbon monoxide and later sensitization of the myocardium suggest that caution of exposure be exercised by cardiac deficient personnel.

1,1,1-Trichloroethane (Methylchloroform)

1,1,1-Trichloroethane must be differentiated from 1,1,2-trichloroethane and 1,1,2,2-tetrachloroethane which are more toxic chemicals (Torkelson et al., 1958). 1,1,1-Trichloroethane is classified as a hazardous waste and priority pollutant by the EPA (Settig, 1985). 1,1,1-Trichloroethane is not a proven carcinogen. However, caution in its use is advised because of the carcinogenicity of structurally related chlorinated hydrocarbons. Solvents such as 1,2-dichloroethane (ethylene dichloride), 1,1,2-trichloroethane, 1,1,2,2-tetrachloroethane, and hexachloroethane (*p*-dichloroethane) are classified as carcinogens by the EPA since they have produced tumors in mice.

1,1,1-Trichloroethane is used primarily as a solvent, degreaser, dry cleaning agent, and propellant. More than 1×10^9 lb of 1,1,1-trichloroethane is produced per year. It is estimated by NIOSH that 2,900,000 workers are exposed to 1,1,1-trichloroethane annually.

1,1,1-Trichloroethane is rapidly absorbed from the lungs and gastrointestinal tract. Absorption of liquid 1,1,1-trichloroethane through the skin has been reported by Stewart and Dodd (1964), while Riihimaki and Pfaffli (1978) observed that no significant toxicological symptoms appeared as the result of dermal exposure to 1,1,1-trichloroethane vapor. Most of the absorbed 1,1,1-trichloroethane is exhaled by the lungs unchanged. Small amounts are metabolized to trichloroacetic acid and trichloroethanol. Accumulation following chronic exposure has not been conclusively demonstrated (Torkelson et al., 1981).

The principal organ affected by 1,1,1-trichloroethane is the CNS with symptoms of depression ranging from headache, disorientation, and drowsiness to convulsions, stupor, coma, and death (Torkelson et al., 1958).

1,1,2,2-Tetrachloroethane (sym-Tetrachloroethane, $Cl_2C=CCl_2$)

1,1,1,2-Tetrachloroethane and 1,1,2,2-tetrachloroethane are positional isomers. Usually tetrachloroethane refers to the 1,1,2,2 isomer. Both isomers are classified as hazardous waste and priority toxic pollutants. However, 1,1,2,2-tetrachloroethane has been classified by the National Cancer Institute as a carcinogen (NCI, 1978). 1,1,1,2-Tetrachloroethane is used as a solvent in several products while 1,1,2,2-tetrachloroethane is used in dry cleaning as a fumigant and in glues and lacquers. It is estimated that some 5000 workers in the U.S. are exposed to 1,1,2,2-tetrachloroethane annually.

Both isomers of tetrachloroethane cause depression of the CNS; however, the toxic effects of the 1,1,2,2 isomer are more defined (Hamilton and Hardy, 1983) and is considered as the most toxic of the chlorinated hydrocarbons used in industry. The toxicity of 1,1,2,2-tetrachloroethane has decreased its use almost to a point of elimination.

The 1,1,2,2 isomer is a potent hepatotoxic agent, 10 to 20 times as toxic as tetrachloroethylene (Parkki, 1986). Damage to the liver is manifest as acute or subacute necrosis and steatosis. Early symptoms of intoxication observed are nausea, drowsiness, irritability, hand tremors, and numbness of the toes. Chronic exposure to the 1,1,2,2 isomer may also result in general malaise, loss of appetite and abdominal discomfort, mental confusion, stupor, and convulsions. Damage to the kidneys from exposure to the 1,1,2,2 isomer is observed in the form of nephritis as is shown by the appearance of albumin and casts in the urine (Sittig, 1985).

Trichloroethylene (Trichloroethene, TCE, $Cl_2C=CCl$)

Trichloroethylene has been widely used as a dry cleaning agent, degreaser, and solvent. It is the primary solvent in the extraction of caffeine from coffee. In fact, trichloroethylene was used in medicine for anesthesia; however, its use as an anesthetic agent has diminished due to the development of more efficacious agents.

Trichloroethylene has been classified as a carcinogen, hazardous substance, hazardous waste, and priority pollutant by the EPA (Sittig, 1985). Trichloroethylene is absorbed primarily by inhalation and ingestion, but absorption through the skin has also been reported (Stewart and Dodd, 1964). Most of the trichloroethylene absorbed is excreted by exhalation, but some is stored in the fat. It is metabolized in the liver to chloral hydrate which in turn is reduced to trichloroethanol. This is later oxidized to trichloroacetic acid, which is excreted in the urine (Waters et al., 1977; Smith, 1966).

Exposure to trichloroethylene vapor can result in irritation of the eyes, nose, and throat. If exposure is severe, the CNS is depressed with symptoms such as headache, dizziness, tremors, nausea and vomiting, lethargy, blurred vision, and intoxication. Damage to the liver and kidneys have been attributed to exposure to trichloroethylene.

Tetrachloroethylene (Perchloroethylene, PCE, $Cl_2C=CCl_2$)

Tetrachloroethylene is widely used in commercial dry cleaning solvents, degreasers, fumigants for grain, and as a veterinary anthelmintic. It is classified as a carcinogen by NCI and as a hazardous waste and priority pollutant by the EPA.

Following inhalation and absorption, tetrachloroethylene is distributed throughout the body and stored in fatty tissues (Stewart et al., 1961). Absorption through the skin may occur to some extent (Stewart and Dodd, 1964). Most of the tetrachloroethylene is exhaled through the lungs, but an appreciable amount is metabolized to trichloroacetic acid (Liebman and Ortiz, 1977).

Exposure to tetrachloroethylene may result in depression of the CNS which is exhibited as lightheadedness and confusion, disorientation, coma, and death. Studies by Hake and Stewart (1977) did not reveal residual neurological or behavioral toxicity upon chronic exposure to tetrachloroethylene (Hamilton and Hardy, 1983). Exposure of skin to tetrachloroethylene has resulted in burns and blisters.

C. SUMMARY

Some scientists believe that the toxicological information concerning solvents has developed too slowly. Perhaps this is true, particularly for some of the solvents. One reason for this is that the technology necessary to analyze low concentrations of solvents has not been available. As a result of the lack of information, the public has adopted the impression that most solvents are virtually harmless.

There is increasing evidence that the chronic, long-term (perhaps a lifetime) intake of all solvents should be studied more closely. Stricter regulations must be based upon increased scientific information, and scientific information must be collected by scientific experimentation. The question of extrapolation of data collected in animals to allowable exposure in man requires additional research data. The elucidation of the mechanism of the neurotoxicity of *n*-hexane and methyl-*n*-butyl ketone is an example of the benefits of research in this area.

In this chapter, the toxicity and mode of action of only a few of the many solvents has been presented.

REFERENCES

Adams, E.M., Spencer, H.C., Rowe, V.K., McCollister, D.C., and Irish, D.D. Vapor toxicity of carbon tetrachloride determined by experiments on laboratory animals, *A.M.A. Arch. Ind. Hyg. Occup. Med.* 6:50,1952.

Aksoy, M. Benzene as a leukemogenic and carcinogenic agent, *Am. J. Ind. Med.* 8:9–20, 1985.

Altenkirch, H., Mager, J., Stoltenburg, G., and Helmbrecht, J. Toxic polyneuropathies after sniffing a glue thinner, *J. Neurol.* 214:137–152, 1977.

Altenkirch, H., Wagner, H.M., Stolenburg-Didinger, G., and Steppart, R. Potentiation of hexacarbon neurotoxicity by methyl-ethyl-ketone (MEK) and other substances: clinical and experimental aspects, *Neurobehav. Toxicol. Teratol.* 4:623–627, 1982.

Astrand, I., Ehiner-Samuel, H., Kibbom, H., and Ovruim, P. Toluene exposure. I. Concentration in alveolar air and blood at rest and during exercise, *Work Environ. Health* 9:119–130, 1972.

Bardodej, Z. and Bardodejova, E. Biotransformation of ethylbenzene, styrene, and alphamethylstyrene in man, *Am. Ind. Hyg. Assoc. J.* 31:206–209, 1970.

Baselt, R.C. *Disposition of Toxic Drugs Chemicals in Man,* 2nd ed., Biomedical Publications, Davis, CA, 1982, 157–159.

Beirne, G.J. and Brennan, J.T. Glomerulonephritis associated with hydrocarbon solvents, *Arch. Environ. Health* 25:365–369, 1972.

Benton, C.D., Jr. and Calhoun F.P., Jr. The ocular effect of methyl alcohol poisoning, *Trans. Am. Acad. Ophthalmol. Laryngol.* 56:875–885, 1952.

Bergman, K. Whole-body-autoradiography and allied tracer techniques in distribution and elimination studies of some organic solvents: benzene, toluene, xylene, styrene, methylene chloride, chloroform, carbon tetrachloride, and trichloroethylene, *Scand. J. Work Environ. Health* 5(Suppl. 1):29–53, 1979.

Bohlen, P., Schlunegger, U.P., and Lauppi, L.S. Uptake and distribution of hexane in rat tissues, *Toxicol. Appl. Pharmacol.* 25:242–249, 1973.

Boor, J.W. and Hurtig, H.I. Persistent cerebellar ataxia after exposure to toluene, *Ann. Neurol.* 2:440–442, 1977.

Brandt, L., Nilsson, P.G., and Mitelman, F. Non-industrial exposure to benzene as leukaemogenic risk factor, [Letter] *Lancet* 2(8047):1074, 1977.

Browning, E. *Toxicity and Metabolism of Industrial Solvents,* Elseiver Publishing Co., New York, 1965, 91–93.

Brugnone, F., DeRosa, E., Perbellini, L., and Bartolucci, G.B. Toluene concentrations in the blood and alveolar air of workers during the workshift and the morning after, *Br. J. Ind. Med.* 43):56–61, 1986.

Carlsson, A. Distribution and elimination of C^{14}-xylene in rat, *Scand. J. Work Environ. Health.* 7:51–55, 1981.

Carlton, F.B., Siger, D., Welch, L.W., and Heath, J.J. Chronic neurological sequelae following acute toluene exposure, *Vet. Human Toxicol.* 31:353, 1989.

Carpenter, C.P., Kinkead, E.R., Geary, D.L. Jr., Sullivan, L.J., and King, J.M. Petroleum hydrocarbon toxicity studies. V. Animal and human response to vapors of mixed xylenes, *Toxicol. Appl. Pharmacol.* 33:543–558, 1975.

Cherry, N., Hutchins, H., Pace, T., and Waldron, H.A. Neurobehavioral effects of repeated occupational exposure to toluene and paint solvents, *Br. J. Indust. Med.* 42:291–300, 1985.

Cohen, R. Reversible subacute ethylene glycol monomethyl ether toxicity associated with microfilm production: a case report, *Am. J. Ind. Med.* 6:441–446, 1984.

Cohr, K.H. and Stokholm, J. Toluene: a toxicologic review, *Scand. J. Work Environ. Health* 5:71–90, 1979.

Cooper, J.R. and Kini, M.M. Biochemical aspects of methanol poisoning, *Biochem. Pharmacol.* 11:405–416, 1962.

Couri, D. and Milks, M. Toxicity and metabolism of the neurotoxic hexacarbons *n*-hexane, 2-hexanone and 2,5-hexanedione, *Ann. Rev. Pharmacol. Toxicol.* 22:145–166, 1982.

DiVincenzo, G.D., Kaplan, C.J., and Dedinas, J. Characterization of the metabolites of methyl *n*-butyl ketone, methyl iso-butyl ketone, and methyl ethyl ketone in guinea pig serum and their clearance, *Toxicol. Appl. Pharmacol.* 36:511–522, 1976.

DiVincenzo, G.D. and Kaplan, C.J. Uptake, metabolism, and elimination of methylene chloride vapors by humans, *Toxicol. Appl. Pharmacol.* 59:130–140, 1981.

Donner, J., Maki-Paakkanen, J., Norppa, H., Sorsa, M., and Vainio, H. Genetic toxicology of xylenes, *Mutat. Res.* 74:171–172, 1980.

Drew, R.J. and Fouts, J.R. The lack of effects of pretreatment with phenobarbital and chlorpromazine on the acute toxicity of benzene in rats, *Toxicol. Appl. Pharmacol.* 27:183–189, 1974.

Dutkiewicz, T. and Tyras, H. Study of the skin absorption of ethylbenzene in man, *Br. J. Ind. Med.* 24:330–332, 1967.

Dutkiewicz, T. and Tyras, H. Skin absorption of toluene, styrene, and xylene by man, *Br. J. Ind. Med.* 25:243, 1968.

Dyro, F.M. Methyl ethyl ketone polyneuropathy in shoe factory workers, *Clin. Toxicol.* 13:371–376, 1978.

Ellenhorn, M.J. and Barceleaux, D.G. *Medical Toxicology and Treatment of Human Poisoning.* Elsevier Science Publishing Company, Inc., New York, 1988, 983–985.

El Masry, A.M., Smith, J.N., and Williams, R.T. The metabolism of alkylbenzenes: *n*-propylbenzene and *n*-butylbenzene with further observations on ethylbenzene, *Biochem. J.* 64:50–56, 1956.

Engstrom, J. and Bjurstrom, R. Exposure to xylene and ethylbenzene. II. Concentration in subcutaneous adipose tissue, *Scand. J. Work Environ. Health* 4:195–203, 1978.

Fishbein, L. An overview of environmental and toxicological aspects of aromatic hydrocarbons. II. Toluene, *Sci. Total Environ.* 42:267–288, 1985a.

Fishbein, L. An overview of environmental and toxicological aspects of aromatic hydrocarbons IV. Ethylbenzene, *Sci. Total Environ.* 44:269–287, 1985b.

Gerarde, H.W. *Toxicology and Biochemistry of Aromatic Hydrocarbons.* Elsevier Publishing Co., New York, 1960, 148–149, 171–180.

Gerarde, H.W. Toxicological studies on aspiration hazard and toxicity of hydrocarbons and hydrocarbon mixtures, *Arch. Environ. Health (Chicago)* 6:329–341, 1963.

Gibson, J.E. and Hardisty, J.F. Chronic toxicity and oncogenicity bioassay of inhaled toluene in Fischer-344 rats, *Fundam. Appl. Toxicol.* 3:315–319, 1983.

Goldstein, B.D. Benzene toxicity: a critical evaluation: hematoxicity in humans, *J. Toxicol. Environ. Health* 2(Suppl.):69–105, 1977.

Goto, I., Matsumra, I., Inoue, N., Murai, Y., Shida, K., Santa, T., and Kuroiwa, Y. Toxic polyneuropathy due to glue sniffing, *J. Neurol. Neurosurg. Psychiat.* 37:848–873, 1974.

Grabski, D.A. Toluene sniffing producing cerebellar degeneration, *Am. J. Psychiat.* 118:461–462, 1961.

Hake, C.L. and Stewart, R.D. Human exposure to tetrachloroethylene: inhalation and skin contact, *Environ. Health Perspec.* 21:231–238, 1977.

Haley, T.J. *Toxicology.* T.J. Haley and W.O. Berndt, Eds. Hemisphere Publishing Corporation, Washington, DC, 1987, 515.

Hamilton, A. and Hardy, H.L. *Industrial Toxicology,* 4th ed. A.J. Finkel, Ed. John Wright Publishers, Littleton, MA, 1983, 271–276.

Hanley, T.R., Yano, B.L., Nitschke, K.D., and John, J.A. Comparison of the teratogenic potential of inhaled ethylene glycol monomethyl ether in rats, mice, and rabbits, *Toxicol. Appl. Pharmacol.* 75:409–422, 1984.

Hansbrough, J.F., Zapata-Sirvent, R., Dominic, W., Sullivan, J., Boswick, J. and Wang, X.W. Hydrocarbon contact injuries, *J. Trauma* 2:250–252, 1985.

Harger, R.N. and Forney, R.B. Aliphatic alcohols [review], *Prog. Chem. Toxicol.* 3:1–61, 1967.

Hersh, J.H., Podruk, P.E., Rogers, G., and Weisskopf, B. Toluene embryopathy, *J. Pediatr.* 106:922–927, 1985.

Ishii, N., Herskowitz, A., and Schaumburg, H. *n*-Hexane polyneuropathy: a clinical and experimental study, *J. Neuropathol. Exp. Neurol.* 31:198–199, 1972.

Jori, A., Calamari, D., DiDomenico, A., Galli, C.L., Galli, E., Marinovich, M., and Silano, V. Ecotoxicological profile of xylenes, *Ecotoxicol. and Environ.* 11:44–80, 1986.

Juntunen, J., Matikainer, E., Antti-Poika, J., Suoranta, J., and Valle, M. Nervous system effects of long term occupational exposure to toluene, *Acta Neurol. Scand.* 72:512–517, 1985.

Knox, J.W. and Nelson, J.R. Permanent encephalopathy from toluene inhalation, *N. Eng. J. Med.* 275:1494–1496, 1966.

Kopelman, P.G. and Kalfayan, P.Y. Severe metabolic acidosis after ingestion of butanone, *Br. Med. J.* 286(6358):21–22, 1983.

Krantz, J.C., Jr., Carr, C.J., and Vitcha, J.F. Anesthesia: study of cyclic and non-cyclic hydrocarbons on cardiac automaticity, *J. Pharm. Exper. Therap.* 94:315–318, 1948.

Kuna, R.A. and Kapp, R.W. The embryotoxic/teratogenic potential of benzene vapor in rats, *Toxicol. Appl. Pharmacol.* 57:1–7, 1981.

Lehmann, H.B. and Flury, F. *Toxicology and Hygiene of Industrial Solvents.* Williams and Wilkins Publishers Baltimore, MD, 1943, 138–145, 191–196.

Lieber, C.S. The metabolism of alcohol, *Sci. Am.* 234:25–33, 1976.

Liebman, K.C. and Ortiz, E. Metabolism of halogenated ethylenes, *Environ. Hlth. Persp.* 21:91–97, 1977.

Longacre, S.L., Kocsis, J.J., and Snyder, R. Influence of strain differences in mice on the metabolism and toxicity of benzene, *Toxicol. Appl. Pharmacol.* 60(3):398–409, 1981.

Low, L.K., Meeks, J.R., and Mackerer, C.R. Health effects of the alkylbenzenes. I. Toluene, *Toxicol. Ind. Health* 4(1):49–75, 1988.

Low, L.K., Meeks, J.R., and Mackerer, C.R. Health effects of the alkylbenzenes. II. Xylenes, *Toxicol. Ind. Health* 5(1):85–105, 1989.

McMartin, K.E., Ambre, J.J., and Tephly, T.R. Methanol poisoning in human subjects — role for formic acid accumulation in the metabolic acidosis, *Am. J. Med.* 68:414–418, 1980.

Mendell, J.R., Saida, K., Ganansia, M.F., Jackson, D.B., Weiss, H., Gardier, R.W., Chrisman, C., Allen, N., Couri, D., O'Neill, J., Marks, B., and Hetland, L. Toxic polyneuropathy produced by methyl *n*-butyl ketone, *Science* 185:787–789, 1974.

Morley, R., Eccleston, D.W., Douglas, C.P., Greville, W.E., Scott, D.J., and Anderson, J. Xylene poisoning: a report on one fatal case and two cases of recovery after prolonged unconsciousness, *Br. Med. J.* 3:442–443, 1970.

National Cancer Institute. Bioassay of 1,1,1,2-tetrachloroethane for possible carcinogenicity. DHEW Publication No. (NIH) 78–827. National Institutes of Health, Bethesda, MD, 1978.

National Institute for Occupational Safety and Health (NIOSH). *Registry of Toxic Effects of Chemical Substances,* Vol. 2. D. Sweet, Ed. National Institute for Occupational Safety and Health, Washington, DC, 1985–1986, 1548.

Parkinson, G.S. Benzene in motor gasoline — an investigation into possible normal transport operations, *Ann. Occup. Hyg.* 14:145–153, 1976.

Parkki, M.G. Biotransformation reactions and active metabolites, in *Safety and Health Aspects of Organic Solvents: Progress in Clinical and Biological Research,* Vol. 220. V. Riihimaki and U. Ulfvarson, Eds. Alan R. Liss, Inc., New York, 1986, 89–96.

Parsons, C.E. and Parsons, M.E.M. Toxic encephalopathy and "granulopenic anemia" due to volatile solvents in industry: report of two cases, *J. Ind. Hyg. Toxicol.* 20:124–133, 1938.

Paulson, G.W. and Waylonis, G.W. Polyneuropathy due to *n*-hexane, *Arch. Intern. Med.* 136:880–882, 1976.

Perbellini, L., Brugnone, F., and Faggionato, G. Urinary excretion of the metabolites of *n*-hexane and its isomers during occupational exposure, *Br. J. Indust. Med.* 38:20–26, 1981.

Perbellini, L., Brugnone, F., Mozzo, P., Cocheo, V., and Caretta, D. Methyl ethyl ketone exposure in industrial workers, *Int. Arch. Occup. Environ. Health* 54:73–81, 1984.

Perbellini, L., Mozzo, P., Brugnone, F., and Zedde, A. Physiologicomathematical model for studying human exposure to organic solvents: kinetics of blood/tissue *n*-hexane concentrations and 2,5-hexanedione in urine, *Br. J. Ind. Med.* 43:760–768, 1986.

Picciano, D.J. Cytogenetic studies of workers exposed to benzene, *Environ. Res.* 19:33–38, 1979.

Pratt, O.E. Alcohol and the developing fetus, *Br. Med. Bull.* 38:48–52, 1982.

Proctor, N.H. and Hughes, J.P. *Chemical Hazards of the Workplace.* J. B. Lippincott Co., Philadelphia, PA, 1978, 282–283.

Reinhardt, C.F., Azar, A., Maxfield, M., Smith, P., and Mullin, L. Cardiac arrhythmias and aerosol "sniffing", *Arch. Environ. Health* 22:265–279, 1971.

Rickert, D.E., Baker, T.S., Bus, J.S., Barrow, C.S., and Irons, R.D. Benzene disposition in the rat after exposition by inhalation, *Toxicol. Appl. Pharmacol.* 49:417–423,1979.

Riihimaki, V. and Savolainen, K. Human exposure to m-xylene. Kinetics and acute effects on the central nervous system, *Ann. Occup. Hyg.* 23:411–422, 1980.

Riihimaki, V. and Pfaffli, P. Percutaneous absorption of solvent vapors in man, *Scand. J. Work Environ. Health* 4:73 (Biol. Abst. HEBP, 79:00239), 1978.

Ritchie, J.M. The aliphatic alcohols, in *The Pharmacological Basis of Therapeutics,* 5th ed. L.S. Goodman and A. Gilman, Eds. Macmillan Publishing Co., Inc. New York, 1975, 137–151.

Ruff, R.L., Petito, C.K., and Acheson, L.S. Neuropathy associated with chronic low level exposure to *n*-hexane, *Clin. Toxicol.* 18:515–519, 1981.

Sabri, M.I., Moore, C.L., and Spencer, P.S. Studies on the biochemical basis of distal axonopathies: I. Inhibition of glycolysis produced by neurotoxic hexacarbon compounds, *J. Neurochem.* 32:683–689, 1979.

Sabri, R., Truhaut, R., and Laham, S. Researches toxicologiques sur les solvants de remplacement de benzene. Etude des xylenes, *Arch Malad. Profess. Med.* 21:301, 1960.

Saida, K., Mendell, J.R., and Weiss, H.S. Peripheral nerve changes induced by methyl *n*-butyl ketone and potentiation by methyl ethyl ketone, *J. Neuropath. Exp. Neurol.* 35:207–225, 1976.

Sandmeyer, E.E. Aliphatic hydrocarbons, in *Patty's Industrial Hygiene and Toxicology,* 3rd ed. G.D. Clayton and F.E. Clayton, Eds. John Wiley & Sons, New York, 1981, 3175–3252.

Savolainen, H. Glial cell toxicity of ethyleneglycol monomethylether vapor, *Environ. Res.* 22:423–430, 1980.

Savolainen, H., Vainio, H., Helojoki, M., and Elovaara, E. Biochemical and toxicological effects of short term intermittent xylene inhalation exposure and combined ethanol intake, *Arch. Toxicol.* 41:195–205, 1978.

Schaumburg, H.H. and Spencer, P.S. Degeneration in central and peripheral nervous systems produced by pure *n*-hexane: an experimental study, *Brain* 99:187–192, 1976.

Schmidt, R., Schnoy, N., Altenkirch, H., and Wagner, H.M. Ultrastructural alteration of intrapulmonary nerves after exposure to organic solvents, *Respiration* 46:362–369, 1984.

Schwetz, B.A., Leong, B.K.L., and Gehring, P.J. Embryo- and fetotoxicity of inhaled chloroform in rats, *Toxicol. Appl. Pharmacol.* 28:442–451, 1974.

Sherwood, R.J. Evaluation of exposure to benzene vapor during the loading of petrol, *Br. J. Ind. Med.* 29:65–69, 1972.

Sittig, M. *Handbook of Toxic and Hazardous Chemicals and Carcinogens,* 2nd ed., Noyes Publications, Park Ridge, NJ, 1985, 194–885.

Smith, G.F. Trichlorethylene: a review, *Br. J. Ind. Med.* 23:249–262, 1966.

Smyth, H.F., Jr., Carpenter, C.P., Weil, C.S., Pozzani, U.C., and Striegel, J.A. Range-finding toxicity data: list VI. *Am. Ind. Hyg. Assoc. J.* 23:95–107, 1962.

Snyder, R. and Kocsis, J.J. Current concepts of chronic benzene toxicity, in *CRC Critical Reviews in Toxicology,* CRC Press, Inc., Boca Raton, FL, 1975, 265–288.

Soderlund, S. Exertion adds to solvent inhalation dangers, *Health Safety* 55:42–43, 1975.

Spencer, P.S. and Schaumburg, H.H. *n*-Hexane and methyl *n*-butyl ketone, in *Experimental and Clinical Neurotoxicology.* P.S. Spencer and H.H. Schaumburg, Eds. The Williams and Wilkins Company, Baltimore, 1980, 456–475.

Spencer, P.S., Sabri, M.I., Schaumburg, H.H., and Moore, C. Does a defect in energy metabolism in the nerve fiber cause axonal degeneration in polyneuropathies?, *Ann. Neurol.* 5:501–507, 1979.

Sterman, A.B. and Schaumburg, H.H. Neurotoxicity of selected drugs, in *Experimental and Clinical Neurotoxicology.* P.S. Spencer and H.H. Schaumburg, Eds. The Williams and Wilkins Company, Baltimore, 1980, 593–612.

Stewart, R.D. and Dodd, H.C. Absorption of carbon tetrachloride, trichloroethylene, tetrachloroethylene, methylene chloride and 1,1,1-trichloroethane through the human skin, *Am. Ind. Hyg. Assoc. J.* 25:439–446, 1964.

Stewart, R.D., Gay, H.H., Erley, D.S., Arbor, A., Hake, C.L., and Schaffer, A.W. Human exposure to tetrachloroethylene vapor, *Arch. Environ. Hlth.* 2:516–522, 1961.

Stoughton, R.W. and Lamson, P.D. Relative anesthetic activity of butanes and pentanes, *J. Pharmacol. Exp. Ther.* 58:74–77, 1936.

Strubelt, O. Interactions between ethanol and other hepatotoxic agents, *Biochem. Pharmacol.* 29:1445–1449, 1980.

Susten, A., Dames, B., Burg, J., and Niemeir, R.W. Percutaneous penetration of benzene in hairless mice: an estimate of dermal absorption during tire-building operations, *Am. J. Ind. Med.* 7:323–335, 1985.

Taher, S. M., Anderson, R. J., McCartney, R., Popovtzer, M., and Schrier, R. Renal tubular acidosis associated with toluene sniffing. *N. Eng. J. Med.* 290:765–768, 1974.

Takeuchi, Y., Ono, Y., Hisanaga, N., Iwata, M., Aoyama, M., Kitoh, J., and Suriura, Y. An experimental study of the combined effects of *n*-hexane and methyl ethyl ketone, *Br. J. Ind. Med.* 40:199–203, 1983.

Tephly, T.R., Makar, A.B., McMartin, K.E., Hayreh, S.S., and Martim-Amat, G. Methanol — its metabolism and toxicity, *Biochem. Pharmacol. Methanol* 1:145–164, 1979.

Tephly, T.R., Tinelly, T., and Watkins, W.D. Alcohol metabolism: role of microsomal oxidation *in vivo, Science* 166:627–628, 1969.

Toftgard, R. and Gustafsson, J.A. Biotransformation of organic solvents. A review, *Scand. J. Work Environ. Health* 6:1–18, 1980.

Toraason, M., Niemeier, R.W., and Hardin, B.D. Calcium homeostasis in pregnant rats treated with ethylene glycol monomethyl ether (EGME), *Toxicol. Appl. Pharmacol.* 86:197–203, 1986.

Torkelson T.R. and Rowe V.K. Halogenated aliphatic hydrocarbons, in *Patty's Industrial Hygiene and Toxicology,* 3rd ed. G.D. Clayton and F.E. Clayton, Eds. John Wiley & Sons, New York, 1981, 3433–3599.

Torkelson, T.R., Oyen, F., McCollister, D.D., and Rowe, V.K. Toxicity of 1,1,1-trichloroethane as determined on laboratory animals and human subjects, *Am. Ind. Hyg. Assoc. J.* 19:353–362, 1958.

Vigliani, E.C. and Saita, G. Benzene and leukemia, *N. Eng. J. Med.* 271:872–876, 1964.

von Oettingen, W.F. The halogenated hydrocarbons: their toxicity and potential dangers, *J. Ind. Hyg. Toxicol.* 19:349–448, 1937.

von Oettingen, W.F. *The Halogenated Hydrocarbons of Industrial and Toxicological Importance.* Elsevier Science Publishing Company, Amsterdam, 1964, 522.

Walsh, W.A., Scarpa, F.J., Brown, R.S., Ashcraft, K.W., Green, V.A., Holder, T.M., and Amoury, R.A. Gasoline immersion burn, *N. Eng. J. Med.* 291:830, 1974.

Waters, E.M., Gerstner, H.B., and Huff, J.E. Trichloroethylene. I. An overview, *J. Toxicol. Environ. Health* 2:671–701, 1977.

Weisburger, E.K. Carcinogenicity studies on halogenated hydrocarbons, *Environ. Health. Perspect.* 21:7–16, 1977.

Wimer, W.W., Russell, J.A., and Kaplan, H.L. *Alcohols Toxicology.* Noyes Data Corp. Park Ridge, NJ, 1983, 27–45.

Winneke, G. and Fodor, G.G. Dichloromethane produces narcotic effect, *Occup. Health Safety* 34:49, 1976.

Yang, R.S.H. The toxicology of methyl ethyl ketone, *Residue Rev.* 97:131–143, 1986.

Yant, W.P. and Schrenk, H.H. Distribution of methanol in dogs after inhalation and administration by stomach tube and subcutaneously, *J. Ind. Hyg. Toxicol.* 19:337–345, 1937.

Zavon, M.R. Methyl cellosolve intoxication, *Am. Ind. Hyg. Assoc. J.* 24:36–41, 1963.

Zimmerman, H.J. *Hepatotoxicity: The Adverse Effects of Drugs and Other Chemicals on the Liver.* Appleton-Century-Crofts, New York, 1978, 349–369.

Chapter

7

Toxicity of Pesticides

Janice E. Chambers

A. INTRODUCTION

The concern generated about environmental toxicology arose more over pesticide-related issues than over any other group of toxicants. The landmark event triggering the environmental movement is undoubtedly the publication in 1962 of *Silent Spring*, a book in which the biologist Rachel Carson described many examples of environmental abuses, most of which involved pesticides (Carson, 1962). Although the book contained some inaccuracies and unwarranted conclusions, it raised the consciousness of many citizens, scientists, and politicians to the issues central to the environmental movement. As a result new regulations and agricultural strategies were developed that changed pesticide use patterns, thus averting the "silent spring" predicted by Carson to result from the pesticide-induced demise of the song birds. Nevertheless, pesticides remain a concern in the environment and this chapter will describe not only the former issues but also the more current concerns of environmental quality.

It should not be surprising that pesticides are potentially harmful. They are designed specifically to kill ("-cide") unwanted species. Considering the commonality present throughout biology in physiological and biochemical processes, it is logical that these same biological mechanisms can be affected in nontarget as well as target organisms. Therefore, toxicity resulting from the primary mode of action of pesticides is a very real possibility. However, other effects, such as reproductive effects or carcinogenicity, may also be manifested, and are more likely to occur as the result of low dose, long-term exposure. The acute and chronic toxicity of exposures to many pesticides from a human and laboratory animal perspective have been described in detail by Murphy (1986). Either short-term or long-term exposure to pesticides can harm an ecosystem by adversely affecting the most sensitive population, which then indirectly affects populations either dependent upon it or competing with it. This would result in population losses or shifts.

B. AGRICULTURAL DEMOGRAPHICS AND PESTICIDE USAGE STATISTICS

Despite their documented or potential harm, pesticides have yielded an enormous economic advantage and have been an important factor in allowing the demographic shifts in human populations seen this century. These economic advantages must be borne in mind as the hazards and concerns about pesticides are discussed.

Before the dawn of the 19th century, about 90% of the American population lived on farms and was involved directly with growing food and fiber crops (Kohn, 1987). By the Civil War, the urban and rural populations were equal in size. However, by the 1960s, only about 5% of the American population was directly involved in supplying food for the U.S. as well as other countries. Currently about 3 to 4% of the U.S. population supplies the food and fiber. A major reason for the demographic shift in the farming population was the development of synthetic organic pesticides and the resulting enhancement in crop success. Naturally, the contribution of fertilizers, healthier and more productive crop varieties, and the development of farm machinery cannot be discounted in the success of modern agriculture. Nonetheless, pesticides are now estimated to increase crop yields by 1.3- to 12-fold by decreasing crop losses by 17 to 78%.

With all these advances, crop losses are still highly significant and are now estimated to be about 37%, with a 13% loss attributed to insect damage, 12% to plant pathogens, and 12% to weeds (Pimentel, 1987). All these types of damage can be overcome with pesticides. Although many lay people equate the terms pesticide with insecticide, this is incorrect, since herbicides and fungicides which are used to a large extent and molluscicides, nematocides, and rodenticides to a smaller extent, are also pesticides. Since insecticides have been responsible for or have been blamed for most of the environmental problems involving pesticides, the emphasis in this chapter will be on insecticides. Pesticides are also important in public health for mosquito, flea, and vermin control and in domestic/urban applications in cockroach and ant control; however, these latter applications are of less relevance than agricultural usage from the standpoint of environmental toxicology. Despite the altered philosophy of pesticide usage as a result of the environmental movement of the *Silent Spring* era, pesticide usage has continued to increase since the introduction of the synthetic organic insecticides, and in fact has increased by about 2.3 times since 1970 to maintain desired agricultural productivity. These increases have occurred although some of the newer pesticides are effective at lower doses than insecticides used earlier. Contrast, for example, the use of the synthetic pyrethroids at application rates of 0.01 to 0.1 lb/acre to 1,1,1-trichloro-2,2-bis(4-chlorophenyl)ethane (DDT, chlorophenothane) at 1 to 5 lb/acre (Kohn, 1987). Some of the newer herbicides are also effective at application rates of oz/acre concentrations. Besides increases in world population and a greater demand for food and fiber production, an increase in pesticide usage has also been required because of the development of resistance in many pest species. Reports document the resistance of over 420 species of insects and mites to various pesticides, and even a few weeds have developed resistance. Despite integrated pest management strategies, pesticide usage is still high. Current pesticide usage worldwide is estimated at almost 1.5×10^9 lb annually, with about 600 basic pesticides in about 50,000 formulated products (Wilkinson, 1987). Thus, pesticides not only have a history of environmental problems, but they are a current concern and will continue to be so in the future.

Although large amounts of pesticides (ton/year) are produced in manufacturing plants, the effluent from these plants is normally not the major source of contamination. One serious exception was the severe contamination of the James River in Virginia by chlordecone (Kepone®), which resulted in major contamination of aquatic organisms (Sheets, 1980). The disregard for rational handling and safety procedures of Kepone® also resulted in serious human health effects, mainly reproductive, in the plant workers (Guthrie, 1980). Another river that has been severely contaminated with pesticides is the Rhine River in Europe (Rand and Barthalmus, 1980).

C. CLASSIFICATION OF PESTICIDES

Chemicals have been used for pest control since the ancient Greek, Roman, and Chinese civilizations about 3 millennia ago when the ability of sulfur dust to control insects was known by philosophers and other learned people (Freed, 1987). Other inorganic chemicals such as saltwater, lead arsenate, and copper salts were also used over the centuries. However, the production of the synthetic organic pesticides in the 20th century has allowed the evolution of a hitherto unseen degree of pesticidal efficacy. Although some plants themselves were known to contain organic chemicals with insecticidal properties such as nicotine in tobacco and pyrethrum in *Chrysanthemum* spp., DDT, as the first highly successful synthetic organic insecticide, heralded a new era of insect control (Kohn, 1987). Although DDT was synthesized in 1874, its insecticidal properties were not reported until 1939.

Paul Mueller received the Nobel Prize in medicine in 1948 for this momentous discovery.

1. Organochlorine Insecticides

DDT is one of a very large group of compounds belonging to the class of organochlorine insecticides that are characterized by high lipophilicity and chemical stability. Metabolic degradation in target and nontarget organisms or environmentally by either chemical, photolytic, or microbial processes is slow. As a result, some organochlorine insecticides persist in the environment for more than 30 years (Pimentel, 1987). On the other hand, many organochlorine insecticides have lower acute mammalian toxicity than many of the new generation insecticides which succeeded them.

Two major groups of organochlorine insecticides exist. The first group is the DDT-type, which includes compounds such as 1,1-dichloro-2,2-bis(4-chlorophenyl)-ethane (DDD, TDE, rhothane), methoxychlor, and dicofol (Kelthane). Although effective as insecticides or acaricides, they have relatively low acute mammalian toxicity. They have been widely used in both agricultural and public health practices. The lethal mechanism of action of these neurotoxic compounds is a persistent opening of the sodium channels in neurons, resulting in repetitive firing of action potentials. Due to its efficacy, its relative low mammalian toxicity, and that it was developed and used for more than two decades before the environmental movement, DDT has been subject to use and abuse. Despite the widespread ecosystem contamination resulting from the heavy use of DDT, it enhanced survival for many soldiers who were exposed to insect vectors of malaria and yellow fever in tropical countries during World War II.

The second major group of organochlorine insecticides is the chlorinated cyclodiene compounds which include aldrin, dieldrin, endrin, heptachlor, and chlordane. The acute toxicity of the cyclodienes vary widely. Their mechanism of action appears to be an inhibition of the inhibitory γ-aminobutyric acid (GABA) receptor-chloride ionophore complex with later hyperactive excitatory responses.

Several other important organochlorine insecticides have been widely used. These include lindane, toxaphene, mirex, and chlordecone (kepone). These do not have the characteristic structure of either of the above two groups and their mode of acute toxicity is less precisely defined, although the same mechanism of action as the cyclodienes has been suggested. These compounds display low acute toxicity.

Some organochlorine insecticides such as the o,p'-isomer of DDT, methoxychlor, and chlordecone have been shown to cause estrogenic effects by binding to estrogen receptors following chronic exposure. Reproductive toxicity has also been reported in birds, rodents, and possibly humans. Carcinogenicity data has also been generated for some of the organochlorine insecticides, but much of the data is unconvincing. If any of these compounds are truly carcinogenic, they are weakly so and act by epigenetic and not genotoxic mechanisms. None of these chronic effects have been well studied in environmentally relevant species. However, these specific chronic effects are of concern in environmental toxicology because of the potential for bioaccumulation and environmental persistence of these chlorinated insecticides. The half-life of DDT in the environment is at least 10 years and one of its metabolites, DDE, persists for decades, as do lindane and heptachlor (Duffus, 1980). At least 2.5 years are required for aldrin and dieldrin to degrade to 95% of their original concentration. Even though DDT was banned in 1972 and virtually all uses of the other organochlorine insecticides have been halted in the U.S., other countries still permit their use, so they continue to accumulate within the environment. However, even within the U.S., hot spots of organochlorine residues exist, either as the result of historical heavy usage in the area or from the use of illegally imported organochlorine insecticides (Hall, 1987). The environmental burden of DDT has been estimated at 1×10^9 lb (Kohn, 1987).

2. Organophosphate and Carbamate Insecticides

Due to the concern raised by Rachel Carson and her contemporaries over the persistence of organochlorine insecticides, more labile insecticides have been developed in recent years. The organophosphorus (OP) and the carbamate insecticides are considerably less stable. The OP insecticides which are comprised of esters of phosphoric and phosphorothioic acid evolved from the chemical technology in which the nerve gases of World War II were developed. Both groups of compounds are neurotoxic, poisoning by inhibiting the enzyme acetylcholinesterase. By inhibiting cholinesterases, the accumulation of the important and widespread neurotransmitter acetylcholine occurs, resulting in later hyperstimulation of cholinergic pathways. Typically, death in mammals is the result of respiratory failure following a series of signs or symptoms involving effects much like those produced by the autonomic nervous system neurotoxins, muscarine and nicotine. Although the range of acute mammalian toxicity levels is wide in these two groups, many of the compounds are highly toxic and pose a threat of accidental poisoning to human and animal populations. Since they are less lipophilic and less persistent than the organochlorine compounds, there is a trade-off in the development of these insecticides, and there is now a greater threat of acute toxicity than was previously present with such compounds as DDT. This is true not only for wildlife but also to humans who come in contact with the insecticides. Those most at risk are pesticide applicators and others directly involved in their manufacture, formulation, and use. There has been an increase in the frequency of agricultural worker hospitalizations from occupational exposures as the organophosphates replaced the organochlorine insecticides in the marketplace (Guthrie, 1980). Currently it is estimated that about 3000 workers are hospitalized in the U.S. annually from pesticide exposure, with the organophosphates being responsible for most of these poisonings.

3. Pyrethroids

A more recently developed group of insecticides are the synthetic pyrethroids, which chemically resemble pyrethrum. Their primary mechanism of action is, similar to DDT, opening of neuronal sodium channels resulting in hyperactivity of nervous pathways. Because pyrethroids are readily detoxified both by enzymatic actions and environmentally, they are not persistent and have not posed severe environmental contamination problems. However, some invertebrates such as crustacea and also some fish species are very sensitive to the lethal effects of the synthetic pyrethroids.

4. Biological Insecticides

Rational synthesis has allowed the development of insecticides which attack specific arthropod targets. Such compounds as diflubenzuron (a chitin synthesis inhibitor) and the insect growth regulators have resulted in the development of insecticides with a greater selective toxicity than was previously possible with the anticholinesterase or sodium channel toxicants which interfere with similar basic mechanisms in both target and nontarget organisms. However these specific inhibitors do represent a potential hazard for nonpest arthropods. Due to their selective toxicity, the fungicides and herbicides are typically of minor concern from the standpoint of environmental toxicology. Finally, four factors are the main determinants of the pesticide's potential to cause adverse effects: inherent toxicity, stability, solubility, and absorptivity (Sheets, 1980).

It is also worth noting that in a few cases pesticide contaminants in formulations are of greater concern than the pesticides themselves. Some of these impurities are highly toxic, and therefore can exert a potential hazard even though they are present in only minor concentrations. Two well-known examples are 2,3,7,8-tetrachlorodibenzo-*p*-dioxin (TCDD, dioxin) in the herbicide 2,4,5-T (Silvex) and the rearrangement/hydrolysis products in malathion.

D. PESTICIDE REGULATION BY THE U.S. GOVERNMENT

Laws about pesticide use were publicized as early as 1910 with the passage of the Insecticide Act (Moore, 1987). However, this was designed primarily to protect consumers from fraudulent product claims. Further regulation became necessary after the advent and expanded use of the synthetic organic insecticides in the 1940s. The Federal Insecticide, Fungicide and Rodenticide Act (FIFRA) was passed in 1947, and in 1952 tolerances of pesticide residues in food, feed, and fiber were set by amendments to the Federal Food, Drug and Cosmetic Act. The Magnuson-Metcalf Bill passed in 1958 required acute and chronic studies of 200 pesticides in fish and wildlife species (Hall, 1987). FIFRA was amended in 1964 to mandate safety labeling on pesticide containers (Moore, 1987). FIFRA was further amended in 1972 to provide protection for the environment and for human health against "unreasonable adverse effects". These FIFRA amendments required information on product chemistry, residue chemistry, environmental fate, mammalian toxicology, and wildlife effects before registration of a pesticide for use on land. In some cases, reentry and spray drift information was also required. Later, further amendments have been added to FIFRA to enhance the database upon which regulatory decisions are based.

Other laws legislated in the U.S. since 1970 which also address environmental quality are the Water Pollution Control Act (colloquially called the Clean Water Act), the Toxic Substances Control Act (TSCA), the Marine Protection Research and Sanctuaries Act (MPRSA), the Resource Conservation and Recovery Act (RCRA), and the National Environmental Policy Act (NEPA).

E. BIOMAGNIFICATION

One of the primary problems arising from the use of the persistent, lipophilic organochlorine insecticides has been the problem of bioconcentration and biomagnification, collectively called bioaccumulation. These compounds, DDT for example, and many of their metabolites such as DDE, are lipophilic and partition into cell membranes and fat stores. Thus, organisms such as algae that live in water polluted with organochlorine compounds accumulate these chemicals since they have a greater affinity for biological lipids than the ambient water. When these organisms are ingested by animals at the next trophic level, such as copepods, the latter accumulate even greater amounts. As these are consumed, even greater levels accumulate. Higher trophic levels of the food chain can thus acquire and store extremely high concentrations of the pollutants, up to 100,000-fold the concentration occurring in the water (Pimentel, 1987). As an example, DDT at a concentration of 0.001 to 0.01 ppb in an aquatic environment, either fresh or salt water, was found at 0.1 ppm in aquatic invertebrates, 0.5 to 2 ppm in fish, and 10 ppm in birds of prey (Edwards, 1973). Also, the pesticide residues can be transferred to progeny via excretion in the egg, so the next generation then starts its life with a body burden of pesticide. Biomagnification has been a significant problem with the organochlorine insecticides and other highly lipophilic xenobiotics such as polychlorinated biphenyls. OP insecticides and carbamates do not bioaccumulate to any extent. However, because of their persistence, residues of these organochlorine compounds can still be found in wildlife samples.

F. AQUATIC TOXICITY

Many fish kills have been attributed to pesticides. In the 1960s and early 1970s, 18% of the reported fish kills were attributable to pesticides, with a category of "unknown causes" being the only category responsible for a larger (29%) percentage of deaths (Nimmo et al., 1987). Among major fish kills in which more than 100,000 fish were killed, pesticides were the sixth ranking cause of death (responsible for 3.7% of the deaths), with municipal wastewater, unknown causes, miscellaneous operations, food processing, and power plants being the categories responsible for more deaths. When selective pressures are severe, resistant populations of fish such as mosquitofish develop (Yarbrough and Chambers, 1979). These resistant populations can bioaccumulate very high levels of pesticides and therefore present a greater threat to their predators than noncontaminated food sources. Additionally, they have become less tolerant of other environmental stresses such as low oxygen stress which could have serious ramifications in overall population survival.

Certain life stages of some fish species are extremely sensitive to the toxic effects of some pesticides (Nimmo et al., 1987). Nonlethal effects such as the development of scoliosis in adult sheepshead minnows *Cyprinodon variegatus* can occur at exposure concentrations of pesticides as low as 4 µg/l for 10 to 17 d, and growth reduction is seen in juveniles at 0.08 µg/l. Effects were observed in experiments over the total life cycle of 141 d at concentrations as low as 0.07 µg/l. Although these effects are nonlethal per se, they have the potential of adversely affecting an organism's competitiveness in the ecosystem, and therefore can result in population changes. Additionally, DDT affects fish behavior (Duffus, 1980).

Invertebrates are also extremely sensitive to some pesticides. The estuarine mysid *Mysidopsis bahia* is adversely affected by diflubenzuron at concentrations above 0.025 µg/l (Nimmo et al., 1987). Of course, an adverse effect by a chitin synthesis inhibitor such as diflubenzuron is predicted in an arthropod.

Unexposed progeny of the organisms exposed to diflubenzuron produced fewer offspring and displayed shorter life spans.

Lastly, a reduction in photosynthesis elicited by DDT has been reported (Duffus, 1980). Such an effect, if profound, could have far-reaching ramifications on ecosystem viability.

G. GROUND WATER

Ground water contamination is of great concern because it is the source of 25% of the fresh water used in the U.S. (Carsel and Smith, 1987). More ground water contamination occurs now than ever before because the new generation pesticides such as carbamates are more water soluble and sorbed less than the older highly lipophilic, strongly sorbed pesticides and thus have a greater tendency to leach into the water. The ground water under certain soil types such as sandy soils is more susceptible to contamination because there is little organic matter in the overlying soil to adsorb the xenobiotics. Aldicarb and ethylene dibromide (EDB) contamination of ground water in Florida is an example (Davies and Doon, 1987). Most contamination is in the low ppb range and are lower for pesticides than for hazardous wastes (Carsel and Smith, 1987). Even though ground water contamination is a much greater problem at present than it was in the 1950s and 1960s, this contamination does not consistently mean that drinking water is contaminated.

H. EFFECTS ON WILDLIFE

Wildlife can suffer effects of pesticides by direct exposure from external contact or ingestion of contaminated food sources, or by shifts in ecosystem composition as a result of toxicity to sensitive ecosystem components. One of the most notorious potential effects on wildlife was the reputed eggshell thinning caused by DDT (Hall, 1987). This effect is controversial, but it was an extremely important factor leading to the banning of DDT. Although eggshell thinning can be shown experimentally, in actuality it was believed to have occurred in the wild only in some raptors and fish-eating birds. DDE, a metabolite of DDT, was probably the primary compound responsible. Of course, these carnivores at the higher trophic levels would have been expected to have bioaccumulated the largest pesticide residue levels. The eggshell thinning was believed to result from alterations in calcium deposition in the eggshell due to interference with estrogen activity (Duffus, 1980). Behavior favoring egg breakage was also believed to be induced after DDT exposure in the birds of prey.

Bird kills which correlated with heavy DDT use in the field were seen in the early 1950s (Hall, 1987). Experimentally, reproductive success in quail and pheasants was diminished following exposure to several organochlorine insecticides, DDD, DDT, endrin, dieldrin, and aldrin. Heptachlor and dieldrin use in the late 1950s in an imported fire ant eradication program caused severe wildlife kills.

More recently, organophosphate and carbamate insecticides have been implicated directly in bird kills, with parathion, diazinon, monocrotophos, phorate, dicrotophos, and carbofuran being the compounds of greatest concern (Hall, 1987). Thus, the high acute toxicity of these anticholinesterases continues to pose a threat to natural populations. Inverse correlations were seen between South Carolina brown pelican populations and DDE residue levels, with population recoveries occurring as DDT use was eliminated. Discharges of pesticides from a manufacturing plant into the Rhine River correlated with a decline in the local sandwich tern population (Rand and Barthalmus, 1980).

Monitoring programs performed in the mid 1960s to the early 1970s revealed residues of organochlorine insecticides and/or PCBs in many wildlife species including eagles, falcons, ducks, marine birds, marine mammals, crocodiles, and sea turtles (Hall, 1987). Although residues have declined since the early 1970s, reflecting the reduction in organochlorine compound use, the residue data cannot be used to substantiate pesticides as the cause of death in wild birds. An encouraging note is most of the studies conducted on wildlife species have shown additive toxic effects between compounds, and not synergism.

Even more encouraging is the fact that only about 25 of the more than 350 chemicals (pesticides, metabolites, and formulation products) tested in wildlife have actually been implicated in causing an adverse effect (Hall, 1987). Most compounds are considered of little hazard to wildlife because of their low toxicity and the low use. Such compounds as the synthetic pyrethroids, methoxychlor, lindane, mirex, toxaphene, and many cholinesterase inhibitors do not constitute a threat to wildlife.

Experiments in wildlife species with some organophosphate and carbamate insecticides have suggested that the compounds affect behavior, hormone balance, salt gland function, cold tolerance, embryonic development, and growth (Hall, 1987). On the other hand, these experiments have also indicated that some suspect compounds were unlikely to affect wildlife adversely.

Most of the information on the effects in wildlife is on birds. Some amphibians are sensitive to pesticides and have died from exposure, but effects on amphibians in general have not been well-researched (Barthalmus, 1980). However, toxic and teratogenic effects have been noted in anuran tadpoles at relatively low, environmentally relevant concentrations of organophosphorus insecticides (Snawder and Chambers, 1989). If such effects are occurring in the field during development, a large impact on frog populations could occur. Reptilian sensitivity to pesticides has also not been as well-documented as toxicity in birds or fish. However, reptiles are less sensitive to pesticide-induced lethality than are amphibians, although pesticide residues have been found in reptiles.

Among mammals, mice and rodents are relatively tolerant (Barthalmus, 1980). However, there has been a suggestion that bats have suffered some lethality from pesticides released from fat stores during the increased energy needs of migrations.

Thus, pesticide effects on wildlife have occurred in the past and have been documented or strongly implicated in the field, and the potential of some pesticides to affect wildlife adversely has been suggested in laboratory studies. However, it is not well understood to what extent wildlife kills can actually be attributed to pesticides, what role synergistic interaction between pesticides and environmental or chemical stresses might play, and what overall harm to wildlife populations can be attributed to pesticide exposures.

I. EFFECTS ON MICROBES

Since microbial metabolism is one of the most significant, if not the major, route of environmental pesticide degradation, it is extremely important to understand both the microorganisms' role in pesticide degradation as well as any effects which pesticides might have on microbial populations. Soil contamination with pesticides is inevitable either as a result of the pesticides applied directly to soils or washing from sprayed plants during rain. Residues bound to plant components will be incorporated into the soils during tillage. In the past, soil contamination has been a very serious problem. As an example, the arsenical insecticides contaminated the soils at sufficient levels to affect subsequent crops (Sheets, 1980). Other pesticides have accumulated in soils to high levels resulting in the contamination of future crops. Contamination of water, from direct application, drift, in effluents, and via runoff, can result in some contamination and thus exposure of microbial populations.

Although microbial metabolism of pesticides leads to detoxication, some of these reactions of microbes are activation reactions, and therefore microbial metabolism can lead to increases in potential toxicity (Stirling, 1980). An example is the conversion of aldrin to dieldrin and later to photodieldrin. Four of the most important types of reactions mediated by microorganisms are dehalogenation, dealkylation, hydrolysis of esters and amides, and ring oxidation followed by cleavage. This diversity of reactions makes most pesticidal chemicals susceptible to some type of microbial degradation. Some of these degradation reactions are by cometabolism, where the pesticide does not serve as the sole nutrient source for the microorganisms. Therefore, cooperative metabolism between two or more strains of microorganisms is possible, resulting in more complete degradation of the pesticide.

However, pesticides can also selectively affect certain strains of microorganisms and thereby alter the composition of the microbiological community. Some pesticides have been shown to inhibit nitrification, which could have far-reaching implications on many other ecological niches as well.

J. SUMMARY

It is clear that pesticides have caused some serious problems in environmental contamination and harm to fish and wildlife. The heavy and indiscriminate use patterns of the past have decreased due to increasing environmental awareness and the development of safer, more effective pesticides. Additionally, integrated pest management practices have reduced automatic and prophylactic spray schedules by employing more nonpesticidal strategies to aid in controlling insects.

Nevertheless, pesticides are still widely used in agriculture as well as public health and urban applications, so they continue to be introduced into the environment. Accidents and manufacturing effluents can also contribute to the environmental load. From point sources, they become widely disseminated by wind drift, animal movement, runoff, and water movement. Thus, a wide variety of nontarget organisms can be exposed to pesticide residues. Perhaps those organisms at greatest risk of exposure are those whose natural habitats have been destroyed by human development and technology and who therefore have been forced to exist closer to human habitation. The most sensitive individuals within a species and the most sensitive species within a community may be adversely affected by pesticide exposures. Such effects may range from growth retardation to physiological or behavioral deficits to death. With substantial effects on the most sensitive individuals or species, population shifts will occur resulting in ecological changes in community composition or predator/prey relationships.

However, the worst case scenarios can be averted with caution and intelligent manufacture, transport, use, and disposal of pesticides, and with a rational and conservative approach to pest control. Perhaps, it is fitting to end this chapter with Rachel Carson's eloquent statement on the rational use of pesticides (Carson, 1962):

> Through all these new, imaginative and creative approaches to the problem of sharing our earth with other creatures there runs a constant theme, the awareness that we are dealing with life — with living populations and all their pressures and counter-pressures, their surges and recessions. Only by taking account of such life forces and by cautiously seeking to guide them into channels favorable to ourselves can we hope to achieve a reasonable accommodation between the insect hordes and ourselves."

ACKNOWLEDGMENTS

The author acknowledges the support of Research Career Development Award ES00190, granted by the National Institutes of Environmental Health Sciences.

REFERENCES

Barthalmus, G.T. Terrestrial organisms, in *Introduction to Environmental Toxicology*. F.E. Guthrie and J.J. Perry, Eds. Elsevier/North Holland, Amsterdam, 1980, 106–119.

Carsel, R.F. and Smith, C.N. Impact of pesticides on ground water contamination, in *Silent Spring Revisited*. G.J. Marco, R.M. Hollingworth, and W. Durham, Eds. American Chemical Society, Washington, DC, 1987, 71–84.

Carson, R. *Silent Spring*. Houghton Mifflin, Boston, 1962, 368.

Davies, J.E. and Doon, R. Human health effects of pesticides, in *Silent Spring Revisited*. G.J. Marco, R.M. Hollingworth, and W. Durham, Eds. American Chemical Society, Washington, DC, 1987, 113–126.

Duffus, J.H. *Environmental Toxicology*. Edward Arnold Ltd., London, 1980, 164.

Edwards, C.A. *Persistent Pesticides in the Environment*, 2nd ed., CRC Press, Cleveland, OH, 1973, 170.

Freed, V.H. Pesticides: global use and concerns, in *Silent Spring Revisited*. G.J. Marco, R.M. Hollingworth, and W. Durham, Eds. American Chemical Society, Washington, DC, 1987, 145–158.

Guthrie, F.E. Pesticides and humans, in *Introduction to Environmental Toxicology*. F.E. Guthrie and J.J. Perry, Eds. Elsevier/North Holland, Amsterdam, 1980, 299–312.

Hall, R.J. Impact of pesticides on bird populations, in *Silent Spring Revisited*. G.J. Marco, R.M. Hollingworth, and W. Durham, Eds. American Chemical Society, Washington, DC, 1987, 85–112.

Kohn, G.K. Agriculture, pesticides, and the American chemical industry, in *Silent Spring Revisited*. G.J. Marco, R.M. Hollingworth, and W. Durham, Eds. American Chemical Society, Washington, DC, 1987, 159–174.

Moore, J.A. The not so silent spring, in *Silent Spring Revisited*. G.J. Marco, R.M. Hollingworth, and W. Durham, Eds. American Chemical Society, Washington, DC, 1987, 15–24.

Murphy, S.D. Toxic effects of pesticides, in *Casarett and Doull's Toxicology*, 3rd ed. C.D. Klaassen, M.O. Amdur, and J. Doull, Eds. Macmillan Publishing Company, New York, 1986, 519–581.

Nimmo, D.R., Coppage, D.L., Pickering, Q.H., and Hansen, D.J. Assessing the toxicity of pesticides to aquatic organisms, in *Silent Spring Revisited*. G.J. Marco, R.M. Hollingworth, and W. Durham, Eds. American Chemical Society, Washington, DC, 1987, 49–70.

Pimentel, D. Is *Silent Spring* behind us?, in *Silent Spring Revisited*. G.J. Marco, R.M. Hollingworth, and W. Durham, Eds. American Chemical Society, Washington, DC, 1987, 175–190.

Rand, G.M. and Barthalmus, G.T. Case history: pollution of the Rhine River, in *Introduction to Environmental Toxicology*. F.E. Guthrie and J.J. Perry, Eds. Elsevier North Holland, Amsterdam, 1980, 236–248.

Sheets, T.J. Agricultural pollutants, in *Introduction to Environmental Toxicology*. F.E. Guthrie and J.J. Perry, Eds. Elsevier/North Holland, Amsterdam, 1980, 24–33.

Snawder, J.E. and Chambers, J.E. Toxic and developmental effects of organophosphorus insecticides in embryos of the South African clawed frog. *J. Environ. Sci. Health* B24:205–218, 1989.

Stirling, L.A. Microorganisms and environmental pollutants, in *Introduction to Environmental Toxicology*. F.E. Guthrie and J.J. Perry, Eds. Elsevier/North Holland, Amsterdam, 1980, 329–342.

Wilkinson, C.F. The science and politics of pesticides, in *Silent Spring Revisited*. G.J. Marco, R.M. Hollingworth, and W. Durham, Eds. American Chemical Society, Washington, DC, 1987, 25–48.

Yarbrough, J.D. and Chambers, J.E. The disposition and biotransformation of organochlorine insecticides in insecticide-resistant and -susceptible mosquitofish, in *Pesticide and Xenobiotic Metabolism in Aquatic Organisms*. M.A.Q. Khan, J.J. Lech, and J.J. Menn, Eds. American Chemical Society, Washington, DC, 1979, 145–159.

Chapter

8

Halogenated Aromatic Compounds

Larry G. Hansen

A. INTRODUCTION

Organohalogen compounds have been introduced in Chapter 6 (solvents) and Chapter 7 (pesticides). Polychlorinated aromatic compounds (PCAs), especially polychlorinated biphenyls (PCBs) and polychlorinated dibenzo-*p*-dioxins (PCDDs) have received enormous attention by scientists, regulators, the press, and the public resulting in an inconceivable volume of printed information in two decades. This is more remarkable because during the first four decades of the five-decade commercial production of PCBs they were dismissed as being biologically inactive but extremely useful commercially. PCBs, PCDDs, and the closely related polychlorinated dibenzofurans (PCDFs) (Figure 8.1) have been examined from nearly every aspect of toxicology and ecotoxicology; yet there is no clear consensus on the past, present, or future impacts of these ubiquitous global contaminants.

Depending on the point of view, the adverse biological effects of PCAs might be viewed as readily generalized, highly specific, or limited to a select group of isostereomers. PCBs, PCDFs, PCDDs, polychlorinated naphthalenes (PCNs), polybrominated biphenyls (PBBs), bromobenzene, hexachlorobenzene, dichlorodiphenyltrichloroethane (DDT), and many others all induce cytochromes P_{450} and other xenobiotic metabolizing enzymes. Many of these compounds are capable of initiating porphyria, chloracne, embryo toxicity, and teratogenesis. Some rather rigid structure/activity relationships have been elucidated so

0-8493-8851-1/94/$0.00+$.50
© 1994 by CRC Press, Inc.

Biphenyl

Dibenzofuran

Dibenzodioxin

Figure 8.1. Structures of closely related bicyclic aromatics.

congeners such as 2,5,2′5′-tetrachlorobiphenyl (PCB 52) and 2,3,6,2′3′6′-hexachlorobiphenyl (PCB 136) are considered "nontoxic" while other isomers such as 3,4,3′4′-tetrachlorobiphenyl (PCB 77) and 3,4,5,3′4′5′-hexachloro-biphenyl (PCB 169) are two of the most potent congeners known.

No single species, and indeed no individual animal, measurably expresses all of the above mentioned manifestations. A complex mixture of 209 possible congeners would result in a similar broad profile of effects. However, many of these effects, including a characteristic "wasting syndrome", can be produced by individual PCDDs, PCDFs, and PCBs which are related to the more toxic 2,3,7,8-tetrachlorodibenzo-*p*-dioxin (2,3,7,8-TCDD). On the other hand, some congeners can potentiate and a few antagonize the action of other congeners. Added to the complexity of the effect of each isomer are species differences, interactions with different classes of contaminants, nutritional factors, and the low levels of very potent contaminants in the commercial preparations. Considering these factors, it is easy to understand why the environmental toxicology of PCBs is rather difficult to define.

Environmental toxicology (or "environmental health") deals with the hazards of exposure to a variety of potentially toxic substances (chemical stress) concurrent with various conditions of nutritional, microbial, physical, and emotional stresses. The goal is the prediction of risk so the most significant challenges can be attenuated. Most halogenated aromatic compounds are toxic hazards and conditions favoring exposure to them have established there are significant health risks associated with the current usage and/or environmental disposition.

To assess risks accurately, exposure and hazards must be determined in some quantitative manner. Exposure must consider concentrations in various media and degree of likely contact with these media, and hazard evaluation depends on the interpretation of direct toxicity measurements within the complex matrix of other stresses, including other chemical stresses. To accomplish the monumental task of accurate risk assessment, elucidation of the mechanisms of action and comparative (structural as well as phylogenetic) toxicities are invaluable in providing the generalizations needed to reduce the information to a manageable matrix. The environmental toxicology of the polychlorinated aromatics must be based on the systematic evaluation of many diverse factors and the careful integration of these factors with observed incidents. The first step is to examine the consequences of halogenating organic compounds.

B. HALOGENATION

Inorganic chloride is a major physiological anion, but halogens covalently bound to carbon are relatively rare in unexposed biological systems. Many exotic organohalogens have been identified from marine species, some from fungi, and a few from higher organisms. Marine organisms have evolved within an environment high in halogens, but terrestrial and freshwater aquatic forms do not usually contain covalently bound halogens as essential components. A notable exception is thyroxine or (3,5-diiodo-4-hydroxy phenoxy)-3,5-diiodophenylalanine, the thyroid hormone in which iodine is covalently bound.

Halogen ions are abundant, inexpensive to produce, and bind easily to carbon atoms, especially unsaturated carbons, to modify the properties of the starting molecule. Thus, halogens have been widely used in the design and manufacture of solvents, other industrial chemicals, pesticides, and medicinal agents. One result of adding halogens is to increase the molecular weight of the compound which in turn frequently increases the specific gravity, melting point, and boiling point and decreases the vapor pressure (Table 8.1). Such a simple modification in these properties has had obvious applications for industrial chemicals. The PCBs, used in a wide variety of applications before concern about environmental contamination mounted, were produced by chlorinating biphenyls in a random fashion until the desired physical properties were achieved.

Another result of halogenating compounds is to increase their stability. The carbon halogen bond (C–X) is stronger than the carbon hydrogen bond, while resonance provides additional stability (double bond character) in vinyl

TABLE 8.1. Some Properties of Halogenated Benzenes

| Halogen | Molecular Weight | Density | Temperature (°C) | | Aqueous Solubility (mmol/l) |
			Melting Point	For 1 atm[a]	
Benzene	78	0.878	5.5	80	24.1
1-F	96	1.022	−42	85	
1-Cl	113	1.107	−45	132	4.30
1-Br	157	1.495	−31	156	2.86
1-I	204	1.831	−28	189	
1-Cl	113	1.107	−45	132	4.30
1,3-Cl	147	1.288	−24	173	0.83
1,4-Cl	147	1.288	−24	174	0.53
1,2,3-Cl	181	1.454	68	218	0.12
1,3,5-Cl	181	1.454	64	208	0.029
1,2,3,4-Cl	216	1.858	46	254	0.036
1,2,4,5-Cl	216		139	245	0.0059
Penta-Cl	250	1.834	84	276	0.0026
Hexa-Cl	285	2.044	231	309	0.000017
Phenol	94	1.071	41	182	708
Penta-Cl Phenol	266	1.978	190	309	30

[a] Temperature at which vapor generates 1 atm.

and aryl halides. These aryl halides are protected from nucleophilic attack through backside protection from the benzene ring. On the other hand, each additional halogen on the benzene ring decreases the strength of the individual C–X bonds thus making them more susceptible to dehalogenation in the environment. Stability is a desirable property in many chemical applications but has the undesirable effect of enhancing the persistence of these chemicals in the environment. The once-common insecticide DDT has aromatic chlorine atoms on the two phenyl rings which are quite resistant to removal, but one of the aliphatic chlorine atoms on the trichloroethane bridge can be removed by dehydrochlorination to yield the persistent metabolite, 1,1-dichloro-2,2′-bis(*p*-chlorophenyl)ethylene (DDE). The resulting vinyl chlorine atoms are much more difficult to remove, providing the stability of DDE.

Organohalides are quite insoluble in water unless there are other polar groups on the molecule. Water solubility decreases with increasing degrees of halogenation (Tables 8.1 and 8.2). They are, however, miscible within each other and with other nonpolar materials such as oils and biological lipids. Organohalides are referred to as being lipophilic because of their relative solubility in water and lipids. They partition out of aqueous media such as blood and urine into fatty tissues and depot fat. This makes them less available for biotransformation and excretion processes; thus, they have a tendency to accumulate in tissue fat (Chapter 3). In aquatic and marine environments,

TABLE 8.2. Properties of Some Chlorobiphenyls

IUPAC Number[a]	Substitution Pattern	Molecular Weight	Melting Point	Water Solubility (μmol/l)	GLC Relative Retention Time (OCN = 1.0)
1	2	189	34	23.4	0.1554
4	2,2′	223	60	3.5	0.2245
15	4,4′	223	148	0.2	0.3387
18	2,5,2′	258	45	0.2	0.3378
28	2,4,4′	258	57	0.3	0.4031
31	2,5,4′	258	67	0.4	0.4024
37	3,4,4′	258	87		0.4858
44	2,3,2′,5′	292	48	0.6	0.4832
52	2,5,2′,5′	292	88	0.052	0.4557
70	2,5,3′,4′	292	104	0.056	0.5407
77	3,4,3′,4′	292	178	0.039	0.6295
95	2,3,6,2′,5′	326	99		0.5464
101	2,4,5,2′,5′	326	77	0.013	0.5816
110	2,3,6,3′,4′	326	oil		0.6314
118	2,4,5,3′,4′	326	109		0.6693
126	3,4,5,3′,4′	326	154		0.7512
128	2,3,4,2′,3′,4′	361	150	0.0012	0.7761
136	2,3,6,2′,3′,6′	361	114	0.017	0.6257
153	2,4,5,2′,4′,5′	361	103	0.0033	0.7036
194	2,3,4,5,2′,3′,5′,5′	430	156	0.00029	0.9620

[a] Ballschmiter et al., 1987.

compounds of low water solubility partition into free swimming organisms or bind easily to particulate matter and thus accumulate in sediments. Bioconcentration of nonpolar compounds can be dramatic, with organisms accumulating concentrations 3 to 6 orders of magnitude greater than that of the ambient water (Metcalf et al., 1975).

Persistence and lipophilicity are very important determinants of bioconcentration potential, while water solubility and volatility are important determinants of environmental transport. Halogenated hydrocarbons tend to be found in lower concentrations in waters and higher concentrations in soils, sediments, and biota. Bioconcentration in wide-ranging species may lead to a unique aspect of ecodisposition which may be termed biodispersion. This may be highly significant for insects, since they have a large collective biomass and many species spend the most active feeding stages in aquatic or subsoil matrices before metamorphosis into the more mobile adult forms.

Highly volatile organohalides rapidly disperse in the atmosphere. Compounds of low and moderate volatility tend to be influenced more by meteorological conditions and may be transported short distances in the atmosphere or retained in other matrices for relatively long periods. Nevertheless, subtle differences in closely related halogenated aromatics, such as positional isomers (Table 8.2), can have profound effects on ecodisposition, bioconcentration, biodispersion, and biological activity. These major differences within subsets of generic properties are crucial in the understanding of the environmental toxicology of halogenated hydrocarbons and will become apparent in later discussions.

C. HALOGENATED HYDROCARBONS

Organic halide or halogenated hydrocarbon are not definitive terms and have different implications for different researchers depending on personal experience and emphasis; thus, some differentiation among the vast number of possible organic halides is necessary.

Hundreds of millions of pounds of alkyl halides such as chloroform and hexachloroethane are produced annually in the U.S. alone. In 1978, the U.S. produced 11×10^9 lb of dichloroethane (Weisburger, 1981). Vinyl halides such as dichloroethylene and vinyl chloride are also produced in massive amounts. Some alkyl halides such as chloral hydrate ("knock-out drops"), halothane (anesthetic), and chloroform can have deleterious effects on the nervous system. Metabolites of alkyl halides and vinyl halides are most often responsible for liver and kidney toxicity as well as carcinogenic effects.

Halogenated alicyclics such as lindane, toxaphene, and mirex in addition to cyclodiene insecticides such as aldrin and heptachlor are frequently included in discussions of halogenated aromatics. These and other halogenated aromatic pesticides such as DDT have been discussed in Chapter 7.

D. HALOGENATED AROMATICS

Halogenated phenols and phenoxy compounds are more polar than other halogenated aromatics and thus behave differently. They are manufactured for specific end uses (pentachlorophenol, 2,4-dichlorophenoxy acetic acid) and as intermediates or byproducts in the production of other chemicals or they may be generated during the degradation of more complex chemicals (e.g., the leaving groups of ronnel and leptophos organophosphorus insecticides). The polar nature of these halogenated aromatics permits them to reach higher concentrations in aqueous media and distribute more evenly among tissues. Significant proportions are excreted without further metabolism.

Halogenated benzenes such as bromobenzene, *p*-dichlorobenzene (PDB), and hexachlorobenzene (HCB) are very lipophilic and, in general, must be metabolized to some degree before they can be excreted. PDB %cand HCB may sublimate so significant aerosol exposure is possible, especially in confined areas. These compounds have a broad spectrum of biological activities, primarily causing hepatic, renal, and nervous system effects. Metabolites include the more bioactive halogenated phenols, some of which are mutagenic and potentially carcinogenic. Some halobenzenes such as PDB have relatively low toxicities, while HCB is a potent microsomal monooxygenase inducer and porphyrinogenic agent.

Joining of two aromatic rings to form naphthalene or biphenyl adds stability to the molecule and permits even greater degrees of halogenation than with compounds having single aromatic rings. Polychlorinated naphthalenes (PCNs) were marketed under the name of Halowax and had applications similar to PCBs. Polybrominated biphenyls (PBBs) were used primarily as fire retardants. Substitution of chlorine by bromine frequently increases the toxic potency of comparable PCB congeners (Safe, 1984).

Some of the major incidents of accidental poisonings by PCAs are presented in Table 8.3. In addition, Table 8.4 lists some of the more publicized areas where PCAs have been implicated as environmental hazards. In most cases other toxicants are also present, which may either exacerbate PCA toxicity or lessen the serious effects of PCAs (e.g., monooxygenase induction) thus resulting in an increase or decrease in the impacts of other toxicants such as polynuclear aromatic hydrocarbons (PNAs) (Hansen, 1987; Winston et al., 1988). The potential hazards to humans are frequently recognized by effects on indigenous species, emphasizing the predictive value of ecotoxicology. Bioconcentration of high concentrations of PCAs in edible species such as fish may enhance the potential hazards to humans.

Most of the remaining discussion will focus on the closely related PCBs, PCDFs, and PCDDs so a more detailed examination of the principles of environmental toxicology can be discussed. Recalling the generic consequences of halogenation will aid in applying the principles discussed in this chapter to other halogenated aromatics, while the specific structure/activity relationships will emphasize the critical importance of the subtle differences among compounds.

E. PCBs, PCDFs, AND PCDDs

1. Production, Usage, and Dispersion

PCBs were first synthesized in the 19th century, but large scale commercial production did not begin until about 1930. PCBs were used in a wide range of applications and worldwide production is estimated at 1.2×10^9 kg. Most commercial production was discontinued in the late 1970s. Some of the commercial preparations are listed in Table 8.5.

TABLE 8.3. Some Poisoning Episodes with Polyhalogenated Aromatics

Date	Incident	Agent	Source
1916	Chloracne	PCN	Gas mask production
1940s/1950s	X-Disease (Cattle)	PCN, etc.	Multiple
1950s	Chick edema	PCDD, PCN, PCB	Heated fats with chlorophenols
1955	Porphyris (Turkey)	HCB	Treated wheat seed used for flour
1968	Yusho (Japan)	PCB, PCDF	Rice oil with heat exchange fluid
1960s	Agent orange (Vietnam)	TCDD	Aerial spraying of herbicides
1973	PBB (Michigan)	PBB	Flame retardant in animal feed
1976	Seveso (Italy)	TCDD	Chemical plant explosion
1979	Yucheng (Taiwan)	PCB, PCDF	Rice oil from heat exchanger

TABLE 8.4. Example of Areas Considered to be Significantly Contaminated with Polychlorinated Aromatic Compounds

General Area	PCAs[a]	Vertebrates Affected[b]
Baltic and North Seas	PCBs > DDT	Fish, marine mammals, piscivorous birds
Great Lakes	PCBs > DDT > cyclodienes	Fish, piscivorous birds
Hudson River, NY	PCBs	Fish
Seveso, Italy	PCDDs	Livestock, wildlife
Sheboygan River, WI	PCBs	Fish, birds
St. Lawrence Seaway	PCBs > DDT	Fish, whales
Times Beach, MO	TCDD	Wildlife (uninhabited)

[a] Most significant, most toxic, or most publicized; other toxic materials, especially heavy metals and polynuclear aromatics, are almost invariably present.

[b] Most prominent species (excluding man) associated with or studied in relation to the area of concern.

TABLE 8.5. Commercial PCB Products and Producers

Trade Name	Producer and Country
Aroclor	Monsanto (USA)
Aroclor, Santotherm	Mitsubishi-Monsanto (Japan)
Chlophen	Bayer (Germany)
Delor	Chemko (Czechoslovakia)
Fenclor	Caffaro (Italy)
Kanechlor	Kanegafuchi (Japan)
Phenoclor, Pyralene	Prodelec (France)
Soval	Sovol (USSR)

Uses of PCBs included dielectric fluids in transformers and capacitors, hydraulic fluids, heat transfer fluids, plasticizers, adhesives, lubricants, flame retardants, and extenders or diluents of other agents. Some microscope immersion oils were pure PCB. PCB-containing mortar coatings in silos once contributed to milk contamination through the feeding of silage to dairy cows. High exposures of residents and workers on these farms also occurred, but most of these sites were closed and dismantled by the late 1970s (Hansen, 1987). Products such as plastics and adhesives in older livestock facilities can result in the introduction of PCBs into the food chain by unexpected routes (Hansen et al., 1989). Widely distributed products such as brake fluids and carbonless copy papers resulted in a very wide distribution of trace amounts of PCAs, but the bulk of environmental contamination originated from electrical applications. Older transformers are the major current source of continued release of PCBs into the environment, and transformer production wastes are responsible for many environmental hot spots. Pre-1972 fluorescent light fixtures and older electrical appliances may be responsible for kitchen air PCB levels that were comparable to that in some research laboratories and which can reach an order of magnitude higher than outdoor air levels (MacLeod, 1981).

PCBs had been considered to be relatively nontoxic and, indeed, were shown to have very low *acute* toxicity (NAS, 1979). There had been some indication of skin and liver pathological changes associated with industrial exposure, but the full extent of the *chronic* health and ecological effects of PCBs is still not completely understood.

The production of 1.2×10^6 t of highly persistent chemicals that were deployed in a wide variety of uses but with little concern for health effects led to the escape of these chemicals into the environment. In 1966, Soren Jensen published a short note identifying PCBs as the troublesome peaks frequently found during the analysis for chlorinated pesticides in Swedish wildlife. Koeman found similar compounds in Dutch samples, and by the time this observation was published in *Nature* in 1969, a major human PCB poisoning episode had occurred in Japan. It soon became apparent that PCBs were widespread and that about one third of the total production had been discharged into the environment.

Unlike PCBs, polychlorinated dibenzofurans (PCDFs) and dibenzodioxins (PCDDs) (Figure 8.1) were not produced intentionally except in very small amounts for analytical standards and other research purposes. Both PCDFs and PCDDs occur as contaminants in the production of other chlorinated aromatics, especially when phenolic intermediates are used in the synthesis, and they are also released during the combustion of chlorinated compounds. Incineration of municipal refuse, which has been shown to generate and release PCDFs and PCDDs during the combustion of most types of organic matter in the presence of inorganic chloride, led to the theory that PCDDs were present long before the industrial revolution. Careful analysis, however, of preserved specimens not exposed to recent pollutants (400-year-old Eskimo and 2800-year-old Chilean mummies) did not detect PCDDs above the levels of modern labora-tory contamination (Ligon et al., 1989).

PCBs subjected to heat over long periods may be converted to PCDFs. The heat exchanger transformer fluid, which leaked into rice oil in the Japanese "Yusho" PCB poisoning incident of 1968 (Kuratsune and Shapiro, 1984), contained high levels of PCDFs which exacerbated the toxicities observed. Soot from fires involving PCB-containing electrical transformers are more toxic than the PCB dielectric fluids alone due to high levels of PCDFs and PCDDs (De Caprio et al., 1983). The optimum temperature range for PCDF and PCDD formation is 250 to 350°C, while virtual destruction occurs at temperatures exceeding 800°C. Thus, carefully controlled high-temperature incineration can be used to destroy PCAs such as "agent orange" without increasing the hazard of the effluent. Uncontrolled overheating at lower tem-peratures such as in the burning of transformers involved in a fire in Binghamton, New York and the Seveso chemical plant explosion (Walsh, 1977) resulted in the generation of highly toxic effluents.

The proportion of individual congeners can have a profound effect on the net biological activity of any residue. The significance of the congener composition of residues is underscored by the possibility that over 400 different PCB, PCDF, and PCDD congeners exist (Table 8.6), and the varying composition of each commercial mixture contributes to the residues discharged (Table 8.7, Figures 8.2 and 8.3). Some of the differences in properties discussed above (Table 8.2) result in unequal effects on individual congeners.

2. Physical, Chemical, and Biotic Factors

High temperature oxidation is only one of the factors that result in the constantly changing composition of complex PCA residues. To understand the

**TABLE 8.6. Numbers of Isomers of Representative
Halogenated Aromatic Compounds**

Number of Halogens	Benzenes	Dioxins	Furans	Biphenyls
1	1	2	4	3
2	3	10	16	12
3	4	14	28	24
4	3	22	38	42
5	1	14	28	46
6	1	10	16	42
7	—	2	4	24
8	—	1	1	12
9	—	—	—	3
10	—	—	—	1
Total	13	75	135	209

TABLE 8.7. Approximate Molecular Composition (w/w%) of Aroclors

Isomer	Aroclor 1242	Aroclor 1254	Aroclor 1260
MonoCl	1.0	<0.1	ND
DiCl	16.0	0.5	ND
TriCl	49.0	1.0	<0.5
TetraCl	25.0	21.0	1.5
PentaCl	8.0	48.0	12.0
HexaCl	1.0	23.0	38.0
HeptaCl	<0.1	6.0	41.0
OctaCl	ND	ND	8.0
DecaCl	ND	ND	<0.1

potential ecological and public health impacts of PCB-PCDF-PCDD residues, an appreciation of the dynamic nature of these residues is required. Media and biota may contain any of the 400 possible PCAs (Table 8.6) in addition to other chlorinated aromatics (e.g., phenols, benzenes, naphthalenes, azobenzenes) as well as brominated counterparts. Patterns of volatility, solubility, and susceptibility to degradation of the PCBs will be used as examples.

With many properties the position of chlorination on the biphenyl ring has an equal or greater effect than modest changes in the degree of chlorination. The biphenyl rings can rotate at the 1,1′ positions and assume orientation within the same plane (coplanar) or at various angles to each other. Chlorine substitution in the *ortho* positions causes steric interference and the degree of coplanarity decreases with increasing *ortho* (2,2′6,6′) chlorination (McKinney et al., 1983). PCDFs and PCDDs are held rigidly in a coplanar orientation by the oxygen bridge(s). The potency of coplanar PCB congeners which mimic

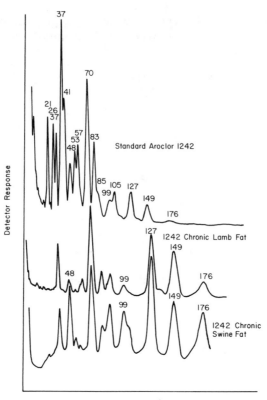

Relative Retention Time p, p'-DDE=100

Figure 8.2. Packed column gas/liquid chromatograms of Aroclor™ 1242 and fat from animals fed Aroclor™ 1242. Numbers refer to retention times relative to *p,p'*-DDE = 100 (total PCB was 15.6 ppm in lamb fat and 10.8 ppm in swine fat) (Hansen, 1987b).

the effects of TCDD (the most potent of the PCAs) depends on their degree of coplanarity and the degree of chlorination at the *para* and *meta* positions (Poland and Knutson, 1982; Safe, 1984, 1987).

Congeners with the highest number of chlorine atoms have the greatest molecular weight and, in general, the highest melting point and the lowest vapor pressure (Table 8.2). However, these properties are also influenced by the position of the chlorine atoms. The retention time of a congener in gas liquid chromatography (GLC) is a rough approximation of the volatility of the congener. The presence of chlorine atoms in the *para* position decreases the volatility and increases the GLC relative retention time, while the presence of *ortho* chlorine atoms decreases the GLC retention time relative to other isomers and even to higher chlorinated *para*-substituted congeners. The retention time of the coplanar PCB 77 is longer than any other tetrachlorobiphenyl and roughly equal to that of the di-*ortho*-substituted PCB 110 (a pentachlorobiphenyl). Moreover,

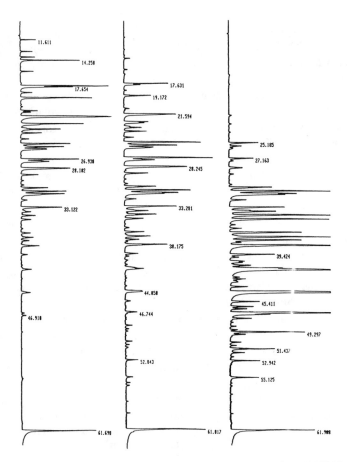

Figure 8.3. Capillary column gas/liquid chromatograms of Aroclors 1260 (top), 1254 (middle), and 1242 (bottom). Numbers are peak retention times in minutes. Peak 61 is octachloronaphthalene internal standard.

PCB 84 , PCB 95 (five chlorine atoms with three in *ortho* position), and PCB 136 (six chlorine atoms with four in the ortho position) have shorter retention times than PCB 77, while the coplanar PCB 126 has a longer retention time than many congeners with six chlorine atoms (Table 8.2). The single property of volatility can have a profound effect on the dispersion of PCB congeners and thus the composition of various residues in the environment.

Water solubility also has a profound effect on individual congener disposition. This effect, however, is more closely correlated with the degree of chlorination, as discussed previously. Although *ortho* substitution also affects polarization and water solubility, the differences among PCB 101, PCB 128, and PCB 136 (Table 8.2) are really rather small. Furthermore, the octanol/water partition coefficients of the above congeners (2,500,000, 5,000,000, and 10,000,000, respectively) represent small differences in large numbers but are much larger than that for PCB 1 (20,000). Thus, the higher chlorinated isomers have a greater tendency to partition into fats or accumulate in biota and/or sediments regardless of substitution pattern.

Within aquatic and marine environments more of the lower chlorinated or *ortho*-substituted congeners may dissolve in the water or volatilize into the atmosphere, while more of the higher chlorinated congeners adhere to particulates and settle out. During periods of low flow in reaches of the Hudson River, 65% of the PCB transported downriver can be accounted for by only three lower *ortho*-substituted chlorinated congeners, namely 2 2,2′ and 2,6 (PCB 1, PCB 4, and PCB 10, respectively) (Bush et al., 1985). During periods of higher flow these congeners constitute a markedly lower proportion of total PCB. This depletion is probably due to a combination of an increase in the dissolution of the higher chlorinated congeners from sediments and volatilization loss of the lower chlorinated congeners. Macroinvertebrates bioconcentrate mainly those congeners containing four, five, or six chlorine atoms (Bush et al., 1985), reflecting the loss of the lower chlorinated congeners and greater availability of the higher chlorinated congeners.

Of course, not all of the PCB congeners bioconcentrated from the media are retained unmetabolized, so the net bioaccumulation reflects these losses. Although the differences in volatility and lipid/water partitioning coefficients result in significant changes, the much more specific and dramatic differences in susceptibility to biodegradation (Jensen and Sundstrom, 1974; Sipes and Schnellman, 1987) might be more important in changing the composition of the residues. However, the small proportion of PCAs found in biota compared to other matrices including soils and sediments (NAS, 1979) may make these changes significant in food chains (Hansen, 1987b) but not in other matrices (Eisler, 1986).

Longer term studies and closer examination of residues have revealed that additional changes were also occurring and that microbial degradation might be responsible. Bacteria isolated from PCB-contaminated soils and sediments seemed to be adapted to PCBs. Strains of aerobic bacteria which can oxidatively degrade lower chlorinated biphenyls have been isolated from these soils. Some of these strains have remarkably broad degradation capacities (Bedard et al., 1986). In more anaerobic media, dechlorination, primarily in the *meta* and *para* positions, is a distinct possibility (Brown et al., 1987). Sediments isolated from areas highly contaminated by PCB, such as Waukegan Harbor (Lake Michigan) and the upper Hudson River, were found to be enriched in the lower chlorinated, *ortho*-substituted congeners and depleted of higher chlorinated congeners. However, their aqueous solubilities would predict the opposite pattern. The concentration of these same congeners (PCB 1, PCB 4, and PCB 10) were also dramatically increased in the sediments (Brown et al., 1987) in spite of their greater tendency to volatilize from the water (resulting in an equilibrium away from the sediments) and be oxidized by aerobes. The net biotransformation of congeners by microbes is inversely related to their partition coefficients (Chen et al., 1988), thus making this sediment enrichment with lower chlorinated congeners even more remarkable and possible anaerobic dechlorination more plausible.

It is likely that dechlorination is an enzymatic process (Brown et al., 1987) by which adapted anaerobic bacteria can take advantage of the greater free energy gradient offered by aromatic dechlorination. This process is apparently very slow (Chen et al., 1988), even slower than the mainly *ortho* photolytic dechlorination which occurs in the upper layers of aquatic systems exposed to sunlight. The anaerobic dechlorination from an environmental perspective may be more significant because PCB concentrations are much higher in sediments, and once chlorine atoms are removed, the products are more susceptible to the 2,3- and 3,4-dioxygenases of aerobic bacteria as well as the P_{450} monooxygenases of higher organisms.

The biotransformation of PCBs in multicellular animals is almost always initiated by a cytochrome P_{450} monooxygenase. The initial step may be through an indirect hydroxylation (addition + keto-enol rearrangement) or an epoxidation (arene oxide formation) with or without displacement of the chlorine (NIH Shift) or even a loss of chlorine atoms (Sipes and Schnellman, 1987). The presence of two adjacent unchlorinated carbons (vicinyl hydrogens) favors, but is not a requirement for, epoxidation. PCBs chlorinated in the *para* (4,4′) and *meta* (3,3′,5,5′) + *para* positions are more resistant to biotransformation. PCBs with four or more chlorine atoms but with hydrogen in either the 4, or especially, in the 4,5 positions (i.e., 2,3- or 2,6-dichloro, or 2,3,6-trichloro) of one or both rings are more rapidly metabolized than other higher chlorinated biphenyls. The P_{450} monooxygenases induced by phenobarbital and noncoplanar PCBs have a greater potential for oxidizing *ortho*-substituted congeners; those induced by coplanar PCBs more readily oxidize the coplanar PCBs (Kaminsky et al., 1981).

Oxidized products may bind to cellular macromolecules but they most frequently are rapidly conjugated, either to glucuronic acid or to glutathione derivatives. Glucuronides are readily excreted while the glutathione conjugates are further metabolized to mercapturic acids (Chapter 3). Figure 8.4 depicts some of the intermediates and metabolites of PCB 15 (4,4′-dichlorobiphenyl).

The mercapturic acids excreted in the bile may be cleaved by intestinal flora to form the sulfhydryl product which is later methylated and reabsorbed (Bakke et al., 1982). The methylthio PCB is oxidized first to a sulfoxide and then to a sulfone and may partition into depot fat and other tissues to become a significant portion of the total PCB residue (Jensen et al., 1979; Haraguchi et al., 1984). The methyl sulfonyl PCBs often show highly specific tissue binding (Brandt and Bergman, 1987) which may be responsible for some of the toxic effects of PCBs. Methyl sulfonyl metabolites of HCB and DDT have also

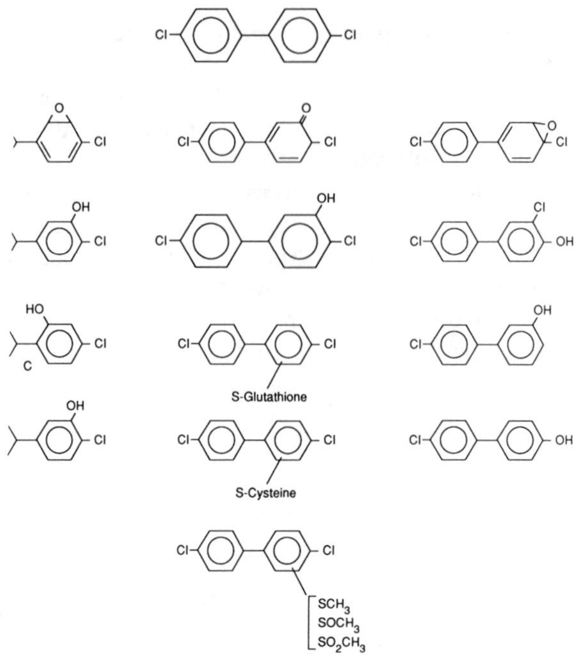

Figure 8.4. Metabolites of 4,4′-dichlorobiphenyl (Sipes and Schnellman, 1987).

been identified (Jensen et al., 1979) and those from chlorobenzenes may have greater biological activity than the parent compound (Kato et al., 1988). Thus, this route of metabolism is a generalized and significant route for PCAs.

PCDFs and PCDDs are metabolized via routes similar to those for PCBs; however, elucidation of metabolites is difficult because the high toxicity of the 2,3,7,8-substituted congeners severely limits the doses that may be administered. Hydroxylation and dechlorination and/or NIH shifts have been demonstrated for TCDD, indicating the formation of arene oxides (Olson, 1986). TCDF also forms hydroxylated products and methoxy derivatives of these products, again with some loss or shift of chlorines. Dihydroxy (2,2′) biphenyls could also arise following oxidation of the 4 position of the dibenzofuran followed by ring opening and reduction. The 2,3,7,8-tetraCDF and 1,2,3,7,8-pentaCDF are metabolized and excreted rather rapidly by the rat, but the 2,3,4,7,8-pentaCDF is very persistent (Brewster and Birnbaum, 1988). PCDFs substituted in the 4 position were also the most persistent in "Yusho" patients, indicating the importance of the 4 and 6 carbons (adjacent to oxygen) in the oxidative metabolism of PCDFs.

PCDFs, PCDDs, and coplanar PCBs have reduced biological activity following oxidative metabolism; however, many noncoplanar PCBs are more bioactive following oxidation. Accumulation of metabolites in the environment, and especially in food chains, may be of great toxicological significance.

Those metabolites generated *in situ*, of course, would have even greater potential for biological activity.

F. COMPOSITION AND ANALYSIS

Environmental toxicology depends heavily on analytical chemistry for obvious reasons; however, it is virtually impossible to understand and assess the impact of PCBs, PCDFs, and PCDDs without an appreciation of the limitations and evolution of the analytical methods employed in the two-decade attempt to define the problem. Frequently, apparent trends in increasing and/or decreasing environmental residues were actually more highly influenced by changes in analytical methodology than by actual changes in concentrations. Moreover, the methods of quantitation and reporting were often grossly misleading and may have had a greater impact than warranted.

Most analytical methods employed nonpolar solvent extractions, some form of cleanup and gas/liquid chromatography with an electron capture detector (ECD-GLC) (Erickson, 1986). Figure 8.2 illustrates a typical packed column isothermal chromatogram of standard Aroclor 1242 and the chromatogram of extracted fat from sheep and swine fed this mixture. Figure 8.3 illustrates the better resolution of Aroclor standards which was achieved with a 50 m capillary column and programmed temperature increases. The ECD is extremely sensitive to halogenated compounds so solvents, reagents, and glassware (and even the wax on the shiny surface of aluminum foil) can interfere if meticulous standards are not maintained in the laboratory (Erickson, 1986).

It is apparent from Figures 8.2 and 8.3 that even when appropriate analytical manipulations were developed quantitation and reporting of results could be quite complex. Results were most frequently quantitated on the basis of the commercial preparation that most closely resembled the chromatographic pattern of the sample extract. Some analysts have attempted to quantitate each peak based on the height or area that would result for a given concentration of the selected "standard" and then to sum these relative concentrations. In the mid 1980s the most common method was to select three or four major peaks characteristic of the closest matching standard mixture and report the results as a concentration of that mixture based on the average of the characteristic peaks. This procedure resulted in an estimation rather than quantitation.

The composition of the residues resulting from a single commercial preparation changes as soon as physical, chemical, and/or biotic forces act on the mixture. The estimations based on a "standard" PCB would be incorrect as the composition changed. Since these changes are known to occur, errors in estimation occurred for environmental samples.

Another error obvious to even the amateur analyst with a very basic knowledge of the properties of PCBs, but overlooked by some analysts or regulators who needed to generate and use comprehensible values, is the method of reporting. The lower chlorinated congeners are more volatile, more water soluble, and more readily biodegraded. Aroclor 1242, for example, contains 42% chlorine by weight, so it has a higher proportion of lower chlorinated congeners than does Aroclor 1254 or Aroclor 1260 (Figure 8.3; Table 8.7). Environmental conditions deplete many lower chlorinated congeners, so plankton and small fish from a pond contaminated with Aroclor 1242 may have a residue pattern more closely resembling Aroclor 1254, while residues more closely resembling Aroclor 1260 would accumulate in the sediment.

Although typical analytical surveys may report 5 ppm PCB as Aroclor 1254 in fish, it is not uncommon for this information to appear somewhere as 5 ppm Aroclor 1254. Since it would take about four times the amount of Aroclor 1242 to result in the same concentration of the penta- and hexachlorobiphenyls found in Aroclor 1254 (Table 8.7), this may also eventually be interpreted that 20 ppm Aroclor 1242 rapidly (within a year) degraded to 5 ppm Aroclor 1254.

Analytical and data processing developments have increased our ability to detect specific congeners (Safe et al., 1987). However, all specific congener analyses are not equal, and even these can sometimes be misleading.

Fortunately, the most toxic PCB congeners and the more toxic PCDFs and PCDDs are present in very low concentrations relative to other congeners. Methods have been developed to concentrate the toxic coplanar PCB 77, PCB 126, and PCB 169 as well as PCDFs and PCDDs from sample extracts so their concentrations can be more accurately determined. Although once thought to be absent or insignificant in PCBs and environmental residues, the enrichment method has confirmed the presence of the highly toxic PCB 77 and PCB 126. Using "dioxin equivalent" factors, in many cases concentrations of some PCB congeners have been found to contribute more to the net toxicity of the residue than do PCDFS and PCDDs (Olafsson et al., 1988; Tanabe, 1988; Kubiak et al., 1989).

G. EXPERIMENTAL TOXICITY

A summary of the adverse effects associated with PCB toxicosis (Table 8.8) vividly illustrates the breadth of biological activity of a single class of PCA. Even though PCBs can elicit a broad spectrum of biological and toxic effects, they are structurally specific toxicants, which means they must combine with and stimulate a receptor or receptors to initiate a reaction which leads to the effects responsible for the observed response.

PCB $\leftrightarrow$ PCB:Receptor(s) $\rightarrow$ Action(s) $\rightarrow$ Effect(s) $\rightarrow$ Responses

Some of the receptors known to be involved in PCA toxicity include those involved with heme and/or cytochromes P_{450} biosynthesis, inhibition of

TABLE 8.8. Common Manifestations of PCB Exposure in Various Animals

Hepatic Effects
        Hepatomegaly, bile duct hyperplasia
        Widespread (e.g., rabbit) or focal (e.g., mouse) necrosis
        Lipid accumulation, fatty degeneration
        Induction of microsomal monooxygenases and other enzymes
        Decreased activity of membrane ATPases
        Depletion of fat-soluble vitamins, especially vitamin A
        Porphyria
Gastrointestinal Effects
        Hyperplasia and hypertrophy of gastric mucosa
        Gastric ulceration and necrosis
        Proliferation and invasion of intestinal mucosa (monkey),
         hyperplasia, hemorrhage, necrosis (hamster, cow)
Respiratory System
        Chronic bronchitis, chronic cough
Nervous System
        Alterations in catecholamine levels
        Impaired behavioral responses
        Developmental deficits
        Depressed spontaneous motor activity
        Numbness in extremities
Skin
        Chloracne (most characteristic human response)
        Edema, alopecia
Immunotoxicity
        Lymphoid involution (spleen, lymph nodes, especially
         thymus)
        Subsequent reduction of circulating lymphocytes
        Suppressed antibody responses
        Enhanced susceptibility to viruses
        Suppression of natural killer cells
Endocrine System
        Altered levels of circulating steroids
        Estrogenic, antiestrogenic, antiandrogenic effects
        Decreased levels of plasma progesterone
        Adrenocortical hyperplasia
        Thyroid pathology, changes in circulating thyroid hormones
Reproduction
        Increased length of estrus
        Decreased libido
        Embryo and fetal effects following *in utero* exposure
Carcinogenesis
        Promoters
        Attenuation of some carcinogens

cytochromes P_{450}, receptors modulating catecholamine production and release, steroid receptors, and the Ah receptor.

The receptor most often linked with halogenated aromatic hydrocarbon toxicity is the Ah receptor, so named because one of its effects is the dramatic induction of aryl hydrocarbon hydroxylase (AHH, a cytochrome P_{450}) activity following its binding by xenobiotics. A constellation of other effects (Tables 8.9 and 8.10) such as thymic atrophy, fatty liver, chloracne (humans), wasting syndrome (guinea pigs), and teratogenesis (cleft palate in mice) also segregate with the Ah genetic locus (which is derepressed by the agonist Ah receptor complex) (Poland and Knutson, 1982). The classical and most potent agonist is 2,3,7,8-TCDD, but other agonists such as polyhalogenated dibenzofurans, azobenzenes, naphthalenes, and biphenyls with at least three halogens are isosteric and fit the 3×10 angstrom coplanar dimensions of TCDD and bind to the AH receptor (Figure 8.5) (Poland et al., 1979, Poland and Knutson, 1982; Safe, 1984, 1987).

The most important characteristics of the toxicity elicited by these compounds (Poland and Knutson, 1982), are

1. The parent compounds, and not metabolites, are responsible for toxicity.
2. The toxic effects produced by these compounds are mediated through their binding to the cytosol ("Ah") receptor.
3. The capacity of a congener to elicit a single toxic response indicates the potential to produce the complete syndrome.

Based on this model, TCDD is about 40 times as potent as the comparable dibenzofuran and 400 times as potent as the comparable PCB (Table 8.10). However, TCDD is only about 2.5 times more potent than the most potent PCB and PCDF. Acute lethality is highly species dependent (Table 8.9), but the LD_{50} for TCDD in guinea pigs and rats is less than 0.05 mg/kg while that for the comparable PCDF and PCB are higher and that for the commercial PCB mixture is at least 20,000 times higher. These dramatic differences in acute toxicity may be an oversimplification since the variety of subacute and chronic effects listed in Table 8.8 suggests a more complex interpretation of the effect of each congener.

Many of the lesions associated with TCDD-like (Ah receptor) effects involve hyperplasia or hypoplasia of epithelial tissues (Table 8.9) (Poland and Knutson, 1982). The most apparent and most studied effect is induction of about a dozen cytochrome P_{450} monooxygenases with aryl hydrocarbon hydroxylase (AHH) being the most directly associated with the Ah receptor. Lipid accumulation and vitamin A depletion are also associated primarily with Ah receptor activity, but hepatomegaly, induction of other P_{450}s, as well as phase II enzymes and porphyria resulting from interference with heme synthesis are rather generic properties of many PCAs.

TABLE 8.9. Acute Toxicity (LD_{50}) and Histopathological Changes Elicited by 2,3,7,8-TCDD

	Monkey	Guinea Pig	Cow	Rat	Hamster
Oral LD_{50}	<70	1	NR[a]	22M[b]	5000
(μg/kg)				45F[b]	
Hyperplasia and/or metaplasia					
Gastric	++	0	+	0	0
Intestinal	+	NR	NR	NR	++
Urinary	++	++	++	0	NR
Biliary	++	NR	+	NR	NR
Alveolar	NR	NR	NR	++	NR
Skin	++	NR	+	0	NR
Hypoplasia, atrophy or necrosis					
Thymus	+	+	+	+	+
Bone marrow	+	+	NR	NR	NR
Testicle	+	+	NR	+	NR
Other					
Liver	+	0	NR	++	+
Edema	+	0	NR	0	+

[a] NR = no evidence reported, 0 = not observed, + = observed, ++ = observed and more severe.

[b] M = male, F = female.

Source: Poland and Knutson, 1982.

Early studies which indicated a similar or mixed action for noncoplanar (conformationally restricted) PCB congeners may be suspect because of the later discovery of PCDF contamination (Goldstein, 1979). In fact, any study of "pure" congeners must be considered tentative until the question of coplanar PCA contamination has been adequately addressed. Nevertheless, a blanket rejection of toxic effects by conformationally restricted congeners based on possible contamination is unjustified.

The later demonstration that many pure congeners do have mixed activity, including a lower potency to mimic the TCDD syndrome (Table 8.10) (Safe, 1984, 1987), has further diminished the importance of congeners that have no Ah receptor activity. The net result has been to discount congeners which (1) have no Ah receptor activity and/or (2) were reported to have effects in studies which did not address coplanar PCA contamination. For example, early studies demonstrated that *ortho*-substituted PCB 1 and PCB 4 were significantly more uterotropic than *para*-chlorinated congeners as well as the Aroclor mixtures (Ecobichon and MacKenzie, 1974). Even though coplanar contamination would therefore not significantly influence the results, these congeners were further discounted, perhaps because of their rapid metabolism (Kaminsky et al., 1981). Thus, anaerobic dechlorination of PCBs which enriched sediments and waters with these

TABLE 8.10. Relative Ah Receptor Potencies of Some PCBs, PCDFs, and PCDDs

Congener	Body (g)	Liver (%)	Thymus (%)	TCDD Equivalent[b]
		Weights[a]		
Control	85	4.6	0.37	0.0
2,3,7,8 TCDD	72[c]	5.4[c]	0.24[c]	1.0
2,3,7,8 TCDF				.025
2,3,4,7,8 PCDF				.38
1,2,3,4,7,8 HCDF				.27
1,2,3,4,7,8 HCDD				.045
1,2,4,7,8 PCDD				.0046
1,2,4,7,8 PCDF				.0009
37 3,4,4'	82	5.4[c]	0.35	ND
52 2,5,2',5'	93	4.7	0.36	0.0
60 2,3,4,4'				.0000085
77 3,4,3',4'	86	6.0[c]	0.26[c]	.0027
101 2,4,5,2',5'	82	5.3[c]	0.32	0.0
105 2,3,4,3',4'	79	7.7[c]	0.26[c]	.0011
118 2,4,5,3',4'	102	7.6[c]	0.38	.0000083
126 3,4,5,3',4'	58[c]	5.3	0.18[c]	.40
128 2,3,4,2',3',4'	89	5.6[c]	0.35	<.0000089
153 2,4,5,2',4',5'	87	5.9[c]	0.36	0.0
169 3,4,5,3',4',5'	52[c]	5.4[c]	0.17[c]	.0016
170 2,3,4,5,2',3',4'	82	5.6	0.33	<.0000089
189 2,3,4,5,3',4',5'	98	6.4[c]	0.43	.0000102
Aroclor™ 1242				.0000137
Aroclor™ 1254				.0000099

[a] Parkinson et al., 1983 (changes in body weight were not always detected during the limited duration of the study).
[b] TCDD equivalent as measured by relative AHH induction (Safe, 1987).
[c] P < 0.01 by Student's t Test.

lower chlorinated congeners was considered detoxication (Brown et al., 1987). However, conformational restriction and hydroxylation appear to be the two most important criteria for PCBs to have estrogenic activity (Korach et al., 1988). Other significant subtle effects may also be overlooked in the futile search for a single unifying mechanism of PCA toxicity.

Heme metabolism, for example, has several vulnerable steps and is regulated by feedback mechanisms. Even though the excess capacity of heme synthesis requires a major interruption before heme levels are compromised, excessive production of intermediate products such as porphyrins can be toxic. The porphyria-related diseases are rather species specific and frequently slow to develop. Barbiturates and heavy metals are also responsible for some porphyrias, but PCA induced porphyria is usually the most serious type (Porphyria cutanea tarda). Polychlorinated benzenes, naphthalenes, dibenzodioxins, and noncoplanar as well as coplanar PCBs and many metabolites induce delta-aminolevulinic

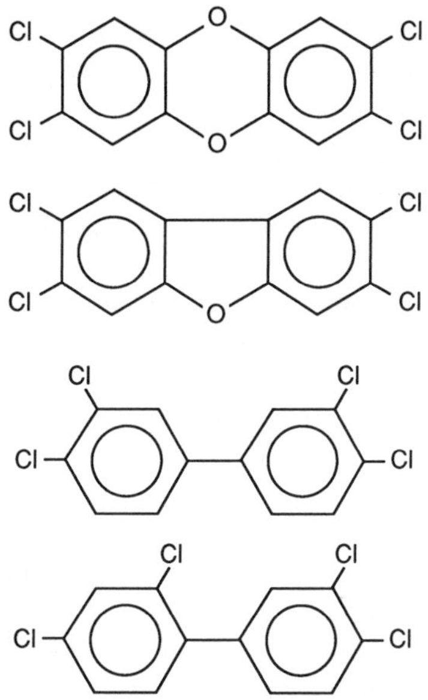

Figure 8.5. Comparison of 2,3,7,8-TCDD, 2,3,7,8-TCDF, 3,4,3′,4′-TCB, and 2,4,3′,4′-TCB.

acid synthetase (Kato et al., 1988). Additional actions such as inhibition of the subsequent decarboxylases exacerbate the condition. Several other environmental toxicants can exacerbate the development of porphyria which is more readily expressed following P_{450} induction and by heavy metals such as lead which inhibits additional synthetic enzymes such as ALA dehydratase and ferrochelatase in the heme synthesis pathway.

Gastrointestinal hyperplasia, ulceration, and hemorrhage have been observed in PCB poisoned swine exposed to moderate doses in rats exposed to higher doses and in cattle exposed to environmentally relevant doses when concomitantly subjected to stress and weight loss (Hansen, 1987b). Thus, these effects are probably not associated exclusively with the Ah receptor and may be more generalized PCA manifestations than Table 8.9 would indicate.

Chloracne is the most characteristic lesion in humans exposed to TCDD. Chloracne is a more specific and serious form of chemical acne and seems to be associated primarily with the Ah receptor in sensitive people. The "skin rash" frequently reported by persons exposed chronically to PCB may be a milder form of chemical acne not necessarily associated with the Ah receptor and should not be considered chloracne unless confirmed by medical diagnosis. Immune suppression is caused by a wide range of chemical toxicants including many types of PCAs. Thymic atrophy appears to be primarily Ah receptor-specific, but generalized responses such as increased susceptibility to infectious agents and suppressed antigen responses are probably outcomes of several different PCA actions, heavy metals, and other toxicants. A significant impact on immune competency may be associated with increased levels of glucocorticoids. Marked adrenocortical hyperplasia is seen in Baltic seals exposed to large amounts of PCAs, especially PCB and DDT, as well as other pollutants. The thymic involution observed in TCDD exposed animals, however, is not dependent on corticosteroids (Poland and Knutson, 1982).

Additional endocrine disturbances may be subtle, complex, and somewhat ambiguous. PCBs and many other toxicants directly alter the histological structure of the thyroid, while TCDD associated morphological changes may be related to effects on thyroid autoregulation resulting in altered thyroid hormone production, circulation, and excretion. PCBs can also compete with endogenous compounds such as thyroid hormones, retinol, and progesterone for transport protein binding sites and, thus, magnify other responses in the organism. On the other hand, this displacement may expose the hormones to more rapid degradation and excretion, expecially when accompanied by induction of catabolic monooxygenases and conjugating enzymes.

TCDD has antiestrogenic (Astroff and Safe, 1990) and antiandrogenic (Bookstaff et al., 1990) effects. Ah agonists such as TCDD and 2,3,4,7,8-PCDF do not bind to steroid hormone receptors and steroid hormones do not bind to the Ah receptor, but the Ah receptor in some way is thought to mediate the antiestrogen effects (Astroff and Safe, 1990). Some other nonsteroidal antiestrogens such as tamoxifen are weak estrogen receptor agonists as well as antagonists in some species. Ah receptor agonists have activities similar to steroidal antiestrogens (e.g., progestins), and decrease uterine nuclear estrogen and cytoplasmic estrogen receptor levels. On the other hand, conformationally restricted PCBs, and especially hydroxylated metabolites, bind to estrogen receptors and are estrogenic (Ecobichon and MacKenzie, 1974; Korach et al., 1988). Some of the effects observed following PCA exposure could also be related to actions on the hypothalamic-pituitary-gonadal axis; for example, Aroclor 1242 can potentiate *in vitro* gonadotropin-releasing hormone (GnRH) actions (Jansen et al., 1993) while TCDD can decrease GnRH action on the pituitary (Bookstaff et al., 1990).

The interactions among PCAs and cytochromes P_{450} also appear to contribute to a lack of uniform responses. Some perturbations may be due to increased steroid catabolism by induced P_{450}s. The cytochrome P_{450}s in steroid processing tissues such as gonads, pituitary, adrenal cortex, corpus lutea, etc. are less uniformly associated with the endoplasmic reticulum and may be induced in different proportions than those of the liver, GI tract, and lung which are more commonly responsible for xenobiotic metabolism. Some of the P_{450}s involved in steroid metabolism can bind xenobiotic substrates, but the turnover is much slower than with the steroids. This results in competitive inhibition of synthetic enzymes (e.g., adrenal), concurrent with the induction of catabolizing enzymes (e.g., liver). Induced P_{450}s can be inhibited by metabolites of PCBs (Hansen, 1987b; Nagayama et al., 1989). Phenobarbital induced P_{450} can enhance the metabolism and excretion of estradiol, testosterone, and thyroid hormones and increase the metabolism of conformationally restricted (noncoplanar) PCBs to estrogenic products. Thus, even though endocrine disturbances are a major effect of TCDD, actions unrelated to the Ah receptor are also significant by themselves and may antagonize or enhance many TCDD actions.

Certain teratogenic effects such as cleft palate in mice are associated with the Ah receptor, but embryotoxicity can result from noncoplanar PCBs and/or metabolites. Prenatal exposure can also result in later biochemical and behavioral changes in offspring which may not become apparent until postnatal development.

Alterations in endocrine function may result in nervous system and behavioral manifestations, but there also appears to be some direct nervous system effects. Both coplanar and noncoplanar PCBs can affect catecholamine levels in various portions of the brain (Hansen, 1987a, 1987b). Such actions can also affect reproductive functions because of the monoamine innervation of neuroendocrine organs. Epidemiological studies have associated prenatal PCB exposure with behavioral and learning decrements in humans, further confirmed by experimental exposures in rodents and primates (Schantz et al., 1989).

H. COMPARATIVE ENVIRONMENTAL TOXICITY

Acute lethality is not a serious consideration for most PCAs and certainly not for PCBs. There are PCB effects on aquatic plankton and invertebrates at water concentrations below 0.1 ppm which can modify populations and cause ecological changes (Hansen, 1987a). These effects may not be readily apparent since exposures have occurred gradually and the ecosystems may have adapted and changed long before PCA cause/effect relationships were considered.

The important effects on human health are those associated with chronic exposure, those which develop slowly after a pulse of exposure, and subtle effects on development which may occur following prenatal exposure. The subtlety and breadth of toxicity that are observed in laboratory experiments illustrate that the hazards from environmental, food chain, and occupational exposures to PCBs may be very difficult to evaluate; in fact, these hazards may only pose a serious risk when accompanied by exposure to other toxicants. Since concurrent or sequential exposure to additional toxicants is likely, we must determine the potential risk for the net adverse effects from various profiles of environmental chemicals, which include significant proportions of PCAs.

These hazards, however, are difficult to establish experimentally and may be impossible to assess accurately and completely by epidemiological methods.

Nevertheless, there is a wealth of information regarding the health status of vertebrates environmentally exposed to various levels of PCAs with various proportions of other common toxicants. These exposures are from environmental sources, insuring that the levels and combinations are relevant to human environmental health. Detailed examination of these situations and integration of net health effects with known experimental effects may be the only way to estimate the real risks posed by environmental PCAs. Guilt by association is less than ideal, but within a mixture of known toxicants the predominant chemicals must be considered to play at least some role until proven otherwise.

Cancer is a major concern in environmental risk assessment. The mechanism of carcinogenicity of PCBs, PCDFs, and PCDDs is still controversial, and although PCBs are established as tumor promoters in rodent liver, there is no firm evidence for a similar action in humans (Hayes, 1987). In some situations PCBs have been shown to antagonize both initiators and promoters of cancer, presumably through their potent induction of a broad spectrum of P_{450} monooxygenases and phase II enzymes. This same action of PCBs, however, may enhance effects of other carcinogens (note that Aroclor 1254 is the standard inducer used to detect promutagens requiring bioactivation in the Ames *Salmonella* mutagenicity assay). Studies of occupational exposure to PCBs have not yet been effectively correlated with cancer incidence (Chemical Manufacturers Association, 1988). Although Aroclor 1254 has been shown to antagonize the tumorigenic action of aflatoxin in trout, it is prudent to consider the possible PCB enhancement of polynuclear aromatic (PNA) carcinogenicity in other fish.

The association of liver lesions in English sole from different sites in Puget Sound with PNAs was greatest in the single site with the highest level of PCB contamination (Hansen, 1987a). Other examples of increased tumor incidence in fish may be due to many factors in addition to concurrent exposure to PCBs and PNAs, but liver enzymes in exposed fish are significantly more active in activating promutagens (Winston et al., 1988).

A high incidence of human bladder cancer has been reported in the St. Lawrence estuary area in which a heavy load of PNAs (estimated as benzo-[a]-pyrene concentration) and PCAs (PCBs and DDT) have been detected in the sediment. Fifteen stranded male beluga whales from the estuary contained 180 ± 75 ppm PCB in their blubber, and a young adult whale with 270 ppm of PCB in its blubber had an extensively metastasized urinary bladder carcinoma (Martineau et al., 1987; Hansen,1987a). This single description of a very rarely observed cancer is by no means a definitive correlation, but follow-up studies (Martineau et al., 1988) revealed six tumors in 15 whales that were examined and further supports a potential link between PCAs and PNAs and cancer.

Female beluga whales had about a fourfold lower concentration of PCB residues in their blubber than did the males, possibly due to the added excretory routes of parturition and nursing in females. However, a juvenile female was found with a blubber concentration of 576 ppm PCB and lesions consistent with PCB poisoning. Additional information concerning the relationship between contamination, pathology, and the lack of recovery of the St. Lawrence population of beluga whales suggests that the high PCB burden may be a contributory factor since the lesions described are consistent with signs of PCB poisoning in other mammals. In addition, the residue concentrations compare with those in the threatened seal populations in the Baltic.

PCB and DDT residues are higher in the threatened Baltic Sea seal populations than in other more stable populations from the same area (Hansen, 1987a). Threatened female seal populations with greater levels of DDT and metabolites in their tissues have pathological changes in their uteri while implantation failure and early abortion were frequently noted in those with relatively higher PCB residues. Other reproductive disorders such as premature births in California sea lions have also been loosely associated with elevated PCB/DDT residues in marine mammals. The high residues are the result of food chain biomagnification and are also associated with other pathological conditions, including immune suppression, gastric and intestinal ulcers, liver lesions, skin lesions, and kidney pathology in seals and beluga whales (Bergman and Olsson, 1985). Marked adrenocortical hyperplasia is common in adult Baltic seals and may be closely associated with other pathological findings (Bergman and Olsson, 1985). A high incidence of benign uterine tumors in Baltic grey seals seemed to be associated with age and not with PCAs.

The nearly total failure of mink ranching around Lake Michigan in the 1960s was eventually linked to PCBs in the fish diet. Mink are extremely sensitive to PCBs compared to other mammals, with reproductive failure occurring at dietary concentrations of 2 ppm Aroclor 1254 and estimated 3 to 9 month dietary LC_{50}s of 8.6, 6.7, and 0.1 ppm for Aroclor 1242, Aroclor 1254 and 3,4,5,3′4′5′-hexachlorobiphenyl, respectively (Aulerich et al., 1985). Signs of toxicity are consistent with the general effects of PCBs and include induction of P_{450} monooxygenases, degenerative kidney changes, anorexia, gastric ulcers, loss of hair, and enlarged liver, pancreas, and adrenal glands.

The PCB devastation of commercial mink production was preceded a couple years by a notable decline in reproduction and abnormalities of hatchlings in colonial and migratory water birds. Fish-eating birds also bioaccumulate PCAs from dietary sources. In the early 1970s it became apparent that the PCBs as well as PCDDs could be linked to low reproductive success and deformed chicks in populations from Long Island Sound, the lower Great Lakes, and Green Bay, Wisconsin (Kubiak et al., 1989). Similar PCB effects in domestic fowl were also demonstrated, with the focus of these studies being on reproductive effects and immune suppression (Hansen, 1987a, b). Behavioral effects and hormonal abnormalities are commonly observed in wild birds from contaminated areas or in those dosed experimentally with PCBs (Eisler, 1986).

Recent detailed analysis of Green Bay Forster's tern egg residues have suggested that a large proportion (>90%) of the embryotoxic activity could be due to the high TCDD equivalent concentrations of coplanar PCBs and a smaller proportion (>7%) attributed to TCDD (Kubiak et al., 1989). Several precautions and qualifications are presented regarding additive, synergistic, and antagonistic effects of other contaminants with PCB, but TCDD equivalents can be correlated with embryotoxicity and the recovery of populations suffering depressed reproduction can be correlated with reductions in TCDD residues. Nevertheless, the TCDD equivalency approach neglects other important effects. Potential impacts of media enriched in readily metabolized and/or conformationally restricted congeners (Bush et al., 1985; Brown et al., 1987) may be seriously underestimated and the real cause/effect relationships totally overlooked. These congeners can also affect neurochemistry, induce steroid metabolizing monooxygenases, and act as effective inhibitors of monooxygenases and some of the more estrogenic compounds and metabolites.

Adverse extrinsic effects on bird reproduction (e.g., nest abandonment or ineffective parental incubation) were also highly significant and may be related to any of the many PCAs found in the tern eggs (Kubiak et al., 1989). Since many combinations of PCAs can also cause aberrant parental behavior as well as changes in hormone levels (Eisler, 1986), it is prudent to consider PCA actions other than those associated with the Ah receptor as causes for the extrinsic effects on reproduction.

The increase in warnings from comparative environmental toxicology may have serious implications regarding the impact of PCAs, and particularly PCBs, on human environmental health. The reproduction problems of mink and Baltic seals were of concern prior to the demonstration of global PCB contamination in 1966. By 1970 effects on fish-eating birds became apparent and PCBs were suspect agents in reproductive problems encountered in other marine mammals. When the St. Lawrence population of beluga whales did not recover (as did other populations) after hunting pressures were removed, chemical pollution in which PCBs played a major role was belatedly demonstrated as an important factor. Adverse human reproductive effects and developmental deficits have been linked to PCB exposures from consumption of Great Lakes fish and exposure on farms with PCB-coated silos. Similar responses have been seen in experimentally exposed mammals, including primates (Hansen, 1987b; Schantz et al., 1989). These effects are not necessarily related to the Ah receptor. Humans exposed to TCDD show a relative resistance or moderate response to many of the Ah locus effects except chloracne.

Recent reconsideration of TCDD carcinogenesis in humans indicates that the risk, especially by the outdated EPA linear risk assessment model, has been greatly overestimated (Roberts, 1991). Adding to the ambiguity of Ah receptor-mediated carcinogenesis is a recent retrospective study of PCB exposed factory workers who showed an increased risk of brain cancer and malignant melanoma (NIOSH, 1991). Based on the nature of the capacitor impregnating process in this factory, workers would have been exposed to the more volatile and less persistent, lower chlorinated, and non-*para*-chlorinated biphenyls, so a lower risk would have been predicted using TCDD equivalency. Thus, even though the high toxicity of TCDD and coplanar PCBs should not be discounted, consideration of the metabolism of the active conformationally restricted PCB congeners may be more relevant than TCDD equivalency in assessing the risks of environmental PCAs to humans.

It is unlikely that environmental PCB burdens will decrease significantly soon (Tanabe, 1988), and it is prudent, in light of effects other than cancer, to decrease whatever exposures can be controlled. On the other hand, dispersive forces, and in some cases degradation, have decreased the concentration of many residues so, except for hot spots, geographical differences in PCB burdens are disappearing. The dispersion makes it more difficult to control moderate exposure, however, so increased attention must be given to reducing further discharges, the containment of hot spots, and avoiding practices which tend to increase exposures. The environmental toxicology of PCAs is certainly not completely understood, but the information is adequate to demand continued efforts toward further studies to understand their mechanisms of action in concert with planned attenuation of the ingress of these compounds into the environment.

REFERENCES

Astroff, B. and Safe, S. 2,3,7,8-TCDD as an antiestrogen: effect on rat uterine peroxidase activity, *Biochem. Pharmacol.* 39:485–488, 1990.

Aulerich, R.J., Bursian, S.J., Breslin, W.J., Olson, B.A., and Ringer, R.K. Toxicological manifestations of 2,4,5,2′4′5′-, 2,3,6,2′3′6′- and 3,4,5,3′4′5′-hexachlorobiphenyl and Aroclor 1254 in mink, *J. Toxicol. Environ. Hlth.* 15:63–79, 1985.

Bakke, J.E., Bergman, A.L., and Larsen, G.L. Metabolism of 2,4′5-trichlorobiphenyl by the mercapturic acid pathway, *Science* 217:645–647, 1982.

Ballschmiter, K., Schafer, W., and Buchert, H. Isomer-specific identification of PCB congeners in technical mixtures and environmental samples by HRGC-ECD and HRGC-MSD, *Fres. Zeitschr. Anal. Chem.* 326:253–257, 1987.

Bedard, D.L., Unterman, R., Bopp, L.H., Brennan, M.J., Haberl, M.L., and Johnson, C. Rapid assay for screening and characterizing microorganisms for the ability to degrade PCBs, *Appl. Environ. Microbiol.* 51:761–768, 1986.

Bergman, A. and Olsson, M. Pathology of Baltic grey seal and ringed seal females with special reference to adrenocortical hyperplasia: is environmental pollution the cause of a widely distributed disease syndrome?, *Finn. Game Res.* 44:47–62, 1985.

Bookstaff, R.C., Moore, R.W., and Peterson, R.E. 2,3,7,8-TCDD increases the potency of androgens and estrogens as feedback inhibitors of luteinizing hormone secretion in male rats, *Toxicol. Appl. Pharmacol.* 104:212–224, 1990.

Brandt, I. and Bergman, A. PCB methyl sulphones and related compounds: identification of target cells and tissues in different species, *Chemosphere* 16:1671–1676, 1987.

Brewster, D.W. and Birnbaum, L.S. Disposition of 1,2,3,7,8-pentachlorodibenzofuran in the rat, *Toxicol. Appl. Pharmacol.* 95:490–498, 1988.

Brown, J.F., Bedard, D.L., Brennan, M.J., Carnahan, J.C., Feng, H., and Wagner, R.E. Polychlorinated biphenyl dechlorination in aquatic sediments, *Science* 236:709–712, 1987.

Bush, B., Simpson, K.W., Shane, L., and Koblintz, R.R. PCB congener analysis of water and caddisfly larvae (Insecta: Trichoptera) in the Upper Hudson River by glass capillary chromatography, *Bull. Environ. Contam. Toxicol.* 34:96–105, 1985.

Chemical Manufacturers Association, PCB Program, 2501 M Street, NW, Washington, DC 20037, 1988.

Chen, M., Hong, C.S., Bush, B., and Rhee, G.Y. Anaerobic biodegradation of polychlorinated biphenyls by bacteria from Hudson River sediments, *Ecotoxicol. Environ. Safety* 16:95–105, 1988.

DeCaprio, A.P., McMartin, D.N., Silkworth, J.B., Rej, R., Pause, R., and Kaminsky, L.S. Subchronic oral toxicity in guinea pigs of soot from a PCB-containing transformer fire, *Toxicol. Appl. Pharmacol.* 68:308–322, 1983.

Ecobichon, D.J. and MacKenzie, D.O. The uterotropic activity of commercial and isomerically-pure chlorobiphenyls in the rat, *Res. Commun. Chem. Pathol. Pharmacol.* 9:85–95, 1974.

Eisler, R. Polychlorinated biphenyl hazards to fish, wildlife and invertebrates: a synoptic review, *U.S. Fish Wildl. Serv. Biol. Rep.* 85(1.7):72, 1986.

Erickson, M.D. *Analytical Chemistry of PCBs.* Butterworth Publishers, Stoneham, MA, 1986, 508.

Goldstein, J.A. The structure-activity relationships of halogenated biphenyls as enzyme inducers, *Ann. N. Y. Acad. Sci.* 320:164–178, 1979.

Hansen, L.G. Food chain modification of the composition and toxicity of PCB residues, *Rev. Environ. Toxicol.* 3:149–212, 1987b.

Hansen, L.G., Environmental toxicology of PCBs, *Environ. Toxin Series* 1:15–48, 1987a.

Hansen, L.G., Sullivan, J.M., Neff, C.C., Sanders, P.E., Lambert, R.J., Beasley, V.R., and Storr-Hansen, E. PCB contamination of domestic turkeys from building materials, *J. Agr. Food Chem.* 37:135–139, 1989.

Haraguchi, H., Kuroki, H., Masuda, Y., and Shigematsu, N. Determination of methylthio and methylsulphone polychlorinated biphenyls in tissues of patients with Yusho, *Food Cosmet. Toxicol.* 22:283–288, 1984.

Hayes, M.A. Carcinogenic and mutagenic effects of PCBs, in *Environ. Toxin Series* 1:77–96, 1987.

Jansen, H.T., Cooke, P.S., Porcelli, J., Liu, T.C., and Hansen, L.G. Estrogenic and antiestrogenic actions of PCBs in the female rat: in vitro and in vivo studies, *Reprod. Toxicol.* 7:237–248, 1993.

Jensen, S., Jansson, B., and Olsson, M. Number and identity of anthropogenic substances known to be present in baltic seals and their possible effects on reproduction, *Ann. N. Y. Acad. Sci.* 320:436–448, 1979.

Kaminsky, L.S., Kennedy, M.W., Adams, S.M., and Guengerich, F.P. Metabolism of dichlorobiphenyls by highly purified isozymes of rat liver cytochrome P-450, *Biochemistry* 20:7379–7384, 1981.

Kato, Y., Kogure, T., Sato, M., and Kimura, R. Contribution of methylsulfonyl metabolites of *m*-dichlorobenzene to the heme metabolic enzyme induction by the parent compound in rat liver, *Toxicol. Appl. Pharmacol.* 96:550–559, 1988.

Korach, K.S., Sarver, P. Chae, K., McLachlan, J.A., and McKinney, J.D. Estrogen receptor-binding activity of polychlorinated hydroxybiphenyls: conformationally restricted structural probes, *Mol. Pharmacol.* 33:120–126, 1988.

Kubiak, T.J., Harris, H.J., Smith, L.M., Schwartz, T.R., Stalling, D.L., Trick, J.A., Sileo, L., Docherty, D.E., and Erdman, T. Microcontaminants and reproductive impairment of the Forster's tern on Green Bay, Lake Michigan — 1983, *Arch. Environ. Contam. Toxicol.* 18:706–727, 1989.

Kuratsune, M. and Shapiro, R. PCB poisoning in Japan and Taiwan, *Am. J. Ind. Med.* 5:1–153, 1984.

Ligon, W.V., Jr., Dorn, S.B., May, R.J., and Allison, M.J. Chlorodibenzofuran and chlorodibenzo-*p*-dioxin levels in Chilean mummies dated to about 2800 years before the present, *Environ. Sci. Technol.* 23:1286–1290, 1989.

MacLeod, K.E. Polychlorinated biphenyls in indoor air, *Environ. Sci. Technol.* 15:926–928, 1981.

Martineau, D., Beland, P., Desjardins, C., and Lagace, A. Levels of organochlorine chemicals in tissues of beluga whales *(Delphinapterus leucas)* from the St. Lawrence Estuary, Quebec, Canada, *Arch. Environ. Contam. Toxicol.* 16:137–147, 1987.

Martineau, D., Lagace, A., Beland, P., Higgtins, R., Armstrong, D., and Shugart, L.R. Pathology of stranded beluga whales from the St. Lawrence Estuary, Quebec, Canada, *J. Comp. Path.* 98:287–311, 1988.

McKinney, J.D., Gottschalk, K.E., and Pedersen, L. A theoretical investigation of the conformation of PCBs, *J. Molec. Struct. Theochem.* 104:445–449, 1983.

Metcalf, R.L., Sanborn, J.R., Lu, P.-Y., and Nye, D. Laboratory model ecosystem studies of the degradation and fate of radiolabeled tri-, tetra- and pentachlorobiphenyl compared with DDE, *Arch. Environ. Contam. Toxicol.* 3:151–165, 1975.

Nagayama, J., Kiyohara, C., Mohri, N., Hirohata, T., Haraguchi, K., and Masuda, Y. Inhibitory effect of methylsulphonyl polychlorinated biphenyls on aryl hydrocarbon hydroxylase activity, *Chemosphere* 18:701–709, 1989.

National Academy of Sciences (NAS). Polychlorinated biphenyls. National Research Council, National Academy of Science Press, Washington, DC, 1979, 182.

NIOSH. Health hazard evaluation report: Westinghouse Electric Corporation, Bloomington, IN. HETA 89-116-2094. National Institute of Occupational Safety and Health, Cincinnati, OH, 1991.

Olafsson, P.G., Bryan, A.M., and Stone, W. PCBs and PCDFs in the tissues of patients with Yusho or Yu-Chen: total toxicity, *Bull. Environ. Contam. Toxicol.* 41:63–70, 1988.

Olson, J.R. Metabolism and disposition of 2,3,7,8-tetrachlorodibenzo-*p*-dioxin in guinea pigs, *Toxicol. Appl. Pharmacol.* 85:263–273, 1986.

Parkinson, A., Safe, S.H., Robertson, L.W., Thomas, P.E., Ryan, D.E., Reik, L.M., and Levin, W. Immunochemical quantitation of cytochrome P-450 isozymes and epoxide hydrolase in liver microsomes from PCB or PBB treated rats, *J. Biol. Chem.* 258:5967–5976, 1983.

Poland, A., Greenlee, W.F., and Kende, A.S. Studies on the mechanism of action of the chlorinated dibenzo-*p*-dioxins and related compounds, *Ann. N. Y. Acad. Sci.* 320:214–229, 1979.

Poland, A. and Knutson, J.C. 2,3,7,8-Tetrachlorodibenzo-*p*-dioxin and related halogenated aromatic hydrocarbons: examination of the mechanism of toxicity, *Ann. Rev. Pharmacol. Toxicol.* 22:517–554, 1982.

Roberts, L. Dioxin risks revisited, *Science* 251:624–626, 1991.

Safe, S., Safe, L., and Mullin, M. PCBs: environmental occurrence and analysis, *Environ. Toxin Series* 1:1–14, 1987.

Safe, S. PCBs and PBBs: biochemistry, toxicology, and mechanism of action, *CRC Crit. Rev. Toxicol.* 13:319–393, 1984.

Safe, S. Determination of 2,3,7,8-TCDD toxic equivalent factors (TEFs): support for the use of the *in vitro* AHH induction assay, *Chemosphere* 16:791–802, 1987.

Schantz, S.L., Levin, E.D., Bowman, R.E., Heironimus, M.P., and Laughlin, N.K. Effects of perinatal PCB exposure on discrimination-reversal learning in monkeys, *Neurotoxicol. Teratol.* 11:243–250, 1989.

Sipes, I.G. and Schnellmann, R.G. Biotransformation of PCBs: metabolic pathways and mechanisms, *Environ. Toxin Series,* 1:97–110, 1987.

Tanabe, S. PCB problems in the future: foresight from current knowledge, *Environ. Pollut.* 50:5–28, 1988.

Walsh, J. Seveso: the questions persist where dioxin created a wasteland, *Science* 197:1064, 1977.

Winston, G.W., Shane, B.S., and Henry, C.B. Hepatic monooxygenase induction and promutagen activation in channel catfish from a contaminated river basin, *Ecotoxicol. Environ. Safety* 16:258–271, 1988.

Chapter

9

Environmental Ionizing Radiation

Lorris G. Cockerham and
Michael B. Cockerham

A. INTRODUCTION

Electromagnetic radiation is divided into nonionizing and ionizing radiation according to the energy required to eject electrons from molecules (Sanders, 1986). Radiations that are nonionizing have lower energy levels and are not capable of producing ionization. Examples of these lower energy level electromagnetic radiations are ultraviolet, visible light, infrared, microwave, radio wave, electric wave, and sound. Ionizing radiation, which exhibits the properties of both waves and particles, has enough energy to produce ionization in matter. Those that exhibit corpuscular properties include alpha particles and beta particles, while those that behave more like waves of energy include X-rays and gamma rays.

Before discussing the effects of ionizing radiation, some of the many factors that influence the toxicity of radiation should be reviewed. One of the major factors related to the exposure is the dose or total amount of radiation received (Table 9.1). The absorbed dose of radiation is the quotient dE/dm where dE is the differential energy deposited into a differential mass, dm (Hall, 1984). The unit of absorbed dose in the CGS (centimeter-gram-second) system is the rad (radiation-absorbed dose) and 1 rad = 100 erg/g; a dose of 1 rad of ionizing radiation has been absorbed when 100 erg of energy have been deposited in each gram of material (Robertson, 1989). Another term commonly used, particularly in the field of radiation protection, is the rem (roentgen

0-8493-8851-1/94/$0.00+$.50
© 1994 by CRC Press, Inc.

TABLE 9.1. Radiation Quantities and Units Used in Radiobiology

Unit or Quantity	Symbol	Application
Becquerel	Bq	SI quantity of radioactivity Bq = 1 disintegration/sec Bq = 2.7×10^{-11} Ci
Curie	Ci	Quantity of radioactivity 1 Ci = 3.7×10^{10} dps 1 Ci = 3.7×10^{10} Bq
Gray	Gy	SI unit of absorbed dose 1 Gy = 100 rad = 1 J/kg
Rad	rad	Unit of absorbed dose 1 rad = 0.01 Gy = 100 erg/g
Rem	rem	Unit of dose equivalent rad $\times$ Q $\times$ other modifying factor 1 rem = 0.01 Sv
Sievert	Sv	SI unit of dose equivalent rad $\times$ Q $\times$ other modifying factor, 1 Sv = 100 rem
Linear energy transfer	LET	Energy deposition per unit of path length; usually in eV/μm
Relative biological effectiveness	RBE	Same effect from same dose of reference radiation, used in radiobiology
Quality factor	Q	Biological effectiveness of radiations
Electron volt	eV	Unit of energy 1 eV = 1.6×10^{-12} erg 1 eV = 1.6×10^{-19} J

Source: BEIR V, 1990. Reprinted with permission from *Health Effects of Exposure to Low Levels of Ionizing Radiation (BEIR V)*. Copyright 1990 by the National Academy of Sciences. Courtesy of the National Academy Press, Washington, D.C.

equivalent man). This unit was developed to enable radiation protection personnel to set standards of exposure (rem = rad $\times$ quality factor $\times$ distribution factor). The quality factor is a unit to equate the relative biological effectiveness (RBE) of one radiation to another, and the distribution factor attempts to compensate for the varying sensitivity of the different parts of the body. The roentgen, that amount of radiation required to produce 1 electrostatic unit of charge per cubic centimeter of air, is an older radiation exposure term still found in the literature. This is a measure of only the actual ionizations produced by X-ray or gamma ray irradiation in air.

In 1980 the International Commission on Radiological Units and Measurement (ICRU) introduced the "System Internationale" or SI units to express radiation dose (Hall, 1984). The Gray (Gy), the SI unit for absorbed dose, corresponds to an energy absorption of 1 J/kg or 100 rad. This concept of energy absorption is useful for determining absorbed doses of X-rays and gamma rays. However, determination of the absorbed dose in tissues exposed to fast neutron radiation involves more elaborate calculations which are beyond the scope of this discussion. Suffice it to say that the absorbed dose of neutron radiation depends on the transfer of energy from neutrons to directly ionizing particles in the tissue and must be described by the kinetic energy released in the material.

For general use, a quantity different from the rad or Gray has been introduced, the dose equivalent. The dose equivalent allows for the relative effectiveness of a particular type of radiation. Gamma rays and X-rays are regarded as the standard and a quality factor of 1 is multiplied by the dose to compute the dose equivalent. Therefore, the dose equivalent (Seiverts) for X-rays and gamma rays is equal to the dose (Grays). However, neutrons are thought to be roughly 10 times more effective in producing tissue damage than X-rays and therefore are assigned a quality factor of 10.

A factor influencing the toxicity of radiation is the dose rate ($D = dD/dt$, where the differential dose varies with respect to time, or where there is no variability in dose, [dE/dm], $D = D/t$). When reviewing experiments in radiation toxicology, these variables (i.e., total dose, dose rate, type of radiation, and variability of the model) must be considered.

The increasing use of radiation in the modern world, recent incidence of massive radiation exposure, and the advent of space exploration dictate that certain basic elements of radiation exposure be addressed. Exposure in the extraterrestrial environment will become especially significant when manned space missions move beyond the present low inclination earth orbits to the projected manned orbital space platform, when working colonies are established on the moon and with extended space missions to Mars (Conklin and Hagan, 1987; Bogo, 1988). A specific concern is the potential for exposure of astronauts to the ionizing radiation associated with solar flares and proton showers. Rust (1982) and Stauber et al. (1983) estimated that astronauts in a shielded area such as the inside of a space vehicle like the shuttle would receive a dose of 1.5 Gy during a major proton shower. An astronaut outside this shielded area during the same time could receive about 10 Gy. The major increase in planned extravehicular activities in space station missions would significantly increase the hazard of space radiation (Conklin and Hagan, 1987). In fact, the biological and medical effects of radiation could be the limiting factor to man's long term presence in space (McCormack et al., 1988).

B. EXPOSURE TO IONIZING RADIATION

The planet earth is essentially a closed system except for the input of cosmic radiation and debris. Most of us limit our concern of environmental

TABLE 9.2. Average Annual Effective Dose Equivalent of Ionizing Radiations

Source	Dose Equivalent		Effective Dose Equivalent	
	mSv	mrem	mSv	%
Natural				
Radon	24.0	2400	2.0	55.0
Cosmic	0.27	27	0.27	8.0
Terrestrial	0.28	28	0.28	8.0
Internal	0.39	39	0.39	11.0
Total natural	—	—	3.0	82.0
Artificial				
Medical				
X-ray diagnosis	0.39	39	0.39	11
Nuclear medicine	0.14	14	0.14	4.0
Consumer products	0.10	10	0.10	3.0
Occupational	0.009	0.9	<0.01	<0.3
Nuclear fuel cycle	<0.01	<1.0	<0.01	<0.03
Fallout	<0.01	<1.0	<0.01	<0.03
Miscellaneous	<0.01	<1.0	<0.01	<0.03
Total artificial	—	—	0.63	18
Total natural and artificial	—	—	3.6	100

Source: BEIR V, 1990. Reprinted with permission from *Health Effects of Exposure to Low Levels of Ionizing Radiation (BEIR V)*. Copyright 1990 by the National Academy of Sciences. Courtesy of the National Academy Press, Washington, D.C.

radiation to its damage to humans, perhaps on the beach or in the mountains. However, the quantity of environmental radiation could pose a risk to ecological balance. The radiation environment on earth could range from the irreducible natural background levels to extremely high levels following global nuclear warfare. Biological damage may be detected at levels slightly above the former, while the latter would make the planet uninhabitable. The sources of environmental radiation may be broken into two major components: natural and artificial or technologically induced radiation (Table 9.2).

1. Natural Background Radiation

Natural background radiation is the greatest contributor to radiation exposure in the world. In most countries natural background radiation contributes slightly more than half of the absorbed radiation dose (Mettler and Moseley, 1985). Relative contributions to the total absorbed dose may range from 42% in highly developed countries to about 94% in most developing countries. Exposure to natural sources of irradiation is unavoidable for the most part, and life has evolved under a continuous exposure to ionizing radiation. This background radiation has three components: (1) cosmic radiation (external), (2) terrestrial radiation (external), and (3) naturally occurring radionuclides (internal).

Cosmic Radiation

Cosmic radiation originates predominately from galactic sources and consists mostly of high-energy protons and alpha particles (BEIR III, 1980; Mettler and Moseley, 1985; Robertson, 1989). At the earth surface, cosmic radiation varies with altitude, geomagnetic latitude, and solar modulation (Hobbs and McClellan, 1986; Fry, 1987). For instance, in the U.S., 48% of the population lives at sea level to 152.5 m and receives a dose rate of about 27 mrem/yr (0.27 μSv/yr), while in Leadville, Colorado (altitude 3200 m), the residents receive about 125 mrem/year.

This effect of altitude becomes increasingly important to passengers and crews of high flying aircraft. It is estimated that cabin attendants and crew members receive about 160 mrem/year above that received at sea level (Eisenbud, 1987). The cosmic rays are attenuated by the earth atmosphere, resulting in a shielding effect which decreases with altitude. Cosmic ray exposure doubles every 1500 m above the earth surface (Mettler and Moseley, 1985).

Above the earth atmosphere radiation consists of two main components. One is the highly energetic cosmic radiation which is geomagnetically trapped in the earth's magnetic field. The second component is beyond the earth's magnetic field and is due to background cosmic radiation consisting of about 85% protons and 14% alpha particles. Astronauts traveling into outer space must traverse two belts of geomagnetically trapped radiation, the primary cosmic radiation, and radiation from solar flares.

Within the U.S. the effect of latitude on cosmic radiation dose rate is less than 10%, with an average dose rate at sea level of about 270 μSv/year (Mettler and Moseley, 1985). However, in the U.K. the annual dose rate varies from about 280 μSv/year in the south of England to 310 μSv/year in the north of Scotland (Fry, 1987). The variation in dose rate with latitude depends primarily upon variations in the earth's magnetic field, with which cosmic radiation interacts (BEIR III, 1980).

Terrestrial Radiation

Terrestrial radiation levels and rates from natural background sources are functions of geographic location and living habits. In most areas on earth the terrestrial radiation level varies within relatively narrow limits, but in certain regions of Brazil, China, France, Italy, Madagascar, and Nigeria the terrestrial radiation levels substantially exceed the normal range (Mettler and Moseley, 1985; Eisenbud, 1987; Luxin et al., 1990). For instance, a person tanning on some beach along the Atlantic coast of Brazil may receive as much as 17.5 cGy/year from the sand alone (Mettler and Moseley, 1985). Meanwhile, exposure from the fine monazite particles of the soil in the Dong-anling and Tongyou regions of China would be between 18 and 20 cGy/year (Luxin et al., 1990).

The conterminous U.S. may be divided into three general radiation regions (BEIR III, 1980). The Atlantic and Gulf coastal plains receive an average of 23 mrems/year while the range in the Colorado plateau area may be as high as 140 mrems/year. The average terrestrial level for the remainder of the U.S. is only 46 mrems/year with an estimated national average of 40 mrems/year.

The terrestrial radiation rate varies with the type of soil in the area and the naturally occurring radionuclide content of the soil. Approximately 70 of the 340 nuclides found in nature are radioactive (Eisenbud, 1987). These radionuclides have existed in the earth's crust since its formation and are known as "primordial radionuclides". These primordial radionuclides have half-lives comparable to the age of the universe and are the source of terrestrial radiation (Mettler and Moseley, 1985).

Three distinct chains of primordial radioactive elements are found in the earth's crust and account for much of the terrestrial radiation exposure (Mettler and Moseley, 1985). These are (1) the uranium series, (2) the thorium series, and (3) the actinium series. Uranium, the origin of the actinium series, is found in various quantities in rocks and soils. The uranium isotopes are alpha emitters and therefore do not contribute to the gamma background radiation. Uranium in soils and in fertilizers can be absorbed by plants and, via the food chain, be subsequently found in animal tissues. At equilibrium, an adult human male may be expected to have a uranium body burden of 100 to 125 μg. The ^{232}Th decay series may also move through the food chain, but due to its relative insolubility and low specific gravity it is present in biological materials only in insignificant amounts (Eisenbud, 1987). Thorium may be found in silty clay and peaty soils and can be absorbed by vegetables such as potatoes, corn, carrots, beans, and squash. However, the principal source of human exposure to thorium is through inhalation of soil particles. Thorium is removed very slowly from bone and its concentration is found to increase with age.

^{226}Ra, an alpha emitter originating in the uranium decay series, is present in varying amounts in all rocks, soils, and water and is of special importance along with its daughter products (Mettler and Mosely, 1985). ^{226}Ra, with a half-life of 1622 years, decays to radon (^{222}Rn), a noble gas radionuclide with a half-life of 3.8 d. Radon also emits alpha particles but adds to the gamma radiation level of the environment through its gamma-emitting daughters.

Radium is very similar to calcium in its chemical properties and is absorbed by plants from the soil in a manner similar to calcium. It then passes through the food chain to humans, where 70 to 90% is concentrated in bone. The amount of ^{226}Ra moving through the food chain depends upon its content in the soils and its rate of absorption by plants, which in turn is dependent on the amount of exchangeable calcium in the soil. Brazil nuts, which have a tendency to concentrate barium (another chemical very similar to radium) may have a ^{226}Ra content about 1000 times greater than the average diet.

Internal Radiation

Internal radiation, the third component of natural background radiation, results from naturally occurring radionuclides contained within the body and contributes about 11 to 17% of the average radiation exposure of the population (Southwood, 1987; BEIR V, 1990). While some of the radionuclides are freely dispersed throughout the body, others are concentrated in specific organs. All of the emitted decay energy from these sources is absorbed locally (Robertson, 1989; Yamamoto et al., 1990). The absorption and deposition of naturally occurring radionuclides of bismuth, carbon, hydrogen, lead, polonium, potassium, radium, radon, thorium, and uranium result primarily from the inhalation and ingestion of these materials in air, food, and water (BEIR III, 1980).

In a terrestrial ecosystem radionuclides that occur naturally in soil, or are deposited in the soil, are incorporated metabolically into plants (Eisenbud, 1987). The absorption of radionuclides from soil depends on the chemical form and distribution coefficient of the radionuclide as well as the metabolic requirements of the plant and physicochemical factors in the soil. In addition to root absorption, plants are contaminated by direct foliar deposition. Foliar deposition is potentially a major source of food chain radionuclide contamination since the radionuclide may be absorbed by the plant or transferred directly to animals consuming or coming in direct contact with the foliage.

Atmospheric radionuclides are eventually deposited on surface waters as well as on soil. Therefore, the atmosphere is coupled to soils, surface waters, and subsurface aquifers. Radionuclides are removed from surface soils by processes including surface runoff and leaching into soil water. They are eventually transported into streams or subsurface aquifers. Radionuclides that leach into deep underground aquifers may eventually reach surface waters and become incorporated into the biosphere again.

Rivers, estuaries, and coastal waters are major receptors of effluent radionuclides from industrial plants and cities. These waters are of special importance because of their high biological activity and productivity. Phytoplankton in these relatively shallow waters convert inorganic compounds in the aquatic environment into food for higher organisms. Zooplankton, the basic food of several higher trophic levels, uses phytoplankton for nourishment. Certain bottom dwelling fish and animals also consume phytoplankton.

The risk of consumption of radionuclides in marine and fresh water organisms depends, in part, on where the radionuclide is located in the organism. A radionuclide is more of a risk if it is concentrated in an organ consumed by higher organisms, such as humans, than if it is deposited in a portion that is not eaten. The radionuclides of cobalt (^{60}Co) and zinc (^{65}Zn) concentrate in edible tissues while those of radium (^{226}Ra) and strontium (^{90}Sr), although concentrated by clams, oysters, scallops, and certain crabs, are stored in the shell which is not ordinarily consumed (Eisenbud, 1987).

Uptake and retention of radionuclides by an organism are influenced by the portal of entry, the chemistry and solubility of the compound, particle size, and metabolism. Uptake normally occurs via three principal routes of entry: inhalation, ingestion, and skin absorption. Of the three, inhalation poses the greatest risk. Particles greater than 10 μm in diameter do not penetrate into the lung. These particles are removed by the nasal hairs or cilia found in the upper respiratory tract. Therefore, particle size will determine the compartment into which the particle is deposited.

Ingestion of contaminated food contributes to exposure. Gastrointestinal (GI) exposure depends upon transit time through the gut, while absorption depends upon the solubility of the radionuclide. Contamination of skin with radionuclides is of less consequence since the skin forms a formidable barrier. However, contamination of an open wound may result, not only in continuous radiation of the surrounding tissue but in the introduction of the radionuclide into the rest of the body.

Regardless of the portal of entry, the radionuclide passes throughout the body of the animal and is deposited into the milk, flesh, and eggs. When the radioactive material enters the body it becomes an internal emitter. It will continue to radiate the body until it is excreted by some physiologic process, mainly in the urine and feces, or until its radioactivity decays (Cerveny and Cockerham, 1986). The time it takes an organism to eliminate half of the radionuclide is known as the *biological half-life* and the time required for a radionuclide to decay to half of its activity is the *physical half-life* (Table 9.3). If the biological and physical half-lives are known for a particular radionuclide, the *effective half-life* may be calculated (Mettler and Moseley, 1985).

2. Radon

Although not mentioned specifically in the above discussion of internal radiation from radionuclides, radon (^{222}Rn), the short-lived decay product of ^{226}Ra, may be the most important internal radionuclide, accounting for about 60% of the effective dose equivalent from internal emitters (Mettler and Moseley, 1985). As seen in Table 9.2, radon and its decay products contribute 55% of the total average annual effective dose equivalent of 3.6 mSv.

^{222}Ra is a naturally occurring, colorless, odorless, tasteless, radioactive gas formed from the decay of uranium and radium. Since uranium has been present since the earth was formed and has a half-life of 4.5×10^9 years, radon has also

TABLE 9.3. Half-Lives of Some Biologically Significant Radionuclides

Nuclide	Physical	Biological[a]	Effective
^{14}C	5.73×10^4y	40 d	40 d
^{137}Cs	30 y	70 d	70 d
^{131}I	8 d	12 d	8 d
^{55}Fe	657 d	2000 d	494 d
^{32}P	14 d	260 d	14 d
^{239}Pu	2.4×10^5y	180 y	180 y
^{24}Na	15 h	11 d	14 h
^{90}Sr	29 y	36 y	16 y
^{3}H	12 y	12 d	12 d
^{235}U	7.1×10^8y	20 d	15 d
^{65}Zn	245 d	400 d	152 d

[a] Whole body.

Source: Mettler and Moseley, 1985; Cerveny and Cockerham, 1986; Sanders, 1986; Eisenbud,
       1987.

been present for some time and will likely exist for many years at the same level as it is now.

By the early part of the 20th century radon was being linked with lung cancer while being used therapeutically. In the 1960s attention was focused on the emission of radon from uranium mill tailings in parts of Colorado and Utah. An enterprising contractor was using these mill tailings for aggregate in the construction of houses. However, since uranium is not mined in many places, concern regarding its use did not remain in the national spotlight for very long.

It was not until December 1984 that the presence of radon became a major national issue. A man named "Wattress" from Boyertown, Pennsylvania, working in a power plant near Philadelphia, set off all the radiation detection alarms one day when he came to work. After an intensive investigation, the "Wattress" house was determined to be the source of the radiation. Radiation in the air inside the Wattress house measured 2700 pCi/L, the highest level that has ever been recorded in a house (Mendelsohn, 1988). Once radon was proven to be the source of the radiation, environmental radon became a national issue. Yet, despite a major effort to characterize indoor radon, there has been no direct comprehensive determination of the concentration of radon in U.S. homes (Nero, 1988).

Production

Production of radon is primarily from the widespread distribution of uranium and its decay products in the soil. Every square mile of surface soil to a depth of 6 in. contains about 1 g of radium, a decay product of uranium. Decay of radium releases radon in small amounts to the atmosphere (Nazaroff et al., 1988a). The radium content of the soil reflects the radium content of the rocks from which it was formed, and radon released to the atmosphere is increased in areas with granite formations and uranium and thorium ore deposits.

Mineral waters were thought to have some curative powers in Roman times, and it is known that many mineral springs contain relatively high concentrations of radium and radon (Eisenbud, 1987). Spas in Europe, Japan, South America, and the U.S. have exploited the alleged curative powers of radioactive waters. Some of the best known mineral springs in the U.S. are at Saratoga, New York and Hot Springs, Arkansas. Visitors to spas not only drink and bathe in the radioactive waters but may, as at Badgastein, Austria, sit in caves where radon emanates from the surrounding rocks.

Potable water supplies may also contain radon due to the deposition of radium isotopes. Groundwater supplies may contribute significantly to indoor radon concentrations. Private groundwater supplies constitute a somewhat greater source of radon than do public supplies (Nazaroff et al., 1988b).

Commercially produced radon has been used by radiopharmaceutical companies and hospitals to produce radioactive seeds or needles which are then implanted into tumors (Cohen, 1979). For the most part, the use of radon in this procedure has been replaced by radionuclides made in accelerators and nuclear reactors.

Use

Each year thousands of people throughout the world use radon in some form for therapeutic purposes. Ailments treated in various ways include acne, allergies, arthritis, asthma, diabetes, gout, hypertension, and ulcers. The National Health System in the U.S.S.R. prescribed radon bath treatments daily. Medical uses of radon to treat malignancies began in the U.S. in the early part of the 20th century and was used in the treatment of dermatological disorders as late as 1950.

The emanation of ^{222}Rn from soil and its concentration in groundwater have been used more recently as good predicators of earthquakes, for the study of atmospheric transport, and for exploration of petroleum and uranium deposits (Cothern, 1987).

Environmental Entry

^{226}Ra in soil is the largest single source of radon in the atmosphere (NCRP, 1984). Radon is continually being formed in the soil and released into the air at about 2×10^9 Ci ^{222}Rn/year (Harley, 1973). The rate varies from one location to another depending on many factors including soil composition, ambient temperature, and barometric pressure. A small amount of ^{222}Rn is absorbed by plants and then released into the air.

When ^{222}Rn enters the atmosphere from the ground it is dispersed into the air and its concentration in open areas is low. However, when ^{222}Rn enters a building, either from the building material, the soil beneath the floor, or around the walls, the concentration builds up because of the restrictions in air movement within the building and lack of air supply from the outside. The ^{222}Rn and its decay products with short half-lives may adsorb to dust particles in the air and be inhaled.

Groundwater that is in contact with rock or soil containing radium will pick up the ^{222}Rn and release it to the atmosphere when the water comes to the surface. With a contribution of 5×10^8 Ci ^{222}Rn/year, groundwater is considered the second largest source of environmental radon. Tailings from uranium mines, coal residues, natural gas, and some building materials release small amounts of radon to the environment.

Environmental Levels

Although the units of radiation and absorbed dose were defined earlier, the measurement of radioactivity was not discussed. The unit of radioactivity for many years has been the *curie* (Ci), which is defined as the quantity of radioactivity producing 3.7×10^{10} nuclear disintegrations per second. The System Internationale introduced in 1980, defined new terms for radioactivity and radiation dose. The newer SI unit being substituted for the curie is the *becquerel* (Bq), with 1 Bq being equal to one disintegration per second.

Not only are radioactivity levels of ^{222}Rn written as pCi/L or Bq/m^3, but as *working level* (WL) or *working level month* (WLM). The WL for radon exposure is defined as any combination of short-lived radon daughter radionuclides in 1 L of air that will result in the emission of 1.3×10^5 MeV of potential alpha energy (BEIR IV, 1988). This corresponds to a ^{222}Rn concentration of 3.7 kBq/m^3. Exposure to one WL for a month of 170 working hours will result in one WLM. However, a 30-d month is equal to 720 h and exposure to one WL for 720 h gives a cumulative exposure of 4.235 WLM. Therefore, remaining at home for 12 h/d for a month at an exposure level of 1 WL would result in an exposure of greater than 2 WLM/month.

Typically, the ^{222}Rn concentration in air over soil is 4 Bq/m^3 (100 pCi/m^3), but this value may vary by a factor of two or more with seasonal and diurnal variations (BEIR IV, 1988). Seasonally, levels are highest in early autumn and lowest in early spring, while the diurnal variations show an early morning peak and a decrease in the afternoon.

Distribution studies of household levels showed that 7% of the U.S. single-family houses have concentrations greater than 150 Bq/m^3 and that possibly a million houses in the U.S. have annual average ^{222}Rn concentrations of 300 Bq/m^3 or more. Average winter concentrations found in houses in the Spokane River Valley of Washington and Idaho and in eastern Pennsylvania were on the order of 500 Bq/m^3. Some houses in eastern Pennsylvania had levels as high as 100,000 Bq/m^3, and in one neighborhood of Clinton, New Jersey, winter concentrations exceeded 7500 Bq/m^3 (Nero, 1988).

^{222}Ra levels in surface water are essentially zero. However, levels in aquifers of igneous and metamorphic rocks are high. Aquifers associated with granite consistently show the highest levels of ^{222}Rn, averaging 100 Bq/L (Cothern et al., 1986).

Radon in soil is usually referred to as soil-gas and the levels in soil are affected primarily by the radium content and its distribution in the soil, soil porosity, moisture, and density. Measurements of ^{222}Rn soil-gas have been minimal but those reported range from 7000 Bq/m^3 in Spokane, Washington to 1,000,000 Bq/m^3 in Reading Prong, New Jersey (Michel, 1987). Plants apparently absorb ^{222}Rn from soil but little information concerning this process is available. Pearson (1967) reported that ^{222}Rn emanated from the leaves of field corn. The reported emanation rate of 9.15×10^{-6} Bq/cm^2/sec was almost twice that reported from the surrounding soil.

Environmental Fate

Regardless of where it is found, the ultimate fate of ^{222}Rn is degradation by radioactive decay. ^{222}Ra, with a half-life of 3.82 d, degrades by alpha-emission to ^{218}Po. The radon daughters are electrically charged and attach to inert dust in the atmosphere, resulting in the formation of radioactive dust. Because most of the daughters are short-lived, with ^{210}Pb having the longest half-life of 22 years, equilibrium is reached in the air in about 22 h (Eisenbud, 1987). The concentration of ^{222}Rn daughters in the atmosphere can be decreased by the passage of air containing dust over the ocean or by a thunderstorm. In any event, the mean life of the atmospheric dust with attached radon daughters is thought to be 15 d. However, radon and radon daughters are transformed only through radioactive decay, emitting alpha particles in the process.

3. Technologically Induced Radiation

Health Sciences

The use of man-made radiation in the health sciences is normally divided into three areas: (1) diagnostic X-ray examination, (2) nuclear medicine, and (3) therapeutic radiation. The use of X-rays in diagnostic examinations, including dental, represents the single largest man-made source of radiation exposure in the U.S. (BEIR III, 1980). In the more highly developed countries such as the U.S. exposure from medical sources may even exceed natural background exposure (Mettler and Moseley, 1985). Dental X-rays are the most common of the diagnostic examinations with outpatient examinations contributing 30% of the exposure.

The use of radiopharmaceuticals in nuclear medicine has almost doubled over a 10-year period. It is estimated that up to 12 million doses of radiopharmaceuticals are dispensed each year in the U.S. for diagnostic purposes

(BEIR III, 1980). However, the per capita effective dose equivalent from these procedures in the U.S. is only about 140 µSv (Mettler and Moseley, 1985).

Radiation therapy has been used almost exclusively for the treatment of malignant neoplasms. The high absorbed dose, 50 to 70 Gy, required in most malignant conditions leads to *nonstochastic* or direct effects such as cell death. Therefore, normal tissue surrounding the neoplasm may also be exposed and incur long range risk. The risk, however, may be eclipsed by the immediate benefits associated with increased life expectancy resulting from the destruction of the neoplasm.

Nuclear Power Production

When radiation exposure from nuclear power production is mentioned, the immediate thought is of nuclear power reactors and the environmental dispersion of radionuclides, particularly ^{85}Kr, ^{3}H, ^{14}C, and ^{129}I. However, exposure from nuclear power production should also include mining, uranium fuel fabrication, and waste storage and disposal (Mettler and Moseley, 1985).

Although uranium mining increases the amount of uranium and its decay products, including radon and its daughters, the environmental risks from the radioactive emissions from uranium mines is insignificant (Hobbs and McClellan, 1986; Eisenbud, 1987). However, mill tailings may represent a significant source of environmental radiation due to the emanation of ^{222}Rn, dispersion of the tailings by wind and water, and by the use of mill tailings in building construction.

About 1000 land-based nuclear reactors have been constructed and are operational throughout the world. Some of the reactors were built for research or the production of radioisotopes and plutonium. Approximately 200 naval vessels throughout the world are powered by nuclear reactors. Yet, the environmental release from nuclear operations in the U.S. results in a dose rate for the average person of <1 mrem/year (BEIR III, 1980).

The discussion has focused primarily on radiation released from nuclear power production during normal operations. These values do not necessarily hold true during accidents or in the handling, storage, and disposal of radioactive waste.

Nuclear Weapons

The first atomic weapon was detonated on a July morning in 1945 on a New Mexico desert north of Alamagordo. Since that day, hundreds of test explosions have been conducted by the U.S., the U.S.S.R., the U.K., India, France, and the People's Republic of China. In 1963 the U.S., the U.K., and the U.S.S.R. signed an agreement to end atmospheric testing. However, France, China, and India did not sign this agreement and continued to conduct atmospheric testing (Eisenbud, 1987). Between 1945 and 1984 the total estimated yield of all atmospheric nuclear explosions was about 546 Mton (Mettler and

Moseley, 1985). A 1 Mton explosion equals the explosive force of 1×10^6 ton of TNT.

The detonation of a nuclear weapon generates enough heat to vaporize the weapon instantly and stop the nuclear reaction. The nuclei created by the reaction are now highly energetic and emit high-energy radiation such as X-rays and gamma rays as they return to a lower energy state. The released radiation heats the surrounding air, forming a shock wave and a fireball. For a nuclear weapon with a yield of 1 Mton, the fireball will rise at a rate of almost 400 ft/sec and reach an altitude of about 60,000 ft. Convective forces initiated by the fireball will result in enormous amounts of air, soil, and debris being sucked upward, creating a crater as much as 400 ft deep and 1200 ft in diameter. As the fireball cools, the newly created, highly unstable nuclei condense onto particles of soil and debris, returning to earth as radioactive fallout (Fetter and Tsipis, 1981).

In a nuclear explosion, over 400 radioactive isotopes are released into the biosphere. Among these, about 40 radionuclides are considered potentially hazardous. Of particular interest are those isotopes whose organ specificity and long half-lives present a danger of irreversible damage or induction of malignant alterations. Both early and delayed fallout result in the deposition of radioactive material in the environment (Cerveny and Cockerham, 1986).

The radioactive fallout from nuclear explosions may be divided into three portions depending on the yield and height of the burst. The larger, intensively radioactive particles fall close to the site within hours. Slightly smaller particles behave somewhat like aerosols and are dispersed into the troposphere where they will stay for several months. The fallout from this portion remains in bands around the earth at the latitude of the detonation. The third portion penetrates the stratosphere and its particles are deposited worldwide over a period of months to years (Eisenbud, 1987). Most of the radioactive fallout is downwind from the explosion and up to 70% is comprised of the largest particles, returning to earth close to the detonation site within hours. The intensity of the radioactivity varies inversely with distance from the site of explosion. With a steady wind, the pattern of accumulated dose of radioactivity assumes the shape of nested cigar-shaped contours, each contour denoting a particular dose.

A 1 Mton thermonuclear weapon, detonating at ground level with a steady wind of approximately 15 mi/h, would produce a radioactive fallout dose rate of 400 rem in 24 h in an area of approximately 400 sq mi. At a dose rate of 2 rem/year, more than 20 times the maximum recommended by the EPA, an area of 1200 sq mi would remain unfit for use for a year and more than 20,000 sq mi would be uninhabitable for a month (Fetter and Tsipis, 1981).

In 1961 the U.S.S.R., without warning, broke a 3-year moratorium on atmospheric testing and detonated 50 nuclear devices. This led to a renewed competition in the detonation of these devices between the U.S. and the

U.S.S.R. for the next 2 years causing an abrupt increase in fallout to an all-time high (Eisenbud, 1987). During this time the U.K. reported a fallout dose of approximately 80 μSv (Fry, 1987). This increase in fallout led to worldwide pressure to halt atmospheric testing, resulting in the signing of a nuclear weapons test ban agreement early in 1963 by the U.S., the U.K., and the U.S.S.R. (Eisenbud, 1987).

After the signing of the agreement and elimination of open atmospheric testing by the signatory nations, the average fallout dose in the U.K. fell to 7 μSv in 1984 and, with one exception, has remained between 6 μSv and 10 μSv since (Fry, 1987). The annual average whole-body fallout rate in the U.S. is now approximately 45 μSv (4.5 mrem) and is projected to stay at this level through the year 2000 (BEIR III, 1980; Mettler and Moseley, 1985).

Accidents

Although the environmental release of radionuclides from nuclear reactor operations is about 1 mrem/year per person, malfunctions can occur and accidents can happen (BEIR III, 1980). On March 28, 1979 the worst accident in the history of U.S. commercial nuclear power generation occurred on Three Mile Island in Pennsylvania (Hobbs and McClellan, 1986). Although radiation exposure to the plant workers and the public was insignificant, the nuclear power industry was set back almost a decade. Orders for the construction of new nuclear plants were canceled (Eisenbud, 1987).

This is not to say that all nuclear power plant accidents will not result in the discharge of radionuclides. Not all of the nearly 800 nuclides produced in the reactor are radioactive and of these only 54 are considered a significant risk (Eisenbud, 1987). With core damage, the severity of the accident and therefore the risk, depends mainly on the release of radioactive isotopes, mainly ^{131}I and ^{137}Cs, into the environment.

Since 1952 there have been 14 reactor accidents that involved core damage. One, the Windscale, U.K. Atomic Energy works accident, was the first time radioactive material was released from a reactor accident. In October 1957, a plutonium production reactor located on the coast of Cumbria in northwest England released about 20,000 Ci [131]I (Eisenbud, 1987). The core in one of the two air-cooled, graphite-moderated, natural uranium reactors was partially consumed by fire, releasing the fission products onto the seashore and foothills southwest of the Cumbrian Mountains. As an aftermath of this accident four fatal leukemia cases in people that were under 20 years of age between 1950 and 1980 were recorded in the nearby village of Seascale. Based on statistical epidemiological calculations, only 0.5 cases would have been expected (Stather et al., 1987).

More recently, at 1:23 a.m. on Saturday, April 26, 1986, an explosion occurred at the water-cooled, graphite-moderated reactor of Unit No. 4 at

Chernobyl power station along the Pripyat River about 90 km north of Kiev, Ukraine. This accident resulted in the release of an estimated 81 MCi of radioactivity into the atmosphere by May 6, 1986 and was the most costly industrial accident in history (Eisenbud, 1987).

The radioactive cloud, because of wind direction, moved northwest from Chernobyl, covering much of Scandinavia by Monday, April 28 (Clarke, 1987). On the Baltic coast of Sweden, the radiation levels were 14 times the normal background level. By Wednesday, April 30, the wind direction at Chernobyl had shifted and part of the radioactive cloud began moving to the southeast. A high pressure also developed over eastern Europe, splitting the cloud into two parts. The western portion of the cloud covered all or part of Scandinavia, Poland, Germany, Austria, Czechoslovakia, Hungary, Switzerland, Italy, and Yugoslavia. By Friday, May 2, the eastern portion of the cloud had covered all or part of Romania, Bulgaria, Yugoslavia, Greece, and Turkey, while the western portion had moved over part of France and the U.K. The Netherlands and Belgium were completely covered by this time.

One week after the accident two masses of contaminated clouds still existed, one over northwestern Europe and the other over southeastern Europe. By Monday, May 5, the largest contaminated cloud lay over southern Germany, most of Italy, Greece, and eastern Europe. The northwestern cloud was then dispersing over the Atlantic.

The accident at Chernobyl released large quantities of various radionuclides in the atmosphere, resulting in widespread fallout throughout the northern hemisphere (Fry, 1987). This, in turn, led to a large increase in measured values of a range of radionuclides in various environmental domains. In all cases the most prominent radionuclides were ^{131}I, ^{134}Cs, and ^{137}Cs. The highest levels of ^{131}I and ^{137}Cs fallout were detected in southern Germany, Austria, northern Italy, and Greece (Clarke, 1987). The radionuclide deposition was dependent upon meteorological conditions during passage of the contaminated air. In the U.K. high deposition occurred in areas where it rained during passage of the contaminated air mass. By a strange twist of fate, these areas included north Wales, southwest Scotland, and the Lake District of England (Fry, 1987). This latter area of England was the site of the 1957 Windscale accident.

In comparing the size of the releases and radiological consequences of the 1957 Windscale accident, the Three Mile Island (TMI) accident in 1979, and the Chernobyl accident in 1986, a striking difference in the impact among the three accidents was noted (Clarke, 1987). While the TMI release was higher than at Windscale, the effective dose was much lower. This was due to the radionuclides released, ^{131}I at Windscale and ^{133}Xe at TMI, and that the fission products at TMI were retained within the vessel. The release from the Chernobyl accident was 10 times higher than from TMI but the doses were hundreds of times higher than from TMI. Again, this was partially due to the radionuclides released at Chernobyl (^{131}I and ^{137}Cs). The total impact of the Chernobyl accident was greater in the U.K. than was the Windscale accident, while the total impact of the TMI accident was 100 times less in the U.S. than the Windscale accident was in the U.K.

Four years after the Chernobyl accident Soviet diplomats acknowledged that the medical, environmental, and political consequences are much greater than previously discussed (Barringer, 1990). Four million people in Ukraine, Byelorussia, and western Russia are said to be living on contaminated ground, and the death rate of children with leukemia in Minsk is 50 times greater than a year before the accident. One Soviet diplomat, Victor Borovikov, said it could take more than 200 years to clean up the damage from the 1986 nuclear accident (*Arkansas Gazette,* 1990).

Waste Management

The disposal of radioactive waste is a dilemma facing a technologically advanced society. One of the basic demands of such a growing society is availability of convenient and inexpensive sources of energy. As the demand for energy increases, the reliance on nuclear power will increase, as will the production of radioactive waste. The long-term storage and disposal of high-level nuclear waste is a problem that has not been adequately resolved.

Radioactive waste is classified according to its physical and chemical properties as well as its source (Eisenbud, 1987). The three general categories of radioactive waste are

1. Low-level waste
   a. Residues from laboratory research
   b. Uranium mill tailings
   c. Waste generated in the cleanup of uranium, radium, and thorium processing plants
2. High-level waste
   a. Spent fuel from civilian nuclear power reactors
   b. Liquid and solid residues from the reprocessing of civilian spent fuel
   c. Liquid and solid wastes from the reprocessing of fuel used for military purposes
3. Transuranic wastes — mainly alpha-emitting residues from military manufacturing

Low-level radioactive waste management is facing a crisis, not because of technical problems and environmental risks involved, but because of the sociopolitical ramifications of the widespread public concern of the perceived risk of shallow land burial (Eisenbud, 1987). Although damage to public health has not been shown to have resulted from operations at six commercial shallow land burial sites and five major government shallow land burial facilities, three of the commercial sites have been closed. Of the three remaining commercial low-level sites in the U.S., the Chem-Nuclear Systems site at Barnwell, South Carolina is the only commercial site in the east and it receives more than half of the low-level waste generated in the U.S. Since the national inventory of low-level radioactive waste is growing at about 10^5 m^3/yr (30% from medical institutions), some other method must be employed to dispose of low-level radioactive waste.

Of greater concern is the storage of large quantities of highly toxic liquid and solid wastes which must be isolated from the environment for thousands of years. The selection of sites to store high-level radioactive waste requires more than technical decisions. Each site that is technically acceptable must also have public support (Hill et al., 1982). There is general agreement by the public that there is a need to store radioactive waste. However, when a site is found which meets the technical qualifications, the local public wants the waste stored somewhere else. Eisenbud (1987) calls this a "not in my backyard" (NIMBY) syndrome.

Much of the high-level nuclear waste is stored at temporary sites in concrete-encased steel tanks a few meters below the surface of the ground. These tanks have a capacity ranging from 15,000 gal to 1×10^6 gal and an expected lifetime of 15 to 40 years (Hobbs and McClellan, 1986; Eisenbud, 1987). Considering the finite lifetime of the steel tanks, it becomes clear that these wastes must be transferred to other containers or more secure sites in the future.

Scientists appointed by the International Council of Scientific Unions believe that interim storage for 50 to 100 years will greatly reduce the problem of thermal loading when the waste is finally transferred to the final disposal site (Harrison, 1984). They believe that current and future technology will allow the safe disposal of the nuclear waste. Several options for the permanent management of radioactive waste have been considered (Gonzales, 1982; Hobbs and McClellan, 1986; Eisenbud, 1987). The major methods being considered for disposal or long-time storage include (1) as solids in salt mines, deep underground caverns, man-made vaults, deep ocean seabeds, deep ocean subseabeds, deep mined cavities, or insertion into Greenland glaciers; (2) as liquids in deep wells, deep underground caverns or in tanks; and (3) as a grout mixture injected into deep rock fissures. The disposal method now in favor is deep underground mined cavities.

As a result of the Nuclear Waste Policy Act signed into law on January 7, 1983, the U.S. Department of Energy (DOE) has proposed five preliminary sites for characterization. These five sites are located either in basalt, bedded salt, domal salt, or tuff geological media found in Richton Dome, Mississippi, Yucca Mountains, Nevada, Deaf Smith County, Texas, Davis Canyon, Utah, or the DOE Hanford Site in Washington. January 31, 1998 has been established as the date when the first repository operation will begin.

C. BIOLOGICAL EFFECTS OF IONIZING RADIATION

Radiation toxicology is a specialized branch of toxicology involving the study of the adverse effects of radiation on living organisms. It is a multidisciplinary science, borrowing freely from several of the basic sciences. The cytopathologic effects of radiation exposure are quite similar to those induced in other types of cellular injury. Radiation-induced cell changes may result in death of the organism, death of the cells, or cancers with no features distinguishing them from those induced by other types of cell injury.

Radiation exposure occurs from many of the sources mentioned previously. *Directly ionizing* radiation carries an electric charge that interacts directly with atoms in the tissue or medium by electrostatic attraction or repulsion. *Indirectly ionizing* radiation is not electrically charged, but results in production of charged particles by which its energy is absorbed. A characteristic of charged particles produced directly or indirectly is *linear energy transfer* (LET), the energy loss per unit of distance traveled, expressed in kiloelectron volts (keV) per micrometer (μm). The LET, depends on the velocity and charge of the particle produced and varies from about 0.2 to >1000 keV/μm.

Many particles spend virtually all their energy at LETs of less than a few keV/μm. The most significant of these particles are the principal components of primary cosmic radiation and high-energy electrons such as those emitted by beta radiation. These high-energy electrons, as well as the indirectly ionizing radiation such as X-rays and gamma rays that produce them, are referred to as low-LET radiation. Low energy electrons, produced by both direct and indirect ionizing radiation, are intermediate in LET.

High-LET radiation also contributes to the environmental radiation load. Alpha radiation emitted by internally deposited radionuclides is probably the most important directly ionizing high-LET radiation. Neutron radiation is the principal indirectly ionizing high-LET radiation.

Irradiation induces formation of reactive chemical products when it enters a biological system (Kennedy et al., 1984). For example, superoxide radicals can be generated in the reaction of hydrated electrons or hydrogen atoms with dissolved oxygen following gamma radiation of aqueous solutions. This radiolytically-formed free radical is involved in oxidative chain reactions (Fridovich, 1976) with the possibility of interconversion and postirradiation generation of other forms of activated oxygen, leading indirectly to further irradiation-induced cellular damage (Greenstock, 1984).

Low-level ionizing radiation of living cells results in the formation of free radicals and hydroxyl radicals (Upton, 1982; Halliwell and Gutteridge, 1985; Johnson, 1986). Much of the cellular DNA damage inflicted by ionizing radiation is due to the formation of free radicals and hydroxyl radicals. Likewise, the formation of hydroxyl radicals by ionizing radiation can initiate lipid peroxidation, leading to cell membrane damage and the formation of cytotoxic aldehydes.

1. Ecotoxicology

Plants and lower animals are much more resistant to the effects of radiation than humans (Eisenbud, 1987). Nevertheless, when a mixed forest on Long Island was subjected to chronic gamma irradiation during the growing season for 12 years, a phenomenon known as retrogression occurred (Grosch, 1980; Moriarity, 1983). Woodwell (1970) reported the results of the irradiation of an oak-pine forest by a radiation source containing 9500 Ci ^{137}Cs. Within 6 months a vegetation gradient developed. What was originally an oak-pine forest became an oak forest, with the elimination of the pine. The oak forest then became a shrub zone, followed by a sedge zone. Near the radiation source, the sedge gave way to a central devastated area, where only mosses and lichens survived. The changes were similar to those seen along gradients of increasing severity of climatic changes, such as increasing altitude on a mountain. The chronic exposure to gamma radiation reversed the ecological succession.

After 12 years of chronic irradiation, only lichens and green algae survived up to a distance of 20 m from the source, an area receiving as much as 3000 roentgen (R)/d. Only sedges and grass lived from 20 to 75 m from the source, where the dose was 20 to 160 R/d. From 75 to 85 m, with a dose of 10 to 20 R/d, blueberry bushes died and no oaks survived. From 85 to 115 m from the source, where 5 to 10 R/d were received, oaks survived but pines did not. Scraggly, stunted pines were seen at a distance of 125 m, where 2 R/d of gamma radiation was received. Even as far away as 150 m from the source, the pine needles were shorter than normal and the diameters of the trees were reduced.

The only animals left on the Long Island experimental plot were insects (Grosch, 1980). Leaf-eating insects attacked the oak trees and shrubs. Bark insects and wood borers developed larger populations in response to an abundance of food and an absence of predators. All other animals died or left the denuded areas.

These observations agree with other studies (Odum, 1971) on the effects of gamma radiation on whole communities and ecosystems in a tropical rain forest of Puerto Rico and in the Nevada desert. The effects of mixed gamma/neutron radiation have been studied on fields and forests in Georgia and at the Oak Ridge National Laboratory in Tennessee. Short-term effects of gamma radiation have been studied at the Savannah River Ecology Laboratory in South Carolina. The Oak Ridge National Laboratory has also conducted a low-level chronic radiation study of a lake bed community.

In all of these studies, the differential sensitivity of a species in an ecosystem is of considerable interest (Odum, 1971). If an ecosystem receives a higher level of radiation than was confronted in its evolution, the elimination of sensitive strains or species may result.

2. Effects on Plants

In gamma source experiments, such as the ones mentioned above, weeds and grasses proved relatively tolerant of radiation and field crops were only slightly more sensitive. Exposure to sustained irradiation from a large gamma source is not an exact simulation of chronic exposure to short-lived radioisotopes, but it does provide an indication of the effects of an acute, high-level radioisotope exposure.

Radionuclides such as ^{210}Ra, ^{226}Ra, and ^{222}Rn that occur or are deposited in the soil are translocated by plants. In addition to root uptake radioisotopes are absorbed following direct deposition on foliar surfaces. Although the mechanism of radionuclide adsorption by plants is not well understood, it is known that individual radionuclides are translocated from the root or the leaves to the remainder of the plant. Mean ^{232}Th concentrations of 0.018 ± 0.022 pCi/kg have been found in the edible portions of 25 vegetables, including beans, carrots, corn, potatoes, and squash. Flora near the summit of the Morro do Ferro, a hill in the state of Minas Gerais, Brazil, have absorbed so much ^{228}Ra that they can be autoradiographed easily (Eisenbud, 1987).

Individual plant species vary widely in their susceptibility to the damaging effects of ionizing radiation. Exposure of the mixed vegetation of a Long Island forest to gamma rays demonstrated the vulnerability of conifers compared to the radiotolerance of grasses. Exposure in early stages of development or during the growing season, when cells are dividing, increases the radiosensitivity of the plant. During cell division, cell death or damage is induced at much lower irradiation levels than those required during the interphase stage of the cell cycle. Investigations of plant damage from gamma ray sources demonstrated that radiation causes a reduction in the number of cells per meristem (growing point). This reduction in cells varies directly with the number of cells displaying chromosome damage.

In higher plants, sensitivity to ionizing radiation is directly proportional to the chromosome volume of the cell nucleus (Odum, 1971). However, certain community attributes such as biomass and diversity also become determinants of species vulnerability. Indeed, the "unshielded" biomass above ground is a major determinant in that plants sprouting from seeds or from shielded underground parts have an increased chance of recovery.

3. Effects on Animals

In higher animals, unlike in plants, there is no simple, direct relationship between nuclear chromosome volume and sensitivity to ionizing radiation.

Rather, the effects of irradiation on specific organ systems are more critical (Odum, 1971).

Developmental Stages

As in plants, proliferating cells are much more sensitive to irradiation than differentiated, nondividing cells. An organism is more radiosensitive during its early stages of development, when most cells are dividing, than at any other stage of its life. The LD_{50} for fish embryos is 16 to 18 times smaller than required in later life. For insects, the LD_{50} in adults is about 1000 times larger than during the developmental stages (Grosch, 1980).

Instead of lethality, morphological abnormalities are associated with irradiation during the middle stages of development. Irradiation-induced anomalies occurring during this time may result in death of the organism or abnormal development of one or more organ systems. Exposure during this period may result in gross malformations, growth retardation at term or as an adult, and structural neuropathology (Mettler and Moseley, 1985). In the human, most major organogenesis occurs during the first trimester of pregnancy, with embryonic death and congenital abnormalities resulting from irradiation exposure during this period (Robertson, 1989).

During late organogenesis and in the perinatal period, just before and just after birth, radiation damage tends to be functional rather than structural. Perinatal irradiation with X-rays and gamma rays (140 to 180 cGy) induces changes in tissue enzyme activity (Andrew and Lytz, 1981). The major effects of perinatal *in utero* exposure in humans is seen in the developing CNS with neurological and behavioral damage not obvious in histological examination (Cockerham and Prell, 1989; Robertson, 1989).

Reproduction

During a period of about 20 years in the early part of this century, radiation was used in an attempt to increase fertility. Exposure normally was to 1.5 to 2.25 Gy over a period of 3 weeks. These levels apparently had little effect on fertility or on any later conceived children (Mettler and Moseley, 1985).

In recent years, the safety of radiation levels is more often questioned since the threshold for the effects of ionizing radiation on male reproduction is difficult to predict (Schrag and Dixon, 1985). Although fully developed sperm cells and primary spermatocytes are relatively radioresistant, the proliferating spermatogonial cells of the testis are highly sensitive to ionizing radiation (Robertson, 1989).

Occupational radiation exposure of the testis produces a significant decrease in serum gonadotropins and significant changes in semen production and morphology. However, the effects of ionizing radiation on spermatogenesis are normally reversible with the recovery of fertility predictable. Although an acute irradiation dose of 6 Gy to the testis is likely to produce permanent sterility, conception has occurred for males after years of either aspermic or hypospermic conditions following absorbed doses between 2.3 and 3.7 Gy (Mettler and Moseley, 1985; Robertson, 1989).

Although there are no proliferating stem cells in the mature female reproductive system, the oocytes in follicles are in various stages of development. Animal experiments indicate that the radiosensitivity of the ova depends on the maturity of the follicle (Robertson, 1989). Irradiation depletion of the radiosensitive mature and intermediate follicles will result in periods of temporary sterility followed by fertility due to maturation of surviving immature follicles. Cases have been reported in which women receiving as much as 6.4 Gy became pregnant up to 2 years later and delivered normal children (Mettler and Moseley, 1985). However, the estimated dose required to produce permanent sterility in the female ranges from 6.25 to 30 Gy, depending on the age of the subject (Robertson, 1989).

Genetic Effects

Mutations are structural changes occurring in genes as a result of exposure to a number of environmental agents including chemicals, heat, and ionizing radiation (Eisenbud, 1987). Partially due to the efforts of the film industry, its post-World War II movies portray inaccurately the genetic effects of irradiation. In fact, the genetic effects of irradiation dominated the philosophy of radiation risk and radiation protection (Mettler and Moseley, 1985). Only during the past two decades has the scientific community increased its appreciation of the somatic effects of ionizing radiation. Data evaluation during this time also suggested that the previous estimates of genetic risk were too high.

Damage to DNA along a low-LET radiation tract is the same kind that occurs spontaneously. It will likely consist of single strand breaks in the double helix and may be repaired by cellular enzymes. The ultimate effects depend more on the effectiveness of the repair than upon the initial break. Chromosomal breaks from low-LET radiations have a small probability of resulting in a translocation during repair of a lesion. In contrast, high-LET radiation lesions have a high probability of interaction and result in a greater number of unrepaired or misrepaired lesions. These lesions may be amplified many times during transcription and translation and are the major contributors to the genetic damage resulting from irradiation.

Current studies indicate that there may have been one possible mutation in the 78,000 children of people who survived the atomic bombings at Hiroshima and Nagasaki. When the mutation rate of the group of exposed parents is estimated, this single mutation has a high probability of being unrelated to radiation exposure.

Available information suggests that the radiation dose required to double the human mutation rate varies between 0.8 and 2.4 Gy (UNSCEAR, 1988).

Using this value, assessments may be made of the contribution of the natural background radiation to the frequency of various types of genetically determined disease. Although these assessments are highly uncertain, they imply that only a small amount (1 to 6%) of all genetic mutations can be attributed to exposure to natural background radiation (Robertson, 1989).

Radiation Carcinogenesis

Radiation carcinogenesis is considered by most radiobiologists as the most important effect of exposure to levels of ionizing radiation below 1 Gy. Ionizing radiation increases the incidence of virtually every type of neoplasm, whether benign or malignant, and the carcinogenic effects of radiation have been observed in practically all species (Upton, 1984).

The development of cancer appears to be a multistage process involving an *initiator* and at least one *promotor*. Promotors may elicit the production of activated forms of oxygen, including superoxide radicals and peroxides, and hydroxyl radicals, which either directly or indirectly affect DNA (Mettler and Moseley, 1985). Certainly, radiation causes strand breaks and modification of DNA bases. It is also well known that radiation of living cells elicits the formation of free radicals, and hydroxyl radicals, and it is well established that hydroxyl radicals are responsible for the radiation-induced DNA damage (Troll and Wiesner, 1985).

At times it becomes difficult to characterize ionizing radiation as either initiator or promotor. For example, cigarette-smoking uranium miners exposed to radon have radiation-induced lung cancer at five times the rate of nonsmoking miners. It is difficult to assign the role of initiator or promotor to either radon or cigarette-smoke since both have been implicated in the etiology of lung cancer.

Acute Radiation Syndrome

There have been fewer than 25 documented fatalities worldwide between 1946 and 1985 that can be attributed to radiation accidents (Mettler and Moseley, 1985). Although exposure of the whole body to lethal amounts of ionizing radiation is very rare, any discussion of the biological effects of ionizing radiation would be incomplete without mentioning acute radiation sickness and acute radiation syndrome (ARS).

Acute radiation sickness is manifest in characteristic clinical sequelae known as ARS, a combination of syndromes determined primarily by the total radiation dose received, the rate of exposure, and the distribution of the radiation in the body (Young, 1987). Signs and symptoms of ARS result from injury to bone marrow, gastrointestinal system, cardiovascular system, CNS, gonads, and skin. The variation in radiation sensitivity of these tissues causes the signs and symptoms of ARS to occur in three successive phases: an initial prodromal phase, a subsequent latent period, and the manifest illness phase.

The length of each phase may vary directly with the radiation dose, and the time between each phase may vary indirectly with the dose, so at an extremely high dose of radiation, the phases will blend with the latent period disappearing completely (Table 9.4).

The initial prodromal phase is characterized by a combination of gastrointestinal and neuromuscular symptoms such as anorexia, nausea, vomiting, diarrhea, apathy, tachycardia, fever, headaches, insomnia, dizziness, and vertigo. The pathogenesis of the prodromal phase is not known, but several causal factors have been suggested, including direct radiation effects on the central and autonomic nervous systems, disturbance of endocrine balance, and the production and release of various chemical mediators (Cockerham et al., 1987; Cockerham et al., 1988; Donlon and Walden, 1988). The latent period, which follows the prodromal phase, is relatively asymptomatic and is believed to be the time between initial cell damage and the interference of radiation with cell renewal in the affected organs.

The manifest illness phase of the ARS is classically divided into three major syndromes traditionally known as the hemopoietic syndrome, gastrointestinal (GI) syndrome, and central nervous system (CNS) syndrome. However, the current view replaces the CNS syndrome with the neurovascular syndrome.

The hemopoietic syndrome occurs following exposure to 200 to 700 cGy. Radiation doses of 100 cGy or more can significantly damage the blood-forming capability of the body. Approximately 50% of individuals exposed to 300 cGy will die within 2 months. The signs and symptoms result from radiation damage to the bone marrow, lymphatic organs, and immune system. The pathophysiological effects of damage to bone marrow include increased susceptibility to infection, bleeding, anemia, and lowered immunity. Death usually results from hemorrhage and infection.

At radiation doses of 7 to 50 Gy, injury to the GI tract inhibits the renewal of the cell lining. The intestinal epithelial stem cell is the target of radiation damage, and the resulting decrease in mitotic activity leads to denudation of the intestinal mucosa, fluid and electrolyte imbalance, and bacteremia (Gunter-Smith, 1987). The symptoms of the GI syndrome include lethargy, diarrhea, dehydration, and sepsis. At doses of 3 to 8 Gy, temporary injury to the tight junctions between epithelial cells of the mucosal lining permits the escape of bacterial endotoxins into the bloodstream. As the dose increases, the epithelial lining is more extensively depleted. With doses of 10 to 15 Gy, denudation of the mucosa exacerbates the loss of fluid and electrolytes. At doses of 12.5 Gy and above, early mortality occurs due to dehydration, and electrolyte imbalance, with death occurring 4 to 5 d after exposure.

ARS is sometimes divided into four subsyndromes: (1) hemopoietic, (2) GI, (3) cardiovascular, and (4) CNS syndromes (Mettler and Moseley, 1985). Limiting the definition of ARS to only three syndromes leads to a disregard of changes in the cardiovascular system at radiation doses between 30 and 50 Gy,

**TABLE 9.4. Radiation Dose Ranges and Associated
Pathophysiological Events**

| | Pathophysiological Events | | |
Dose Range (cGy)	Prodromal Effects	Manifest-Illness Effects	Survival
75–150	Mild	Slight decrease in blood cell count	Virtually certain
150–300	Mild to moderate	Beginning symptoms of bone marrow damage	Probable (>90%)
300–530	Moderate	Moderate to severe bone marrow damage	Possible — bottom third of range: $LD_{5/60}$ middle third: $LD_{10/60}$ top third: $LD_{50/60}$
530–830	Severe	Severe bone marrow damage	Death within 3.5 to 6 weeks bottom half: $LD_{90/60}$ top half: $LD_{99/60}$
830–1100	Severe	Bone marrow pancytopenia and moderate intestinal damage	Death within 2 to 3 weeks
1100–1500	Severe	Combined gastrointestinal and bone marrow damage, hypotension	Death within 1 to 2.5 weeks
1500–3000	Severe gastrointestinal damage upper half of range: early transient incapacitation (ETI), gastrointestinal death		Death within 5 to 12 d
3000–4500	Gastrointestinal and cardiovascular damage		Death within 2 to 5 d

Source: Baum et al., 1984.

because this radiation level falls within the range that produces GI syndrome (Cockerham and Hawkins, 1987; Hawkins and Cockerham, 1987; Hawkins and Forcino, 1988). The effect of radiation on intestinal microcirculation has received little attention, although this response may be an important factor in the development of GI radiation syndrome. Irradiation-induced hypotension and a decreased cerebral blood flow have been shown to have a temporal relationship with early transient incapacitation (ETI) and performance decrement (Cockerham and Forcino, 1988). A reduction in systemic blood pressure can reduce the driving force required to maintain cerebral blood flow, leading to cerebral hypoxia, which in turn could result in decreased performance and eventually incapacitation, two of the symptoms found in CNS syndrome and neurovascular syndrome.

Neurovascular syndrome is the least understood of the radiation-induced deaths. The syndrome is unique in that death occurs very quickly before damage to the GI and hemopoietic systems becomes apparent. Readily obvious CNS signs and symptoms include disorientation, loss of muscular coordination, respiratory distress, apathy, prostration, convulsive seizures, and coma associated with death. Exposure of 50 Gy is thought to be necessary for neurovascular syndrome and doses above 100 Gy are thought to be required for direct damage of the nervous system. However, ionizing radiation exposure modifies electroencephalographic activity in a variety of brain areas following exposure to 4 to 10 Gy (Tolliver and Pellmar, 1987), and 4 Gy will alter *in vitro* neuronal firing patterns in the hippocampus (Bassant and Court, 1978). In *in vivo* exposure to 20 Gy, γ-radiation decreased hippocampal synaptic transmission and spike generation at 3 d postradiation (Hollinden and Pellmar, 1989).

D. CONCLUSION

To assess the average exposure of residents of the U.S. to ionizing radiation, the National Council on Radiation Protection and Measurements determined the collective effective dose equivalent from each of six main radiation source categories (NCRP, 1987). The collective effective dose equivalent is calculated by multiplying the average per capita effective dose equivalent by the estimated number of people exposed (BEIR V, 1990). The average effective dose equivalent was then calculated by dividing the collective effective dose equivalent by the total U.S. population. The dose equivalent accounts for differences in relative biological effectiveness by multiplying the absorbed dose by the quality factor while the effective dose equivalent relates the dose equivalent to risk.

As seen in Table 9.2, natural radiation sources contribute 82% of the total average annual effective dose equivalent of 3.6 mSv. By far the largest contributor is radon and its decay products (55%). Radon in domestic water supplies is also the chief contributor to radiation exposure from consumer products (BEIR V, 1990). While much is written about radiation exposure from nuclear power production and nuclear weapons testing fallout, their contributions are negligible compared to environmental radon, the largest source of human exposure to ionizing radiation.

REFERENCES

Andrew, F.D. and Lytz, P.S. Biochemical disturbances associated with developmental toxicity, in *Developmental Toxicology*. C. Kimmel and J. Buelke-Sam, Eds. Raven Press, New York, 1981, 145–165.

Arkansas Gazette. Little Rock, Arkansas, June 20, 1990, A1.

Barringer, F. Soviets admit Chernobyl disaster worse, *The News and Observer*, Raleigh, North Carolina, April 28, 1990.

Bassant, M.H. and Court, L. Effects of whole-body irradiation on the activity of rabbit hippocampal neurons, *Radiat. Res.* 75:593–606, 1978.

Baum, S.J., Anno, G.H., Young, R.W., and Withers, H.R. Nuclear weapon effect research at PSR-1983: Vol. 10. Symptomatology of acute radiation effects in humans after exposure to doses of 75 to 4500 rads (cGy) free-in-air. DNA-TR-85-50. Defense Nuclear Agency, Washington, DC, 1984, 66.

BEIR III. The effects on populations of exposure to low levels of ionizing radiation. Committee on the Biological Effects of Ionizing Radiations, National Research Council. National Academy Press, Washington, DC, 1980.

BEIR IV. Health risks of radon and other internally deposited alpha-emitters. Committee on the Biological Effects of Ionizing Radiations, National Research Council. National Academy Press, Washington, DC, 1988.

BEIR V. Health effects of exposure to low levels of ionizing radiation. Committee on the Biological Effects of Ionizing Radiations, National Research Council. National Academy Press, Washington, DC, 1990.

Bogo, V. Radiation: behavioral implications in space, *Toxicology* 49:299–307, 1988.

Cerveny, T.J. and Cockerham, L.G. Medical management of internal radionuclide contamination, *Med. Bull. U.S. Army, Europe* 43:24–27, 1986.

Clarke, R.H. Dose distributions in Western Europe following Chernobyl, in *Radiation and Health. The Biological Effects of Low-Level Exposure to Ionizing Radiation.* R. Jones and R. Southwood, Eds. John Wiley & Sons Ltd., New York, 1987, 251–264.

Cockerham, L.G. and Hawkins, R.N. Radiation injury and the splanchnic circulation, in *Pathophysiology of the Splanchnic Circulation, Vol. II.* P. Kvietys, J. Barrowman, and D.N. Granger, Eds. CRC Press, Inc., Boca Raton, FL, 1987, 55–66.

Cockerham, L.G. and Forcino, C.D. Post-irradiation alterations in cerebral blood flow, in *Terrestrial Space Radiation and It's Biological Effects.* P. McCormack, C. Swenberg, and H. Bucker, Eds. Plenum Press, New York, 1988, 495–507.

Cockerham, L.G. and Prell, G.D. Prenatal radiation risk to the brain, *NeuroToxicology* 10:467–474, 1989.

Cockerham, L.G., Forcino, C.D., Pellmar, T.C., and Smart, S.W. Effect of methysergide on postirradiation hypotension and cerebral ischemia, presented at the Cerebral Hypoxia and Stroke Symposium, Budapest, Hungary, August 22–24, 1987.

Cockerham, L.G., Arroyo, C.M., and Hampton, J.D. Effects of 4-hydroxypyrazolo (3,4-d) pryimidine (allopurinol) on postirradiation cerebral blood flow: implications of free radical involvement, *Free Rad. Biol. Med.* 4:279–284, 1988.

Cohen, B. Radon: characteristics, natural occurrence, technological enhancement, and health effects, *Progress in Nuclear Energy.* 4:1, 1979.

Conklin, J.J. and Hagan, M.P. Research issues for radiation protection for man during prolonged spaceflight, *Adv. Rad. Biol.* 13:215–284, 1987.

Cothern, C.R. History and uses in, *Environmental Radon.* C. Cothern and J. Smith, Eds. Plenum Press, New York. 1987, 31–58.

Cothern, C., Lappenbusch, W., and Michel, J. Drinking-water contribution to natural background radiation, *Health Phys.* 50:33–47, 1986.

Donlon, M.A. and Walden, T.L., Jr. The release of biologic mediators in response to acute radiation injury, *Comments Toxicology* 2:205–216, 1988.

Eisenbud, M. *Environmental Radioactivity. From Natural, Industrial, and Military Sources.* 3rd ed. Academic Press, Inc., New York, 1987.

Fetter, S.A. and Tsipis, K. Catastrophic releases of radioactivity, *Sci. Am.* 244:41–47, 1981.

Fridovich, I. Oxygen radicals, hydrogen peroxide, and oxygen toxicity, in *Free Radicals in Biology, Vol. 1.* Academic Press, New York, 1976, 239–277.

Fry, F.A. Doses from environmental radioactivity, in *Radiation and Health. The Biological Effects of Low-Level Exposure to Ionizing Radiation.* R. Jones and R. Southwood, Eds. John Wiley & Sons Ltd., New York, 1987, 9–17.

Gonzales, S. Host rocks for radioactive-waste disposal, *Am. Sci.* 70:191–200, 1982.

Greenstock, C.L. Oxy-radicals and radiobiological oxygen effect, *Israel J. Chem.* 24(1):1–10, 1984.

Grosch, D.S. Radiations and radioisotopes, in *Introduction to Environmental Toxicology.* F. Guthrie and J. Perry, Eds. Elsevier/North Holland, Inc., New York, 1980, 44–61.

Gunter-Smith, P.J. Effect of ionizing radiation on gastrointestinal physiology, in *Military Radiobiology* J. Conklin and R. Walker, Eds. Academic Press, Inc., San Diego, 1987, 135–151.

Hall, E.J. *Radiation and Life,* 2nd ed. Pergamon Press, New York, 1984, 4–20.

Halliwell, B. and Gutteridge, J.M.C. *Free Radicals in Biology and Medicine.* Clarendon Press, Oxford, England, 1985, 28, 135, 159, 169.

Harley, J. Environmental radon, in *Noble Gases.* R. Stanley and A. Moghissi, Eds. U.S. Energy Development and Research Agency, National Environmental Research Center, Washington, DC, 1973, 109–114.

Harrison, J.M. Disposal of radioactive wastes, *Science* 226:11–14, 1984.

Hawkins, R.N. and Cockerham, L.G. Postirradiation cardiovascular dysfunction, in *Military Radiobiology.* J. Conklin and R. Walker, Eds. Academic Press, New York, 1987, 153–163.

Hawkins, R.N. and Forcino, C.D. Effects of radiation on cardiovascular function, *Comments Toxicology* 2:243–252, 1988.

Hill, D., Pierce, B.L., Metz, W.C., Rowe, M.D., Haefele, E.T., Bryant, F.C., and Tuthill, E.J. Management of high-level waste repository siting, *Science* 218:859–864, 1982.

Hobbs, C.H. and McClellan, R.O. Toxic effects of radiation and radioactive materials, in *Casarett and Doull's Toxicology,* 3rd. ed. C. Klaassen, M. Amdur, and J. Doull, Eds. Macmillan Publishing Company, New York, 1986, 669–705.

Hollinden, G.E. and Pellmar, T.C. Attenuation of synaptic transmission in hippocampal slices following whole animal exposure to ionizing radiation, *Soc. Neurosci. Abstr.* 15:134, 1989.

Johnson, K.J. Neutrophil-independent oxygen radical-mediated tissue injury, in *Physiology of Oxygen Radicals.* A. Taylor, S. Matalon, and P. Ward, Eds. American Physiology Society, Bethesda, MD, 1986, 152–153.

Kennedy, A.R., Troll, W., and Little, J.B. Role of free radicals in the initiation and promotion of radiation transformation *in vitro, Carcinogenesis* 5:1213–1218, 1984.

Luxin, W., Yongru, Z., Zufan, T., Weihue, H., Deqing, C., and Yongling, Y. Epidemiological investigation of radiological effects in high background radiation areas of Yangjiang, China, *J. Radiat. Res.* 31:119–136, 1990.

McCormack, P.D., Swenberg, C.E., and Bucker, H. Preface, in *Terrestrial Space Radiation and Its Biological Effects*. P. McCormack, C. Swenberg, and H. Bucker, Eds. Plenum Press, New York, 1988, v–vi.

Mendelsohn, M.L. Introduction to the radon problem. Presented at The Toxicology Forum 1988 Annual Summer Meeting, July 18–22, 1988, Given Institute of Pathobiology, Aspen, CO.

Mettler, F.A., Jr. and Moseley, R.D., Jr. *Medical Effects of Ionizing Radiation*. Harcourt Brace Jovanovich, New York, 1985, 288.

Michel, J. Sources, in *Environmental Radon*. C. Cothern and J. Smith, Eds. Plenum Press, New York, 1987, 81–130.

Moriarity, F. *Ecotoxicology. The Study of Pollutants in Ecosystems*, 2nd ed. Academic Press, San Diego, CA, 1983, 68–69, 159, 164–165.

National Council on Radiation Protection and Measurements (NCRP). Evaluation of occupational and environmental exposures to radon and radon daughters in the United States. NCRP Report No. 78. National Council on Radiation Protection and Measurements, Washington, DC, 1984.

National Council on Radiation Protection and Measurements (NCRP). Ionizing radiation exposures of the population of the United States. NCRP Report No. 93. National Council on Radiation Protection and Measurements, Washington, DC, 1987.

Nazaroff, W.W., Moed, B.A., and Sextro, R.G. Soil as a source of indoor radon: generation, migration, and entry, in *Radon and Its Decay Products in Indoor Air*. W. Nazaroff and A. Nero, Jr., Eds. John Wiley & Sons, New York, 1988, 57–112.

Nazaroff, W.W., Doyle, S.M., Nero, A.V., Jr., and Sextro, R.G. Radon entry via potable water, in *Radon and Its Decay Products in Indoor Air*. W. Nazaroff and A. Nero, Jr., Eds. John Wiley & Sons, New York, 1988, 131–157.

Nero, A.V., Jr. Radon and its decay products in indoor air: an overview, in *Radon and Its Decay Products in Indoor Air*. W. Nazaroff and A. Nero, Jr., Eds. John Wiley & Sons, New York, 1988, 1–53.

Odum, E.P. Radiation ecology, in *Fundamentals of Ecology*. 3rd ed. W.B. Saunders Company, Philadelphia, 1971, 451–467.

Pearson, J. Natural environmental radioactivity from radon-222. Publication No. 999-RH-26, U.S. Department of Health, Education, and Welfare, Rockville, Maryland, 1967.

Robertson, J.B. Toxicology of ionizing radiation, in *A Guide to General Toxicology*. 2nd ed. J. Marquis, Ed. Karger, New York, 1989, 141–156.

Rust, D.M. Solar flares, proton showers and the space shuttle, *Science*. 216:939–946, 1982.

Sanders, C.L. *Toxicological Aspects of Energy Production*. Battelle Press, Columbus, OH, 1986, 253–284.

Schrag, S.D. and Dixson, R.L. Occupational exposures associated with male reproductive dysfunction, *Ann. Rev. Pharmacol. Toxicol.* 25:567–592, 1985.

Southwood, R. Opening remarks, in *Radiation and Health. The Biological Effects of Low-Level Exposure to Ionizing Radiation.* R. Jones and R. Southwood, Eds. John Wiley & Sons Ltd., New York, 1987, 3–6.

Stather, J.W., Dionian, J., Brown, J., Fell, T.P., and Muirhead, C.R. Assessing risks of childhood leukaemia in seascale, in *Radiation and Health. The Biological Effects of Low-Level Exposure to Ionizing Radiation.* R. Jones and R. Southwood, Eds. John Wiley & Sons Ltd., New York, 1987, 65–80.

Stauber, M.C., Rossi, M.L., and Stassinopoulos, E.G. An overview of medical-biological radiation hazards in earth orbits, in *Space Safety and Rescue, Vol. 58, Science and Technology Series*, IAA 83-256. G. Heath, Ed. American Astronautical Society, Univelt, San Diego, 1983, 267–300.

Tolliver, J.M. and Pellmar, T.C. Ionizing radiation alters neuron excitability in hippocampal slices of the guinea pig, *Radiat. Res.* 112:555–563, 1987.

Troll, W. and Wiesner, R. The role of oxygen radicals as a possible mechanism of tumor promotion, *Ann. Rev. Pharmacol. Toxicol.* 25:509–528, 1985.

UNSCEAR. Sources, effects and risks of ionizing radiation. United Nations Scientific Committee on the Effects of Atomic Radiation, United Nations, New York, 1988, 647.

Upton, A.C. The biological effects of low-level ionizing radiation, *Sci. Am.* 246:41–49, 1982.

Upton, A.C. Biological aspects of radiation carcinogenesis, in *Radiation Carcinogenesis: Epidemiology and Biological Significance.* J. Boice, Jr. and J. Fraumeni, Jr., Eds. Raven Press, New York, 1984, 9–19.

Woodwell, G.M. Effects of pollution on the structure and physiology of ecosystems, *Science* 168:429–433, 1970.

Yamamoto, M., Ueno, K., Igarashi, Y., Shiraishi, K., and Kawamura, H. Determination of low-level Ra-226 in human bone by α-spectrometry, *J. Radiat. Res.* 31:85, 1990.

Young, R.W. Acute radiation syndrome, in *Military Radiobiology* J. Conklin and R. Walker, Eds. Academic Press, Inc., San Diego, 1987, 165–190.

Chapter

10

Animal and Plant Toxins

Jason S. Albertson and
Frederick W. Oehme

A. INTRODUCTION

The environmental toxicologist must be aware of a vast array of potentially toxic substances produced by animals and plants. Venomous or poisonous animals are represented in every phylum and class of animals except birds. These classes of animals occupy every continent and nearly every body of water on the globe. Approximately 1200 individual species are considered to be toxic (Russell, 1984). Numerous poisonous plants grow on our planet and impact both human and animal populations. Hundreds of people are affected by exposure to plant toxins, and agricultural animals cause significant economic losses annually due to ingestion of toxic plants. The apparent ubiquity of animal and plant toxins is of major concern to the world population.

B. ANIMAL TOXINS (ZOOTOXINS)

Venomous animals are those creatures which produce a poison in a highly-developed gland or group of cells and inject this substance into another animal or plant by a bite or sting. In contrast, a poisonous animal cannot directly inject its poison but the poison is generally delivered via ingestion of the poisonous organism by the victim.

Envenomation may be used by an organism to accomplish one or more goals. It is used as an offensive weapon as in the immobilization and digestion of prey. It can also be an effective deterrent against aggression or predation.

0-8493-8851-1/94/$0.00+$.50
© 1994 by CRC Press, Inc.

Poisonous animals, on the other hand cannot rely on their poisons as agents of offense or defense. Their poisons are usually a metabolic product and are toxic by coincidence. Intoxication may also come about by the ingestion of an organism that is toxic in itself (Russell, 1965). Envenomations and bites represented 3.5% of all human poisonings in the U.S. in 1988, some 47,829 cases (Litovitz et al., 1989).

1. Characteristics of Animal Toxins

Animal toxins vary considerably in their complexity, consisting of a mixture of large and small molecular weight compounds. They include a diverse number of polypeptides, enzymes, and amines which are essentially involved in local soft tissue and/or neurologic injury and are probably responsible for most of the damage. Phospholipase A, a widely distributed component of venoms, acts by disrupting cell membranes, uncoupling oxidative phosphorylation, inhibiting cellular respiration, and stimulating the release of histamines, serotonin, and kinins. *L*-amino-oxidase, a ubiquitous component of snake venom, activates tissue peptidases and causes hypotension and cardiovascular alterations. Hyaluronidase causes tissue liquefaction and aids in the spread of the venom extravascularly. Numerous proteases involved in the breakdown of tissues and production of hemorrhage have been isolated from snake venom (Russell, 1967). Other constituents of animal toxin include various anticoagulants, aminopolysaccharides, lipids, steroids, glycosides (found in toad secretions), and formic acid, in the venom of wasps and bees and in ant saliva (Russell and Gertsch, 1983).

The majority of venoms exert their effects directly on the cells and tissues they contact. The degree of damage is dependent on how much of a specific component accumulates at a site where it causes an effect. Most venoms damage a variety of tissues and produce a plethora of toxicologic effects. Creatures whose venom is neurotoxic include the Elapid snakes (coral snakes, cobras, mamba), the black widow spider, and scorpions. Venom components are also known to be hemolytic, hemorrhagic, thrombogenic, proteolytic, cardiotoxic, and allergenic.

Poisonous Snakes

Pit Vipers. The poisonous snakes of the U.S. belong to three groups: pit vipers, elapids, and colubrids (Table 10.1). Pit vipers constitute the largest group and include five subspecies of copperheads (*Agkistrodon contortrix*), three subspecies of cottonmouth water moccasins or cottonmouth (*A. piscivorus*), three subspecies of pygmy rattlesnake (*Sistrurus miliaris*), three subspecies of massassauga (*S. catenatus*), as well as at least 26 species of rattlesnakes (*Crotalus* spp.). The primary function of snake venom is the immobilization of the victim and predigestion of the victim's tissue.

**TABLE 10.1. Classification and Distribution of
Poisonous Snakes of the U.S.**

Classification	Distribution
Pit vipers	
Rattlesnake (*Crotalus* spp.)	Entire continental U.S.
Copperhead (*Agkistrodon mokasen*)	Eastern, southern U.S.
Cottonmouth (*Agkistrodon piscivorus*)	Eastern, southern U.S.
Elapids	
Coral snake (*Micrurus* spp.)	Eastern, southern U.S.
California lyre snake (*Trimorphodon* *vanderbrughii*)	Southwestern U.S.
Colubrids	
Mangrove snake (*Boiga dendrophila*)	Southwestern U.S.

Pit viper venom contains two significant fractions: enzymatic and nonenzymatic polypeptide fractions. The major enzymes are hyaluronidase and phospholipase A, which respectively disrupts membranes and uncouples oxidative phosphorylation. Major coagulopathies (lack of blood clotting) occur which affect different enzymatic steps of the clotting cascade depending upon species of the pit viper involved. The nonenzymatic fraction directly affects the cardiovascular and respiratory systems. Rattlesnake venom causes either pooling of blood within the hepatosplanchnic bed or the major vessels of the thorax, depending upon the species of the victim. The net result of the venom in all species is hypotension. Most pit viper venom causes destruction of blood vessel walls, resulting in the leakage of red blood cells and plasma. Up to one third of the total volume of circulating blood may be lost into tissue space within several hours if the envenomation is severe. Although bites from pit vipers are most commonly associated with hematologic abnormalities, some species such as the Mojave rattlesnake (*Crotalus scutulatus scutulatus*) also contain neurotoxins.

Several factors influence the severity of a pit viper bite. The degree of toxicity varies among these species. Probably the most dangerous species is the coral snake. Death ensues within hours of envenomation, however death is rarely reported following the bite of a copperhead. The size of the snake is an important factor in predicting the resulting toxicity as larger snakes are usually able to inject a proportionally greater volume of venom. The quantity and quality of venom is also influenced by temporal factors. Younger snakes have higher concentrations of peptides in their venom. The peptide fraction is also known to increase during the spring of each year. The time since the last bite and the volume previously injected into a victim contributes to the volume subsequently injected since the venom lost must be regenerated. The motivation of the snake also makes a difference in that offensive strikes are usually more severe than defensive ones. Agonal strikes are perhaps the most severe (a decapitated head is dangerous for up to 1.5 h). Factors associated with the response of the victim are its size, site of the bite, elapsed time since the bite occurred, and the victim's physical activity since envenomation.

The most common clinical sign associated with a rattlesnake bite is marked edema (swelling) and erythema (redness) at the site of the bite. Edema and erythema may extend over the entire head, limb, or even over the entire body. There is immediate local pain, locomotion becomes painful and stiff, and generalized severe muscle pain develops. The victim becomes excessively thirsty. In small victims nausea, vomiting, and diarrhea can result. Shock is an important complication in young patients. Anaphylaxis has been reported but is a rare clinical syndrome. If the victim is bitten on the face or throat, intense soft tissue edema and inflammation can cause blockage of upper airways resulting in dyspnea.

Elapids (Coral Snakes). A bite from a coral snake is much more severe than that of a rattlesnake and thus warrants immediate medical attention. Coral snakes are not likely to attack unless cornered. Young children are often attracted to the snake's bright colors and hence are more frequent victims of coral snake bites. Pain from the bite is transient. Coral snake venom is primarily neurotoxic; however, the mechanism of action is poorly understood. Central nervous system signs, such as numbness of limbs, disorientation, and skeletal muscle and respiratory muscle paralysis, predominate. The only treatment is immediate antivenin therapy (Peterson and Meerdink, 1989).

Lizards

The only two known venomous lizards are the Gila monster (*Heloderma suspectum*) and the Mexican beaded lizard *(H. horridum)*. These lizards have grooved teeth with four venom glands on each side that supply the teeth with venom from the dorsal surface. All the teeth are dangerous rather than only the rear teeth, as has been reported. Gila monsters and Mexican beaded lizards have been known to bite with a chewing action thereby envenomating their prey more effectively than with a single bite. *Heloderma* venom has not been well-characterized but one neurotoxin, hyaluronidase, vasoactive peptides, and two arginine esterases have been isolated. The venom also contains a heat-stable smooth muscle stimulating factor. Recent work has uncovered a kallikrein-like hypotensive enzyme from the venom of the Mexican beaded lizard (Peterson and Meerdink, 1989). Pain and swelling are observed at the site of the bite. These symptoms may move progressively toward the body proper, resulting in shock, vomition, and CNS depression by the victim.

Amphibians

Many amphibian species produce glandular secretions of the skin that prevent desiccation, discourage predators, and control the growth of microorganisms on the surface. These secretions have cytotoxic and hemolytic effects. Although all toads and some frogs secrete repulsive substances from their skin, only a few species secrete toxic substances. These include the Colorado River toad *(Bufo alvarius)*, marine toads *(B. marinus)*, and the Golden arrow frog *(Dendrobates* spp.) from Central America. These particular amphibians are found in Florida, the Colorado River drainage system in Arizona and California, and in Hawaii. The parotid glands (skin glands behind the eye) secrete a complex venom that produces symptoms similar to those elicited by cardiac glycosides. The clinical symptoms are profuse salivation, prostration, cardiac arrhythmias, convulsions, and death in as little as 15 min.

The poisons of newts and salamanders have been studied in great detail. Taritotoxin has been isolated from three species of newts: the California newt *(Taricha torosa)*, the European newt *(Triturus* spp.), and the unk *(Bombia variegota)*. These toxins exert their effects on the CNS, resulting in hypertension and contact anesthesia.

Marine Animals

Coelenterates. Among the Coelenterates are the venomous hydroids (Portuguese man-o-war), jellyfish, corals, and sea anemones. Coelenterates have venomous stinging cells known as nematocytes which may be used in defense and/or food gathering. Clinical signs associated with a nematocyte sting are a tingling, stinging sensation, shooting, throbbing pain, prostration, dermal reddening, pruritus (itching), and urticaria, shock, muscle cramps, nausea, vomiting, backache, loss of speech, ptyalism (excessive salivation), paralysis, delirium, convulsions, and death (Russell, 1984).

Mollusca. Mollusks are unsegmented invertebrates which are often covered by a calcareous shell and have a ventral muscular foot used for locomotion. Of the some 80,000 species, about 85 are known to be poisonous to humans. The snails of the class Conus are the most dangerous. The venom is used to paralyze their prey and rarely for defense. A sophisticated venom delivery system consisting of a venom duct (radula) and hollow teeth (radular teeth) is used. A bite from a gastropod can result in a sharp, stinging pain. Symptoms including localized ischemia, cyanosis, numbness, paralysis, and coma often result.

Cepholopoda. The many genera of the class Cepholopoda including the octopus, squid, cuttlefish, and nautilus are venomous. The venom gland is part of the digestive system of these animals. It is unclear which components of the Cephalopod saliva constitute the venomous fraction. The bite consists of two small superficial punctures resulting in a burning or tingling sensation, hemorrhage from the wound, swelling, and inflammation. These wounds are painful but rarely fatal.

Echinodermata. Another class of invertebrates with species toxic to humans are the Echinodermata, more commonly known as sea stars, sea urchins, and sea cucumbers. Tiny spines (pedicellariae) composed of calcium carbonate cover their bodies. These pedicellariae function in food-gathering, grooming, and self-defense. The spines release a toxin directly onto the skin. Usual clinical signs include an immediate and intense burning sensation and inflammation at the site of the wound, followed by numbness and muscular paralysis. These injuries are usually not serious (Russell, 1984).

Fish. Venomous fish inflict wounds via spines which are either mechanically injurious and/or associated with a venom-producing apparatus. The spines are frequently associated with the dorsal or pectoral fins. More than 200 species of fish are considered hazardous. Envenomation from rat fish, elephant fish, and catfish (Oriental variety) causes an instantaneous, throbbing pain which can be incapacitating. Localized inflammation may be severe, leading to gangrene and shock. The sting of the weaver fish (*Trachinidae* spp.), scorpion fish (*Scorpaenidae*), lion fish (*Pterosis* spp.), and stone fish (*Syraceja*) initially result in signs similar to a catfish sting, but may lead to swelling and paralysis of the affected limb, systemic illness, and death. The sting of the hound or spring shark and sting ray are less severe, causing intense pain and less commonly hypotension, vomiting, diarrhea, and sweating. The toad fish and the surgeon fish inflict mechanical wounds with sharp, lance-like movable spines located near the base of their tails.

Several types of fish are poisonous if eaten. Fish that have consumed dinoflagellates can be hazardous if eaten. Also carnivorous fish that have eaten herbivorous fish which have consumed poisoned plant material can be toxic if eaten. Several types of fish are inherently poisonous. Shark livers may be toxic, and the tropical form of the moray eel should not be eaten. Scromboid fish, such as tuna, mackerel, and bonito can be poisonous if any spoilage occurs. Puffer fish, prized in Japan as a delicacy, should be prepared in a special manner. Signs of poisoning include tingling of the lips and tongue, ataxia within 10 to 45 min following ingestion, nausea, vomiting, diarrhea, paralysis, and convulsions resulting in a 60% mortality rate.

Fish producing ciguatoxin or ichthyosarcotoxin should not be eaten. Any marine fish can become toxic, probably as a result of their feeding habits. These toxins are water soluble, so the meat must be soaked, the toxin allowed to leach out, and the water discarded before it is used. Roe (fish eggs) should not be eaten. Signs of ciguatoxicosis are tingling of the lips, tongue, and throat leading to numbness, nausea, vomiting, abdominal cramping, diarrhea, headache, nervousness and convulsions, extreme weakness and paralysis, muscle pain, visual disturbance, and dermatitis. Mortality is about 7%. Ciguatoxin inhibits the enzyme acetylcholinesterase and allows the accumulation of acetylcholine at the postsynaptic membrane, followed by disruption of nerve function and respiratory paralysis.

Sea Snakes. Several varieties of cobra-like marine elapid snakes are known. The bite of a sea snake produces signs similar to a cobra strike. Within 20 min to several hours the victim senses that his or her tongue is thickening. Sensations of muscle stiffness, progressive ascending paralysis and lockjaw are characteristic symptoms. Convulsions may ensue and the mortality rate is approximately 25%.

Shellfish. Shellfish in many oceans have been associated with serious poisonings. Dinoflagellate unicellular algae that contain specific toxins are filtered through their digestive and respiratory mechanisms of these shellfish. Although no harm befalls the shellfish, humans and other animals eating the shellfish are poisoned. Reports of toxicoses are seasonal (usually in the summer) and are attributable to "algal blooms", often referred to as "red tides". Shellfish species eliciting toxicosis are cockles, mussels, clams, and oysters. The toxin is located within the digestive organs (the "dark meat"), the gills, and the siphon. The toxin is heat and water soluble. Poisoned individuals suffer from nausea, vomiting, diarrhea, and abdominal cramping approximately 12 h after ingestion. (These symptoms may also result from bacterial or microorganism contamination of the shellfish.) An allergic condition has been described in those individuals previously sensitized to the toxin.

Fulminating shellfish poisoning is the result of saxitoxin produced by the dinoflagellate. Tingling and burning of the lips, gums, tongue, and face occurs, which then spreads to other areas of the body. Numbness, facial weakness, arthralgia (joint pains), ptyalism, increase thirst, difficulty in swallowing, and paralysis leading to death may occur.

Turtles. Sea turtles have at times been known to be toxic, particularly those caught in the Indo-Pacific region. The liver should be eaten only with caution. Signs of toxicosis are throbbing or dull frontal headache, nausea, vomiting, diarrhea, abdominal pain, dizziness and drowsiness, irritability, photophobia, and convulsions.

Arthropoda/Arachnida

Only a relatively small number of arthropods are sufficiently venomous to be a health hazard to humans. Nevertheless, arthropods are blamed for more poisonings in humans than all other phyla combined. Of the nearly 30,000 species of spiders considered venomous, only a small number have fangs capable of penetrating the intact skin. The bite of the black widow spider (*Latrodectus* spp.) produces muscle fasciculations, sweating, and malaise. The tarantula (*Aphonepelma* spp.), although appearing to be highly aggressive and dangerous, inflicts wounds with only mild effects.

The bite of the brown recluse or violin spider (*Lovosceles reclusa*), a ubiquitous creature living in homes, barns, and outdoors, produces little initial discomfort. However, the bite area often becomes necrotic and ulcerative, producing massive sloughing of soft tissue which leaves a disfiguring scar. Systemic signs that can occur include weakness, nausea, convulsive seizures, vomiting, hemolytic anemia, and thrombocytopenia (Peterson and Meerdink, 1989).

Scorpions (*Centroides* spp.), distant cousins of the spider are well-known for the painful wound they can inflict with the spine of their tail. The venom of *Centroides sculpturatus* contains five separate neurotoxins that activate sodium channels, resulting in prolonged action potentials and depolarization of presynaptic terminals. This causes an excitatory neurotoxicity. Thousands of cases of scorpion stings are reported each year in Mexico. The initial sting produces little pain. Later signs differ between children and adults. In children, signs range from listlessness and abnormal head movements to tachycardia (rapid heart beat), tachypnea (rapid breathing), ataxia (incoordination), and respiratory paralysis. If death does not occur, the child may be asymptomatic within 36 h. The syndrome in adults is less severe, and the victim may be asymptomatic within 12 to 24 h (Keegan, 1980).

Insecta

The California harvester ant (*Pogonomyrmex californicus*) and the fire ant (*Solenopsis xyloni*) produce painful local bites due to formic acid in their saliva. These injuries can be serious if multiple bites are inflicted simultaneously. The stings of wasps, hornets, and bees are responsible for approximately 200,000 anaphylactic-like reactions each year in the U.S. A sting can elicit local swelling and cause anaphylaxis in a previously sensitized victim. Multiple stings result in severe swelling and an allergic response which may kill in a few minutes to hours. The bite of butterflies have been known to cause local inflammation.

Blister beetle (*Epicuata* spp.) ingestion causes significant effects in live-stock, especially horses. The beetles inhabit alfalfa in swarms. Although they are killed in significant numbers by modern methods of hay processing that include crimping and baling, they remain in the baled hay. The toxic principle, cantharidin, produces inflammation of the digestive and urinary tracts. If severe renal involvement occurs, death can result.

Mammals

Only a few species of mammals are venomous. The male platypus (*Ornithorphynchos anathmus*) has a movable horny spur on the inner side of each hind limb near the heel. This spur is connected to a crural or poison gland. Attacks on humans have been few and none have been fatal. There is considerable swelling and pain at the site of the wound and the victim may suffer some degree of shock. The apparatus of the venom gland and the action of the venom of the spiny anteater (*Tachyglossus* spp. and *Zaglossus* spp.) are similar to those of the platypus. Shrews have submaxillary salivary glands with toxin-producing cells. The toxin is secreted with the saliva and is injected when the shrew bites its prey.

The livers of polar bears and pinnipeds, such as seals and walrus, bio-concentrate vitamin A, making the liver potentially toxic, thus it should not be eaten.

C. POISONOUS PLANTS

The ingestion of toxic plants by humans and animals is a common occurrence. In 1988, plants were the fourth most frequent substance involved in human poisonings in the U.S. Almost 94,000 episodes occurred during this period (6.9% of all exposures). While some of these patients did not show any effects, others had symptoms that required treatment or hospitalization, and some even died (Litovitz et al., 1989).

With certain species only very small amounts of some plant parts (flowers, fruits or berries, leaves, stems, barks, roots) need to be ingested to produce dramatic effects. Some plants are poisonous if chewed or swallowed while others are allergenic, or cause dermatitis or mechanical injury. Some plants are harmful only if eaten or chewed at certain stages of their growth; others are toxic at all stages. The species of plant and its stage of growth or the season of the year are important in determining the potential hazards following ingestion.

It is very difficult to prepare meaningful statistics on serious or fatal poisonings by plants because of incomplete reporting and the limited information available to physicians to obtain credible plant identification. The general toxic effects and causative plant-chemical agents in plants associated with the most frequent poisonings are reviewed below.

1. Gastroenteritis-Inducing Plants

The majority of plant intoxications in the U.S. occur from the ingestion of plants that contain gastroenteric irritants. Symptoms range from burning of the mouth and throat, when toddlers chew on the leaves of such common houseplants as the vine growing on the window sill (*Philodendron* sp.) or the dumbcane (*Dieffenbachia*) in the big pot by the front door, to severe vomiting, intestinal cramping, and purging diarrhea from the fresh roots and stems of the pokeweed (*Phytolacca*), seeds of wisteria, berries of spurge laurel (*Daphne* spp.), or leaves of buttercups (*Ranunculus* sp.). The onset of irritation is variable, depending on the nature of the toxin, the necessity for biological activation of the irritant, and the various mechanisms by which emesis or gastroenteritis is elicited. Many plants cause only minor symptoms in adults but can produce profound emesis and diarrhea in a small child, resulting in dehydration and electrolyte imbalance.

Serious intoxications are produced by plants containing toxalbumins, such as the rosary pea (*Abrus precatorius*) and the castor bean (*Riccinus communis*). Following chewing and ingestion of the seeds of these plants, a considerable delay may ensue before onset of signs. This delay is dependent on the quantity of material swallowed. The delay in symptoms may be as short as 1 to 2 h or as long as 1 d. Hemorrhagic gastroenteritis commonly results in signs of nausea, emesis, abdominal cramps, and diarrhea (Table 10.2).

2. Digitalis-Containing Plants

Humans have been fatally poisoned by eating the berries, chewing on leaves or flowers, or drinking water from vases in which digitalis-containing plants such as lily-of-the-valley (*Convallaria majalis*), foxglove (*Digitalis purpurea*), or oleander (*Nerium oleander*) are being displayed. Initially there is a local irritation to the mouth followed by vomiting. In contrast to the effects of pure cardiac glycosides, poisoning by these plants is associated with diarrhea and abdominal pain due to the presence of saponins and other irritants. Acute intoxication is manifest as a defect in cardiac conduction, sinus bradycardia, and hyperkalemia (Table 10.3).

3. Nicotine, Cystisine, and Coniine Containing-Plants

These alkaloids all exert similar actions. Several fatal poisonings in the western U.S. have resulted from the ingestion of these plants, including wild flora in salads and wild tobacco leaves (*Nicotininia* sp.). Cystisine intoxication occurs when the seeds from the pea-like pods of the golden chain tree (*Laburnum anagyroides*) are eaten by humans. Coniine poisoning follows ingestion of the parsley-like leaves or the seeds of the poison hemlock (*Conium maculatum*). Socrates was allegedly executed by being forced to drink a solution of the extract of poison hemlock. Vomiting occurs within the hour (most commonly within 15 min), accompanied by profuse salivation. Abdominal cramping is minimal and diarrhea is rare. Headache, confusion, muscle weakness, incoordination, pyrexia, pupillary dilation, and elevated heart rate ensue. Death results from respiratory arrest.

4. Atropine-Containing Plants

Several plants in the U.S. contain atropine or atropine-related alkaloids. The most important of these is jimsonweed (*Datura stramonium*). Humans are poisoned by eating the seeds or sucking on the flowers of jimsonweed. Hallucinations frequently result. Early symptoms are pupillary dilation with blurred vision due to a lack of accommodation and dryness of the mouth followed by

TABLE 10.2. Plants That Cause Digestive Tract Irritation, the Toxic Agent Involved, and the Clinical Signs Produced

Plant	Toxic Agent	Clinical Signs
Rosary pea seed	Phytotoxin (related to toxalbumin)	Nausea, colic, diarrhea, weakness, trembling, anemia, uremia
Poinsettia (*Euphorbia pulchenia*)	Irritating sap	Emesis, delirium
Philodendron	Calcium oxalate	Renal dysfunction, listlessness
Dumbcane (*Dieffenbachia*)	Oxalate crystals	Oropharyngeal irritation
Ivy (*Hedera helix*)	Saponic glycoside in berries	Purgative, excitement
Lantana	Alkaloid in ripe berries, clippings	Low dose — GI irritation, photosensitization; High dose — muscular weakness, circulatory collapse
Daffodil (*Narcissus*)	Alkaloid in bulb	Emesis, GI irritation, convulsions, anuria
Rhubarb (*Rheum chaponticum*)	Oxalates in leaves	Coma, death, colic, emesis
Pokeweed (*Phytolacca*)	Alkaloid all parts, leaves most toxic	GI irritation, spasms, convulsions
Death camus (*Zygadenus*)	Alkaloid in bulb	GI irritation
Mayapple (*Podophyllum pelatum*)	Resinoid in root	Violent diarrhea
Castor bean (*Riccinus communis*)	Phytotoxin in seed	GI irritation, convulsions
Elderberry (*Sambucus*)	Unconfirmed principle (all parts)	GI irritation
Daphnea (*Daphne*)	Irritating glycosides (all parts)	GI irritation, bloody diarrhea
Wisteria	Unidentified toxin	GI irritation, collapse
Black Locust (*Robinia*)	Heat labile phytotoxin and glycosides (all parts, especially seeds, bark)	GI irritation, nervous signs, depression

dysphagia and difficulty in speaking. The skin becomes hot, dry, and flushed. A rash may appear about the face, ears, and neck. Severe intoxications produce pyrexia, delirium, and hallucinations — convulsions may appear. Atropine poisoning mimics a variety of pathologic states such as encephalitis, meningitis, uremia, concussion, and overt psychosis. Children are especially sensitive to atropine poisoning. Fortunately, fatalities are uncommon, and recovery occurs within 24 h (Table 10.4).

**TABLE 10.3. Plants That Contain Digitalis-Like Principles
and the Clinical Signs of Their Effects**

Plant	Toxic Principle	Clinical Signs
Mistletoe *(Phoradendron serotinum)*	Toxic amines in berries	Acute GI irritation, cardiovascular collapse
Lily-of-the-valley *(Convalaria majalis)*	Glycoside in leaves	Irregular heart rate, vomiting
Purple Foxglove *(Digitalis purpurea)*	Digitalis glycosides in leaves	GI irritation, irregular beats, tremors
Oleander *(Nerium oleander)*	Digitalis-like glycosides	Nausea, dizziness, bloody diarrhea, circulatory irregularities, respiratory arrest

5. Convulsion-Producing Plants

The principle plant in the U.S. which produces convulsions as its primary toxic sign following ingestion is water hemlock (*Cicuta* sp.). Water hemlock is only found in wet, swampy lands and may be mistaken for wild parsnips. Within 15 min to 1 h following ingestion the victim experiences nausea, salivation, emesis, and tremors. This is followed rapidly by multiple grand mal seizures. Death occurs secondary to prolonged anoxia during the tonic contractions of the respiratory muscles (Table 10.5).

6. Cyanide-Containing Plants

Several plants contain cyanide or various cyanide precursors that release cyanide upon consumption. Cyanide interrupts the electron transport chain and halts cellular respiration. Thus cyanide causes cellular hypoxia or anoxia. Classical acute signs are rapid respirations, cherry-red blood, and terminal seizures. Death ensues in 10 to 30 min if the patient is not treated promptly. Cyanide toxicosis is commonly seen in livestock ingesting forages such as *Sorghum* species (milo, Sudan grass, Johnson grass) and corn (Table 10.6).

7. Nitrate-Accumulating Plants

Nitrates accumulate in the vegetative tissues especially in the stalk of certain plants. Nitrate toxicosis from ingestion of plant material is most common in ruminants, and nitrate is reduced to nitrite by rumin microorganisms. Nitrite is approximately 10 times more toxic to an animal than is the nitrate ion. The nitrite ion oxidizes the ferrous (+2) iron of hemoglobin to the ferric (+3) state, resulting in the formation of methemoglobin which cannot transport oxygen to the tissues. If enough methemoglobin is formed, a severe oxygen deficiency may occur in the tissues. Acute poisoning usually occurs .5 to 4 h after ingestion of a high-nitrate feed. Characteristic signs include those attributable to tissue anoxia such as anxiety, polypnea (increased respiration), dyspnea (labored breathing), and a rapid, weak pulse. Blood high in methemoglobin is typically dark brown or chocolate in color, imparting a cyanotic appearance to the mucous membranes. Death usually ensues within a few hours if appropriate treatment is not initiated.

8. Dermatoses-Causing Plants

In general, dermatotoxic plants cause irritation, blistering, pain, and itching. Systemic reactions may also occur if dermatotoxic plants are eaten, but these cases are rare. Treatment consists of washing the skin with soapy water within 5 min of exposure to remove the toxic principle and to prevent further spread of the toxin over the skin (Table 10.7).

D. MUSHROOMS

Unlike vascular plants, mushrooms may be difficult to identify. However, as they produce relatively only a few different toxins, a provisional diagnosis based upon history and clinical signs can be made. Mushrooms that produce signs within 2 h of ingestion may be subdivided into four major categories. These include mushrooms causing gastrointestinal irritation, those producing hallucinations without drowsiness, those causing delirium, sleep, or coma, and those causing a disulfiram-like effect (inhibition of hepatic biotransformation of alcohol) or intoxication. Genera of mushrooms in these categories are *Coprinus* and *Amanita*, with *Amanita* being the most toxic. The toxic effect of a fifth class of mushrooms is characterized by a prolonged latent period of 6 h or more before the onset of clinical signs. This latter group is further subdivided into those that produce sensations of abdominal fullness and severe headache 6 h after ingestion, those that induce severe emesis and diarrhea 12 h after ingestion, and those that cause polydypsia (intense thirst) and polyuria (excessive production of urine) 3 d following ingestion. Mushrooms of the *Cortinarius* species are known to cause severe polydypsia after a latent period of up to 17 d. Other signs include nausea, headaches, muscular pain, and chills (Lampe, 1979).

TABLE 10.4. Plants That Contain Atropine and Similar Agents, and the Clinical Signs Produced Following Ingestion

Plant	Toxic Agent	Clinical Signs
Belladonna *(Atropa belladonna)*	Atropine alkaloid	GI irritation, nervous signs
Potato, Tomato *(Solanum)*	Solanum alkaloid	GI irritation, nervous signs, hot skin, pupil dilation
Jimsonweed *(Datura stramonium)*	Alkaloids, hyoscyamine, scopolamine	Thirst, delirium, convulsions, coma

TABLE 10.5. Toxic Agents of Plants That Affect the Nervous System and the Clinical Signs Produced

Plant	Toxic Agent	Clinical Signs
Health family (rhododendron, azalea)	Resinoid carbohydrate (andromedotoxin)	GI irritation (laurel, depression, paralysis, coma)
Crocus *Colchicum autumnale)*	Alkaloid	Vomiting, nervous signs
Water hemlock *(Cicuta)*	Unsaturated alcoholic resinoid	Violent convulsions, acute nervous signs, respiratory depression
White snakeroot *(Eupatorium rhugosum)*	Trematol	Ketosis, nervous signs, muscular tremors
Yew *(Taxus)*	Alkaloid (taxine)	Acute death, GI irritation, weakness, vital center depression, convulsions

TABLE 10.6. Plants That Contain Cyanide Glycosides

Plant	Toxic Agent
Hydrangea *(Hydrangea macrophylla)*	Cyanogenic glycoside
Cherry *(Prunus)*	Cyanogenic glycoside in leaves, pits
Apple *(Malus)*	Cyanogenic glycoside in seeds
Peach, apricot *(Prunus)*	Cyanogenic glycoside in leaves, pits

TABLE 10.7. Plants Causing Dermal Irritation, the Toxic Agent Involved, and the Resulting Clinical Signs

Plant	Toxic Agent	Clinical Signs
Poison Ivy *(Toxicodendron radicans)* Poison sumac *(Toxicodendron vernix)*	Allergic sap	Vesicant dermatitis
Snow-on-the-mountain *(Euphorbia)*	Irritant vesicant sap	Local irritation, GI irritation if eaten
Nettles *(Urtica)*	Hairs contain irritant	Local irritation

Gastroenteric upset of rapid onset may be produced by many species of mushrooms that need not be identified for adequate therapy to occur. Profuse sweating is caused by mushrooms containing muscarine. Hallucination-provoking mushrooms contain psilocybin, a dose-responsive psychoactive substance. Young children are particularly susceptible to this toxin, but adults rarely require treatment.

Ingestion of mushrooms containing muscinol and ibotenic acid, compounds found in the *Amanita* species, causes delirium associated with sleep or coma. Children are again at greater risk than adults. A metabolite of coprine from *Coprinus* species is responsible for the toxic effects. Originally it was thought that the toxin exerted its affect by inhibiting aldehyde dehydrogenase, but more recent studies have suggested that it inhibits the synthesis of release of the neurotransmitter norepinephrine (Ellerhorn and Barceloux, 1988). *Gyromitra exculenta* produces a toxin resulting in headache and emesis 6 h after ingestion. Fatal hepatic necrosis may occur.

Most fatalities associated with mushroom ingestion in North America are caused by *Anamita phalloides*. It produces a thermostable toxin (anatoxin) which selectively affects the liver and kidney. Symptoms include vomiting, sweating, diarrhea, and abdominal pain.

E. MYCOTOXINS

A mycotoxin is a toxic secondary fungal metabolite found in foods as a result of contamination of food components such as grains and peanuts by certain fungi. The mycotoxin is thus consumed in the diet. Mycotoxicoses are not readily identified due to their insidious and often nonspecific disease manifestations. Treatment with drugs and antibiotics usually have no effect, and antigenic (antibody) stimulation is minimal. Outbreaks may be seasonal and associated with a particular feed or food component. Direct microscopic examination of the suspected foodstuff may reveal the presence of a fungus, and testing can detect the presence of a specific mycotoxin. Toxigenic molds are common feed and food contaminants and are viable within a wide range of environmental conditions.

Fungal spoilage of feeds and foods depends upon specific environmental and substrate characteristics. Potentially toxigenic fungi do not always produce mycotoxins. Fungal growth stressors such as a decrease in available substrate, moisture content of the grain, fungal viability and physical state, and insecticide treatment of the stored product can affect mycotoxin production. Grain damage due to insects or harvesting methods and environmental conditions can enhance fungal growth. Fungi in stored grains require a suitable carbohydrate substrate, grain moisture of 10 to 18%, at least 70% relative humidity in the storage atmosphere, a specific temperature, and oxygen content.

1. Occurrence of Mycotoxins

Mycotoxin formation in the field is contingent upon particular climactic conditions common to certain geographic locales. Five major fungal groups producing mycotoxins occur in North America (Figure 10.1). Moldy grains may inadvertently be incorporated into foods and feeds. In some areas milling and processing may kill or diminish the fungal spore population, making mold contamination inapparent. However, the mycotoxins formed under field conditions are generally heat-stable and may persist in the processed product without visible mold growth.

Fungal spores present in processed foodstuffs can proliferate if growth criteria are met, resulting in mycotoxin production and contamination. These grains may then be shipped to distant locations not generally associated with a particular mycotoxin. At least 100 molds are known which produce the same mycotoxin, while in other instances more than one mycotoxin can be produced by a single fungal species.

2. Biological Effects of Mycotoxins

Mycotoxicoses is most frequently recognized in livestock. Effects range from acute to chronic, depending upon the mycotoxin involved and the amount ingested. Heavy mold growth decreases the palatability and energy content of feeds and can result in feed refusal and poor performance in production.

Hepatotoxins

Aflatoxins, produced by *Aspergillus* and *Penicillium* species of fungi, are the most prominent liver-damaging mycotoxins. Many hepatotoxic mycotoxins initially cause minor lesions which may later develop into hepatic neoplasms. Histologic lesions are characterized by cellular degeneration, fatty change, hemorrhages, and hepatocellular necrosis. Hepatocytes contain

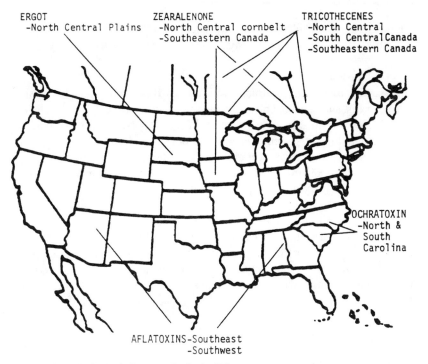

ERGOT
-North Central Plains

ZEARALENONE
-North Central cornbelt
-Southeastern Canada

TRICOTHECENES
-North Central
-South Central Canada
-Southeastern Canada

OCHRATOXIN
-North &
South
Carolina

AFLATOXINS-Southeast
-Southwest

Figure 10.1. Distribution of mycotoxins in the U.S.

enlarged nuclei, and bile duct hyperplasia is common. Acute toxicosis results in icterus (jaundice), hemolytic anemia, hemorrhages, and elevated serum liver enzyme activities. Chronic toxicosis results in poor performance, reduced protein synthesis, hepatic fibrosis, and cirrhosis. Infected poultry have low plasma carotene levels, a reduction in the absorption of fats, and steatosis. A secondary photosensitization may occur since the damaged liver cannot detoxify many photodynamic plant pigments.

Nephrotoxins

Oxalic acid and other nephrotoxins can be produced by two fungal genera *Aspergillus* and *Penicillium* spp. These products cause renal tubular nephrosis. Ochratoxin and citrinin produce renal damage that results in polyuria and polydypsia. Mycotoxin nephropathy is a common problem in Europe, but a major disease outbreak in the U.S. has not been reported. Animals, especially swine, with chronic kidney disease seem to be more susceptible to mycotoxin nephropathy.

Hematopoietic and Coagulation Conditions

Fusarium toxins (trichothecenes such as T-2 and diacetoxyscirpenol) have been associated with hemorrhage, hematoma, weakness, anemia, granulocyto-penia, and increased susceptibility to infections. T-2 and aflatoxins can interfere with the normal blood clotting pathway. Dicumarol, from moldy sweet clover hay, causes acute fatal coagulopathy.

Direct Irritation

Fusarium toxins cause dermonecrotic effects, oral ulceration and necrosis, gastroenteritis, and intestinal bleeding. Diacetoxyscirpenol (DAS) causes hemorrhagic bowel lesions in swine; deoxynivalenol (DON, vomitoxin) induces vomiting and loss of appetite in swine, poultry, and cattle. Some of the trichothecenes produce oral ulceration and necrotic lesions of the small intestine and colon. Certain mycotoxins, especially trichothecenes, suppress appetite and are more likely to elicit symptoms if the dietary requirements are not met.

Reproduction and/or Endocrine Disorders

High levels of zearalonone cause signs of hyperestrogenism in female swine and decreased reproductive capacity in young boars. It can cause anestrus, retained corpora lutea, infertility, and nymphomania in females. Ergotized grain in late gestation induces agalactia in sows because it inhibits the release of prolactin.

Respiratory Effects

Mold-damaged sweet potatoes produce impomearone, associated with hyaline membrane formation in the lungs and pulmonary adenomatosis in cattle.

Central Nervous System Disorders

Ergot and several *Penicillium* toxins (tremorgens) affect the nervous system. Acute ergotism is the result of a lysergic acid-producing principle. Other mycotoxins cause hyperexcitability, incoordination, hypermetria, and tremors. Equine "moldy corn poisoning" or leukoencephalomalacia is a destructive lesion of the brain resulting in somnolence and death.

Immune System Dysfunction

Aflatoxin and rubratoxin impair the immune response causing increased susceptibility to infectious diseases. Aflatoxins cause thymic involution and diminished cell-mediated immunity. Trichothecene mycotoxins (T-2, DAS) depress the bone marrow, causing lymphopenia and thymic involution in cattle, cats, and poultry and resulting in impaired humoral and cell-mediated immunity.

3. Detailed Description of Selected Mycotoxins

Aflatoxins

Eighteen different aflatoxins are known, some produced by fungi and others produced as by-products of animal metabolism following ingestion.

Aflatoxins are stable compounds and not easily destroyed by the common feed-processing methods. Dried grains may contain considerable amounts of aflatoxins. Aflatoxins do not adversely affect palatability but are hepatotoxic to mammals, poultry, and fish. Centrilobular hepatic necrosis is a common pathological change. Aflatoxins induce hepatic neoplasia in laboratory animals and decrease the liver stores of vitamin A in all species.

Ochratoxin and Citrinin

Ochratoxin A is nephrotoxic following chronic exposure, but at higher levels of ingestion may affect other organ systems as well, such as the liver, intestine, lymphoid tissue, and leukocytes. Citrinin, a metabolite of *Penicillium*, is another nephrotoxin that commonly occurs with ochratoxin in the feed.

Sporidesmin

Sporidesmin is a hepatotoxin produced only by the fungus *Pithomyces chartarum*. It mainly affects the biliary tree, thus impeding bile flow. It alters the metabolism of plant pigments which then accumulate, react with sunlight, and cause photosensitization.

Trichothecenes

More than 50 trichothecenes are known to exist. They are stable compounds that form on grain when it is exposed to wet weather and fluctuating temperatures. Several members of this group of mycotoxins affect livestock. Deoxynivalenol (DON, vomitoxin) suppresses appetite in swine and results in decreased weight gain — death is uncommon. T-2 toxin, diacetoxyscirpenol (DAS), and the macrocyclic trichothecenes are far more toxic. Signs, which depend upon the dose consumed, range from feed refusal and hemorrhage to cardiovascular shock. There may also be local cytotoxic effects from direct contact resulting in necrotic oral lesions.

Zearalonone

Zearalonone is produced by *Fusarium roseum*, found in corn-containing feeds. It has a high estrogenic activity, with swine and dairy cattle being the most sensitive. The reproductive organs are affected, resulting in mammary enlargement, vulvar edema, and vaginal prolapse. Male animals suffer a decrease in reproductive capacity.

Slaframine ("Slobber Factor")

Excessive salivation results from the consumption of legume forages, especially red clover parasitized by the fungus *Rhizoctonia leguminicola*. The response from a single exposure to a toxic amount of slaframine is excessive salivation, anorexia, polyuria, and diarrhea, with periodic excessive lacrimation. Symptoms usually subside within 96 h.

Tremorgenic Toxins

Six tremorgenic mycotoxins produce neurologic diseases in cattle characterized by sustained tremors that can result in convulsions and death. Most animals show no signs unless disturbed or excited. Animals with mild to moderate signs recover uneventfully if the feed containing the mycotoxin is changed. Severely affected animals may never recover and can die during a seizure.

Ergot

Ergot (*Claviceps* spp.) is a parasitic fungus which attacks the developing ovary of the grass flower, especially rye grass. The parasitic fruiting body (the sclerotium) replaces or invades the grass ovary. Growth of the fungus is promoted by warm, moist conditions. Three major groups of alkaloids produced by the ergot fungus are ergotamine, ergotoxine, and ergometrine. Ergot alkaloids enhance smooth muscle contraction causing constriction of blood vessels and the muscles of the female reproductive tract. The toxicity of ergot is usually characterized by necrosis and gangrene of the extremities, mimicking frostbite. Ergot alkaloids also inhibit prolactin secretion, resulting in noninflammatory agalactia. Nervous egotism is characterized by hyperexcitability and tremors, which may progress to clonic convulsions upon induced stress and excitement.

Rubratoxin

The fungal species of *Penicillium* produce a hepatotoxin which has a similar effect to the aflatoxins, but it is not known to be a carcinogen. Rubratoxin interferes with the development of immunity in young swine.

Stachbotryotoxicosis

This mycotoxic condition can occur in most animal species, but has been most commonly reported in horses. The fungus *Stachybotrys alternans* produces a toxin that causes anorexia, diarrhea, and reduced milk production, as well as oral hemorrhages and ulcerations. Stachbotryotoxicosis has not been reported as a toxicosis in North America.

F. CONCLUSION

The aim of this chapter has been to acquaint the toxicology student with the most important animal and plant toxins present in our environment. There are texts devoted to many of the toxins discussed above, as well as to other environmental toxins. An in-depth review of the effects of these compounds is available in the quoted texts and references listed.

REFERENCES

Ellerhorn, M.J. and Barceloux, D.G. *Medical Toxicology*. Elsevier Science Publishing Company, Amsterdam, 1988, 1200–1351.

Keegan, H.L. *Scorpions of Medical Importance*. University Press of Mississippi, Jackson, MS, 1980.

Lampe, K.F. Toxic fungi, *Ann. Rev. Pharmacol. Toxicol.* 19:85–104, 1979.

Lampe, K.F. Toxic effects of plant toxins in *Casarett and Doull's Toxicology: The Basic Science of Poisons*, 4th ed. M.O. Amdur, J. Doull, and C.D. Klaassen, Eds. Pergamon Press, Elmsford, New York, 1991, 804–815.

Litovitz, T.L., Schmitz, B.F., and Holm, K.C. 1988 annual report of the American Association of Poison Control Centers National Data Collection System, *Am. J. Emerg. Med.* 7:495–545, 1989.

Peterson, M.E. and Meerdink, G.L. Bites and stings of venomous animals, in *Current Veterinary Therapy X Small Animal Practice*. R.W. Kirk and J.D. Bonagura, Eds. W.B. Saunders Co., Philadelphia, 1989, 177–186.

Russell, F.E. Marine toxins and venomous and poisonous marine animals, in *Advances in Marine Biology*, Vol. III. F.S. Russell, Ed. Academic Press, London, 1965, 255–384.

Russell, F.E., Comparative pharmacology of some animal toxins, *Fed. Proc.* 26:1206–1224, 1967.

Russell, F.E. Marine toxins and venomous and poisonous marine animals, in *Advances in Marine Biology*. J.H.S. Blaxter, F.S. Russell, and C.M. Yonge, Eds. Academic Press, London, 1984, 60–217.

Russell, F.E. and Gertsch, W.J. Letter to the editor (arthropod bites), *Toxicon.* 21:337–339, 1983.

Section III

Chapter

11

Toxicology of Air Pollution

Donald E. Gardner and Susan C. M. Gardner

A. INTRODUCTION

1. Historical Perspective of Atmospheric Pollution

The nature and degree of air pollution have varied over time and the attitude of man towards it has also changed. As the human population grew, so did the levels of environmental pollution. There is little evidence that suggests that air pollution was perceived as a major health problem up to the end of the 13th century. By that time coal was used extensively, and gases emitted from the burning became the most important air pollutant. As early as 1900 the irritating combination of coal smoke and fog, commonly known as smog, became associated with significant increases in urban death rates. The potential hazards associated with the exhaust gases of the internal combustion engine were reported as early as 1915. During the second world war the levels of air pollution in industrial and urban areas increased dramatically but it has only been in the past few years that substantial evidence has been accumulated on the adverse effects of air pollution on human health and well being. It is now understood that air pollution can be disseminated over vast areas by contaminating the atmosphere, resulting in adverse human health effects, and contaminating the environment far away from the original source.

The Environmental Protection Agency is the regulatory agency that has jurisdiction for controlling chemical substances that are emitted into the environment. This includes the contamination by the chemical itself as well as potential degradation products that may be formed within the environment. The federal government has assigned responsibility for generating the needed information on chemicals for their control by organizations in different ways.

For example, for pervasive air pollutants originating from many sources the responsibility for gaining scientific information rests with the federal government. For chemicals that are intentionally used in our daily lives for some specific purpose, such as pesticides, the proof of safety is the responsibility of the industry. To control air pollution we must know (1) how much of, as well as where, when, and how, the pollutant will reach the environment, (2) what will happen to the pollutant in the atmosphere as a result of physical, chemical, and biological reactions, (3) what effects the chemical causes on man, animals, and plants, and (4) will it interact with other substances to enhance or reduce its toxic effect.

Toxicology is one of the fastest moving scientific fields, and the field of inhalation toxicology continues to grow as the public and the scientific community seek more information on the biological hazards of airborne materials. This interest has resulted in an explosion of new research activities and publications that have significantly helped to improve our understanding of the effects and mechanisms of response associated with air pollution. The aim of this chapter is to discuss the basic principles of inhalation toxicology and to relate these principles to the toxicology of air pollution.

2. Primary Emissions and Secondary Pollutants

Air pollution occurs when either chemical or physical agents are present in the air in quantities that can produce significant effects on either man, animals, vegetation, or abiotic materials. These substances may be emitted from certain man-made sources (xenobiotic) and from natural events. Natural air pollutants includes forest fires, bacteria, pollen, and dust particles. Sources of man-made air pollution are those associated with emissions from transportation, industrial factories, electric power plants, coal and oil combustion, and the burning of waste. The list of chemicals emitted into the atmosphere from these sources is very large and may be different from one area of the country to another and from year to year.

In general, air pollutants are divided into two groups, those emitted directly into the atmosphere from specific sources (primary pollutants) or those that result from the interaction between two or more primary pollutants (secondary pollutants) through complex chemical reactions that occur in the atmosphere under certain conditions.

The primary pollutants may be classified in groups depending on their chemical and physical properties. They include organic compounds, compounds containing nitrogen, sulfur, carbon, and halogens, radioactive compounds, and particles ranging in size from 0.1 to 100 μm in size.

The organic group includes saturated and unsaturated aliphatic and aromatic hydrocarbons together with a variety of oxygenated and halogenated derivatives. These pollutants may be emitted as vapors, liquid droplets, or solid particles. Outdoor stationary sources such as chemical plants and petroleum refineries are usually identified as the greatest contributor of these pollutants. Individuals may also be exposed to significant amounts of volatile organic compounds such as benzene when filling their automobile gas tanks. Exposure to organic chemicals such as formaldehyde and vinyl chloride may be associated with an increased risk of cancer.

The most abundant nitrogen compounds in the atmosphere include nitric oxide (NO), nitrogen dioxide (NO_2), and ammonia (NH_3). The first two substances are produced by combustion at high temperatures and in other industrial operations. Ammonia gas is normally present in trace amounts in the atmosphere and is not considered to be a major health hazard. However, at high concentrations it produces immediate irritation of the skin, eyes, and upper and lower respiratory tract, which can cause pulmonary edema and enhance the risk of secondary bacterial infections. High atmospheric concentrations of ammonia are usually associated with an accidental release from an industrial source such as a fertilizer manufacturer or in the production of other chemical products.

Compounds such as sulfur dioxide (SO_2) and sulfur trioxide (SO_3) are released in the environment from the combustion of sulfur containing fuels and from many industrial processes. Emission of high concentration of sulfur compounds and smoke was responsible for thousands of deaths during the air pollution disasters in London in 1952 and 1962. Waste disposal practices can generate hydrogen sulfide (H_2S) and the nauseous odors of mercaptans, unwanted by-products of petrochemical plants and the kraft pulp process used by paper manufacturing industries. In urban areas the oxidation of SO_2 is a major source of particulate sulfates. Airborne acidic products of sulfur dioxide include sulfurous and sulfuric acid. These sulfur-containing acids in the atmosphere account for more than 60% of the acid found in low pH rain that acidifies lakes and rivers and also causes damage to buildings, vegetation, and wildlife.

Large amounts of carbon monoxide (CO) and carbon dioxide (CO_2) are produced from the combustion of carbonaceous fuels such as gasoline. While CO_2 is a normal component of air, high concentrations may cause adverse environmental effects. Worldwide atmospheric concentrations of CO_2 have been increasing significantly. There are concerns over what effects these elevated levels of CO_2 will have on altering the world's temperature. The highest concentration of CO are found where vehicular traffic is the heaviest. Measurements of these levels indicates a bimodal pattern during the day which is associated with the fluctuation of automobile traffic.

Pollution of the atmosphere by particulates is complex and depends on many factors such as the chemical and physical properties of the particles, their interaction with other airborne chemicals, and the emission source. Airborne particles vary in type, size, and amount. The chemical composition depends on the nature of the original source, such as the type of fuel being used, and the industrial activities. For example, in areas with several metallurgical industries one could expect to find higher levels of iron, zinc, manganese, and sulfate in the air. On the other hand the composition of the material emitted from the stacks, chimneys, or furnaces burning fossil fuel (fly ash) will vary with the types of fuel used, method of firing, and distance from the stack that the emissions are being measured. The bulk of this material usually consist of silica, alumina, iron oxide, and carbon.

Airborne particulates are classified into two groups according to their size: fine (<2.5 μm) and coarse (>2.5 μm). The largest particles do not remain airborne for any length of time due to their large mass and rapid rate of sedimentation. The smallest particles are electrically charged and often become adsorbed to other particles. The fine and coarse particles are quite different in composition and their removal from the atmosphere. Fine and course particles differ substantially in their deposition pattern in the respiratory tract, an important factor in understanding the health effects of inhaled particulate matter.

Metallurgical and other industrial sources emit certain inorganic halogen compounds such as hydrogen fluoride and hydrochloric acid. These compounds are corrosive and irritating and produce deleterious effects on vegetation, livestock, and wildlife. The burning of large amounts of chlorine containing substances such as municipal solid waste and plastics, especially polyvinyl chloride, is an important source of emissions of hydrochloride acid. Volcanic gases are the only known natural source of this acid in the atmosphere.

Environmental radiation originates from natural and man made sources. The most serious potential risk of contamination of the environment is the accidental release from nuclear reactors. The type and quantity of this atmospheric waste depends on the type of reactor, the fuel type used (uranium, plutonium, or thorium), and the coolant (air, light or heavy water, gas, or liquid metal). Radioactive compounds can also arise from the combustion of fossil fuels. Natural radioactivity in the atmosphere is due to radionuclides which are derived from radioactive minerals (radon and thoron) in the earth and from the interaction of cosmic rays with gases in the atmosphere producing radioactive species such as tritium and ^{14}C. When these compounds decay, their short lived daughter products attach to existing airborne dust.

In addition to these primary pollutants, scientists are concerned about new products formed in the atmosphere as a result of various chemical and physical reactions that take place. These secondary pollutants may be more toxic than the original compounds from which they are derived. The resulting products and the rate of their formation are influenced by the concentration of the substrates, degree of photochemical activation, and the amount of moisture in the atmosphere. The resulting concentrations of these compounds are also dependent on meteorological dispersive conditions and the local topography. Examples of secondary pollutants of major concern include ozone, organic hydroperoxides, aldehydes, peroxyacetyl nitrates, and certain short lived free radicals. The chemistry and mechanisms of formation of many of these pollutants are very complex. For example, the evaporation of gasoline alone, when mixed with other components in the air, can account for hundreds of species of organic molecules found in the atmosphere. Following the formation of these secondary pollutants some may then be altered further through various oxidation and photochemical reactions.

Traditionally, regulatory strategies have emphasized the control of outdoor sources of air pollution, but only recently has concern been raised over the potential harmful health effects of indoor air pollution from such substances as formaldehyde, radon, cigarette smoke, asbestos, and nitrogen dioxide. Exposure to these compounds is extremely important since it is estimated that individuals may spend 1 or 2 hours a day in the outdoor environment and 22 to 23 h indoors. Of this, <40% is spent in an occupational indoor environment. Measurements indicate that exposure levels to some indoor pollutants may be greater than outdoor concentrations. Thus, with longer exposure, such substances may pose a serious health risk. Air quality standards for air inside dwellings do not exist, although testing and evaluation of radon has been initiated.

Indoor emissions consist of compounds emitted from room surfaces and carpets and from activities taking place in the rooms. Degassing from building fixtures, furniture, paints, and textiles depends on the age of the material and conditions in the room. Contaminants are also derived from heating and cooking devices and from cleaning substances. The size of the room, rate of production of the pollutant, and ventilation of the room air determines the concentration of chemicals in the air.

3. Difficulties in the Study of Air Pollution

Chronic and acute lung disease remains one of the major causes of disability and death in all age groups. The lung is an organ that is continuously exposed to the outside environment and is highly susceptible to damage by inhaled substances. The average adult breathes about 15 kg of air each day, and when compared to the daily intake of about 1.5 kg of food and 2 kg of water, it becomes clear that inhaled air is a major source of exposure to environmental chemicals. This becomes even more important when individuals are at work or exercising. Such activities will increase the pulmonary ventilation rate by approximately 20% and cause a shift from nasal breathing to mouth breathing. Such changes significantly increases overall dose and deposition of inhaled substances in the respiratory system and thus increases the individual's health risk.

Establishing air health standards for the general population is extremely complex since the individual chance of acquiring a disease due to inhaled air pollutants depends on the susceptibility of the individual to that chemical. The susceptibility of individuals may differ based on genetic factors (airway size, normal clearance mechanisms, alpha-1 antitrypsin levels), acquired factors (age, state of health), and personal habits. A small percentage of individuals may actually be adversely affected by exposure to any level of pollutant. The most susceptible in this high risk group includes infants, elderly people, and individuals with severe cardiac and pulmonary disease. Safe environmental standards for a community must consider the susceptibility of these sensitive groups.

The ultimate goal of experimental studies is to estimate the toxicity in animals and to define the mechanism of action of the compound. Extrapolation of data from such studies to man has always been implied, but only qualitative extrapolation is possible. If a certain chemical causes a response in several animal species, most toxicologists would predict that a similar alteration would occur in humans since animals and man have similar morphological, chemical, and physiological systems. However, the specific exposure level required to produce a similar response in humans may be quite different. At present it is not possible to predict with certainty the exact concentration at which these effects would occur. Animals are used as models to estimate risk that man might encounter following exposure. To perform meaningful risk assessment and develop risk management practices for establishing a scientifically defensible standard for an airborne chemical, we must be able to perform quantitative extrapolation.

Quantitative extrapolations require the determination of (1) the amount and location of the inhaled toxicant that reaches the specific target tissue in both animals and humans after exposure to a given concentration (dosimetry) and (2) whether a specific response differs across species (species sensitivity).

Research on air pollution has focused primarily on the outdoor environment, and relatively little attention has been paid to the problems of indoor pollution. Inside buildings the rate of air exchange is low in comparison to that in the outdoor environment, and a buildup of pollutants produced by human activities and building materials can take place. Very little information exists about the indoor home atmosphere. Effective research to determine the overall health risk from indoor pollution and establishing of priorities for different abatement strategies will likely require methodologies and approaches different from those used for the outdoor environment. People are often less likely to suggest any association between indoor air quality and moderate health effects. To improve our understanding of the health risk associated with living indoors will require extensive studies on (1) the physical factors affecting indoor air quality such as emission rates and decay rates, (2) the development of sensitive biological markers as indicators of human exposure, (3) the identification and protection of those individuals who may be uniquely sensitive to chemical exposure even at very low concentrations, and (4) the development of an understanding of the possible interactions that may occur between atmospheric pollutants present outdoors when combined with those indoors.

There are repeated reports on the toxic effects of air pollution to wildlife. Both gaseous and particulate emissions from industrial accidents, urban pollution, and natural sources such as volcanoes have caused significant effects on the wildlife population, ranging from death and injury to more subtle physiological and behavioral effects such as the change in the nesting habits of birds. Such exposures may be enhanced through certain meteorological conditions, possible synergistic interaction of airborne mixtures, bioaccumulation, biomagnification, and enhanced bioavailability due to certain environmental conditions.

Adverse effects have been attributed to airborne trace metals such as lead, arsenic, zinc, and cadmium; to gases including hydrogen sulfide, oxidants, sulfur dioxide; and to airborne fly ash and cement dust (Newman and Schreiber, 1988). Effects are observed in rabbits, field mice, and birds exposed to levels below EPA ambient air quality standard, indicating that these wildlife species may be uniquely sensitive to such air pollutants. Animal sensitivity varies with specific chemical and species exposed.

B. AIR POLLUTION REGULATIONS AND STANDARDS

1. National Ambient Air Quality Standards

The original Clean Air Act (CAA), passed in 1955, authorized the U.S. Public Health Service to limit air pollution. In 1972, a new Clean Air Act provided the necessary authority to establish National Ambient Air Quality Standards (NAAQS) for human health and the environment. Since that time there have been several major efforts by Congress to improve on this legislation. The standards are based on the scientific knowledge that there are thresholds for air pollutants below which no adverse effects are experienced by humans or the environment. When the air pollution in an area exceeds the proposed standard (i.e., nonattainment area), there must be an effort to reduce the air emissions to comply with mandated standards.

The two types of standards are referred to as primary and secondary standards. The former defines the air quality necessary to prevent any adverse effect on human health. The secondary standards are aimed at preventing effects on property, vegetation, and other elements of the environment. The standards vary and depend on length of exposure and concentration. Short term standards are set for 1, 8, or 24 h and serve to protect individuals from slightly higher levels of pollutants that may be tolerated for a brief period. Longer term standards (annual basis) protect the population from repeated low levels of exposure over a prolonged period. Six standards have been established (Table 11.1).

This standard setting process includes a comprehensive summary of the available scientific literature on expected effects which are published in various EPA criteria and control technology documents. Using this information,

TABLE 11.1. National Ambient Air Quality Standards (NAAQS)[a]

| Pollutant | Average Time | Standards | |
		Primary	Secondary
Ozone	1 h	0.12 ppm	Same as primary
Nitrogen dioxide	Annual	0.053 ppm	Same as primary
Carbon monoxide	8 h 1 h	9.0 ppm 35.0 ppm	Same as primary
Particulate matter (PM_{10})[b]	Annual 24 h	50.0 $\mu g/m^3$ 150.0 $\mu g/m^3$	Same as primary Same as primary
Sulfur dioxide	Annual 24 h	0.03 ppm 0.14 ppm	None None
Lead	Calendar quarter	1.5 $\mu g/m^3$	Same as primary

[a]   Code of federal regulations, 40 CFR 59.4-50.12.
[b]   Subscript indicates the particulate mean aerodynamic diameter.

the EPA administrator selects an appropriate standard which, in his/her judgement, is adequate to protect the health of normal and most susceptible subpopulations with an adequate margin of safety. Varying degrees of uncertainty exist about the adequacy of these standards; therefore, as the health data base is improved these standards are appropriately reviewed and revised.

While the federal government is responsible for establishing the NAAQS, the CAA places considerable responsibility on the states for achieving these standards. The prime vehicle for the implementation of the CAA by the states is known as the State Implementation Plan (SIP). The development of SIPs began by dividing each county into geographical air sheds, known as air quality control regions. Each region determines if the air standards are met or exceeded. When the ambient level exceeds the mandated limits, certain control requirements are imposed to reduce emissions to achieve compliance.

2. National Emission Standards for Hazardous Air Pollutants

In addition to setting the NAAQS, the CAA authorizes the EPA to establish National Emission Standards for Hazardous Air Pollutants (NESHAP). Hazardous air pollutants are those for which no ambient air quality standard is applicable and to which exposure may result in or contribute to an increase in mortality or in serious irreversible or incapacitating reversible illnesses. These standards must be established with an ample margin of safety to ensure the protection of public health. Because of environmental damage and health impacts from accidental toxic exposures, states have developed air regulations for over 800 chemicals ranging from common elements such as iron and nickel to complex chemicals such as polynuclear aromatic hydrocarbons. Over the past two decades, the EPA has prescribed emission limitations for seven substances: arsenic, asbestos, benzene, beryllium, mercury, radionuclides, and vinyl chloride.

In establishing these emission standards the EPA takes into consideration factors such as (1) expected severity of the associated human diseases; (2) length of time between exposure and disease, with the longer periods considered especially dangerous; (3) portion of the human intake that is related to airborne substances; and (4) linkage between sources of emissions and reported cases of diseases attributed to the pollutant (Tabler, 1984).

It has been difficult for the EPA to fully satisfy the statutory requirement of NESHAP. The fundamental problem that has prevented an effective regulatory program to control emissions of these pollutants was that it was too stringent. For example, since carcinogenicity is a significant health effect associated with hazardous air pollutants, it become impossible for the EPA to relate the strategy of absolute protection with the emerging knowledge of cancer etiology. The key point in the analysis of carcinogenicity is the traditional position taken by the EPA: there is no "safe" level of exposure for carcinogens. Therefore, it has become virtually impossible for the EPA to write regulations as mandated by the CAA.

3. New Clean Air Act Amendments of 1990

On November 15, 1990 new Clean Air Act Amendments were signed into law, substantially expanding the initial statute. The scope of these environmental regulations in reducing air pollution will have major impacts on both governmental agencies and the private sector. The projected costs to meet these new regulations are in the range of $25 billion per year, making this action one of the biggest undertakings in the history of environmental regulations. While there are many key features of this new law that will have a significant impact on regulations, three examples are discussed that are most closely related to the theme of this text.

Air Toxics

The new amendments will aid the EPA in imposing tighter controls on the emissions of toxic air pollutants and should solve some of the problems experienced under the old NESHAPs. There will be a shift in the focus of regulations away from the previous risk-based approach toward a technology-based control. Based on using the best technology to minimize the release of these substances into the ambient air, companies will be required to install Maximum Achievable Control Technology (MACT) on all stacks.

Congress has directed the EPA to provide a statutory list of 189 substances to be regulated as air toxics. Changes in this list can be made by the EPA if the scientific data demonstrate that such a change is appropriate. The EPA will not establish control requirements on the basis of a specific chemical but will identify and publish a list of categories of major industrial sources that are expected to emit substantial quantities of each of the air toxics. A major source is defined as any facility that emits 10 ton/year of any single air toxic or 25 ton/year of any combination of air toxics. The regulatory goal is to ensure that 90% of the emissions of the 30 most serious chemicals are regulated within 10 years. After each standard is promulgated, industry must comply within 3 to 4 years. While the initial emphasis will be on control technology, the EPA must also determine if still more stringent standards are necessary to protect public health. This would require the EPA to establish carcinogenic risk assessment methodology to be applied to such air toxic exposure.

Acid Rain

Before 1977, there was little concern about the phenomenon of acid rain; however, evidence has proven that emissions of SO_2 and nitrogen oxides (NO_x) are causing damage to lakes, forests, and other ecosystems. Power plants account for about 80% of the SO_2 and 33% of the NO_x emissions. To control the formation of acid rain the new amendments will focus on power plant emissions. The goal is to reduce SO_2 emissions by 10^7 ton annually and NO_2 by 2×10^6 ton. The reductions in SO_2 will be achieved by allocating to the power plants certain "emission allowances". This unique aspect of the acid rain provisions will require plants to reduce their emissions through appropriate controls or be allowed to acquire certain allowances from other plants to come into compliance. Plants that have substantially reduced their levels below the required levels will able to sell their "unused" allowances to other utilities. Such allowance or emission trading should reduce the total SO_2 emissions. Industries now spend over $4 billion a year on compliance with acid rain regulations. It is estimated that this new "pollution credit" approach may actually save industry 30 to 40% in meeting these regulations.

In determining the relative contribution of these emissions to acidic deposition, meteorological variability including differences for rain and the buffering capacity of the soil and water must be considered.

Current knowledge about the formation and transport of acidic substances indicates it has a detrimental effect on terrestrial and aquatic systems, drinking water, agriculture, natural terrestrial ecosystems, and materials. There are relatively few studies regarding the potential human health effects from inhalation of low levels of these acidic compounds.

Chlorofluorocarbons

Based on available evidence, chlorofluorocarbons (CFCs) are considered the major contributor to the depletion of stratospheric ozone (O_3) over the antarctic. The destruction of O_3 is brought about by the action of cosmic radiation on the CFCs releasing chlorine. The chemical stability of the CFCs leads to their desirable safety characteristics but also contributes to their ability to deplete atmospheric O_3. There is no known destructive mechanism for CFCs in the troposphere. Predictions are that CFC concentration in the atmosphere has reached levels high enough to ensure nearly total destruction of O_3 in the lower stratosphere. Such a depletion of the O_3 layer will result in increased exposures of the terrestrial population and marine ecosystems to the damaging effects of ultraviolet radiation.

The CAA amendments seek to develop and manufacture substitutes to replace these CFCs. Stopping the production of CFCs would be a major problem because of their importance in modern technology and society. The EPA will begin phasing chlorofluorocarbons out by the year 2000, and a complete production ban will be effective in 2030. The identification of suitable substitutes will be difficult because toxicity, flammability, and environmental impacts must also be considered. Potential substitutes for CFCs include hydrofluorocarbons (HFC) and hydochlorofluorocarbons (HCFC) which should theoretically result in significantly lower stratospheric chlorine levels.

C. BIOLOGICAL EFFECTS OF AIR POLLUTION

Environmental toxicologists are confronted with the formidable task of interpreting the results from several different types of studies on the health effects of air pollutants. These analyses permit a sound assessment of the relevance and implications of the pollutant levels to which humans can be safely exposed.

In the evaluation of these biological responses, three sources of information can be used to estimate the potential adverse effects of air pollutants on public health. Well-designed and controlled experimental animal studies provide an opportunity (1) to investigate effects of the test substance over a lifetime, uncomplicated by the presence of other pollutants, (2) to utilize an unlimited number of test chemicals at wide ranges of exposure and concentrations, and (3) to use a wide range of sensitive biological indicators to predict hazards. Since animals and man have many physiological similarities, it is assumed that if a particular effect occurs in several animal species it is likely that a similar effect could occur in man, although the exposure levels and duration of the exposure necessary to induce these effects may be quite different. Usually the toxicological data base represented by human studies is quite small compared to the data available from animal studies. This imbalance stems from the nature and inherent limitations of epidemiological and human clinical studies. Acute clinical human studies using volunteers provide data under highly controlled conditions and the concentrations and time of exposure can be directly related to the predicted effects in man. However, chronic exposures cannot usually be performed, the expected effects must be reversible, and the endpoints are limited to nonevasive measurements such as clinical chemistry and pulmonary function endpoints. Epidemiological studies are useful since the test population was exposed under natural conditions. However, in such studies it is extremely difficult to relate a particular exposure to a specific health effect since community studies must cope with a host of complex variables, are restricted to a limited range of natural exposure conditions, and often provide information too late for preventive measures. The best data base for establishing scientifically sound and defensible environmental health standards is a careful integration of all three disciplines.

While many animal studies have been conducted to examine the health effects associated with exposure to air pollutants, most of these studies were conducted at concentrations significantly higher than is normally found in the ambient environment, making it difficult to extrapolate such responses to relatively low levels found in the air. In assessing such risk there are several factors that may influence both the type and magnitude of the individuals susceptible to a given exposure. Individuals differ in both their sensitivity and responsiveness to any chemical exposure. Such sensitivity is often pollutant specific. The response can be affected by the concentration, the duration of the exposure, preexisting disease, sex, age, exercise, and the experimental subject.

This chapter is not intended to review completely all available data on every chemical tested. Instead it will provide the reader with a representative overview of the major studies which will illustrate the effects of various airborne chemicals and relate these observed changes to the potential health risk of the population. An extensive bibliography is included for additional information.

1. Criteria Pollutants

Oxidants

The major chemicals that are important in the formation of oxidant air pollution are nitrogen oxides, aldehydes, carbon monoxide, and hydrocarbons. In areas of abundant sunlight, new substances are formed from complex chemical reactions occurring among these gases, resulting in photochemical oxidant pollution. The most toxic of these oxidant gases is ozone (O_3). Unlike the protective nature of O_3 in the upper atmosphere, O_3 in the earths lower atmosphere can be quite toxic.

Toxicity Profile. An array of health effects is caused by exposure to O_3. Although the precise mechanism of action of O_3 at the molecular level is still unknown, it is agreed that O_3 toxicity is related to its oxidative capacity. Oxidants can react with virtually every class of biological substance. It is likely that the actual mechanism of O_3 toxicity may involve a combination of several interacting mechanisms; however, it appears that the specific site of injury is the cell membrane. Thus, the toxic effects expected would be an alteration in the cell permeability and leakage of essential intracellular enzymes from the cell. If the exposure is severe, the damaged cell may not recover.

The magnitude of the O_3 health effects is highly dependent on concentration, duration of exposure, and the activity level of the subject. Pulmonary functional change and symptoms of toxicity have been measured in both clinical and epidemiological studies. Within the normal population some subgroups respond differently. Differences in response may be due to age, race, gender, disease status, and the specific endpoint measured. Exposed adults report symptoms including chest tightness, irritation, and cough (Folinsbee et al., 1988). Exposure can cause decrement in pulmonary function including rapid, shallow breathing, decreases in the ability to inspire maximally, and decrements in forced expiratory function (Folinsbee and Horvath, 1986). Prolonged exposure can elicit effects at concentrations as low as 0.08 ppm (Horstman et al., 1989).

Since exercise increases both the depth and frequency of breathing, the total volume of inhaled air will also be increased, thus increasing the total dose of O_3 to the lung. In exercising individuals the duration of the exercise is more important than the concentration of the exposure (Folinsbee et al., 1988).

Examining pulmonary lavage fluid can provide useful markers of responses to O_3. Normally, the lung epithelium is a protective barrier controlling molecular and cellular flux across this membrane. Human and animal studies have indicated that O_3 causes a disruption of this barrier resulting in an influx of proteins and blood cells into the alveoli (Alpert et al., 1971; Kehrl et al., 1987). Seltzer et al. (1986) reported increases in pharmacological active substances (prostaglandins and thromboxanes) in lavage fluid from individuals exposed to O_3. Similar changes have been seen in animals. With this inflammatory reaction there may occur increases of other potentially harmful substances such as fibronectin and proteolytic enzymes. These effects may be associated with causing irreversible chronic lung disease. In humans these effects have been observed at concentrations as low as 0.1 ppm and can persist for periods up to 18 h after the exposure.

Numerous other types of responses (i.e., structural, biochemical, and physiological) have been identified in animal studies. Acute exposure to O_3 concentrations <1.0 ppm produces evidence of significant structural changes in cells that line the respiratory tract and the alveoli. With chronic exposure further structural changes may occur which can develop into chronic obstructive lung disease (i.e., fibrosis) which is not reversible.

Changes in various biochemical parameters have been used as markers of pulmonary response to O_3. High concentrations cause decreases in antioxidant metabolism while repeated low levels cause increases (Plopper et al., 1979; Chow et al., 1981; Mustafa et al., 1983). Biochemical studies give evidence that O_3 alters cytoplasmic and lysosomal enzyme activity of cells (Castleman et al., 1973; Chow et al., 1981). Finding increased levels of proteolytic enzymes in lavage fluid indicates lung tissue breakdown (Last et al., 1984).

Several host defense systems are normally operative in protecting the body from infectious and neoplastic diseases. In the upper respiratory tract, ozone reduces the effectiveness of the mucociliary escalator by increasing mucus secretions and reducing cilia beating activity (Kenoyer et al., 1981).

Within the gaseous exchange region of the lung the first line of defense is the alveolar macrophage. O_3 reduces the ability of these cells to engulf (phagocytize) microorganisms and nonviable particles deposited in the lung (Gardner and Graham, 1977; Driscoll et al., 1987; Devlin et al., 1989), reduce their ability to migrate (McAllen et al., 1981), disrupt macrophage membrane integrity (Gardner and Graham, 1977), and reduce lysosomal enzyme and bactericidal activity (Hurst et al., 1970; Kimura and Goldstein, 1981; Van Loveren et al., 1988). Numerous investigators have presented evidence that these effects cause increases in the incidence of respiratory infections (Gardner and Graham, 1977; Miller et al., 1978).

Certain extrapulmonary effects occur indicating that O_3 or some reactive product can cross the blood/gas barrier. O_3 affects both the humoral and cellular mediated immune system (Aranyi et al., 1983; Orlando et al., 1988). The health consequences from such immunosuppression effects of O_3 are not well understood.

Changes in the central nervous system and behavioral pattern in animals at concentrations less than 1.0 ppm have been reported (Tepper et al., 1985). Hematological effects seen in animals include an increase in Heinz bodies in RBC (Menzel et al., 1975), decrease in RBC acetylcholinesterase activity (Moore et al., 1980), and a decrease in RBC survival rate (Moore et al., 1981). Changes have been detected in the serum of animals exposed to O_3 indicating increases in cholesterol and decreases in total lipoprotein triglycerides (Vaughan et al., 1984; Mole et al., 1985). Elevation in serum cholesterol is a risk factor in coronary heart disease.

Pregnant animals and developing fetuses may be at risk from photochemical oxidants. Effects reported have included decreased average maternal weight gain, increased fetal resorption rate, and delays in certain behavioral responses (Kavlock et al., 1980).

Nitrogen Dioxide

Of the oxides of nitrogen which have been studied and occur in the atmosphere through combustion of fossil fuels and a later conversion process, nitrogen dioxide (NO_2) is the most toxic. Exposure to NO_2 can lead to a wide variety of respiratory effects in both man and animals. Like O_3, both reversible and irreversible effects may be caused and the responses depend on the concentration, duration, specific pattern of the exposure, and the species studied. While the magnitude of the effect increases with both concentration and time, concentration has the more pronounced influence. Previous reviews on the health consequences of NO_2 exposure are available (NAS, 1977; WHO, 1977; U.S. EPA, 1982).

Toxicity Profile. NO_2 reacts with the high humidity and temperature of the respiratory tract resulting in formation of chemicals such as HNO and HNO_3. Like O_3, NO_2 has strong oxidative potential. Nitrates have been detected in the blood and urine of animals exposed to NO_2 indicating that NO_2 reacts with various cellular constituents producing nitrates.

Experimental studies with man have indicated that the odor thresholds are about 0.1 ppm. The perception of odor is enhanced by high humidity and is attenuated by increased duration of the exposure.

Clinical studies indicate that healthy subjects at rest do not show difficulties in breathing at concentration <2.0 ppm suggesting that acute exposure to outdoor ambient levels of NO_2 may not acutely impair healthy individuals (Von Nieding and Wagner, 1979; Koenig et al., 1987). Morrow and Utell (1989) found that individuals with chronic lung disease had an adverse response after a short exposure to 0.3 ppm. Von Nieding and Wagner, (1979) reported increases in airway resistance in patients exposed to 1.5 ppm. However, other investigators did not find such responses at levels of 2.0 ppm (Kerr et al., 1979; Linn et al., 1985).

When exercising, patients with asthma may be more sensitive to NO_2 exposure. They exhibited significant increases in airway responsiveness when challenged with nonspecific stimuli (cold air, methacholine). This suggests that the airway passages are constricted, reducing their ability to draw air into the lungs (Mohsenin, 1987; Avol et al., 1988; Bylin et al., 1988). Other studies have not seen such an effect (Linn et al., 1986).

Epidemiological studies have not been able to demonstrate clearly a significant difference in lung function in individuals exposed within the community. The existing data indicates possible diminished lung function and increases in both lower and upper respiratory disease in school children at ambient concentrations (Lebowitz et al., 1985; Vedal et al., 1987). These effects may not be solely associated with NO_2 since other chemicals were present in the environment.

Changes in pulmonary function have been seen in a variety of animal species. Effects reported include increases in respiratory rates and residual volume and decreases in end-expiratory volume and vital capacity (Lafuma et al., 1987; Miller et al., 1987).

Lung biochemistry studies have focused on either identifying putative mechanisms of response or on early indicators of tissue damage. At lower concentrations significant increases occur in lactic dehydrogenase levels and in lung collagen synthesis rate (Sherwin et al., 1972; Last and Warren, 1987);

pulmonary phospholipid synthesis is altered and lecithin synthesis is depressed (Sagai et al., 1984). At elevated exposure levels other biochemical changes take place. These are similar to those seen with O_3 including altered levels of GSH reductase, succinate oxidase, glutathione peroxidase, lung collagen, prostaglandin, and thromboxane (Elsayed and Mustafa, 1982; Azoulay-Dupuis et al., 1983). Following NO_2 exposure there is a significant change in the epithelial tight junctions allowing an increase of protein to influx into the lungs (Selgrade et al., 1981; Hatch et al., 1986).

The epidemiological studies that have shown an association between NO_2 exposure and an increase in respiratory disease are supported by animal studies (Gardner, 1984; McGrath and Oyervides, 1985; Graham et al., 1987; Miller et al., 1987). NO_2 alters host defense mechanism to an extent that the lung is unable to defend itself against bacterial and viral infections. This effect is not only dependent on NO_2 concentration and duration but also on the specific microbial pathogen. Since humans have similar pulmonary defenses this data would suggest that NO_2 could also affect human defenses.

The specific pulmonary defenses affected include structural alterations in the cells making up the mucociliary escalator and effects on the pulmonary macrophage including (1) decreasing bactericidal and viricidal activity, (2) decreasing phagocytic ability, and (3) a variety of morphological and enzymatic changes (Gardner et al., 1969; Greene and Schenider, 1978; Azoulay et al., 1978; Rombout et al., 1986; Frampton et al., 1989; Jakab and Hmieleski, 1988). NO_2 also suppresses primary antibody responses and causes other changes in both the cell-mediated and humoral immune system (Maigetter et al., 1976; Richters and Damji, 1988).

NO_2 exposure causes numerous morphological effects including hypertrophy/hyperplasia and an increase in blood cells in the upper respiratory tract (Rombout et al., 1986; Hayashi et al., 1987). At low concentrations these changes will resolve but more permanent damage occurs at concentrations >2.0 ppm.

NO_2, or its by-products, produces several systemic responses: (1) reduction in body weight (Richters et al., 1987; Stephens et al., 1987); (2) changes in white and red blood cells, platelet, and hemoglobin (Case et al., 1979); (3) altered clinical chemistries (Menzel et al., 1977); (4) cardiovascular system effects (Tsubone and Suzuki, 1984); (5) hepatic changes (Miller et al., 1980; Graham et al., 1982); and (6) effects on the kidney and urine contents (Takahashi et al., 1986). A few studies have indicated developmental or reproductive effects such as possible decreases in liter size, increases in mortality rates of neonates, delays in opening of the eyes, and uncontrolled incisor growth (Tabacova et al., 1985).

Carbon Monoxide

Carbon monoxide (CO) is an odorless, colorless, tasteless gas which is formed as the result of incomplete combustion of carbonaceous material.

Man has experienced CO poisoning since the first discovery of fire. Early medical writings described the seriousness of CO exposure. The important relationship between CO, hemoglobin, and their saturated product carboxyhemoglobin (COHb) on the human physiological process was first described in 1895.

The basic mechanism governing the interaction of CO and O_2 with hemoglobin indicates that CO combines with the reduced hemoglobin molecule much more rapidly than does O_2, thus decreasing the O_2 transport capability of the blood. The tenacity of binding between CO and hemoglobin is more than 200 times that of O_2. Thus, a small amount of inhaled CO can usurp a large proportion of hemoglobin in the blood, seriously impairing the transport of O_2 to extrapulmonary tissue. The alveolar CO levels can be crudely converted to COHb by dividing the CO in the air in ppm by 7 (Kuller and Radford, 1983). The half-life of COHb in the blood is about 4 h. The EPA Air Quality Criteria Document for CO provides an excellent review and detailed evaluation of the scientific data base on the health effects of this gas (U.S. EPA, 1990). Non-smoking human adults normally have <1% COHb in the blood, but for heavy smokers the median value may be as high as 5%.

Toxicity Profile. The early miners recognized the hazards of CO and carried canaries into the mines to monitor CO hazards. The swaying of the bird on its perch before falling provided the workers with an early and sensitive indicator of small quantities of CO. This small bird is significantly more susceptible to CO than larger birds, mice, or man indicating certain species sensitivity.

The effects of CO inhalation have been extensively documented in animal models and human exposure. A unique feature of CO exposure is the biological marker of the dose that an individual has received. By measuring the blood level of COHb the CO exposure can be calculated. In as much as the major effect of CO on the body is related to the capacity for transporting O_2 to the tissue, all organs that require a continual high O_2 supply for normal functions are critical targets. The most important of these are the heart and the central nervous systems (CNS).

Acute exposure to CO levels of 200 to 1200 ppm can cause headaches at 10 to 20% blood COHb levels, mental confusion and incoordination at 40%, convulsions at 50 to 60%, and death at COHb levels above 70%. Normal recovery from a coma can occur if tissue anoxia is not too severe. Adaptation to chronic low levels of CO can occur through an increased hematocrit, increased hemoglobin, and increased blood volume. Community population studies on CO in ambient air have not associated this gas with changes in normal pulmonary function or disease (Lebowitz et al., 1987).

Since the heart requires a continuous supply of O_2 to maintain electrical and contractile integrity, the critical importance of the cardiovascular system during exposure to CO is evident. During periods of tissue hypoxia, the heart compensates by increasing both its rate and its output to meet normal O_2 demands of the body. Alteration of electrical activity and decline in contractile force and ventricular fibrillation soon follow the induction of myocardial hypoxia. The threshold for this latter effect seems to be about 100 ppm CO (9% COHb) (Kuller and Radford, 1983). The most susceptible individuals at risk are those with clinical heart disease or chronic obstructive pulmonary disease.

Studies have shown that exposure to 50 ppm CO can cause chest pain in patients with angina. These individuals had a COHb level of 2 to 4.5%. Whether such effects can be translated to increased risk of heart attacks and sudden death cannot be determined (Aronow, 1981). Animal studies have indicated that CO exposure may promote atherosclerosis, especially in individuals on high fat diets (Klein et al., 1980; Kuller and Radford, 1983).

Exercise capacity is reduced with CO exposure. When the COHb exceeds 45%, human subjects are unable to perform slight physical exertion. The critical concentration at which COHb effects maximal exercise capacity is about 5% (Horvath et al., 1975; Klein et al., 1980).

COHb levels of 7% have been associated with decrements in attention span and learning ability in humans. Similar effects have been reported in studies with laboratory animals exposed to CO (Horvath et al., 1971; Putz et al., 1979).

While adults may have a certain capacity to adjust to moderately high COHb levels without irreversible effects, the fetus is much more susceptible to decreases in tissue oxygen supply so even a moderate CO exposure could have a deleterious effect during fetal development (NRC, 1985). Exposure of pregnant rabbits to CO (10% COHb) resulted in lower birth weights and increases in neonatal mortality and limb deformities (Alstrup et al., 1972). The offspring of female rats exposed to CO (16% COHb) during gestation had significant learning deficits, indicating the susceptibility of the fetus to CO (Flechter and Annau, 1980; Mactutus and Flechter, 1984).

Particulate Matter

Airborne particles exist in the atmosphere in many sizes and vary widely in physical and chemical composition depending on the source emitting the particulates and meteorological conditions. In general, particulates can be divided into two groups according to size, with coarse particles being >2.5 μm in diameter and fine particles >2.5 μm. The fine particles contain sulfates, carbon, ammonium salts, lead, and nitrates. The larger particles consist of oxides of silicon, aluminum, calcium, and iron, as well as pollen, sea salt, and airborne substances from certain man-made products such as tires. Primary particles are discharged directly from natural or xenobiotic sources while secondary particles result from chemical and physical reactions in the atmosphere. Once airborne, particles grow and are chemically transformed through various interactions with other particles and gases.

Toxicity Profile. Potential health effects from exposure to such dusts have been recognized as early as the 16th century when high incidences of respiratory disease were noted in miners and stonecutters. Deposition of inhaled particles is complex and depends on many factors. Once airborne, numerous environmental factors must be considered in determining the toxicity of such aerosols including inhalation patterns, disease state, chemical reactivity, corrosiveness, surface charge, and porosity (Lippmann and Schlesinger, 1984). These factors may affect the behavior of the particles during exposure or may influence the response of the body to the deposited particle.

Once these particles are deposited in the respiratory tract a wide variety of toxic effects may result. The adverse effect elicited depends not only on the specific chemical composition of the particle itself but also on its combination with common gases such as NO_2 and with other particulates present in the aerosol. The toxicity from these exposures and the underlying mechanisms of action may be chemical specific, resulting in a specific respiratory response, or they may also involve systemic toxicity by affecting other organs and their associated functions. If the chemical is corrosive or irritating, acute tissue injury will be produced regardless of the site of deposition.

It is important to understand the general mechanism involved with the toxicity of particles and this discussion will focus on this aspect rather than on the specific toxic effect associated with a specific chemical particle.

In the tracheobronchial (TB) region of the respiratory tract the deposition of fine and relatively small (<2 µm) particles is unlikely and is nearly complete for particles greater than 10 µm (Miller et al., 1979). Bronchoconstriction is a common response to particles deposited in this region. The irritant effects of these particles on the receptors of the TB region may be similar to the result observed in many epidemiological studies, e.g., aggravation of respiratory diseases such as bronchitis, asthma, and emphysema and an increased susceptibility to acute infections (U.S. EPA, 1986).

Inhaled particles induce changes in the distribution and activities of various cell types in the respiratory tract. Acute exposure to high levels initially stimulates clearance by increasing mucus secretion and mucociliary activity. However, with repeated exposure there are increases in the number of secretory cells and the mucous layer becomes thicker, resulting in a depression in the ability of the host to clear particles from the upper respiratory tract (Lippmann and Schlesinger, 1984).

If the particles reach alveolar regions of the lung, the clearance from this area is also slowed and the deposited particles then can interfere with the transfer of oxygen to blood. This may produce irreversible damage especially in sensitive individuals.

With mouth breathing, the TB and pulmonary deposition is enhanced. In subjects undergoing moderate to heavy exercise, oronasal breathing occurs which, in turn, alters the specific site and amount of deposited particles. One would expect deeper penetration of particles into the respiratory tract when breathing through the mouth or oronasally than during nasal breathing.

Some data indicate that children may be at a greater risk to airborne particulates. Since minute ventilation is approximately linear with body mass, children may receive a higher TB dose of particles than adults since they inhale more air per unit body mass and are at a greater risk.

Sulfur Dioxide

Sulfur dioxide (SO_2) is the principle form of sulfur oxide emitted from stationary sources. In the ambient air SO_2 may be inhaled either as a gas or bound to airborne particulates. Depending on their physical and chemical properties, these inhaled substances are deposited in various regions of the respiratory tract where several chemical reactions may take place. These reactions include oxidation of SO_2 to sulfate which is excreted in the urine, formation of sulfurous acid which may be absorbed into the bloodstream through the pulmonary capillaries, reversible binding to proteins, or irreversible autoxidation with the formation of free radicals. These chemical reactions may be expected to alter the toxicity of SO_2. Due to its high water solubility SO_2 is deposited primarily in the upper respiratory tract where the relative humidity is high; however, SO_2 can penetrate more deeply into the lung during mouth breathing and with exercise. Because SO_2 often coexists with airborne particulate matter, there has been a strong association between these two pollutants. The particulates may act as carriers of SO_2 by delivering the SO_2 to the more sensitive areas (alveoli) of the lung.

Toxicity Profile. The ability to smell SO_2 is highly variable, but there are reports that concentrations as low as 0.5 ppm can be detected. Acute inhalation of <1.0 ppm produces mild respiratory symptoms and small changes in the tracheobronchial region, including bronchoconstriction and altered air flow in both normal and asthmatic subjects. Changes in respiratory function in mild asthmatics has been reported at concentrations as low as 0.5 ppm SO_2 after only a 10 min exposure (Jaeger et al., 1979; Schachter et al., 1984; Folinsbee et al., 1985; Horstman et al., 1986; Kulle et al., 1986). These effects are transient and reversible and usually return to normal within 1 h. A variety of factors can modify this response such as dry or cold air, oral breathing, and exercise (Linn et al., 1983; Bethel et al., 1984). Such effects have been duplicated in animals exposed to 2.0 ppm SO_2. An enhancement of the effects of SO_2 has been reported with concomitant exposure to particulates such as sodium chloride or ferrous sulfate.

In addition to changes in lung function, SO_2 also alters normal respiratory defense systems (mucociliary clearance and nasal mucus flow) in both normal and asthmatic subjects at concentrations as low as 1.0 ppm (Hirsch et al., 1975; Carson et al., 1987). It is known that in diseases characterized by retarded clearance from the tracheal epithelium, e.g., chronic bronchitis, there is a predisposition to respiratory infections.

There is no evidence that long term continuous exposure to relatively high concentrations (5.0 ppm) of SO_2 produces any significant pathological changes in the lungs of exposed animals; however, at twice that concentration certain histopathological lesions become evident (Lewis et al., 1973; Alarie et al., 1975).

Lead

Although ubiquitous in nature, the ambient lead concentrations can, to a large extent, be attributed to the increase in industrialization. Lead concentrations measured in urban areas are about two to three times higher than the levels found in remote landscapes. The Greenland snowpack has shown a 200-fold increase in lead content over the last 3000 years (Ng and Patterson, 1981). Murozumi et al. (1969) have studied samples of polar glaciers and showed the long range transport of airborne lead particles. This study demonstrates the tendency of lead to be transported great distances by weather systems.

Airborne lead can have a health impact not only through inhalation but also through the ingestion of particulate lead which can be deposited on food or fur of organisms and be taken in orally during eating or grooming. Both of these routes of exposure can significantly increase the total body burden and should be considered simultaneously when assessing exposure of humans or other organisms.

The respiratory absorption of lead is dependent on two processes: the deposition of particulates of lead into the lung and the absorption of lead into the bloodstream. Deposition is dependent on particulate size, ventilation rate, and respiratory morphology. Approximately 30 to 50% of the lead inhaled by an adult will be deposited deep into the respiratory tract (Gross, 1981). Systemic absorption from the alveoli of the lung occurs directly, while lead deposited in the upper respiratory tract can be absorbed by swallowing and absorption from the gut. Administration of [204]Pb tracers to human volunteers indicates that nearly all the lead deposited in the lower respiratory region is absorbed (Rabinowitz et al., 1977). Experimental animal data support this conclusion. Absorption of lead varies significantly with the age, sex, and nutritional status of the organism. Lead uptake is much greater in children than adults which is thought to reflect a higher deposition rate. The chemical form of the lead does not seem to be a major factor in absorption.

Toxicity Profile. The toxicity of lead is due to its ability to bind to biologically important molecules, disrupting physiological functions. Hematological, neurological, and renal effects have been associated with lead toxicity at low lead exposures. Lead toxicity is due in part to depletion of heme because of the direct inhibition of delta-aminolevulinic acid dehydratase and other enzymes involved in heme biosynthesis (Dresner et al., 1982). The resulting anemia is due to a shortening of erythrocyte survival time. Neurological effects have been demonstrated at blood lead levels as low as 40 to 60 µg/dl in adults and may be responsible for altered behavior and decreases in IQ (Lilis et al., 1977; Irwig et al., 1978; Hammond et al., 1980). Morphological changes in renal mitochondria are an early response to lead exposure (Goyer and Rhyne, 1973). Lead induced nephropathy has been seen in humans at blood levels between 40 and 100 µg/dl. Reproductive and developmental processes can also be adversely affected by lead exposures (Lane, 1949). Lead is readily transferred across the placenta and in rodent studies fetotoxicity has occurred at a concentration of 10 µg/m³ (U.S. EPA, 1986). Epidemiologic studies indicate that fetal exposure to lead may have undesirable effects on mental development of the newborn and the length of the gestation period. In the male, lead may have an adverse effect on the development of sperm and the seminal vesicles.

Animal studies have reported that lead exposure increases susceptibility of the lung to inhaled infectious agents (U.S. EPA, 1986). This indicates that, once inhaled and deposited, lead may be cytotoxic to the pulmonary macrophage.

The role that lead plays in the induction of human cancer is controversial. At present, lead has been classified by the EPA as a probable human carcinogen based on information that significant increases in renal tumors have been observed in bioassays with animals exposed either to soluble lead salts in their diets or by injections. However, there is no epidemiological evidence that workers exposed to lead have a higher incidence of renal cancer than the normal population. Lead has been reported to induce cell transformation and possible mutagenicity (U.S. EPA, 1986).

The EPA has established that at lead exposures of 3.0 µg/m³ or less there appears to be no effect on blood levels. It is at concentrations of 10.0 µg/m³ or more that blood levels increase significantly with increasing exposure. These studies, however, have been done only on male adults and might be different in females and children.

2. Air Toxics (Hazardous Air Pollutants)

Assessing the risk of the many hazardous air pollutants is highly uncertain because it requires using a variety of untested assumptions to compensate for the absence of data. Scientific evidence indicates that, in some cases, a given chemical is hazardous at high levels of exposure but has no effect below a certain level. For cancer causing chemicals, scientists have not agreed on the definition of thresholds below which these effects would not occur. For hazardous pollutants, it is considered that any level of exposure may pose some risk of adverse effects with the risks increasing as the exposure increases. Consequently, risk assessment analysis has been the primary element in the implementation of laws on air toxics.

Under the new Clean Air Act, the emphasis will be on the technological control of air emissions of these toxic substances rather than on the risk based approach. However, the decision on what are these threshold quantities is still based primarily on the chemical toxicity of the substance as well as its reactivity, volatility, and dispersibility within the environment.

Over the past several years the EPA has listed seven compounds as hazardous. These include benzene, asbestos, vinyl chloride, inorganic arsenic, radionuclides, mercury, and beryllium. Of these, only the latter two are listed for noncarcinogenic reasons.

Several epidemiological investigations have established a causal association between the inhalation of benzene and leukemia in humans. Evidence for benzene carcinogenicity is also sufficient in experimental animals. The EPA has classified benzene as a known human carcinogen.

Lung cancer and mesothelioma are the most important asbestos related causes of death among asbestos exposed individuals. Animal studies confirm these human epidemiological results. The three major varieties of asbestos (amosite, crocidolite, and chrysotile) can cause lung cancer with only limited differences in carcinogenic potency. The EPA classifies asbestos as a known human carcinogen.

A strong association exists between exposure to vinyl chloride and the occurrence of tumors of different types in different organs including the liver, lung, and brain and the hematopoietic system. There is also sufficient evidence that vinyl chloride is carcinogenic in a variety of animal species (mice, rats, and hamsters), and for this reason the EPA has classified vinyl chloride as a known human carcinogen.

Several arsenical compounds, especially the trivalent inorganic forms, have been reported to cause tumors in the lung. Most animal studies have not been able to confirm these findings. Inorganic arsenic has been listed as a known human carcinogen by the EPA.

When radioactive materials in particulate form are inhaled and deposited in the lungs, a radiation dose is delivered primarily to the pulmonary parenchyma. However, with radon gas most of its radiation is absorbed by cells lining the airways with very little reaching the alveoli. All of these radiations can cause tumors in test animals and man.

Mercury has not been classified by the EPA as a carcinogen because of inadequate evidence. The information on the effects of inorganic mercury in humans is very limited, but such exposures may be associated with nervous system damage. All of the occupational limits for mercury are based on its potential neurotoxicity, the most sensitive endpoint. Data on mutagenicity are lacking, but methylmercury has been shown to be a potent teratogen in humans (Harada, 1966).

The pulmonary carcinogenicity of beryllium in rats has been confirmed, but human evidence continues to be questionable. Inhalation of beryllium has been found to cause acute chemical pneumonitis and/or delayed pulmonary granulomatosis disease. These conditions have been reproduced in animal experiments.

To control these and other hazardous chemicals in our environment, more information on the interaction of these chemicals with the environment and their impact on human health will be required. In addition, a data base needs to be developed that is capable of tracking the amounts of these compounds being produced, pinpointing the specific location of production, and giving an estimation of the number of individuals exposed to various concentrations of these compounds. A comprehensive characterization of the potential health effects induced by these chemicals including cancer, birth defects, neurological and pulmonary effects, and the mechanisms of action underlying these responses is necessary for future abatement plans.

D. PRINCIPLES OF RISK ASSESSMENT FOR INHALED CHEMICALS

All substances are toxic under certain conditions of exposure. The important question is not only of toxicity but rather of risk. Everyone accepts some degree of risk in their daily activities. However, we must determine the probability that an activity will cause an adverse effect under actual conditions of human exposure. Determining this risk for airborne chemicals requires more extensive data and evaluation than merely the characterization of the toxic properties of the chemical. It also involves an understanding of the conditions of human exposure to ascertain both the likelihood that exposed humans will be adversely affected and to characterize the nature of the effects they may experience.

The common usage of the term "safe" often means "without risk", but in technical terms this is misleading since science cannot ascertain the conditions under which an exposure is likely to be absolutely without risk. Science cannot prove the existence of what is essentially a negative condition. However, science can establish conditions under which risks are so low that they would be considered to be of no practical consequence. These conditions usually incorporate large safety factors so even elevated exposures can be defined as safe with extremely low risks. The essential steps needed for assessing risk are included in the following sections.

1. Hazard Identification

A proper hazard evaluation includes a critical review of the pertinent data base from animal experiments, epidemiological investigations, and clinical studies. Other information involving effects on isolated organs, cell lines, and subcellular components are also considered, but these are less certain indicators of toxicity potential. This information is then used to determine if exposure can increase the incidence of health injury or disease in people and to identify conditions of exposure associated with this effect. It often includes a characterization of the behavior of the chemical within the body and the interactions it undergoes within the target cell.

2. Dose Response Assessment

This involves defining the quantitative relationship between the exposure to the chemical and the extent of the toxic response. To do this one must distinguish between two types of exposure measurements: measurement of the actual amount received by the subject, either man or animal, in the target tissue and the measurement of the amount of the substance in the specific medium, i.e., air. The first measure is usually expressed as dose and is critical for risk assessment since risk is better related to the actual dose received by the cells and tissues that exhibit biological effects. The second measure is concentration which is used to establish dose. A major obstacle in determining exposure/dose relationships is in defining the most appropriate expressions of dose. Most human health risk assessments are developed on the basis of risk per unit of exposure to the toxic agent. In such cases, exposure is used as a surrogate for dose. Using animal inhalation data to estimate human dose may be difficult since animals are usually exposed to high doses and effects must be extrapolated to low doses; animals and humans, due to differences in size and metabolism, often differ in susceptibility, and the human population is heterogeneous, making some individuals more susceptible than others. In determining dose response relationships one considers the intensity of the exposure, concentration/time relationships, thresholds, metabolism, and the shape of the dose response curve.

3. Human Exposure Evaluation

This involves estimating the number of people exposed and the magnitude, duration, and timing of their exposure. These estimates may incorporate data from past, current, or anticipated exposures in the future. This involves collecting data on the quantities of the chemical released into the atmosphere, the ultimate fate of the substance in the environment, which depends on transport, persistence, chemical interactions, and degradation, and estimates of amount inhaled. Once this information is available, then human exposure can be estimated. Unfortunately, data are frequently very limited in exposure assessment. Models exist to determine the transport of chemicals, and certain assumptions can be made regarding deposition and bioavailability of the inhaled chemical. This information is then used to estimate the dose acquired by humans.

4. Risk Characterization

This final step involves combining all of the above data to predict the frequency and severity of health effects in a human population assuming the specific conditions of exposure. Each of the above steps has an associated uncertainty and the final risk characterization must describe the biological and statistical uncertainty. The possibility of zero risk cannot be excluded. Because the degree of uncertainty varies considerably among risk assessments for different chemicals, the lack of consideration of this uncertainty can lead to inappropriate levels of concern for different chemicals. Thus, uncertainty associated with such assessment needs to be clearly defined. Once this final step in risk assessment is completed, the regulatory agencies can evaluate alternative regulatory options incorporating such information as political, social, economic, and engineering technology in their decision process.

REFERENCES

Alarie, Y.C., Krumm, A.A., Busey, W.M., Ulrich, C.E., and Kantz, R.J., II. Long-term exposure to sulfur dioxide, sulfuric acid mist, fly ash, and their mixtures: results of studies in monkeys and guinea pigs, *Arch. Environ. Health* 30:254–262, 1975.

Alpert, S.M., Gardner, D.E., Hurst, D.J., Lewis, T.R., and Coffin, D.L. Effects of exposure to ozone on defensive mechanisms of the lung, *J. Appl. Physiol.* 31:247–252, 1971.

Alstrup, P., Olson, H., Trolle, D., and Kjeldsen, K. Effects of moderate carbon monoxide exposure on fetal development, *Lancet* 2:1220–1222, 1972.

Aranyi, C., Vana, S.C., Thomas, P.T., Bradof, J.N., Fenters, J.D., Graham, J.A., and Miller, F.J. Effects of subchronic exposure to a mixture of O_3, SO_2, and $(NH_4)_2SO_4$ on host defenses of mice, *J. Toxicol. Environ. Health* 12:55–71, 1983.

Aronow, W.S. Aggravation of angina pectoris by two percent carboxyhemoglobin, *Am. Heart J.* 101:154–157, 1981.

Avol, E.L., Linn, W.S., Peng, R.C., Valencia, G., Little, D., and Hackney, J.D. Laboratory study of asthmatic volunteers exposed to nitrogen dioxide and to ambient air pollution, *Am. Ind. Hyg. Assoc. J.* 49:143–149, 1988.

Azoulay, E., Soler, P., and Blayo, M.C. The absence of lung damage in rats after chronic exposure to 2 ppm nitrogen dioxide, *Bull. Eur. Physiopathol. Respir.* 14:311–325, 1978.

Azoulay-Dupuis, E., Torres, M., Soler, P., and Moreau, J. Pulmonary NO_2 toxicity in neonate and adult guinea pigs and rats, *Environ. Res.* 30:322–339, 1983.

Bethel, R.A., Sheppard, D., Epstein, J., Tam, E., Nadel, J.A., and Boushey, H.A. Interaction of sulfur dioxide and dry cold air in causing bronchoconstriction in asthmatic subjects, *J. Appl. Physiol.: Respir. Environ. Exercise Physiol.* 57:419–423, 1984.

Bylin, G., Hedenstierna, G., Lindvall, T., and Sundin, B. Ambient nitrogen dioxide concentrations increase bronchial responsiveness in subjects with mild asthma, *Eur. Respir. J.* 1:606–612, 1988.

Carson, J.L., Collier, A.M., Hu, S.-C., Smith, C.A., and Stewart, P. The appearance of compound cilia in the nasal mucosa of normal human subjects following acute, *in vivo* exposure to sulfur dioxide, *Environ. Res.* 42:155–165, 1987.

Case, G.D., Dixon, J.S., and Schooley, J.C. Interaction of blood metalloproteins with nitrogen oxides and oxidant air pollutants, *Environ. Res.* 20:43–65, 1979.

Castleman, W.L., Dungworth, D.L., and Tyler, W.S. Cytochemically detected alterations of lung acid phosphatase reactivity following ozone exposure, *Lab. Invest.* 29:310–319, 1973.

Chow, C.K., Plopper, C.G., Chiu, M., and Dungworth, D.L. Dietary vitamin E and pulmonary biochemical and morphological alterations of rats exposed to 0.1 ppm ozone, *Environ. Res.* 24:315–324, 1981.

Devlin, R., Graham, D., and Koren, H. Changes in RNA accumulation and protein synthesis in macrophages from humans exposed to ozone, *Am. Rev. Respir. Dis.* 139(Suppl.):A47, 1989.

Dresner, D.L., Ibrahim, N.G., Mascarenhas, B.R., and Levere, R.D. Modulation of bone marrow heme and protein synthesis by trace elements, *Environ. Res.* 28:55–66, 1982.

Driscoll, K.E., Vollmuth, T.A., and Schlesinger, R.B. Acute and subchronic ozone inhalation in the rabbit: response of alveolar macrophages, *J. Toxicol. Environ. Health* 21:27–43, 1987.

Elsayed, N.M. and Mustafa, M.G. Dietary antioxidants and the biochemical response to oxidant inhalation, *Appl. Pharmacol.* 66:319–328, 1982.

Flechter, L.D. and Annau, Z. Prenatal carbon monoxide exposure alters behavioral development, *Neurobehavioral Toxicol.* 2:7–11, 1980.

Folinsbee, L.J. and Horvath, S.M. Persistence of the acute effects of ozone exposure, *Aviat. Space Environ. Med.* 57:1136–1143, 1986.

Folinsbee, L.J., Bedi, J.F., and Horvath, S.M. Pulmonary response to threshold levels of sulfur dioxide (1.0 ppm) and ozone (0.3 ppm), *J. Appl. Physiol.* 58:1783–1787, 1985.

Folinsbee, L.J., McDonnell, W.F., and Horstman, D.H. Pulmonary function and symptom responses after 6.6-hour exposure to 0.12 ppm ozone with moderate exercise, *J. Air Pollut. Control Assoc.* 38:28–35, 1988.

Frampton, M.W., Smeglin, A.M., Roberts, N.J., Jr., Finkelstein, J.N., Morrow, P.E., and Utell, M.J. Nitrogen dioxide exposure *in vivo* and human alveolar macrophage inactivation of influenza virus *in vitro, Environ. Res.* 48:179–192, 1989.

Gardner, D.E. Oxidant-induced enhanced sensitivity to infection in animal models and their extrapolations to man, *J. Toxicol. Environ. Health* 13:423–439, 1984.

Gardner, D.E. and Graham, J.A. Increased pulmonary disease mediated through altered bacterial defenses, in *Pulmonary Macrophage and Epithelial Cells: Proceedings of the 16th Annual Hanford Biology Symposium; September 1976.* C.L. Sanders, R.P. Schneider, G.E. Dagle, and H.A. Ragan, Eds. Energy Research and Development Administration, Washington, DC, 1977, 1–21.

Gardner, D.E., Holzman, R.S., and Coffin, D.L. Effects of nitrogen dioxide on pulmonary cell population, *J. Bacteriol.* 98:1041–1043, 1969.

Goyer, R.A. and Rhyne, B.C. Pathological effects of lead, *Int. Rev. Exp. Pathol.* 12:1–77, 1973.

Graham, J.A., Miller, F.J., Gardner, D.E., Ward, R., and Menzel, D.B. Influence of ozone and nitrogen dioxide on hepatic microsomal enzymes in mice, *J. Toxicol. Environ. Health* 9:849–856, 1982.

Graham, J.A., Gardner, D.E., Blommer, E.J., House, D.E., Menache, M.G., and Miller, F.J. Influence of exposure patterns of nitrogen dioxide and modifications by ozone on susceptibility to bacterial infectious disease in mice. *J. Toxicol. Environ. Health* 21:113–125, 1987.

Greene, N.D. and Schneider, S.L. Effects of NO$_2$ on the response of baboon alveolar macrophages to migration inhibitory factor, *J. Toxicol. Environ. Health* 4:869–880, 1978.

Gross, S.B. Human oral and inhalation exposures to lead: summary of kehoe balance experiments, *J. Toxicol. Environ. Health* 8:333–377, 1981.

Hammond, P.B., Lerner, S.I., Gartside, P.S., Hanenson, I.B., Roda, S.B., Foulkes, E.C., Johnson, D.R., and Resce, A.J. The relationship of biological indices of lead exposure to the health status of workers in a secondary lead smelter, *J. Occup. Med.* 22:475–484, 1980.

Harada, Y. Congenital (or fetal) Minamata disease, in *Minamata Disease*, M. Katsanuma, Ed. Study group of Minamata disease, Kunamato University, Japan, 1966, 93.

Hatch, G.E., Slade, R., Stead, A.G., and Graham, J.A. Species comparison of acute inhalation toxicity of ozone and phosgene, *J. Toxicol. Environ. Health* 19:43–53, 1986.

Hayashi, Y., Kohno, T., and Ohwada, H. Morphological effects of nitrogen dioxide on the rat lung, *Environ. Health Perspect.* 73:135–145, 1987.

Hirsch, J.A., Swenson, E.W., and Wanner, A. Tracheal mucous transport in beagles after long-term exposure to 1 ppm sulfur dioxide, *Arch. Environ. Health* 30:249–253, 1975.

Horstman, D., Roger, L.J., Kehrl, H., and Hazucha, M. Airway sensitivity of asthmatics to sulfur dioxide, *Toxicol. Ind. Health* 2:289–298, 1986.

Horstman, D., McDonnell, W., Folinsbee, L., Abdul-Salaam, S., and Ives, P. Changes in pulmonary function and airway reactivity due to prolonged exposure to typical ambient ozone (O$_3$) levels, in *Atmospheric Ozone Research and Its Policy Implications: Proceedings of the 3rd US-Dutch International Symposium, May 1988, Nijmegen, The Netherlands.* T. Schneider, S.D. Lee, G.J.R. Wolters, and L.D. Grant, Eds. Elsevier Science Publishers, Amsterdam, 1989, 755–762.

Horvath, S.M., Dahms, T.E., and O'Hanton, J.F. Carbon monoxide and human vigilance, *Arch. Environ. Health* 23:343–347, 1971.

Horvath, S.M., Raven, P.B., Dahms, T.E., and Gray, D.J. Maximal aerobic capacity at different levels of carboxyhemoglobin, *J. Appl. Physiol.* 38:300–303, 1975.

Hurst, D.J., Gardner, D.E., and Coffin, D.L. Effect of ozone on acid hydrolases of the pulmonary alveolar macrophage, *J. Reticuloendothel. Soc.* 8:288–300, 1970.

Irwig, L.M., Harrison, W.O., Rocks, P., Webster, I., and Andrew, M. Lead and morbidity: a dose-response relationship, *Lancet* 2(8079):4–7, 1978.

Jaeger, M.J., Tribble, D., and Wittig, H. Effects of 0.5 ppm sulfur dioxide on respiratory function of normal and asthmatic subjects, *Lung*, 156:119–127, 1979.

Jakab, G.J. and Hmieleski, R.R. Reduction of influenza virus pathogenesis by exposure to 0.5 ppm ozone, *J. Toxicol. Environ. Health* 23:455–472, 1988.

Kavlock, R.J., Meyer, E., and Grabowski, C.T. Studies on the developmental toxicity of ozone, *Toxicol. Letters*. 5:3–9, 1980.

Kehrl, H.R., Vincent, L.M., Kowalsky, R.J., Horstman, D.H., O'Neil, J.J., McCartney, W.H., and Bromberg, P.A. Ozone exposure increases respiratory epithelial permeability in humans, *Am. Rev. Respir. Dis.* 135:1124–1128, 1987.

Kenoyer, J.L., Phalen, R.F., and Davis, J.R. Particle clearance from the respiratory tract as a test of toxicity: effect of ozone on short and long term clearance, *Exp. Lung Res.* 2:111–120, 1981.

Kerr, H.D., Kulle, T.J., McIlhany, M.L., and Swidersky, P. Effects of nitrogen dioxide on pulmonary function in human subjects: an environmental chamber study, *Environ. Res.* 19:392–404, 1979.

Kimura, A. and Goldstein, E. Effect of ozone on concentrations of lysozyme in phagocytizing alveolar macrophages, *J. Infect. Dis.* 143:247–251, 1981.

Klein, J.P., Forster, H.V., Stewart, R.D., and Wu, A. Hemoglobin affinity for oxygen during short term exhaustive exercise, *J. Appl. Physiol.* 48:236–242, 1980.

Koenig, J.Q., Covert, D.S., Marshall, S.G., VanBelle, G., and Pierson, W.E. The effects of ozone and nitrogen dioxide on pulmonary function in healthy and in asthmatic adolescents, *Am. Rev. Respir. Dis.* 136:1152–1157, 1987.

Kulle, T.J., Sauder, L.R., Hebel, J.R., Miller, W.R., Green, D.J., and Shanty, F. Pulmonary effects of sulfur dioxide and respirable carbon aerosol, *Environ. Res.* 41:239–250, 1986.

Kuller, L.H. and Radford, E.P. Epidemiological bases for the current ambient carbon monoxide standards, *Environ. Health Perspect.* 52:131–139, 1983.

Lafuma, C., Harf, A., Lange, F., Bozzi, L., Poncy, J.L., and Bignon, J. Effect of low-level NO_2 chronic exposure on elastase-induced emphysema, *Environ. Res.* 43:75–84, 1987.

Lane, R.E. The care of the lead worker, *Br. J. Ind. Med.* 6:125–143, 1949.

Last, J.A. and Warren, D.L. Synergistic interaction between nitrogen dioxide and respirable aerosols of sulfuric acid or sodium chloride on rat lungs, *Toxicol. Appl. Pharmacol.* 90:34–42, 1987.

Last, J.A., Reiser, K.M., Tyler, W.S., and Rucker, R.B. Long-term consequences of exposure to ozone. I. Lung collagen content, *Toxicol. Appl. Pharmacol.* 72:111–118, 1984.

Lebowitz, M.D., Holberg, C.J., Boyer, B., and Hayes, C. Respiratory symptoms and peak flow association with indoor and outdoor air pollutants, *J. Air Pollut. Control Assoc.* 35:1154–1158, 1985.

Lebowitz, M.D., Collins, L., and Holberg, C.J. Time series analysis of respiratory responses to indoor and outdoor environmental phenomena, *Environ. Res.* 43:332–341, 1987.

Lewis, T.R., Moorman, W.J., Ludmann, W.F., and Campbell, K.I. Toxicity of long-term exposure to oxides of sulfur, *Arch. Environ. Health* 26:16–21, 1973.

Lilis, R., Fischbein, A., Eisinger, J., Blumberg, W.E., Diamond, S., Anderson, H.A., Rom, W., Rice, C., Sarkozi, L., Kon, S., and Selikoff, I.J. Prevalence of lead disease among secondary lead smelter workers and biological indicators of lead exposure, *Environ. Res.* 14:255–285, 1977.

Linn, W.S. Shamoo, D.A., Spier, C.E., Valencia, L.M., Anzar, U.T., Venet, T.G., and Hackney, J.D. Respiratory effects of 0.75 ppm sulfur dioxide in exercising asthmatics, *Am. Rev. Resp. Dis.* 127:278–283, 1983.

Linn, W.S., Solomon, J.C., Trim, S.C., Spier, C.E., Shamoo, D.A., Venet, T.G., Avol, E.L., and Hackney, J.D. Effects of exposure to 4 ppm nitrogen dioxide in healthy and asthmatic volunteers, *Arch. Environ. Health* 40:234–239, 1985.

Linn, W.S., Shamoo, D.A., Avol, E.L., Whynot, J.D., Anderson, K.R., Venet, T.G., and Hackney, J.D. Dose-response study of asthmatic volunteers exposed to nitrogen dioxide during intermittent exercise, *Arch. Environ. Health* 41:292–296, 1986.

Lippmann, M. and Schlesinger, R.B. Interspecies comparison of particle deposition and mucociliary clearance in tracheobronchial airways, *J. Toxicol. Environ. Health* 13:441–469, 1984.

Mactutus, C.F. and Flechter, L.D. Prenatal exposure to carbon monoxide, *Science,* 223:409–411, 1984.

Maigetter, R.Z., Ehrlich, R., Fenters, J.D., and Gardner, D.E. Potentiating effects of manganese dioxide on experimental respiratory infections, *Environ. Res.* 11:386–391, 1976.

McAllen, S.J., Chiu, S.P., Phalen, R.F., and Rasmussen, R.E. Effect of *in vivo* ozone exposure on *in vitro* pulmonary macrophages mobility, *J. Toxicol. Environ. Health* 7:373–381, 1981.

McGrath, J.J. and Oyervides, J. Effects of nitrogen dioxide on resistance to *Klebsiella-pneumoniae* in mice, *J. Am. Coll. Toxicol.* 4:227–231, 1985.

Menzel, D.B., Slaughter, R.J., Bryant, A.M., and Jaurequi, H.O. Heinz bodies formed in erythrocytes by fatty acid ozonides and ozone, *Arch. Environ. Health* 30:296–301, 1975.

Menzel, D.B., Abou-Donia, M.B., Roe, C.R., Ehrlich, R., Gardner, D.E., and Coffin, D.L. Biochemical indices of nitrogen dioxide intoxication of guinea pigs following low level long term exposure, in *Proceedings of an International Conference on Photochemical Oxidant Pollution and Its Control.* B. Dimitriades, Ed. USEPA Report No. EPA-600/3-77-001B. U.S. Environmental Protection Agency, Washington, DC, 1977, 577–587.

Miller, F.J., Illing, J.W., and Gardner, D.E. Effect of urban ozone levels on laboratory-induced respiratory infections, *Toxicol. Lett.* 2:163–169, 1978.

Miller, F.J., Gardner, D.E., Graham, J.A., Lee, R.E., Wilson, W.E., and Bachmann, J.D. Size consideration for establishing a standard for inhalable particles, *J. Air Pollut. Control. Assoc.* 29:610–615, 1979.

Miller, F.J., Graham, J.A., Illing, J.W., and Gardner, D.E. Extrapulmonary effects of NO_2 as reflected by pentobarbital-induced sleeping time in mice, *Toxicol. Lett.* 6:267–274, 1980.

Miller, F.J., Graham, J.A., Raub, J.A., Illing, J.W., Menache, M.G., House, D.E., and Gardner, D.E. Evaluating the toxicity of urban patterns of oxidant gases. II. Effects in mice from chronic exposure to nitrogen dioxide, *J. Toxicol. Environ. Health* 21:99–112, 1987.

Mohsenin, V. Airway responses to nitrogen dioxide in asthmatic subjects, *J. Toxicol. Environ. Health* 22:371–380, 1987.

Mole, M.L., Stead, A.G., Gardner, D.E., Miller, F.J., and Graham, J.A. Effect of ozone on serum lipids and lipoproteins in the rat, *Toxicol. Appl. Pharmacol.* 80:367–376, 1985.

Moore, G.S., Calabrese, E.J., and Grinberg-Funes, R.A. The C57L/J mouse strain as a model for extrapulmonary effects of ozone exposure. *Bull. Environ. Contam. Toxicol.* 25:578–585, 1980.

Moore, G.S., Calabrese, E.J., and Labato, F.J. Erythrocyte survival in sheep exposed to ozone, *Bull. Environ. Contam. Toxicol.* 27:126–138, 1981.

Morrow, P.E. and Utell, M.J. Responses of susceptible subpopulations to nitrogen dioxide. Research Report No. 23, Health Effects Institute, Cambridge, MA, 1989, 1–38.

Murozumi, M., Chow, T.J., and Patterson, C. Chemical concentrations of pollutant lead aerosols, terrestrial dusts and sea salts in Greenland and antarctic snow strata, *Geochim Cosmochim. Acta* 33:1247–1294, 1969.

Mustafa, M.G., Elsayed, N.M., Graham, J.A., and Gardner, D.E. Effects of ozone exposure on lung metabolism: influence of animal age, species, and exposure conditions, in *International Symposium on the Biomedical Effects of Ozone and Related Photochemical Oxidants; March 1982, Pinehurst, NC*. S.D. Lee, M.G. Mustafa, and M.A. Mehlman, Eds. Princeton Scientific Publishers, Inc., Princeton, NJ, 1983, 57–73.

National Academy of Science (NAS). Medical and biological effects of environmental pollutants: nitrogen dioxide. National Academy of Science, Washington, DC, 1977, 1–333.

Newman, J.R. and Schreiber, R.K. Air pollution and wildlife toxicology: an overlooked problem, *Environ. Toxicol. Chem.* 7:381–390, 1988.

Ng, A. and Patterson, C. Natural concentrations of lead in ancient arctic and antarctic ice, *Geochim. Cosmochim. Acta* 45:2109–2121, 1981.

Orlando, G.S. House, D., Daniel, E.G., Koren, H.S., and Becker, S. Effect of ozone on T-cell proliferation and serum levels of cortisol and beta-endorphin in exercising males, *Inhalation Toxicol.* 1:53–63, 1988.

Plopper, C.G., Chow, C.K., Dungworth, D.L., and Tyler, W.S. Pulmonary alterations in rats exposed to 0.2 and 0.1 ppm ozone: a correlated morphological and biochemical study, *Arch. Environ. Health* 34:390–395, 1979.

Putz, V.R., Johnson, B.L., and Setzer, J.V. A comparative study of the effects of carbon monoxide and methylene chloride on human performance, *J. Environ. Pathol. Toxicol.* 2:97–112, 1979.

Rabinowitz, M.B., Wetherill, G.W., and Kopple, J.D. Magnitude of lead intake from respiration by normal man, *J. Lab. Clin. Med.* 90:238–248, 1977.

Richters, A. and Damji, K.S. Changes in T-lymphocyte subpopulations and natural killer cells following exposure to ambient levels of nitrogen dioxide, *J. Toxicol. Environ. Health* 25:247–256, 1988.

Richters, A., Damji, K., and Richters, V. Immunotoxicity of nitrogen dioxide, *J. Leukocyte Biol.* 42:413–414, 1987.

Richters, A., Richters, V., and Sherwin, R.P. Influence of ambient level NO_2 exposure on newborn and adult mice body weights, *J. Environ. Pathol. Toxicol. Oncol.* 7:65–72, 1987.

Rombout, P.J.A., Dormans, J.A.M.A., Marra, M., and van Esch, G.J. Influence of exposure regimen on nitrogen dioxide-induced morphological changes in the rat lung, *Environ. Res.* 41:466–480, 1986.

Sagai, M., Ichinose, T., and Kubota, K. Biochemical effects of nitrogen dioxide, *Toxicol. Appl. Pharmacol.* 73:444–456, 1984.

Schacter, E.N., Witek, T.J., Jr., Beck, G.J., Hosein, H.R., Colice, G., Leaderer, B.P., and Cain, W. Airway effects of low concentrations of sulfur dioxide: dose-response characteristics, *Arch. Environ. Health* 39:34–42, 1984.

Selgrade, M.K., Mole, M.L., Miller, F.J., Hatch, G.E., Gardner, D.E., and Hu, P.C. Effect of NO_2 inhalation and vitamin C deficiency on protein and lipid accumulation in the lung, *Environ. Res.* 26:422–437, 1981.

Seltzer, J., Bigby, B.G., Stulbarg, M., Holtzman, M.J., Nadel, J.A., Ueki, I.F., Leikauf, G.D., Goetzl, E.J., and Boushey, H.A. O_3-induced change in bronchial reactivity to methacholine and airway inflammation in humans, *J. Appl. Physiol.* 60:1321–1326, 1986.

Sherwin, R., Dibble, J., and Weiner, J. Alveolar wall cells of the guinea pig: increase in response to 2 ppm nitrogen dioxide, *Arch. Environ. Health* 24:43–47, 1972.

Stephens, M.A., Menache, M.G., Crapo. J.D., Miller, F.J., and Graham, J.A. Pulmonary function in juvenile and young adult rats exposed to low levels of NO_2 with diurnal spikes, *Toxicol. Environ. Health* 23:79–90, 1987.

Tabacova, S., Nikiforov, B., and Balabaeva, L. Postnatal effects of maternal exposure to nitrogen dioxide, *Neurobehav. Toxicol. Teratol.* 7:785–789, 1985.

Tabler, S.K. EPA's program for establishing national emission standards for hazardous air pollutants, *J. Air Pollut. Control Assoc.* 34:532–543, 1984.

Takahashi, Y., Mochitate, K., and Miura, T. Subacute effects of nitrogen dioxide on membrane constituents of lung, liver, and kidney of rats, *Environ. Res.* 41:184–194, 1986.

Tepper, J.S., Weiss, B., and Wood, R.W. Alterations in behavior produced by inhaled ozone or ammonia, *Fundam. Appl. Toxicol.* 5:1110–1118, 1985.

Tsubone, H. and Suzuki, A.K. Reflex cardiopulmonary responses by stimulation to type J receptors in rats exposed to NO_2, *J. Toxicol. Environ. Health* 13:905–917, 1984.

U.S. Environmental Protection Agency (EPA). Air quality criteria for nitrogen oxide. Air Pollution Control Office, U.S. Environmental Protection Agency, Washington, DC, 1982, 14.1–15.17.

U.S. Environmental Protection Agency (EPA). Air quality criteria for lead, Vol. IV. EPA-600-8-83/028. Air Pollution Control Office, U.S. Environmental Protection Agency, Washington, DC, 1986, 12.1–12.211.

U.S. Environmental Protection Agency (EPA). Second addendum to air quality criteria for particulate matter and sulfur oxides. EPA/600/8-86/020F. Air Pollution Control Office, U.S. Environmental Protection Agency, Washington, DC, 1986, 3.1–4.37.

U.S. Environmental Protection Agency (EPA). EPA air quality criteria for carbon monoxide. EPA/600/8-90/045A. Air Pollution Control Office, U.S. Environmental Protection Agency, Washington, DC, 1990, 10.1–10.188.

Van Loveren, H., Rombout, P.J.A., Wagenaar, Sj.Sc., Walvoort, H.C., and Vos, J.G. Effects of ozone on the defense to a respiratory *Listeria monocytogenes* infection in the rat. Suppression of macrophage function and cellular immunity and aggravation of histopathology in lung and liver during infection, *Toxicol. Appl. Pharmacol.* 94:374–393, 1988.

Vaughan, W.J., Adamson, G.L., Lindgren, F.T., and Schooley, J.C. Serum lipid and lipoprotein concentrations following exposure to ozone, *J. Environ. Pathol. Toxicol. Oncol.* 5:165–173, 1984.

Vedal, S., Schenker, M.B., Munoz, A., Samet, J.M., Batterman, S., and Speizer, F.E. Daily air pollution effects on children's respiratory symptoms and peak expiratory flow, *Am. J. Public Health* 77:694–698, 1987.

Von Nieding, G. and Wagner, H.M. Effects of NO_3 on chronic bronchitis, *Environ. Health Perspect.* 29:137–142, 1979.

World Health Organization (WHO). Oxides of nitrogen. World Health Organization and U.N. Environmental Program, World Health Organization, Geneva, 1977, 32–60.

Chapter

12

Soil Toxicology

K.C. Donnelly, Cathy S. Anderson, Gary C. Barbee, and Donat J. Manek

A. INTRODUCTION

Soil is the ultimate receptor for most pollutants released into the environment. Chemicals adsorbed to particulate matter that is released into the atmosphere are eventually deposited onto soils or surface water. In surface water, particulate matter will ultimately settle onto sediments (i.e., water-saturated soil). Accidental spills, pipeline leaks, or leaking underground storage tanks may release pure chemicals to the surface or subsurface soil. In addition, land disposal has been the method of choice for centuries, mainly due to economic reasons, for most of our municipal waste and much of our industrial waste. Land disposal units include pits, ponds, lagoons, landfills, septic lines, land treatment facilities, and open dumps. While these disposal practices differ greatly, each procedure may result in the placement of hazardous constituents in direct contact with the soil. Under proper management, these procedures take advantage of the tremendous capacity of the soil to retain and degrade toxic chemicals. However, poor management practices have resulted in the severe contamination of many sites in the U.S. and abroad. The presence of hazardous chemicals in the soil represents a potential source of a continuing release to other environmental media including air, surface water, and groundwater.

It is extremely rare for soil contamination to consist of a single pure chemical. Indeed, when an accidental spill does release a pure product to the soil, adsorption and degradation may rapidly change the composition of the spilled product. In most cases, soil contamination consists of a complex mixture

0-8493-8851-1/94/$0.00+$.50
© 1994 by CRC Press, Inc.

of organic and inorganic chemicals. The constituent composition and concentration of this mixture will exert a significant influence on the ultimate environmental fate of the toxic constituents. Conjunctively, the environmental fate of hazardous constituents will exert a significant influence on their toxicological properties. The following review will consider the influence of physical and chemical properties on the environmental fate of several classes of organic and inorganic chemicals, as well as some of the interactions of complex mixtures of chemicals in the soil.

B. ORGANIC CHEMICALS IN THE SOIL ENVIRONMENT

1. The Soil Medium

Soil is unique in that it is the dynamic natural body that forms, exists, and acts at the interface of the lithosphere, atmosphere, and biosphere. It is a medium in which many chemical, biochemical, biological, and microbiological processes occur. These processes may immobilize, transform, or degrade organic chemicals in the soil.

The micro and macro properties of soil develop over time due to the combined effects of climate, biotic activity, and topography acting on the parent material. On a volumetric basis, soil is about 45% minerals, 5% organic matter, 20 to 30% air, and 20 to 30% water (Brady, 1974). On a dry weight basis, soils of the U.S. range in composition from 93 to 99% minerals and 0.1 to 7% organic matter (Brady, 1974).

The primary particles in soil, those that resist further breakdown, are sand, silt, and clay (Table 12.1). The relative proportions or size distribution of these primary particles is known as the soil texture. When soil primary particles flocculate or are cemented together into three dimensional secondary particles or aggregates this is referred to as soil structure. In fine textured soils, such as silts and clays, structural development can cause natural macropores to develop between the planes of the aggregates which may allow dissolved contaminants or spilled chemicals to bypass most of the soil particles and move rapidly and extensively (Barbee and Brown, 1986). A thorough review of macropore flow in soils has been written by Beven and Germann (1982) and White (1985).

Human exposure to organic chemicals released from the soil environment can occur by several pathways: (1) direct contact with the contaminated soil, (2) inhalation of dust or volatilizing chemicals, (3) ingestion of soil, (4) contact with soil runoff water, and (5) ingestion of food crops or drinking water contaminated from soil pollutants. This discussion will review the soil parameters and chemical properties that control the dissipation of a contaminant between the different environmental compartments, including air, water, soil, and biota.

TABLE 12.1. Characteristics of Primary Soil Particles

Particle	Size	Predominant Elemental Composition	Surface Area (m^2/g)	CEC (meq/100 g)	Chemical Reactivity
Sand	0.05–2.0	SiO_2	0.1	2	Very active
Silt	0.50–0.002	SiO_2	1.0	~9	Inactive
Clay	<0.002				
Kaolinite		Al,Si, Fe,Mg	5–20	3–15	Low activity
Mica		Al,Si,Fe, Mg,K	~100	15–40	Active
Montmorillonite[a]		Al,Si, Fe,Mg	~750	80–100	Very active

[a] Montmorillonite crystal units expose a very large internal surface area which greatly exceeds its external surface area.

Water Solubility

The water solubility (S_w) of an organic chemical is a critical property affecting its environmental fate and transport in soils. Some organic chemicals exist in the soil pore water as ionic species (cationic or anionic). For ionic species, their solubility in the soil pore water depends on the pH of the soil solution. Other organic chemicals (e.g., many organic solvents and hydrocarbons) do not exist in the soil pore water as ionic species but as neutral solutes with different degrees of polarity (Griffin and Roy, 1985). Those chemicals that are polar are electrostatically attracted to the highly polar water molecules and will remain in the soil solution (highly water soluble). Nonpolar organic chemicals have little or no electrostatic attraction to the dipolar water molecules which will lessen their contact with water (hydrophobic) and thus partition out of the soil pore water onto the soil organic matter or clay. Nonpolar organic chemicals may also partition into the soil gas phase and volatilize from the soil.

Soil Adsorbents

Adsorption of organic chemicals by soil particles is an important mechanism for their removal from the soil solution and for the inhibition of their leaching to the groundwater or volatilization from the soil surface. In soil, clay and organic matter are the predominant adsorbents while silt and sand play a negligible role.

Clays are composed of aluminum octahedral and silica tetrahedral sheets bound together by shared oxygen atoms to form 1:1 or 2:1 layers, and are thus commonly known as alumino-silicates. As clays form, silicon atoms (Si^{4+}) are replaced to varying degrees by aluminum (Al^{3+}), or Al^{3+} atoms may be replaced by Mg^{++}, Zn^{++}, or Fe^{++} atoms. This substitution leaves unsatisfied negative charges in the layers from the oxygen atoms (O^{4-}). As a result, clay has a net negative charge, and thus the capacity (i.e., cation exchange capacity [CEC]) (Table 12.1) to adsorb and immobilize cationic species or repulse and enhance the mobility of anionic species. In addition, clay can have a very high specific surface area (Table 12.1) which also affects the physical sorption of organic chemicals. Because field soils are a mixture of minerals, their surface area and net negative charge will be proportional to the amount and type of clay present.

The general types of organic materials that have been identified in fractionated soil organic matter are humic and nonhumic substances. Humic substances (well decomposed materials) make up about 85 to 90% of soil organic matter. Humic substances consist of aromatic polymers with aliphatic side chains and carboxyl, phenolic, and alcoholic hydroxyl, carbonyl, methoxy, and nitrogenated functional groups (Hayes and Swift, 1978; Schulten and Schnitzer, 1990). Senesi et al. (1983) also observed relatively high concentrations of transition metal ions (Fe = Ni > Cu > Zn > Co > Mn > Cr) in the humic fraction, suggesting that these metals are an integral part of the humic complex. Nonhumic substances (recognizable plant debris) are composed of complex, heterogeneous aliphatic structures (e.g., fats, waxes, lignins, terpenes, sterols, etc.) and make up about 10 to 15% of the soil organic matter.

Studies of the structure of humic substances indicate they are not composed of condensed, rigid molecules, but rather of flexible, open, random coils capable of molecular expansion which have many charged sites along the length of the polymer (Hayes and Swift, 1978). Oxygenated functional groups are mainly responsible for imparting the highly negative charged (CEC = 150 to 300 meq/100 g) acidic nature to the humic fraction. Due to the presence of aliphatic and aromatic groups, soil organic matter can have both a nonpolar and a polar nature.

2. Partitioning

The fate of an organic chemical in soil depends upon its distribution or partitioning between the soil and the other environmental compartments — air, water, and biota. A series of equilibrium expressions or partition coefficients (Table 12.2) have been developed to describe the tendency of an organic chemical to migrate from one environmental compartment to another.

Soil — Soil/Water Distribution

The soil sorption constant (K_d) is the basic expression describing the partitioning of an organic chemical between the soil and soil pore water (Table 12.2). K_d is the same constant as defined by the Freundlich adsorption equation. K_d relates the amount of chemical sorbed to soil or sediment to the amount in the soil pore water at equilibrium. K_d is soil specific and must be determined for each soil material. Typically, either batch equilibrium (mechanical mixing

TABLE 12.2. Partition Coefficients Used to Estimate the Distribution of Organic Chemicals Between Soil and Other Environmental Compartments

Partition Coefficient	Equilibrium Definition
K_d	(μg chem in soil/g soil)/(μg chem in water/g water)
K_{oc}	(μg chem/g soil oc)/(μg chem/g water)
K_{ow}	(μg chem/ml n-octanol)/(μg chem/ml water)
K_w	(μg chem/cm^3 water)/(μg chem/cm^3 air)

of the soil, water, and chemical) or column displacement (chemical transport through saturated soil columns) methods are used to determine K_d (Siegrist and McCarty, 1987; Lee et al., 1988).

Soil Organic Matter — Soil/Water Distribution

The primary adsorbent of organic chemicals in soil is the organic fraction. Therefore, to describe the partitioning characteristics of a chemical between the soil pore water and soil organic fraction, the organic carbon (oc) normalized soil sorption coefficient (K_{oc}) was developed (Table 12.2). K_{oc} is essentially independent of the soil mineral properties and expresses organic chemical sorption on the basis of the organic carbon content of the soil.

Empirical relationships were developed which show that K_{oc} has a highly correlated linear relationship with the water solubility (S_w) of the chemical. Hassett et al. (1983), using the water solubility data of 107 nonpolar chemicals, developed the following linear equation:

$$\log K_{oc} = 3.95 - 0.62 \log S_w \qquad (12.1)$$

where S_w is in units of mg/l.

The partitioning of an organic chemical between soil organic matter and soil pore water can be estimated by its partitioning between water and an immiscible organic solvent. Octanol, having a relatively low water solubility (~300 to 540 mg/l), best imitates the soil organic matter and thus the octanol/water (ow) partition coefficient (K_{ow}) has been developed (Table 12.2). Octanol is also considered to best imitate the fatty tissue in fish and mammals and the fatty structure in plants (Kenaga and Goring, 1980) and thus K_{ow} can also be used to estimate the bioconcentration factor (BCF) or bioaccumulation potential of organic chemicals in fish, mammals, and biota.

Because K_{oc} and K_{ow} both describe the partitioning of an organic chemical between the soil pore water and organic matter in soil, it is not surprising that they are highly correlated. Griffin and Roy (1985) and Dragun (1988) summarize the different empirical linear equations developed relating K_{oc} and K_{ow} (Table 12.3).

TABLE 12.3. Empirical Equations Developed Relating K_{ow} and K_{oc}

Equation	Chemicals Studied
$\log K_{oc} = \log K_{ow} - 0.21$ [a]	Aromatic and chlorinated hydrocarbons
$\log K_{oc} = -0.317 + \log K_{ow}$ [b]	Nonpolar and polyaromatics
$\log K_{oc} = 0.49 + 0.72 \log K_{ow}$ [c]	Halogenated alkenes and benzenes
$\log K_{oc} = 0.088 + 0.909 \log K_{ow}$ [d]	Nonpolar organics
$\log K_{oc} = 0.884 \log K_{ow} - 0.199$ [e]	Chlorophenols

[a] Karickhoff et al., 1979.
[b] Means et al., 1980.
[c] Schwarzenbach and Westall, 1981.
[d] Hassett et al., 1983.
[e] Schellenberg et al., 1984.

K_{ow} was originally measured by "shake flask" methods and more recently by reverse phase, high performance liquid chromatography (RP-HPLC). However, Sabijic (1987) suggests that K_{ow} can also be calculated using the molecular structure of the chemical (i.e., the number and types of atoms and bonds and their adjacent environment) or molecular connectivity indices. Sabijic (1987) found that this method is capable of accounting for the soil sorption properties of nearly 95% of all organic chemicals whose soil sorption coefficients have been measured.

Soil Water — Plant Distribution

To estimate the partitioning of an organic chemical between the soil pore water and a plant, Ryan et al. (1988) developed the stem concentration factor (SCF_{soil}) which relates the organic chemical concentration within the plant to the total soil concentration (Table 12.3). SCF_{soil} is defined as

$$SCF_{soil} = \frac{S_b}{S_b K_{oc} f_{oc} + \theta} * SCF \qquad (12.2a)$$

and

$$SCF = \left[10^{(0.95 \log K_{ow} - 2.05)} + 0.82\right] * \left[(0.784)10^{-0.434[(\log K_{ow} - 1.78)^2 / 2.44]}\right] \quad (12.2b)$$

where SCF_{soil} is the stem concentration factor (unitless), S_b is the soil bulk density (g/cm³), f_{oc} is the fraction of soil organic carbon ($f_{oc} = K_d/K_{oc}$), θ is the volumetric soil water content (ml/cm³), and SCF is the stem concentration factor (g/ml), which is based upon the concentration of organic chemical in the stem (µg/g fresh weight) divided by the concentration of organic chemical in the external solution (µg/ml). Equation 12.2a shows that plant uptake of organic chemicals decreases as the organic carbon content of the soil increases and that increases in the soil water content decrease SCF and therefore plant uptake.

Soil Water — Air Distribution

The distribution of a chemical between the soil pore water and soil air can be described by Henry's Law constant (Hc):

$$Hc = \frac{16.04(V_p)(M)}{(T)(S_w)} \qquad (12.3)$$

where Hc is Henry's Law constant (atm-m^3/mol), V_p is the vapor pressure of the pure chemical (atm), M is the molecular weight of the chemical (g/mol), T is the absolute temperature (°K), and S_w is the chemical water solubility (g/m^3). If V_p and S_w are not determined at the same temperature, their values must be temperature corrected.

The reciprocal of Henry's Law constant (1/Hc) is used to describe the partitioning of an organic chemical between the soil pore water and soil atmosphere under equilibrium conditions and is known as the air/water partition coefficient (K_w) (Table 12.2). K_w is appropriate only for the dilute solutions that are typically observed in the environment. Chemicals that have low values for K_w have a greater tendency to partition towards the soil air and volatilize from the soil.

3. Volatilization from the Soil Surface

Organic vapors are capable of moving considerable distances both vertically and horizontally in soil. Exposure to chemicals volatilized from the soil surface can occur from (1) vapor loss from land disposal facilities such a surface impoundments, landfills, or land treatment units (Hwang, 1982; Dupont, 1986; Wood and Porter, 1987), (2) ingestion of plants contaminated by volatilized organics (Ryan et al., 1988), (3) vapor loss from chemical spills on the soil surface (Aurelius et al., 1987), (4) vapor loss from underground storage tanks and pipes (Marrin and Kerfoot, 1988), and (5) vapor contamination of groundwater from volatile organic chemicals migrating out from landfills or subsurface reservoirs of chemicals (Smith et al., 1989).

Volatilization of chemicals from soil involves the desorption of the chemical from soil, movement to the soil surface, and vaporization into the atmosphere. The factors that influence the rate of volatilization have been discussed in detail by Spencer et al. (1982) and include the vapor pressure of the chemical in soil, the chemicals rate of movement to the soil surface, the interaction of the chemical with the soil due to adsorption, and the concentration of the chemical in the soil with depth. Soil properties such as porosity, bulk density, water content, organic matter, and clay content, as well as atmospheric conditions such as air temperature, air flow, and turbulence will also influence volatilization. Spencer et al. (1982) present an equation for predicting the steady state vapor flux of organic chemicals from the soil surface as follows:

$$J = D_o P_a^{10/3} (C_2 - C_s) / P_T^2 L \qquad (12.4)$$

where J is the vapor flux from the soil surface (μg cm^{-2}/d^{-1}), D_o is the vapor diffusion coefficient in air (cm^2/d), P_a is the soil air-filled porosity (cm^3/cm^3), C_2 is the concentration or vapor density of the volatilizing chemical in air at the surface of the soil (μg/l), C_s is the concentration of the volatilizing chemical in air at the bottom of the soil layer, P_T is the total soil porosity, and L is the soil depth (cm). The validity of Equation 12.4 has been experimentally verified using hexachlorobenzene (Spencer et al., 1982) and petroleum industry hazardous wastes (Dupont, 1986). This equation is quite effective for estimating volatile organic emissions from the soil surface under field conditions.

To determine the concentration of volatile organic chemicals in soil, soil gas surveying has been successfully used (Marrin and Kerfoot, 1988; Tillman et al., 1989). Because an organic chemical is partitioned between the soil pore water and soil organic matter, Brown et al. (1990) have successfully used K_{oc} as the criterion for deciding when soil pore water or soil core samples should be taken to detect organic chemicals in soil.

4. Soil Runoff and Erosion

Organic chemicals may be released from soils into surface waters by dissolution into runoff water or by adsorption onto soil particulate matter that is lost by erosion. Chemical losses by these pathways can occur from both agricultural and industrial activities. Each year an estimated 360 ton of pesticides are carried away from U.S. agricultural lands by wind and water (Clark et al., 1985). Weber et al. (1980) reported that pesticide losses from agricultural fields were about 0.5 to 2.2% of that applied. However, from 5 to 16% of the applied pesticides could be lost if the pesticide application is followed by intense rains (Weber, 1988). Industrial activities may also contribute to chemical contamination by erosion or runoff. Runoff water from soils amended with hazardous industrial wastes have been shown to contain significant amounts of mutagenic constituents (Brown and Donnelly, 1984; Davol et al., 1989).

5. Fate Modeling

This section provides a summary of the common input parameters needed to model the fate of an organic chemical in the soil environment. Before entering information into a model, ensure that adequate and relevant data has been collected. First, establish that the quantity and quality of the data is adequate. Second, determine if the spatial distribution of the data are adequate to describe the extent of contamination. Finally, determine if the data should be steady state or nonsteady state.

Information required to assess the health risk associated with soil contamination include (1) type and duration of likely exposures (chronic or intermittent), (2) potential exposure pathways (e.g., dermal contact, inhalation and ingestion), (3) exposure points (e.g., surface water affected by soil runoff and waste sites), and (4) concentration of the chemical at the exposure point. The fate processes that can be modeled include mass balance, dilution, dispersion, and equilibrium partitioning. Other fate processes such as chemical reaction, biodegradation, and photolysis are difficult to predict quantitatively without a significant amount of site specific data and model calibration/validation. The partition coefficients discussed individually can be combined to create models on the ecosystem level which estimate the tendency of an organic chemical to migrate from the soil into other compartments of the environment (Jury et al., 1983).

The soil properties that are required as model inputs include: bulk density/porosity, chemical properties (CEC, pH), topography, moisture relationships (hydraulic conductivity, water holding capacity), organic carbon content, redox potential, structure, temperature relationships, and texture (% sand, silt, clay). Brown et al. (1983) provide an in-depth discussion of the soil factors which influence the fate of organic chemicals in soil. Other references that present information on soil properties and how they are measured include Fluker (1958), Page et al. (1982), Rawls (1983), Klute (1986), and Brady (1974).

Information about the characteristics of a particular soil can be found in the county specific *Soil Survey,* published by the Soil Conservation Service (SCS, U.S. Dept. of Agriculture).

Several sources of information are available from the U.S. EPA as guidance for taking soil samples for risk assessments (U.S. EPA, 1987; 1988a; 1989a,b), and for estimating the migration of contaminants in the soil environment (U.S. EPA, 1987; 1988b; 1989c; and National Technical Guidance Studies—Volume 4). The EPA Center for Exposure Assessment Modeling (CEAM, Environmental Research Laboratory, Athens, GA) can also provide assistance.

Two data bases are available which provide information on the plant uptake of organic chemicals. The first, PHYTOTOX (Royce et al., 1984), deals with the direct effect organic chemicals have on the growth and development of terrestrial plants. As of 1985, data on 3500 chemicals and 700 species had been included. The second data base, UTAB, contains information about the **U**ptake, **T**ransport, **A**ccumulation, and **B**iotransformation of organic chemicals by plants. This data base includes 3900 papers with information on 700 chemicals and 250 species. UTAB is available through the University of Oklahoma (John Fletcher, Dept. of Botany, Univ. of Oklahoma, Norman, OK, 73019).

6. Plant Uptake

Major sources of soil pollutants which may contaminate food crops and forages include pesticides and industrial or municipal waste disposed on agricultural land. Plant uptake of organic chemicals is a complex phenomenon which is influenced by a variety of different mechanisms. The first group of mechanisms influencing uptake are time related and include the chemical half-life and the growth period of the plant. Chemicals with a half-life $T_{1/2}$ of <10 d have low potential for plant uptake. As the chemical $T_{1/2}$ becomes longer, and the growth period of the plant increases, the potential for plant uptake of the chemical also increases. Several studies have provided data on this subject (U.S. EPA, 1979; Smith and Dragun, 1984; Ryan et al., 1988).

Environmental factors, such as ambient temperature, soil water content, pH, and organic carbon content, will also affect plant uptake. For example, the phytotoxicity of some pesticides increases with increasing ambient temperature and soil water content but decreases with increased soil organic carbon content (Lichtenstein et al., 1967; Walker, 1971; and Sklarew and Girvin, 1987). Fredrickson and Shea (1986) observed that plant uptake of the pesticide chlorsulfuron is twice as much when the soil pH is 5.9 than when the pH is 7.5. For this acidic chemical, plant uptake is controlled by its polarity and dissociation constant.

One of the least understood variables effecting plant uptake of organic chemicals is plant species effect. Although differences in uptake have been found between plant species and within the same species, no general conclusions can be drawn from the data (Lichtenstein and Shulz, 1965; Harris and Sans, 1967; and Chaney, 1985).

The basic chemical properties which influence the behavior of organic chemicals and their "availability" to the plant include V_p, K_{ow}, S_w, and Hc. The effect of these coefficients on plant uptake depends on the mechanism of uptake. There are four possible mechanisms of plant uptake (Topp et al., 1986): (1) root uptake and transport to the higher plant parts, (2) vegetative uptake of chemical vapors from the air, (3) external contamination by solids (i.e., soil or dust) followed by penetration into the plant and, (4) transport through oil cells in oil-containing plants. Chemical penetration of the plant will usually proceed through at least two of these pathways, but root uptake and vegetative uptake of organic vapors are often the main routes of plant contamination.

Some relationships between the mechanism of plant uptake and chemical properties have been developed (Ryan et al. 1988):

1. Root uptake of organic pollutants is determined by a series of partition equilibrium reactions:
   soil solids ↔ soil water ↔ plant roots ↔ transpiration stream ↔ plant stem
2. Chemicals with high log K_{ow} values, between 4 and 7 (e.g., TCDD, PCBs, some phthalates, and PAHs) are easily sorbed by soil and/or roots and are not readily taken up by plants.
3. Chemicals with low log K_{ow} values, between 1 and 2 (e.g., halogenated aliphatic hydrocarbons, monocyclic aromatics, and many of the pesticides) can be taken up by the roots and easily transported within the plant to its above-ground parts (McFarlane et al., 1987). The rate of transport will be highly dependent on the transpiration rate of the plant and the size and polarity of the transported molecule (Connor, 1984).
4. Chemicals with Hc $\geq 10^{-4}$ (most of the common pollutants) can be easily absorbed by the plant from the vapor phase. The rate and intensity of this process increases with increasing chemical concentration and vapor pressure, soil moisture content, relative humidity, and air temperature and rate of movement (Harris and Lichtenstein, 1961; Guenzi and Beard, 1970; Farmer et al., 1972; Quistad and Menn, 1983).

7. Summary

A survey of 546 National Priority List (NPL) sites across the U.S. revealed that the most common classes of chemicals identified (Table 12.4) include volatile chlorinated alkanes and alkenes, aromatic solvents, and metals (Hazardous Waste News, 1984).

The volatile chlorinated alkanes and alkenes are very mobile in soil, both in the soil air phase due to their high vapor pressure (V_p) and in the soil water phase due to their moderate water solubility (S_w) and low K_{ow} and K_{oc} coefficients, which indicate their low sorption in soils. The aromatic solvents (benzene and xylene) commonly found at NPL sites (Table 12.4) have lower V_p and S_w and higher K_{ow} and K_{oc} coefficients than the volatile chlorinated aliphatics (Table 12.4), and thus have intermediate mobility in soils. A unique property of the chlorinated organics listed in Table 12.4 is that they have a density greater than water, and therefore often move as separate, dense, nonaqueous phase liquids (DNAPL) in both soils (Schwille, 1988) and groundwater (Feenstra and Cherry, 1988; Hunt et al., 1988).

Aliphatic and polynuclear aromatic hydrocarbons (PNAs) are commonly found in petroleum based wastes. Hydrocarbons have low water solubility (S_w) and vapor pressures (V_p) but high K_{ow} and K_{oc} coefficients, therefore they are immobile in soils and have low potential for volatilization. However, some of the lower molecular weight hydrocarbons may volatilize from soil despite their

Table 12.4. Physical and Chemical Properties of the Organic Chemicals Most Frequently Detected at 546 National Priority List (NPL) Sites

Class	Constituent	NPL Sites Detected	Density (g/cc)	S_w^a (mg/l)	Hc (atm-m³/mol)	K_{oc} (ml/g)	Log K_{ow}	VP (mmHg)
Chlorinated organics	Trichloroethylene	179	1.47	1,100	0.0091	126	2.38	58
	PCBs	121	0.031		0.0011	5.3E + 5	6.04	7.7E − 5
	Chloroform	111	1.49	8,200	0.0028	31	1.97	151
	Tetrachloroethylene	90	1.63	200	0.026	364	2.60	17.8
	1,1,1-Trichloroethane	79	1.35	720	0.016	152	2.50	123
	Methylene chloride	63	1.33	20,000	0.0032	9	1.30	362
	trans-1,2-Dichloroethylene	59	1.26	600	0.0066	59	0.48	324
	1,2-Dichloroethane	44	1.26	8,690	0.00098	14	1.48	64
	Vinyl chloride	44	0.91	1.1	0.082	57	1.38	2660
	Chlorobenzene	42	1.11	488	0.0038	330	2.84	11.7
	1,1-Dichloroethane	42	1.17	5,500	0.0043	30	1.79	182
	Carbon tetrachloride	40	1.59	785	0.023	110	2.64	90
Aromatic organics	Toluene	153	0.87	515	0.0063	300	2.73	28
	Benzene	143	0.88	1,780	0.0056	83	2.12	95
	Phenol	84	1.06	93,000	4.5E − 7	14	1.46	0.34
	Ethylbenzene	73	0.87	152	0.0065	1100	3.15	7
	Xylene	71						
	- *ortho*		0.87	175	0.0057	363	3.04	10
	- *meta*			130	0.0025	588	3.20	10
	- *para*			198	0.0025	552	3.16	10

^a Not applicable — see below.

a S_w = water solubility, Hc = Henry's Law constant, K_{oc} = organic carbon normalized soil sorption coefficient, K_{ow} = octanol/water partition coefficient, vp = chemical vapor pressure.

TABLE 12.5. Relationship Between K_d, K_{oc} Values,[a] and the Soil Mobility of an Organic Chemical

K_d[b]	K_{oc}	Mobility Class
>10	>2000	Immobile
2–10	500–2000	Low mobility
0.5–2	150–500	Intermediate mobility
0.1–0.5	50–150	Mobile
<0.1	<50	Very mobile

[a]   From Dragun, 1988.
[b]   Based on values derived from soil with 2.5% organic matter using a variety of pesticides.

low V_p due to their correspondingly low S_w which results in low values for K_w (Table 12.2). PNAs are strongly adsorbed by soil ($K_{oc} > 2000$; Table 12.5), but in the presence of organic solvents they may become more soluble which may greatly enhance their mobility (Villaume, 1985).

Chlorinated aromatic hydrocarbons such as PCBs and 2,3,7,8-tetrachloro-dibenzo-*p*-dioxin (TCDD) are very stable chemicals with very low S_w and V_p (Table 12.5) and high soil sorption coefficients (K_{ow}, K_{oc}). Usually they will not be transported in the environment except by physical means such as with sediment in surface runoff during storms. Because of their persistence in soil, and their high lipophilicity, chlorinated aromatic hydrocarbons and many other chlorinated organics readily bioaccumulate in the fatty tissues of fish and mammals.

Another class of chlorinated aromatic chemicals of environmental importance are the chlorophenols, including pentachlorophenol (PCP), which for many years has been used for wood preservation. As chlorine substitution increases on the phenol ring, S_w decreases significantly (e.g., phenol S_w = 93,000 mg/l and PCP S_w = 14 mg/l) and conversely soil sorption increases (phenol K_{oc} = 14.2 and PCP K_{oc} = 53,000). However, chlorine substitution can also enhance the mobility of the higher chlorophenols due to the negative charge imparted to the chemical (Barbee et al., 1990). Chlorophenols are considered semi-volatile, but vaporization from the soil is probably not an important mechanism for their transport. The sorption of chlorophenols on soil is pH dependent due to their ionizable hydroxyl group. For example, Lagas (1988) determined that the sorption of the ionized PCP molecule at pH > 6 was 15 to 30 times lower than the sorption of the unionized species.

Pesticides fall within several diverse classes of organic chemicals including chlorinated hydrocarbons, cations, substituted phenylureas, nitroanilines, phenoxy and acidic compounds, phenylcarbamates, ethylenebisdithiocarbamates, and organophosphates (neutral esters of phosphoric and carbamic acids). Several investigators (Jury et al., 1987; Gustafson, 1989) have developed methods for estimating the fate (mobility, degradation, etc.) of pesticides in soil and groundwater. The U.S. EPA (1984) has developed a computer algorithm, Pesticide Root Zone Model (PRZM), to determine the effect of different environmental factors on the fate of pesticides in the soil root zone.

C. INORGANIC CHEMICALS IN THE SOIL ENVIRONMENT

The major categories of inorganic pollutants include heavy metals, nitrogen, phosphorous, some acids and bases, salts, and halides. Sources of common inorganic soil contaminants are listed in Table 12.6.

1. Metals

The categories of heavy metals that are environmentally important soil pollutants include: (1) carcinogenic metals (e.g., arsenic, chromium (chromate), beryllium, and nickel are known carcinogens (Norseth, 1977), (2) metals which are mobile in soil at neutral pH, and (3) metals which move through the food chain. The levels of heavy metals naturally occurring in soils and those levels which are recommended in soils and drinking water by regulatory agencies are provided in Table 12.7.

Coarse textured soils or fine textured soils with macropores (Shaffer et al., 1979) may enhance the mobility, thus leaching metals into permeable subsoils or aquifers. Other factors that affect the migration of metals in soils include organic matter content, CEC, soil moisture, and pH (Jones and Burgess, 1984). Of these factors, pH and CEC will usually predominate. Under soil conditions of pH ≥ 7, molybdenum and selenium can leach from soil (Brown et al., 1983).

Plants have a wide range of tolerance to heavy metals, and the mechanism for this tolerance varies between species. Some plants exclude metal uptake at the root soil interface. Others can chelate the metal once it is within the root. Still other plants (accumulator plants) may take up high levels of metals with no apparent deleterious effect because they have mechanisms to prevent metals from reaching sensitive plant parts. In some cases accumulator plants could be used to lower metal concentrations in soil.

For arsenic, copper, nickel, and zinc (Jones and Burgess, 1984) the food chain is protected because the toxic concentration of these metals in plants is less than that for animals. On the other hand, cadmium, selenium, and molybdenum are not toxic to plants at high concentrations and can be accumulated in plants at levels that may be toxic to animals (Brown et al., 1983). Other metals such as lead (Chamberlain, 1983), cobalt, and mercury (Brown, et al., 1983) can also enter the food chain via plant uptake, but to a lesser extent.

TABLE 12.6. Sources of Common Inorganic Soil Contaminants

Contaminant	Source
Aluminum	Paper coating pretreatment sludge and deinking sludge
Antimony	Paint formulation, textile mills, and organic chemical producers
Arsenic	Production of pesticides and veterinary pharmaceuticals, and wood preservatives
Barium	Manufacturing plants
Beryllium	Smelting industries and atomic energy projects
Boron	Decomposition of organics
Bromide	Agricultural fumigants, industrial wastes, photographic supplies, and pharmaceuticals
Cadmium	Cd-nickel battery production, pigments for plastics and enamels, fumicides, and electroplating and metal coatings
Cesium	Thermoionic power conversion and ion propulsion research, nuclear fallout
Chloride	Chlorinated hydrocarbon and chlorine gas production
Chromium	Corrosion inhibitor, dyeing and tanning industries, plating operations, alloys, antiseptics, defoliants, and photographic emulsions
Cobalt	Steel and alloy production, paint and varnish drying agent and pigment and glass manufacturing
Copper	Textile mills, cosmetics manufacturing, and hardboard production sludge
Fluoride	Phosphatic fertilizer, hydrogen fluoride and fluorinated hydrocarbon productions, and refinery waste
Gallium	Smelter or coal processing plants
Gold	Medical isotopic waste
Iodide	Pharmaceutical and analytical chemistry wastes
Lead	Pb battery manufacture, fuel additives, manufacturing of ammunition, caulking compounds, solders, pigments, paints, herbicides, and insecticides
Lithium	Carbonates of calcareous parent material
Manganese	Iron and steel industries, disinfectants, paint, and fertilizers
Mercury	Electrical apparatus manufacture, electrolytic production of Cl and caustic soda, pharmaceuticals, paints, plastics, paper products, Hg batteries, pesticides, and burning of coal and oil
Molybdenum	Steel and alloy production, pigment, filament, lamp, and electronic tube production
Nickel	Production of stainless steel, alloys, storage batteries, spark plugs, magnets,and machinery
Nitrogen	Sewage sludge, wastewaters, and animal wastes
Palladium	Platinum extraction, dental alloys, electrical contacts, and jewelry
Phosphorous	P quarries, fertilizer, and pesticides
Radium	Uranium processing
Rubidium	Superphosphate fertilizer and coal
Selenium	Coal power plant fly ash
Silver	Photographic, electroplating, and mirror industries
Strontium	Atomic fallout
	Kraft mills, sugar refining, petroleum refining, and copper and iron extraction
Thallium	Fertilizer and pesticide manufacture, sulfur and iron refining, and cadmium and zinc processing

Table 12.6. (Continued)

Contaminant	Source
Tin	Tin can production
Tungsten	Nuclear waste
Uranium	Radioactive waste
Vanadium	Steel and nonferrous alloys
Zinc	Brass and bronze alloys production, galvanized metal production, pesticides, and ink

TABLE 12.7. Comparison of Levels of Heavy Metals Naturally Occurring in Soils and Levels Recommended by Regulatory Agencies for Soil and Drinking Water

Contaminant	Common Range[a] (ppm)	Average Level[a] (ppm)	Regulatory Level[b] (ppm)	Drinking Water[a] (ppm)
Ag	0.01–5.0	0.05	5.0	0.05
As	1–50	5.00	5.0	0.05
Ba	100–3000	430.00	100.0	1.00
Cd	0.01–0.7	0.06	1.0	0.01
Cr	1–1000	100.00	5.0	0.05
Hg	0.01–0.3	0.03	0.2	NA
Pb	2–200	10.00	5.0	0.05
Se	0.1–2	0.30	1.0	0.01

[a]   Brown et al., 1983.
[b]   Freeman, 1989.

Soil lead contamination can occur from paint, automobile emissions, or improper disposal of car batteries (Tetta, 1989). Hutton et al. (1988) found soil cadmium contamination resulting from fly ash depositions downwind from a refuse incinerator. Lower et al. (1985) observed elevated concentrations of lead and zinc in the soil downwind of a smelter. The elevated metal concentrations were correlated with increased levels of genetic damage in *Tradescantia* sp. which had been planted in the contaminated soil. Consequently, metals may also enter the food chain through the deposition of particulate matter.

2. Plant Nutrients

Nutrients essential to plant growth may be a hazard to the environment if they occur in excess quantities or in nonsuitable forms. These elements may be deleterious to both plants and mammals. The main plant nutrients that can be soil pollutants are nitrogen and phosphorous.

Inorganic nitrogen (N) occurs in soils in various states such as ammonium (NH_4^+), ammonia (NH_3), molecular nitrogen (N_2), nitrate (NO_3^-), and nitrite (NO_2^-). Ammonia, a gas, can be produced if ammonium is exposed to a high pH. Ammonium may also be adsorbed to soil particles at cation exchange sites due to its positive charge. Nitrate and nitrite are anions which are repelled by soil and thus are readily leachable. Although nitrite is toxic to plants in low concentrations, nitrates can be used by plants and microorganisms.

Wastes applied to soils that are high in nitrogen should be analyzed to determine their form(s) of nitrogen. Nitrogen transformation or plant uptake, leaching, and volatilization can affect the various forms of nitrogen in soil. Nitrate is the form of nitrogen of major concern because of its high mobility in the soil and water. Also, microbes can convert nitrate to nitrite which can cause methemoglobinemia. The maximum allowable level of nitrate in drinking water is 10 ppm.

Phosphorous does not represent a direct human health hazard, but it may cause a deterioration of surface water if released from soils. Phosphate (PO_4^-) is a major cause of eutrophication in lakes and ponds (Thomas et al., 1973). Phosphorous (P) concentrations typically found in soil range from 0.03 to 3.0 mg/l, and the lower value is usually associated with soils near groundwater. Phosphorous applied to the soil as fertilizer is released primarily via erosion because surface soils tend to strongly adsorb phosphorous. With time, mineralization of soil organic matter results in the release of P which is available for plants, microbes, or leaching. Mineralization rates, soil pH, CEC, clay content, and mineralogical composition will affect the efficiency of the soil for retaining P.

3. Acids and Bases

Acids and bases disposed of in soil usually should be neutralized (pH = 7) before it is mixed into the soil. The capacity of a soil to buffer an acidic or basic waste can influence the loading rate of the waste. Should the soil buffering capacity be exceeded, the soil pH can be adjusted by adding lime or a weak acid. If both acidic and basic wastes are to be coapplied, the basic waste should be incorporated into the soil first, then the acidic waste can be applied (Dragun, 1988). If this sequence is not followed, the soil may partially dissolve and metals in the soil minerals may leach from the soil. Before any waste with an extreme pH is applied to soil, bench scale tests should be conducted to determine the effect of the waste on the physical and chemical properties of the soil.

4. Salts

Salts include many substances that produce ions (other than hydrogen or hydroxyl ions) when dissolved. Salts commonly found in soils include calcium, magnesium, sodium, potassium, chloride, sulfate, bicarbonate, and sometimes nitrate. Salt concentrations in soil may increase due to fertilizer or waste applications, soil moisture evaporation, or irrigation.

The salinity of a soil is determined by measuring the electrical conductivity of the soil water (EC), total dissolved solids (TDS), osmotic pressure, percent salt by weight, or the normality. Elevated levels of salt affect plant growth and the physical structure of soil. Elevated levels of sodium in soil can cause soil clay to disperse, thus causing the soil to become impermeable to water movement both into and through the soil. In fine textured soil, inadequate drainage may prevent the leaching of the salts and thus increase the salinity of the soil. Salts can be sorbed by soil colloids or precipitate as insoluble compounds, or they may be soluble and leach from the soil or leave the soil in surface runoff. Excess salts may be removed from soil by leaching it with solutions of $CaSO_4$.

5.   Halides

Highly reactive halogens have stable anions that are referred to as halides. These include fluoride (F^-), chloride (Cl^-), bromide (Br^-), and iodide (I^-). While halides naturally occur in soils, excessive levels may threaten the well being of animals, cover crops, and microbes. Cl^- is essential to both plants and animals while I^-, and perhaps F^-, is essential to only animals.

F^- occurs naturally in soils at concentrations ranging from 30 to 990 mg/kg. F^- is mobile in soil, but applications of lime will decrease its plant uptake and leachability. F^- levels are monitored due to its health hazard in drinking water. The allowable level of F^- in drinking water is dependent on the annual average maximum daily air temperature. With air temperatures $\leq 12°C$, the maximum F^- level can be 2.4 mg/l, and with air temperatures from 26.3 to 32.5°C the maximum F^- level is 1.4 mg/l.

Cl^- is common in almost every waste stream as either a by-product or as a contaminant from the water source. Soil concentrations of Cl^- vary considerably but average about 100 mg/kg. Most forms of Cl^- are soluble, hence, Cl^- is readily leachable. Soil Cl^- should be monitored by measuring the plant uptake and the concentration in soil leachate. Cl^- should not exceed the 250 mg/l drinking water standard in the soil leachate.

Br^- concentrations in the soil typically range from 2 to 100 mg/kg. Br^- naturally occurs as the bromide ion but can be found in smaller concentrations as bromate (BrO_3^-) and bromic acid. Br^- salts of calcium, magnesium, sodium, and potassium are readily leachable from soils. Therefore, the Br^- that is found in soils is usually that chelated by organic matter. Before a waste containing Br^- is applied to a soil, the natural soil Br^- concentration, plant uptake, and Br^- leachability should be considered to maintain its concentration below phytotoxic levels.

The concentration of I⁻ in soils ranges from 0.1 to 10 mg/kg. Since it is only slightly water soluble, I⁻ is mainly retained in soil because it is chelated by soil organic matter or forms insoluble precipitates with phosphate and sulfate minerals. Animal uptake of I⁻ can be controlled via plant uptake even though I⁻ is not essential to plants. Phytotoxic levels of I⁻ in plants (5 mg/kg) may result from excess salts in the soil. Waste applications should not cause the concentration of I⁻ in the soil to reach phytotoxic levels.

D. TOXICITY OF COMPLEX MIXTURES IN SOIL

Until recently, most studies investigating the toxicological properties of contaminated soil addressed specific chemical(s) or classes of chemicals. However, studies using short term microbial bioassays have begun to investigate the various interactions of the components of a complex mixture. Microbial mutagenicity assay uses bacteria to detect chemicals that cause cell mutations as an indication of the potential of the chemical to induce cancer in mammals. The test is conducted with and without a liver microsomal preparation (S9) to mimic the metabolism which a chemical undergoes in a mammalian system. The results can be expressed as specific activity (revertants per milligram of residue) which is a qualitative measure of toxicity. This value can be used to determine if any toxic chemicals reside in the soil regardless of concentration. Alternatively, the results can be expressed as weighted activity (revertants per gram of original material) which provides a more quantitative measure of the toxicity of a specific unit of soil.

A variety of bioassays may be used to detect a variety of endpoints. Acute toxicity assays include the Microtox test (Beckman Instruments, 1982), the algal toxicity assay (Porcella, 1983), and the *Daphnia magna* assay (Horning and Weber, 1985). Chronic toxicity assays include the *Salmonella*/microsome assay (Ames et al., 1973), the Bacillus DNA repair assay (Kada et al., 1974), an assay using haploid or diploid forms of *Aspergillus nidulans* (Kafer et al., 1982; Scott et al., 1982), as well as plant assays using *Tradescantia* (Nauman et al., 1976; Ma et al., 1981) and soybeans (Vig, 1975). These and many additional biological tests can be used to measure the acute or chronic toxic potential of aqueous or solvent extracts of soil. This information, when coupled with a comprehensive chemical analysis, provides valuable information which can be used when conducting a risk assessment.

The initial studies evaluating the genotoxic risk of complex environmental mixtures focused on cigarette smoke condensate (Kier et al., 1974) and air particulate (Tokiwa et al., 1977). More recent research has extended these investigations to include soils (Donnelly et al., 1988), municipal incinerator ash (Silkowski and Plewa, 1990), and hazardous waste (DeMarini et al., 1987; Donnelly et al., 1987a,b), as well as other environmental media (Heartlein et al., 1981; Claxton et al., 1990). Each of these studies addresses the toxic potential of particulate matter which have become contaminated with a complex mixture of organic and inorganic chemicals. However, it is also important to evaluate the toxic potential of soil which has not been affected by industrial activities.

The mutagenic potential of agricultural soils was examined by Brown et al. (1985). Using the solvent extract of three agricultural soils, the bacterial mutagenicity (as measured in strain TA98) ranged from 35 revertant/mg to 434 revertant/mg; while the weighted activity ranged from 2 to 99 revertant/g of soil (Brown et al., 1985). The analysis of chemical constituents of the three soils did not disclose the source of the mutagenic activity, although it was determined that the soil with the maximum level of activity had received applications of the herbicide 2,4-D for the control of mesquite. Comparable levels of activity have been observed in agricultural soils by Goggelman and Spitzauer (1982) and Withrow (1982), while Heartlein et al. (1981) detected mutagens in the runoff water from an agricultural area.

In a comparison of uncontaminated soils from 18 superfund sites, Brown and Donnelly (1989) observed specific activities ranging from below detection to a maximum level of 93 revertant/mg. Background soils for 5 of the 18 sites induced a positive mutagenic response in the bioassay. However, each of these soils were collected on-site, and further investigation has revealed that some of the soils may have been affected by waste management activities. Natural sources cannot be ruled out as being responsible for the elevated mutagenic activity of background soils. However, the presence of elevated levels of industrial or agricultural chemicals does appear to produce a significant increase in the potential toxicity of soil.

The presence of agricultural or industrial chemicals in the soil does not represent a threat to human health or the environment unless these chemicals are volatilized into the atmosphere, leached into groundwater, runoff into surface water, or translocated into plants. While ingestion or adsorption of chemicals may be an exposure route at some locations, ingestion of contaminated groundwater is usually the primary route of exposure at most Superfund sites (Greene, 1989). The properties of the waste and receiving soil will influence each of these release mechanisms. In addition, changes or degradation of waste constituents brought about by biological, chemical, or physical actions of the soil will also influence the potential risk associated with the disposal of toxic chemicals in the soil.

The soil application of the bottom sediment from a lagoon at a wood preserving plant has been shown to alter the mutagenic potential of the waste amended soil over time. In a Norwood soil, the mutagenic potential of the basic fraction extracted from the waste amended soil was less than 100 net revertants immediately after application (Figure 12.1). However, the mutagenicity with or without metabolic activation increased on each subsequent sampling date.

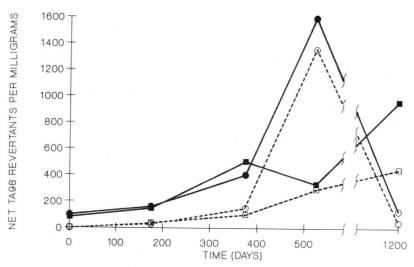

Figure 12.1. Mutagenic potential of wood preserving waste amended Bastrop (circles) or Weswood (squares) soil as measured with (solid line) and without (dashed line) metabolic activation.

The maximum activity observed in the Norwood soil was 962 net revertants per milligram for the basic fraction collected 1200 d following waste application (Donnelly et al., 1987). Similar increases were noted without metabolic activation. For the Bastrop soil, the mutagenic activity reached a maximum 540 d after application and was significantly decreased in the sample collected 1200 d after application (Figure 12.1). The maximum mutagenic response was 1561 net revertants per milligram for the basic fraction collected 540 d after application (Donnelly et al., 1987). The chemical analysis of the extracts of these soils did not disclose the reason for the change in toxicity. Indeed, the constituent composition of samples collected immediately following, and 540 d after, waste application was almost identical. Chemical analysis identified a total of 14 different compounds in the extract of the soil collected immediately after waste application while only eight compounds were identified in the sample collected 360 d following waste application (Donnelly et al., 1987c). The results from biological analysis indicate that degradation and/or transformation may have reduced the mutagenic potential of equivalent weights of waste amended soils. However, the combined results from biological and chemical analysis detected the presence of a highly mutagenic and persistent residue in the soil.

Comparable results were obtained when oily sludges from a refinery were applied to the soil. For a storm water runoff impoundment in a Norwood soil, the mutagenic potential with metabolic activation increased from 70 net revertants per milligram immediately following application, to 868 and 121 net revertants per milligram 360 and 1000 d following waste application, respectively (Brown et al., 1986). In the Bastrop soil, the mutagenic potential displayed a comparable increase 360 d after application and was decreased appreciably 1000 d after application (Brown et al., 1986). However, the mutagenic activity per gram of soil was decreased to about one third and one fourth the initial level in the Norwood and Bastrop soils, respectively.

The impact of soil application on the mutagenic potential of a combined API Separator waste and a Slop Oil Emulsion Solids waste has also been evaluated. The mutagenic activity of this waste displayed an increase to about five to eight times the initial level 360 d after waste application, with a later decrease 1000 d after waste application (Brown et al., 1986). In the Norwood soil the weighted activity was reduced to about 1/10 the initial level while in the Bastrop soil the weighted activity was reduced by about one fifth in the sample collected 1000 d following waste application. Thus, similar changes were observed for six treatments encompassing two soils and three waste streams. Short term bioassays have also been used to assist in the selection of appropriate bioremediation procedures for the 1989 Alaskan oil spill. Claxton et al. (1990) observed that both the Prudhoe Bay crude oil and the weathered oil were weakly mutagenic in strain TA100. These data indicate that biodegradation of industrial waste in soil does not always result in detoxification of waste constituents.

The mutagenic potential of municipal sewage sludge amended soils has also been measured. Angle and Baudler (1984) monitored the persistence and degradation of mutagens in sludge amended soil for 28 d. They observed that the mutagenicity of the sludge amended soil increased during the initial 7 d of the study and rapidly decreased after seven more days. However, the sludge and sludge amended soil did not induce a response that was greater than twice the concurrent solvent control, and no dose response data was provided for the one mutagenic sample that was obtained from this study. Thus, it is difficult to determine the true importance of this data. In a separate study, Johnston et al. (1986) applied sewage sludge to soil and noted that more than 7 weeks were required for the mutagenic activity of the sludge amended soil to reach a maximum and that the activity of the soil returned to background levels rapidly thereafter. In contrast, Hopke et al. (1987) found that the mutagenic potential of the methanol extract of wastewater from a municipal sewage treatment facility was decreased 48 h after addition of the residue to the soil. The mutagenic potential of two soils amended with a municipal sewage sludge was monitored for 2 years by Donnelly et al. (1989). The data from this study indicate that mutagens may persist in municipal sewage sludge amended soil, and that at higher application rates more than 2 years may be required for the mutagenic potential of the soil to return to background levels. Thus, the data obtained using either hazardous industrial waste or municipal sewage sludge suggests that although complex mixtures will be degraded in the soil, the initial products of degradation may be more toxic than the parent compound.

Biodegradation is assumed to be a helpful tool for the remediation of contaminated soil. Certainly, the complete mineralization of organic chemicals to carbon dioxide and water is beneficial to the environment. However, the aerobic oxidation of organic chemicals frequently produces metabolites which have a greater toxic potential than the parent compound (Sims, 1982; Donnelly et al., 1987c; Symons and Sims, 1988). In addition, the hydroxylated metabolites of biodegradation may also be more water soluble than the parent compound. Mutagenic chemicals were detected in both runoff and leachate water from three different soils (sand, loam, and clay soil) which had been amended with an oily sludge from a refinery or a petrochemical plant (Brown and Donnelly, 1984). The mutagenic activity of leachate water and the runoff water from two of the three soils decreased with time following waste application. The activity in the third soil did not decrease over the three years of observation. While mutagenic chemicals were detected in both the runoff water and the leachate, the activity of the runoff water was typically greater than that of the leachate (Brown and Donnelly, 1989). Davol et al. (1989) also detected mutagens in runoff water from soils amended with hazardous industrial wastes. The maximum specific activity was 783 net revertants per milligram of residue which was induced by the runoff water collected from a storm water runoff impoundment sludge amended Weswood soil. The data suggest that 3 years or more may be required for the mutagenic activity of runoff water to return to background levels and that different soils will differ significantly in their capacities to retain mutagenic chemicals during rainfall events (Davol et al., 1989).

Mutagenic chemicals have also been detected in the leachate water from soils amended with municipal sewage sludge. Eight of 27 leachate water extracts from two different soils (a clay and a loam soil) induced a positive mutagenic response (Donnelly and Brown, 1990). The maximum specific activity was 207 net revertants per milligram of residue induced by a Weswood soil which received 150 Mg/ha of sewage sludge. The data from the analysis of leachate and runoff water from contaminated soils indicate that soil properties may have a significant influence on the migration of toxic chemicals. While clay particles and organic matter can adsorb organic chemicals, macropores and cracks in fine textured soils, or the natural permeability of coarse textured soils (i.e., sands) may permit rapid migration of chemicals.

The acute toxicity of an oily sludge amended soil was measured by Symons and Sims (1988). They observed that the toxic potential of the aqueous extract of soils receiving a 2% application of sludge reached nontoxic levels within 180 d following sludge application. The degradation of individual polynuclear aromatic hydrocarbons (PNAs) in the soil was directly correlated with the decrease in acute toxicity at the lowest application rates. At higher application rates (4 and 8% sludge wt/wt), the soil extract continued to induce a toxic response 180 d after waste application, although the toxicity was reduced from levels noted immediately after sludge application (Symons and Sims, 1988).

Various environmental influences can also affect the genotoxic potential of a complex mixture. Both McCoy et al. (1979) and de Flora et al. (1981) have observed activation of mutagens by visible light. Claxton and Barnes (1981) found that various environmental conditions can influence the mutagenicity of air pollutants. The biodegradation of specific PNAs (Sims, 1982), hazardous waste (Brown et al., 1986; Donnelly et al., 1987c), and municipal sewage sludge (Donnelly et al., 1989; Johnson and Baker, 1986) in soil has also been found to result in an increased level of mutagenic activity. In all four studies, there was an initial time-dependent increase in mutagenic activity followed by an eventual decrease in mutagenicity. The mutagenicity of the biological transformation products of 7,12-dimethylbenzanthracene have been monitored in a soil system. Although the initial metabolites of degradation displayed an increased mutagenic potential, the mutagenicity of the metabolites was eventually decreased with time (Park et al., 1988). In addition, metabolic activation of pesticides by plants was demonstrated by Gentile et al. (1977). Thus, environmental influences can act upon complex environmental mixtures to modify their genotoxic potential.

To accurately assess the genotoxic potential of a complex mixture in soil, several factors must be considered. For each compound in the mixture, we must understand the toxicokinetics or the rate of distribution, adsorption, penetration into the cell, metabolism, and excretion. For the mixture of chemicals, these same factors will influence the overall toxicity as well as the interactions of each of the components. Organic chemicals in a complex mixture appear to have one of several potential influences. The potential interactions of the components of a complex mixture include modification of cell uptake, modification of metabolic activation, modification of DNA repair, or competitive antagonism (DiGiovanni and Slaga, 1981; Schoeny et al., 1987). Inhibition occurs when a compound competes for activating enzymes, thus decreasing the rate at which promutagens are converted into their ultimate mutagenic form, while enhancement occurs when metabolic competition reduces the rate of detoxication. Inorganic agents and solid-state carcinogens also influence the toxicity of a complex mixture. The synergism of cadmium with nitrosamines appears to result from its effect on repair mechanisms, while asbestos synergism is related to metabolic effects which result in increased retention times for genotoxic compounds. Finally, environmental parameters can also serve to activate and/or detoxify organic compounds. These factors must be considered when short term bioassays are utilized to evaluate the genotoxicity of complex mixtures. While it may be necessary to fractionate the mixture to isolate and detect the genotoxic compounds, it is also important to evaluate the genotoxic effects of the mixture.

REFERENCES

Ames, B.N., Durston, W.E., Yamasaki, E., and Lee, F.D. Carcinogens are mutagens: a simple test system combining liver homogenates for activation and bacteria for detection, *Proc. Natl. Acad. Sci. U.S.A.* 70(8):2281–2285, 1973.

Angle, J.S. and Baudler, D.M. Persistence and degradation of mutagens in sludge-amended soil, *J Environ. Qual.* 13:143–146, 1984.

Anon. *Hazardous Waste News,* December 24, 1984, p. 415.

Aurelius, M.W. and Brown, K.W. Fate of Spilled xylene as influenced by soil moisture content, *Water, Air Soil Pollut.* 36:23–31, 1987.

Barbee, G.C. and Brown, K.W. Movement of xylene through unsaturated soils following simulated spills, *Water, Air, Soil Pollut.* 29:321–331, 1986.

Barbee, G.C., Brown, K.W., Thomas, J.C., and Murray, H.E. Mobility of organic chemicals from three hazardous wastes through laboratory soil columns, in review.

Beckman Instruments, Inc. Beckman Microtex System Operating Manual, Section 11. Microbics Corporation, Carlsbad, CA. 14 , 1982.

Beven, K. and Germann, P. Macropores and waterflow in soils, *Water Resour. Res.* 18:1311–1325, 1982.

Brady, N.C. *The Nature and Properties of Soils.* Macmillan Publishing Co., Inc., New York, 1974, 621.

Briggs, G.G., Bromilow, R.H., and Evans, A.A. Relationship between lipophilicity and root uptake and translocation of non-ionized chemicals by barley, *Pestic. Sci.* 13:495–504, 1982.

Briggs, G.G., Bromilow, R.H., Evans A.A., and Williams, M. Relationships between lipophilicity and the distribution of non-ionized chemicals in barley shoots following uptake by the roots, *Pestic. Sci.* 14:492–500, 1983.

Brown, K.W., Evans, G.B., Jr., and Frentrup, B.D. *Hazardous Waste Land Treatment.* Butterworth Publishers, Woburn, MA, 1983, 692.

Brown, K.W. and Donnelly, K.C. Mutagenic activity of runoff and leachate water from soils amended with a refinery and a petrochemical waste, *Environ. Pollut.* 35(3):229–246, 1984.

Brown, K.W., Donnelly, K.C., Thomas, J.C., Davol, P., and Scott, B.R. Mutagenicity of three agricultural soils, *Sci. Total Environ.* 41:173–186, 1985.

Brown. K.W., Donnelly, K.C., Thomas, J.C., Davol, P., and Scott, B.R. Degradation of mutagenic compounds in soils amended with two refinery waste, *Water, Air, Soil Pollut.* 29:1–13, 1986.

Brown, K.W., Barbee, G.C., Thomas, J.C., and Murray, H.E. Detecting organic contaminants in the unsaturated zone using soil and soil-pore water samples, *Hazard. Waste Hazard. Mat.* 7:151–168, 1990.

Bryant, D.W. Genotoxic compounds associated with respirable urban air particulate. chemical fractionation and bioassay of complex mixtures, *Environ. Mol. Mutag.* 15(17):10, 1990.

Chamberlain, A.C. Fallout of lead and uptake by crops, *Atmos. Environ.* 17:693–706, 1983.

Chaney, R.L. Potential effects of sludge borne heavy metals and toxic organics on soils, plants, and animals, and related regulatory guidelines. Final report of the Workshop on the International Transportation, Utilization or Disposal of Sewage Sludge Including Recommendation. Annex 3, Workshop 9, 1–56, 1985.

Clark, E.H., Haverkamp, J.A., and Chapman, W. *Eroding Soils: The Off-Farm Impacts.* The Conservation Foundation, Washington, DC, 1985, 252.

Claxton, D.D. and Barnes, H.M. The mutagenicity of diesel-exhaust particle extracts collected under smog-chamber conditions using the *Salmonella typhimurium* test system, *Mutat. Res.* 88:255–272, 1981.

Claxton, L.D., Houk, V.S., and Kremer, F. Mutagenicity and the bioremediation of the Alaskan oil spill. U.S. Environmental Protection Agency, Research Park, NC and U.S. EPA, Cincinnati, OH. *Environ. Mol. Mutag.* 15(17):13, 1990.

Connor, M.S. Monitoring sludge-amended agricultural soils, *Biocycle* 25:47–51, 1984.

Davol, P., Donnelly, K.C., Brown, K.W., Thomas, J.C., Estiri, M., and Jones, D.H. Mutagenic potential of runoff water from soils amended with three hazardous industrial wastes, *Environ. Toxicol. Chem.* 8:189–200, 1989.

de Flora, S. Study of 106 organic and inorganic compounds in the Salmonella/microsome test, Carcinog. 2(4):283–298, 1981.

DeMarini, D.M., Inmon, J.P., Simmons, J.E., Berman, E., Pasley, T.C., Warren, S.H., and Williams, R.W. Mutagenicity in *Salmonella* of hazardous wastes and urine from rats fed these wastes, *Mut. Res.* 189:205–216, 1987.

DiGiovanni, J. and Slaga, T.J., Modification of polycyclic aromatichydrocarbon carcinogenesis, in Polycyclic Hydrocarbons and Cancer, H.V. Gelboin and P.O.P. Ts´o, Eds. (1981), pp. 259–292.

Donnelly, K.C., Brown, K.W., and Kampbell, D. Chemical and biological characterization of a wood-preserving bottom sediment waste. I. Prokaryotic bioassays and chemical analysis, *Mut. Res.* 180:31–42, 1987a.

Donnelly, K.C., Brown, K.W., and Scott, B.R. Chemical and biological characterization of a wood-preserving bottom sediment. II. Eukaryotic bioassays. *Mutat. Res.* 180, 43–53, 1987b.

Donnelly, K.C., Davol, P., Brown, K.W., Estiri, M., and Thomas, J.C. Mutagenic activity of two soils amended with a wood preserving waste, *Environ. Sci. Technol.* 21(1):57–64, 1987c.

Donnelly, K.C., Brown, K.W., and DiGuillio, D.G. Mutagenicity and chemical analysis of soil from superfund site, *Nucl. Chem. Waste and Manage.* 8:135–141, 1988.

Donnelly, K.C. and Brown, K.W. Mutagenic potential of plants grown on municipal sewage sludge amended soil, *Final Report 5867* 1–25, 1989.

Donnelly, K.C., Brown, K.W., and Thomas, J.C. Mutagenic potential of municipal sewage sludge amended soils, *Water, Air and Soil Pollut.* 48:435–449, 1989.

Donnelly, K.C., Brown, K.W., and Thomas, J.C. Bacterial mutagenicity of leachate water from municipal sewage sludge-amended soils, *Environ. Toxicol. Chem.* 9:443–451, 1990.

Donnelly, K.C., Thomas, J.C., Anderson, C.S., and Brown, K.W. The influence of application rate on the bacterial mutagenicity of soil amended with municipal sewage sludge, *Environ. Pollut.* 68:147–159, 1990.

Dragun, J. *The Soil Chemistry of Hazardous Materials.* Hazardous Materials Control Research Institute, Silver Spring, MD, 1988, 458.

DuPont, R.R. and Reineman, J.A. Evaluation of volatization of hazardous constituents at hazardous waste land treatment sites. EPA/600/S2-86/071, 1986.

Farmer, W.J., Ique, K., Spencer, W.F. and Martin, J.P. Volatility of organochlorine residues from soil; effect of concentration, temperature, air flow rate and vapor pressure, *Soil Sci. Soc. Am. Proc.* 36:443–447, 1972.

Feenstra, S. and Cherry, J.A. Subsurface contamination by dense non-aqueous phase liquid (DNAPL) chemicals, in *Proceedings of the IAH International Groundwater Symposium, Canadian National Chapter, Atlantic Region, Halifax, Nova Scotia, May 1–5, 1988.* 61–70, 1988.

Fiedler, D.A. Mutagenic potential of plants grown on a soil amended with mutagenic municipal sewage sludge, Ph.D. Thesis, Texas A&M University, College Station, 1988, 132.

Fluker, B.J. Soil temperatures, Soil Sci. 88(1):35–46, 1958.

Fluker, B.J. Soil temperatures, *Soil Sci.* 86:35–46, 1958.

Fredrickson, D.R. and Shea, P.J. Effect of soil pH on degradation, movement and plant uptake of chlorsulfuron, *Weed Sci.* 34:328–332, 1986.

Freeman, H.M. *Standard Handbook of Hazardous Waste Treatment and Disposal.* McGraw-Hill Book Co., New York, 1989, 48.

Gentile, J.M., Wagner, E.D., and Plewa, M.J. The detection of weak recombinogenic activities in the herbicides alachlor and propachlor using a plant-activation bioassay, *Mutat. Res.* 48:113–116, 1977.

Goggelman, W. and Spitzauer, P. Mutagenic activity, content of polycyclic aromatic hydrocarbons (PAH) and humus in agricultural soils. *Mutat. Res.* 89:187–196,1982.

Greene, J.C. Biological assessment of the toxicity caused by chemical constituents eluted from site soils collected at the Drake Chemical Superfund site. Lock Haven, Clinton Co., PA. U.S. Environmental Protection Agency, Environmental Research Laboratory, Corvallis, OR, 1989, 77.

Griffin, R.A. and Roy, W.R. Interaction of organic solvents with saturated soil-water systems. Open File Report No. 3, Environmental Institute for Waste Management Studies, The University of Alabama, Tuscaloosa, 86, 1985.

Guenzi, W.D. and Beard, W.E. Volatilization of lindane and DDT from soil, *Soil Sci. Soc. Am. Proc.* 34:443–447, 1970.

Gustafson, D.I. Groundwater ubiquity score: a simple method for assessing pesticide leachability, *Environ. Toxicol. Chem.* 8:339–357, 1989.

Harris, C.R. and Lichtenstein, E.P. Factors affecting the volatilisation of insectidal residues from soil, *J. Econ. Entomol.* 54:1038–1045, 1961.

Harris, C.R. and Sans, W.W. Absorption of organochlorine insecticide residues from agricultural soils by root crops, *J. Agr. Food Chem.* 15:861–863, 1967.

Hassett, J.J., Banwart, W.L., and Griffin, R.A. Correlation of compound properties with sorption characteristics of nonpolar compounds by soils and sediments: concepts and limitations, in *Environment and Solid Wastes: Characterization, Treatment and Disposal.* E.C. Francis and S. Auerbach, Eds. Butterworth, Woburn, MA, 1983, 161–178.

Hayes, M.H.B. and Swift, R.S. The chemistry of soil organic colloids, in *The Chemistry of Soil Constituents* D.J. Greenland, and M.H.M. Hayes, Eds. John Wiley & Sons, Inc., New York, 1978, 179–320.

Hazardous Waste News, December 24, 1984, p. 415.

Heartlein, M.W., DeMarini, D.M., Katz, A.J., Means, J.C., Plewa, M.J., and Brockman, H.E. Mutagenicity of municipal water obtained from an agricultural area, *Environ. Mutagen.* 3:519–530, 1981.

Hopke, P.K., Plewa, M.J., and Stapleton, P. Reduction of mutagenicity of municipal wastewaters by land treatment. *Sci. Total Environ.* 66:193–202, 1987.

Horning, W.B., II and Weber, C. I., Eds. Probit analysis of algal growth test data, in *Short-Term Methods for Estimating the Chronic Toxicity of Effluents and Receiving Waters to Freshwater Organisms*. U.S. Environmental Protection Agency, Environmental Monitoring and Support Laboratory, Cincinnati, OH. EPA/600/4-85/014, 1985, 162.

Houk, V.S. and DeMarini, D.M. Use of the microscreen phage-induction assay to assess the genotoxicity of 14 hazardous industrial wastes, *Environ. Mol. Mutag.* 11:13–29, 1988.

Hunt, J.R., Sitar, N., and Udell, K.S. Nonaqueous phase liquid transport and cleanup. I. Analysis of mechanisms, *Water Resource Res.* 24:1247–1258, 1988.

Hutton, M., Wadge, A., and Milligan, P.J., Environmental levels of cadmium and lead in the vicinity of a major refuse incinerator, *Atmos. Environ.* 22:411–416, 1988.

Hwang, S.T. Toxic emissions from land disposal facilities, *Environ. Prog.* 1:46, 1982.

Johnson, B.D. and Baker, D.E. Mutagenicity of sewage sludge amended soil, in *EPA Symposium on the Application of Short-term Bioassays in the Analysis of Complex Environmental Mixtures*. October 20–23, 1986. U.S. Environmental Protection Agency Health Effects Research Laboratory, Research Triangle Park, NC, 1986, (abstr.).

Johnson, J.B., Larson, R.A., Grunau, J.A., Ellis, D., and Jone, C. Identification of organic pollutants and mutagens in industrial and municipal effluents, *Final Report to the Illinois Environmental Protection Agency*, Project FW-38. The Institute for Environmental Studies. University of Illinois at Urbana-Champaign, 1983, 103.

Jones, R. and Burgess, M.S.E. Zinc and cadmium in soils and plants near electrical transmission (hydro) towers, *Environ. Sci. Technol.* 18:731–734, 1984.

Johnston, W.S., Hopkins, G.F., and Maclachlan, G.K. Salmonella in sewage effluent and the relationship to animal and human disease in the north of Scotland, *Vet. Record.* 119:201–203, 1986.

Jury, W.A., Spencer, W.F., and Farmer, W.J. Behaviour assessment model for trace organics in soil. I. Model description, *J. Environ. Qual.* 12:558–564, 1983.

Jury, W.A., Focht, D.D., and Farmer, W.J. Evaluation of pesticide groundwater pollution potential from standard indices of soil-chemical adsorption and biodegradation, *J. Environ. Qual.* 16:422–428, 1987.

Kada, T., Moriha, M., and Shirasu, Y. Screening of pesticides for DNA interactions by REC-assay and mutagenic testing and frameshift mutagens detected, *Mut. Res.* 26:243–248, 1974.

Kafer, E., Scott, B.R, Dorn, G.L., and Stafford, R. *Aspergillus nidulans*: systems and results of tests for chemical induction of mitotic segregation and mutation. I. Diploid and duplication assay systems. A report of the United States Environmental Protection Agency Gene-Tox Program, *Mut. Res.* 98:1–48, 1982.

Karickhoff, S.W., Brown, D.S., and Scott, T.A. Sorption of hydrophobic pollutants on natural sediments, *Water Res.* 13:241–248, 1979.

Kenaga and Goring. Relationship between water solubility, soil sorption, octanol-water partitioning, and concentration of chemicals in biota, in *Aquatic Toxicology*, J.G. Eaton, P.R. Parrish, and A.C. Hendricks, Eds. ASTM STP 707 (Philadelphia: American Society for Testing and Materials, 1980), pp. 78–115.

Kerfoot, H.B. Soil-gas measurement for detection of groundwater contamination by volatile organic compounds, *Environ. Sci. Technol.* 21:1022–1024, 1987.

Kier, L.D., Ames, B.N., and Yamasaki, E. Detection of mutagenic activity in cigarette smoke condensates, *Proc. Natl. Acad. Sci. U.S.A.* 71(10):4159–4163, 1974.

Klute, A., Grismer, M.E., and McWhorter, D.B. Determination of diffusivity and hydraulic conductivity in soils at low water contents from nondestructive transient flow observations, *Soil Sci.* 141:10–19, 1986.

Lagas, P. Sorption of chlorophenols in soil, *Chemosphere* 17:205–216, 1988.

Lee, L.S., Rao, P.S.C., Brusseau, M.L., and Ogwada, R.A. Nonequilibrium sorption of organic contamiants during flow through columns of aquifer materials, *Environ. Toxicol. Chem.* 7:779–793, 1988.

Lichtenstein, E.P. and Shulz, K.R. Residues of aldrin and heptachlor in soils and their translocation into various crops, *J. Agr. Food Chem.* 13:57–63, 1965.

Lichtenstein, E.P., Fuhremann, T.W., Scopes, N.E.A., and Skrent, R.F. Translocation of insecticides from soils into pea plants; effects of the detergent LAS on translocation and plant growth, *J. Agric. Food Chem.* 15:864–869, 1967.

Lower, W.R., Thompson, W.A., Drobney, V.K., and Yanders, A.F. Mutagenicity in the vicinity of a lead smelter, *Teratog., Carcinog., Mutagen.* (3):231–253, 1985.

Marrin D.L. and Kerfoot, H.B. Soil-grass surveying techniques, *Env. Sci. Technol.* 22:740–745, 1988.

Ma, T.H., Sandhu, S.S., and Anderson, V.A. A preliminary study of the clastogenic effects of diesel exhause fumes using the Tradescantia micronucleus bioassay, *Environ. Sci. Res.* 351–358, 1981.

McCoy, E.C., Hyman, J., and Rosenkranz, H.S. Conversion of environmental pollutants to mutagens by visible light, *Environ. Mut.* 68(2):729–734, 1979.

McFarlane, C., Nolt, C., Wickliff, C., Pfleeger, T., Shimabuku, R., and McDowell, M. The uptake, distribution and metabolism of four organic chemicals by soybean plants and barley roots, *Environ. Toxic. Chem.* 6:847–856, 1987.

Means, J.C., Wood, S.G., Hassett, J.J., and Banwart, W.L. Sorption of polynuclear aromatic hydrocarbons by sediments and soils, *Environ. Sci. Technol.* 14:1524–1528, 1980.

Nauman, C.H., Sparrow, A.H., and Schairer, L.A. Comparative effects of ionizing radiation and two gaseous chemical mutagens on somatic mutation induction in one mutable and two non-mutable clones of Tradescantia, *Mutat. Res.* 38:53–70, 1976.

Norseth, T. Industrial viewpoints on cancer caused by metals as an occupational disease, in *Origins of Human Cancer. Book A. Incidence of Cancer in Humans.* Cold Spring Harbor Laboratory, Cold Spring Harbor, New York, 1977, 159–167.

Page, A.L., Logan, T.J., and Ryan, J.A. *Land Application of Sludge: Food Chain Implications.* Lewis Publishers, Inc., Chelsea, MI, 1987,143.

Park, K.S., Sims, R.C., Doucette, W.J., and Matthews, J.E. Biological transformation and detoxification of 7,12-dimethylbenz(a)anthracene in soil systems, *J. Water Pollut. Control Fed.* 60(10):1822–1825, 1988.

Porcella, D.B. Protocol for bioassessment of hazardous waste sites. EPA/600/3-83-054. U.S. Environmental Protection Agency, Environmental Research Laboratory, Corvallis, OR, 1983, 139.

Quistad, G. B. and Menn, J.J. The disposition of pesticides in higher plants, *Res. Rev.* 85:173–197, 1983.

Rao, T.K., Young, J.A., Weeks, C.E., Slaga, T.J., and Epler, J.L. Effect of the co-carcinogen benzo(e)pyrene on microsome mediated chemical mutagenesis in *Salmonella typhimurium, Environ. Mutagen.* 1:105–112, 1979.

Rawls, W.J. Estimating soil bulk density from particle size analysis and organic matter content, *Soil Sci.* 135:123–125, 1983.

Royce, C.L., Fletcher, J.S., Risser, P.G., McFarlane, J.C., and Benenati, F.E. PHYTOTOX: a database dealing with the affect of organic chemicals on terrestrial vascular plants, *J. Chem. Infor. Comput. Sci.* 24:7–10, 1984, (Available through: Fein-Marquart Associates, 7215 York Rd., Baltimore, MD, 21212).

Ryan, J.A., Bell, R.M., Davidson, J.M., and O'Connor, G.A. Plant uptake of non-ionic organic chemicals from soils, *Chemosphere.* 17:2299–2323, 1988.

Sabijic, A. On the prediction of soil sorption coefficients of organic pollutants from molecular stucture: application of molecular topology model, *Environ. Sci. Technol.* 21:358–366, 1987.

Sawhney, B.L. and Brown, K. Eds *Reactions and Movement of Organic Chemicals in Soils.* SSSA Special Publication No. 22. American Society of Agronomy, Inc., Madison, WI, 1989, 474 .

Schellenberg, K., Levenberger, C., and Schwarzenbach, R. Sorption of chlorinated phenols by natural sediments and aquifer materials, *Environ. Sci. Technol.* 18:652–657, 1984.

Schoeny, R. and Warshawsky, D. Mutagenicity of 7H-Dibenzo(c,g)carbazole and metabolism in *Salmonella typhimurium, Mutat. Res.* 188:275–286, 1987.1987.

Schulten, H.R. and Schnitzer, M. Aliphatics in soil organic matter in fine-clay fractions, *Soil Sci. Soc. Am. J.* 54:98–105, 1990.

Schwarzenbach, R.P. and Westall, J. Transport of nonpolar organic compounds from surface water to groundwater. Laboratory Sorption Studies, *Environ. Sci. Technol.* 15:1360–1367, 1981.

Schwille, F. *Dense Chlorinated Solvents in Porous and Fractured Media — Model Experiments.* Translated by J.F. Pankow. Lewis Publishers Inc., Chelsea, MI, 1988, 146.

Scott, B.R., Dorn, G.L., Kafer E., and Stafford, R. *Aspergillus nidulans*: systems and results of tests for induction of mitotic segregation and mutation. Haploid assay systems and overall response of all systems. A report of the U.S. EPA Gene-Tox Program, *Mut. Res.* 98:49–94, 1982.

Senesi, N., Testini, C., and Polemio, M. Chemical and spectroscopic characterization of soil organic matter fractions isolated by sequential extraction procedure, *J. Soil Sci.* 34:801–813, 1983.

Shaffer, K.A., Fritton, D.D., and D.E. Baker. Drainage water sampling in a wet, dual-pore soil system, *J. Environ. Qual.* 8:241–246, 1979.

Shone, M.G.T. and Wood, A.V. A comparison of the uptake and translocation of some organic herbicides and a systematic fungicide by barley. I. Adsorption in relation to physico-chemical properties, *J. Exp. Bot.* 25:390–398, 1974.

Siegrist, H. and McCarty, P.L. Column methodologies for determining sorption and biotransformation potential chlorinated aliphatic compounds in aquifers, *J. Contam. Hydrol.* 2:31–50, 1987.

Silkowski, M.A. and Plewa, M.J. Analysis of the genotoxicity of municipal incinerator ash, *Environ. Mol. Mutag.* 15(17):55, 1990.

Sims, R.E. Fate of Polynuclear Aromatic Hydrocarbons in Soil and Plant Systems. Ph.D. Dissertation. Biological and Agricultural Engineeering Department. pp. 1–174.

Sklarew, D.S. and Girvin, D.C. Attenuation of polychlorinated biphenyls in soil, *Rev. Environ. Contam. Toxicol.* 98:1–41, 1987.

Smith, K., Martin, J., and Als, E. Field investigation to characterize relationship between ground water and subsurface gas contamination at a municipal landfill, in *Superfund '89, Proceedings — 10th National Conference.* The Hazardous Materials Control Research Institute, Silver Spring, MD, 1989, 695.

Smith, L.R. and Dragun, J. Degradation of volatile chlorinated aliphatic priority pollutants in groundwater, *Environ. Int.* 10:291–298, 1984.

Smith, M.G.T., Clarkson, D.T., Sanderson, J., and Wood A.V. *A Comparison of the Uptake and Translocation of Some Organic Molecules and Ions in Higher Plants. Ion Transport in Plants.* Academic Press, New York, 1973, 571–630.

Spencer, W.F., Farmer, W.J., and Jury, W.A. Behavior of organic chemicals at soil, air, water interfaces as related to predicting the transport and volatilization of organic pollutants, *Environ. Toxicol. Chem.* 1:17–26, 1982.

Symons, B.D. and Sims, R.C. Assessing detoxification of a complex hazardous waste, using the microtox bioassay, *Arch. Environ. Contam. Toxicol.* 17:497–505, 1988.

Tetta, D.A. Recycling of battery casings at a superfund site, in *Superfund '89, Proceedings — 10th National Conference.* The Hazardous Materials Control Research Institute, Silver Spring, MD, 1989, 695.

Thibodeaux, L.J. *Chemodynamics, Environmental Movement of Chemicals in Air, Water, and Soil.* John Wiley & Sons, Inc., New York, 1979, 501.

Thomas, R.E., Bledsoe, B., and Jackson, K. Overland flow treatment of raw wastewater with enhanced phosphorus removal. EPA-600/2-76-131, USEPA, Washington, DC (1973).

Tillman, N., Ranlet, K., and Meyer, T.J. Soil-gas surveys: procedures (Part II), *Pollut. Eng.* 21:79, 1989.

Tokiwa, H. Takahashi, K., Morita, K., Takeyoshi, H., and Ohnishi, Y. Detection of mutagenic activity in particulate air pollutants, *Mutat. Res.* 48:237–248, 1977.

Topp, E., Scheunert, I., Attar, A., and Korte, F. Factors affecting the uptake of ^{14}C labelled organic chemicals by plants from soil, *Ecotoxicol. Environ. Safety* 11:219–228, 1986.

U.S. Environmental Protection Agency. Water related environmental fate of 129 priority pollutants. EPA-440/4-79 029B. Vol. 1,2. Office of Water Regulations and Standards, U.S. Environmental Protection Agency, Washington, DC, 1979, 600.

U.S. Environmental Protection Agency. Users manual for the pesticide root zone model (PRZM). EPA/600/3-84/109. Environmental Research Laboratory, Office of Research and Development, U.S. Environmental Protection Agency, Athens, GA, 1984, 216.

U.S. Environmental Protection Agency. A compendium of field operations methods. EPA/540/P- 87/001. Office of Emergency and Remedial Response, U.S. Environmental Protection Agency, Washington, DC, 1987, 644.

U.S. Environmental Protection Agency. Exposure assessment methods handbook (Draft Report Prepared by Versar, Inc.) Office of Health and Environmental Assessment, U.S. Environmental Protection Agency, Washington, DC, 1988a.

U.S. Environmental Protection Agency. Superfund exposure assessment manual (SEAM). EPA/540/1-88/001. Office of Remedial Response, U.S. Environmental Protection Agency, Washington, DC, 1988b, 157.

U.S. Environmental Protection Agency. Risk assessment guidance for superfund, Vol. 1. Human health evaluation manual. EPA/540/1-89/002. Office of Emergency and Remedial Response, U.S. Environmental Protection Agency, Washington, DC, 1989a, 179.

U.S. Environmental Protection Agency. Soil sampling quality assurance user's guide. EPA/600/8-89/046, Office of Research and Development, Environmental Monitoring Support Laboratory, U.S. Environmental Protection Agency, Las Vegas, NV 1989b, 209.

U.S. Environmental Protection Agency. Transport and fate of contaminants in the subsurface. EPA/625/4-89/019. CERI (Cincinnati, OH) and Robert S. Kerr Environmental Research Laboratory, Ada, OK, 1989c, 148.

Vig, B.K. Soybean (glycine max): a new test system for study of genetic parameters as affected by environmental mutagens, *Mut. Res.* 31:49–56, 1975.

Villaume, J.F. Investigations at sites contaminated with dense non-aqueous phase liquids (DNAPL), *Groundwater Monitor. Rev.*, Spring, 1985, 60–74.

Walker, A. Effects of soil moisture content on the availability of soil applied herbicides to plants. *Pestic. Sci.* 2:49–55, 1971.

Weber, J.B. Pesticide dissipation in soils as a model for xenobiotic behavior, in *Pesticides: Food and Environmental Implications. International Symposium Proceedings, International Atomic Energy Agency (IAEA) and Food and Agriculture Organization (FAO), Neuherberg FRG*, November 24–27, 1987, 331.

Weber, J.B., Shea, P.J., and Strek, H.J. An evaluation of nonpoint sources of pesticide pollution in runoff, in *Environmental Impact of Nonpoint Source Pollution*. M.R. Overcash and J.M. Davidson, Eds. Ann Arbor Science, Ann Arbor, MI, 1980, 449.

White, R.E. The influence of macropores on the transport of dissolved and suspended matter through soil, in *Advances in Soil Science,* Vol. 3. Springer-Verlag, New York, 1985, 95–120.

Withrow, W.A. Mutagenicity of roadside soils. Master's Thesis, Illinois Institute of Technology, Chicago, IL, 1982, 118.

Wood, J.A. and Porter, M.L. Hazardous pollutants in class II landfills, *Hazard. Waste Mgmt.* 37:609–610, 1987.

Chapter

13

Environmental Aquatic Toxicology

Kevin M. Kleinow and Mark S. Goodrich

A. INTRODUCTION

In context of the global community, water comprises nearly three fourths of the earth's surface. Although tremendous diversity of life exists in aquatic ecosystems, a commonality among organisms is their close association with the environment. In many regards the intimate nature of organisms with the environment defines the essence of aquatic toxicology as a discipline. While by definition, aquatic toxicology is the qualitative and quantitative study of the toxic effects of chemicals on aquatic organisms, the scope of aquatic toxicology includes much more by including the influences of both biotic and abiotic processes in aqueous systems. From this perspective aquatic toxicology addresses not only all the facets related to the toxicity of chemicals to aquatic organisms, but also the chemical interaction with the life process of an ecosystem. This chapter will deal with unique aspects of the aquatic organism-xenobiotic interaction, as well as those interactions determining environmental fate and toxicity.

The toxicity of chemicals in the aquatic environment is determined by an interplay of interorganismic, intraorganismic, and environmental factors. On the organismic level, species, age, sex, health status, trophic level, ecological niche, toxicant load, inductive status, and physiology are determinant factors in the assessment of risk. On an environmental basis, water quality, distribution, temperature, light, sorption, and solubility are major influencing considerations. The aquatic environment, while often relatively stable on the short term, can vary tremendously between sites and can also be in dynamic flux. In this regard toxicity may also vary between sites, within sites (with depth for example), and with time. The play of environmental factors upon the physiochemical nature of toxicants, availability of toxicants and physiology of aquatic organisms provides a sliding scale with countless variations adding to the complexities of real world hazard assessment.

B. THE AQUATIC ENVIRONMENT

Water bodies possess characteristics which provide an identity much like characteristics define an animal or plant species. Influential in the presentation of and interaction with environmental toxicants, these assemblages of characteristics can be informative indicators of toxicant behavior in aquatic systems. Aquatic ecosystems, nevertheless, are widely varied. Lakes, rivers, marshes, swamps, and bogs are commonly used classifications. However, within a category such as lakes, further delineations according to lake origin, nutrient status, chemical composition, thermal stratification, topography, and water source provide further insight into the disposition of toxicants. Bodies of water are in continual flux. Lakes, for example, pass through various stages of maturity or senescence. This aging presents changing characteristics with the passage of time. In addition, anthropogenic input such as erosion, agricultural runoff, effluent discharges, acid rain, and urbanization can alter the normal aging process.

Lakes can be divided for discussion purposes into various regions which have definitive characteristics and community structure. These zones are presented in Figure 13.1. The well lit, shallow littoral zone along the periphery of a lake is largely populated by rooted aquatic plants, invertebrates, and small fish. Extending lakeward from the littoral zone is the sublittoral zone. Dimly lit, this zone contains fewer organisms than littoral areas. In deeper lakes the sublittoral zone serves as a transition to a third zone known as the profundal zone. It is this latter region which is characterized in midlatitude summer by temperature stratification. The upper strata, the epilimnion, delineated from the lower strata (hypolimnion) by temperature driven density differences, is warm water, which is well lit and mixed by prevailing winds. This strata is trophogenic with high levels of photosynthesis and elevated levels of oxygen. The hypolimnion below the thermal transition zone (thermocline) is tropholytic. In this zone there is limited light, little photosynthesis, minimal mixing, low oxygen, cold temperatures, and high levels of respiration/decomposition.

Open water beyond the marginal vegetation provides habitat to the bulk of the free floating planktonic organisms. Phytoplankton in the epilimnion are the major primary producers in all but the shallowest lakes (Figure 13.1). Linking

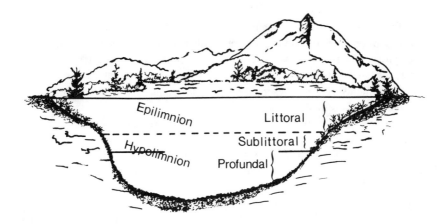

Figure 13.1. Zonation of a temperate lake based on depth, temperature, and productivity.

the primary producers to the rest of the food chain are a variety of invertebrates and fish which serve as primary consumers. Paramount in this capacity are the zooplankton. All of these steps interface with the terminal consumers which are most often carnivorous fish. The benthic or bottom zone which transverses all the foregoing areas provides for additional organismic diversity. Character as well as the quantity of the benthic communities may change with bottom substrate composition, particle size of the sediments, amount of substrate, and chemical environment. Much of the superficial benthic substrate is renewed due to continued sedimentation of organic materials or particulates from the trophogenic zone, sediment/substrate focusing by wave or current action and bioturbation.

Water chemistry may differ significantly between sources, within water bodies according to depth, within water bodies according to season, and with general nutrient status. Nutrient status has a profound effect on the presentation of a lake. Oligotrophic lakes, those relatively poor in nutrients, generally exhibit low plant and phytoplankton productivity, as well as a pronounced floral and faunal species diversity. However, the large number of organismic types are represented in quantitatively low numbers. Topographically these lakes most often have small epilimnion volumes compared to the hypolimnion. Visually these lakes appear clear with blue green waters. Eutrophic lakes on the other extreme are nutrient rich with abundant phytoplankton and littoral plants. These lakes often have shallow broad littoral zones with epilimnion/hypolimnion volume ratios greater than one. Species diversity is less evident than in oligotrophic lakes, but biomass/unit volume is often much greater. The water is often colored yellow, green, or brown with limited transparency.

Lotic environments (running water) present a very different aquatic environment from that of lentic systems (lakes). Currents and shallow depth provide for a species diversity which is longitudinally based, rather than with depth as in lentic systems. Currents vary with the season, depth, and longitudinal profile. These currents provide input of materials in faster waters and depositional effects under slow water conditions. By the nature of flowing water, chemicals may be presented to stationary aquatic organisms in a continuous or pulsatile fashion.

The community structure and community distribution demonstrated in aquatic ecosystems are influential, not only in determining which species will be subjected to a toxicant, but along with physical dynamics and water chemistry will determine the overall significance of toxicant introduction. Modulation of toxicity, toxicant form, and availability by environmental factors have only recently received more than passing attention.

The reader is referred to Chapter 14 on estuarine systems for a similar discussion on marine environments and a variety of limnology and oceanography textbooks for a more in-depth treatment of aquatic ecosystems.

1. Sources of Toxicants

Toxicants enter the aquatic environment by a variety of routes from many different sources. Industrial discharges, municipal inputs, agricultural runoff, groundwater contamination, riverine inputs, aerial deposition, and sediment release are frequent avenues for introduction of toxicants into the water column. Additionally, contaminants may originate from natural physiological and/or physiochemical processing. Biotoxins from aquatic algae, biogeochemical cycling of endogenous copper, zinc, and cadmium (Boyle et al., 1976; Bruland et al., 1978), as well as biogenic generation of ammonia (excretion and degradation of nitrogenous organic matter) are examples of natural sources of aquatic toxicants.

Xenobiotic input into waters may originate from discrete (point) or diffuse (nonpoint) sources. Point sources such as industrial and municipal discharges may provide clear delineations of affected waters in the near field. These demarcations are commonly defined by water quality, incurred chemical residues of affected species, and/or epidemiological factors associated with varying incidences in morbidity and mortality. The degree of mixing, size of the water body, temporal considerations, and flow dynamics provide the determinants for gradient formation, locale of contaminants, and distribution of detrimental effects in the far field.

Nonpoint sources by nature of their diffuse input are not as clearly delineated in the near field as point sources. In actuality few xenobiotic inputs into a watershed, if examined closely or broadly enough, are nonlocalized inputs. Nonpoint sources due to their diffuse nature can provide relatively uniform exposure concentrations. The accuracy of this statement is modulated by the relative size of the waterbody, the distribution of toxicant sources, and the aquatic environment itself. Agricultural runoff of pesticides and fertilizers, as

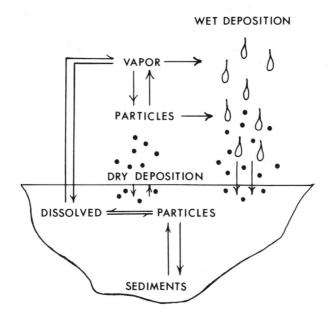

Figure 13.2. Contaminant interactions in the air, water, and air/water interface.

well as aerial deposition of mercury, pesticides, polychlorinated biphenyls, and polycyclic aromatic hydrocarbons are often cited examples of toxicant introduction into aquatic systems by nonpoint sources.

Movement of toxicants into aquatic environments may proceed through water/water, air/water, soil/water, biotic/water, and particle/water interactions. The nature of the participating media and processes involved determine the amounts transferred, the rates of transfer, and the nature of the toxicants which are transferred.

2. Aerial Deposition

Movement of organic contaminants from the atmosphere to the aquatic environment is dependent in part on the distribution of the compound between the vapor and particle phases (Figure 13.2). The atmospheric distribution between these two phases depends on size, surface area, and organic carbon content of the particle phase as well as the contaminant's vapor pressure. In general the lower the volatility the greater the tendency to be associated with the particle phase and vice versa.

Organic contaminants may be removed from the atmosphere from either the vapor or particle phase. While aerial movement appears to occur predominately in the vapor phase (90%), deposition into aquatic environments relies heavily on particle deposition (Eisenreich, 1987). Wet removal of contaminants, by which vapor partitions into rain/snow and particles are scavenged from the atmosphere, appears to far outweigh dry particle deposition in importance. For most compounds examined, particle scavenging appears to dominate both wet and total fluxes.

Transfer of contaminants between the vapor phase and water is governed at the air/water interface by molecular diffusion as driven by the respective concentration gradients. Further modulation is exerted by the resistance to mass transfer between phases. This latter point depends on the water solubility of the compound and its tendency to volatilize. For certain compounds such as volatile, slightly soluble PCBs, water concentrations afforded by particle deposition may act as a source for localized aerial concentrations.

3. Groundwater Input

Contamination of groundwater has, in recent years, received press as an important source of water contamination. It has been estimated that 1% of the producible groundwaters in the U.S. are contaminated (Gass, 1980; OTA, 1984). A large proportion of these waters are in high use agricultural, industrial, or urban areas.

Groundwater contamination comes from many sources including septic tanks, landfills, hazardous waste sites, pipelines, irrigation, pesticide application, wells, and natural sources such as saltwater intrusion or natural leaching. Commonly, introduction of contaminants occurs by leachate formation resulting from water percolation through a contaminated soil matrix or by direct migration from a source buried in a saturated groundwater zone (septic tanks). Other toxicant introductions occur through mechanisms such as interaquifer exchanges (as experienced with drilling) or flow reversals from contaminated surface waters.

Once in the groundwater, contaminants move by advective flow and by the irregularities in mixing accompanying groundwater flow (dispersion). The hydraulic conductivity of a geologic formation depends on factors such as particle size, porosity, and fracture profile. These factors, as well as variances in flow velocity in the flow field, will determine, in part, the extent and spread of contaminants within the groundwater.

Several processes beyond hydraulic factors are responsible for contaminant movement or retardation. Flocculation of colloidal material, trapped particulates, and dissolved organics, besides blocking hydraulic movement, may filter and retain contaminants. Processes such as ion exchange by biotic and abiotic matrices are important in removing constituents from groundwater as well as releasing others. Ions will bind until another ion of greater affinity becomes available or until pH conditions allow release. Geologic materials have a definitive ion exchange capacity. Materials such as clayey soils exhibit the greatest cation exchange capacities. Anion exchange, while less defined for groundwater conditions, appears less influential than cation exchange, as most mineral surfaces in natural water systems are negatively charged. The ion exchange capacity of soils may retard movement of contaminants for long periods of time; however, events such as pH or ionic alterations may allow contaminant release.

Precipitation and complexation are also very common events in groundwater. The formation of ligands between metal ions and either organic or inorganic species may influence movement of contaminants by hindering flow or in some cases facilitating flow by formation of soluble complexes (Griffin and Chou, 1980).

Much like other modalities of contaminant transfer, volatilization may play a role in groundwater contaminant retention and loss. Organic compounds with high vapor pressures, low aqueous solubilities, and low molecular weights may volatilize from groundwater. This process is enhanced by low soil moisture and high soil porosity.

4.   Chemical Limnology of Contaminants

Central to the uniqueness of aquatic toxicology are the many properties of water which influence the physiology and behavior of animals as well as the behavior of toxicants within it. Included on this list are high specific heat, high heat of vaporization, high heat of fusion, expansion on freezing, and universal solvent properties. These characteristics in large measure are related to the structural make-up of water, H-bonding, and polarity of the molecule. Of these properties solubility is perhaps the most critical in regards to aquatic toxicology. Water dissolves more substances and dissolves them more completely than any other liquid. Nevertheless, some substances are relatively insoluble while others are very soluble. In general, water is a poor solvent for nonpolar compounds, that is, those with hydrogen, hydrocarbon, and oxygen. Compounds may be more soluble if constituent components include polar alcohols or NH_4^+. Since nonpolar molecules have little attraction for water "they are squeezed" out by the attraction of water for other water molecules. It is this solubility or insolubility which provides the first overriding determinant in the distribution and availability of contaminants to aquatic organisms. As shown in Figure 13.3 chemicals partition in part in aquatic systems according to their relative polarity and their volatility. Polar compounds (water soluble) which are nonvolatile will be distributed primarily in the water column. Conversely, if the compound is nonpolar and nonvolatile the compound will distribute to organic matrices in the sediments and accumulate in the biota. If the compound is nonpolar and volatile it will, with time, volatilize off the surface of the water into the air. Of course these conditions are idealized. Often chemicals exhibit properties which may be intermediate on the scale of polarity, solubility, and volatility. In such cases the distributional equilibrium will be established which reflect these characteristics.

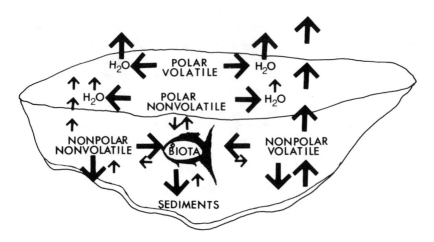

Figure 13.3. Partitioning of chemicals in the aquatic environment.

Many other considerations influence the disposition of toxicants in aquatic systems. In part, these considerations ultimately address the issue of solubility and polarity, but by the guises of complexation, binding, chemical alterations, and precipitation. Many compounds are influenced by the water media in which they are contained. Water, in the sense of natural systems and aquatic toxicology, is not just H_2O, but a composite of H_2O, the materials dissolved in it, and the materials suspended in it. It is the latter two components which modify water to the extent that water is not a generic commodity. This is an important concept as water quality, as it is referred to, is an important modulator of disposition, availability, and toxicity of many compounds.

Water quality as it relates to aquatic toxicology includes biotic and abiotic components. Biotic water quality includes the contribution of living organisms and organically derived substances to the general character of water. On the other hand, numerous physiochemical characteristics comprise the abiotic nature of water. Included on this list are factors such as pH, hardness, alkalinity, salinity, dissolved oxygen, temperature, conductivity, and redox potential. Empirically, if of significant magnitude, biotic and some abiotic characteristics may be grossly detected in water as turbidity, color, odor, or texture. In large measure, abiotic factors are inorganically derived from the environment; however, organically derived substances may also contribute to some of these characteristics of water. Collectively, water quality can influence the availability, chemical form, and ultimately the toxicity of chemicals in the aquatic environment. Often the influence of water quality on toxicity is a collective action, although each characteristic can exert independent as well as interdependent effects.

5. Select Water Quality Parameters

pH

The pH of natural waters falls in the range of 4 to 9, with most waters somewhat alkaline. In water, bicarbonate in equilibrium with carbon dioxide and carbonate provide the primary source of buffering capacity. Shifts to the carbonate side of this equilibrium are largely responsible for the alkaline nature of most waters.

Low water pHs can result from both natural and anthropogenic sources. Organic acids collectively referred to as humic acids can lower pH in aquatic systems. These compounds occur naturally as a result of decomposition of plant material. In regions such as bogs where organic loads are high and buffering capacity low, these acids may provide pH values near 4. Similarly, under natural conditions atmospheric carbon dioxide, if allowed to equilibrate with rain water without the action of other factors, would provide a pH of approximately 5.6 in renewable waters. Other gases such as SO_2 available from volcanic emissions and industrial discharges readily react with water to form sulfurous acid. This gas among others provides the basis for acid rain. For some bodies of water with poor buffering capacity, acid rain, has increased H^+ concentrations such that pHs as low as 2 have been reported. At this end of the pH scale toxicity may result from H^+ concentration alone. Existing populations of aquatic plants and animals often show poor growth and lack of reproductive success. These effects may result in community compositional shifts to more acid tolerant species. Low pH in addition to direct toxic effects can mobilize acid soluble contaminants from otherwise unavailable depots such as the sediments or anthropogenic sources. A variety of compounds including aluminum are thought to be mobilized by this action.

The pH of the water medium relative to the pKa of the contaminant determines the degree of ionization or nonionization of the compound. For organic contaminants, this relationship is described by the Henderson Hasselbach equation:

$$pH = pK_a + \log \frac{\text{ionized}}{\text{nonionized}}$$

If the water pH is lower than the compound's dissociation constant (pK_a) for a weak organic acid, the compound will be preferentially nonionized; for weak organic bases — ionized. The converse is true if the pH is higher than the pK_a. This pH relationship thus influences both water solubility (ionization favors water solubility) and transport through biological membranes (favored by nonionization). This is not to say that all ionic compounds are soluble in water, nor that nonpolar are insoluble. Ionic compounds have differential solubilities. In general, the larger the charge on both ions of an ionizable compound the greater the insolubility of the compound. In addition, solubility appears to be favored by differences in sizes of the oppositely charged ion.

Environmental pH may have an operative influence on environmental fate and processing of toxicants. Studies by Fagerstrom and Jernelov (1972) demonstrated that the amount of methylmercury formed by biogenic methylation of mercury was greater at low pH (5 to 7). Conversely, formation of dimethylmercury was greater at high pHs (>7). This relationship, while not altering the total amounts of mercury methylated, shifted the form produced. This induced change can influence fish mercury content since monomethylmercury is less volatile than dimethylmercury and thus is retained more efficiently in lakes.

Other studies with metals have demonstrated an influence of pH on accumulation potential. Studies with tri-*n*-butyltin and triphenyltin have indicated that increases in bioconcentration factors occurred in carp with increasing environmental pH (Tsuda et al., 1990). This was thought to be associated with a change to the hydroxylated derivatives.

Besides altering bioaccumulation and chemical form, pH has been reported to vary the subsequent toxicity of chemicals in water. Copper is one of the best known examples. Copper demonstrates a multitude of speciation shifts within the pH range of natural waters. At pH 5, for example, a large proportion of the copper is present as Cu^{+2}, the free copper ion, but at higher pHs there are a variety of hydroxides and carbonates formed. Changes in copper toxicity with pH appear to undulate widely when total copper levels are measured but are fairly stable when only forms known to be toxic are examined. It is of interest to note that higher concentrations of these forms are required to elicit toxic effects at low pHs than at higher pHs. These results suggest that additional mechanisms may be operative.

Water Hardness

In everyday life variances in water hardness can be related to the ability to precipitate soap. Hard waters are much better at performing this function than their soft water counterparts. Water hardness as used in the aquatic sciences refers to the water concentration of calcium ions and magnesium ions as well as other polyvalent metals such as manganese, iron, and aluminum. For most fresh waters the major contribution to total hardness is that associated with calcium and magnesium. Such hardness is often referred to as carbonate hardness due to the natural association of ions with carbonate. Waters with a total hardness of <75 mg/l are considered soft, while concentrations ranging from 75 to 300 mg/l encompass the categories of moderately hard, hard, and very hard.

Water hardness is influential in the disposition and toxicity of a variety of aquatic contaminants. While some effects of hardness have been noted for select organic compounds (Slonin, 1977), the most profound effects upon toxicity have been demonstrated for the metals. In general, most heavy metals are more toxic to aquatic life in soft rather than hard water. Hardness alone, however, rarely appears as the sole determinant of metal toxicity. Hardness, pH, and alkalinity (buffering capacity of water) are intimately interactive. For example, when the total hardness of a water sample is greater than the total alkalinity, some of the Ca^{+2} and Mg^{+2} normally associated with HCO_3^- or CO_3^{-2} would be associated with Cl^-, SO_4^{-2}, or NO_3^-. Conversely, when the total alkalinity exceeds the total hardness, HCO_3^- and CO_3^{-2} are free to associate with components other than Ca^{+2} or Mg^{+2} such as K^+ or Na^+. These types of processes have a direct bearing on those interactions affecting metal toxicity. Several postulated mechanisms for the influence of hardness on metal toxicity have been put forth, including reductions in membrane permeability for metals and complexation of hardness ions with metals in water. It is likely that the latter is of minor consequence as added hardness ions have been shown to have little effect on metal toxicity (Zitko and Carson, 1976).

6. Biotic Water Quality

Organic substances are important modulators of solubility, availability, mobility, persistence, and toxicity of contaminants in the aquatic environment. These organic substances may result from decomposition of dead organisms as well as elaboration from or association with live organisms. Such substances may be as small as a single molecule or as large as a particle, whose composition encompasses many chemical classes.

One of the largest categories of organic materials in the aquatic environment as well as in sediments and soils are those substances referred to as humic substances. These plant and animal decomposition products are ubiquitous compounds of poorly defined composition often found as colloidal suspensions in aquatic environments. Because of their poorly defined nature, their concentration is often expressed as dissolved organic carbon (DOC) and are classified in three subfractions based on their pH-dependent solubility in water. Natural waters vary dramatically in both humic substances and dissolved organic carbon. Dissolved organic carbon varying from <2 to >55 mg/l of carbon in uncolored and colored waters may contain anywhere from 50 to 90% humic substances, respectively. Humic substances in the aquatic environment are continually being altered by physicochemical and biological processes which change not only individual component characteristics but intermolecular interactions responsible for properties affecting toxicant behavior.

The effect of humic substances on toxicant solubility, disposition, availability, and toxicity is greatly influenced by the character of the humic substances, the character of the toxicant, and the predominate processing events in the aquatic environment. Humic substances are capable of interacting with metal ions by complexation and ion exchange as well as bind with nonionic material such as nonpolar hydrophobic compounds. Such interactions may lead to the solubilization or immobilization of contaminants. Binding of marginally soluble compounds, for example, may bring more of the compound into the water column (Wershaw et al., 1969). While this action initially facilitates toxicant introduction and mobility, it may also facilitate compound removal through sedimentation. This latter action is largely dependent on particle size. Conversely, binding of highly nonpolar contaminants to soluble dissolved organic carbon has been observed to decrease sorption to sediments largely as a function of contaminant maintenance in the water column by dissolved organic carbon association. Many other interactions have been observed, including declines in contaminant volatility rate (Griffin and Chou, 1980; Hassett and Milicic, 1985) and the bioavailability to aquatic organisms (Leversee et al., 1983; Landrum et al., 1985). Events integral to the environmental processing of contaminants may also be modulated by dissolved organic carbon. Evidence suggests that sunlight may react directly with many organic contaminants and dissolved organic carbon to elicit photochemical changes allowing the production of photoreactant intermediates (Cooper, 1989). Alternatively, dissolved organic carbon has been shown to alter decomposition rates of certain pesticides (Perdue and Wolfe, 1982).

7. Disposition of Contaminants in the Water Column

Particle contaminant interactions are important processes in aquatic ecosystems. Like many other phenomena in the aquatic environment, a periodicity exists for particle type, location, and mobility. Under isothermic conditions in lakes, wind action can act through the water column to resuspend sedimentary material and mix contaminants. Such isothermic conditions are seasonal in temperate lakes with winter and spring being primary periods of this activity. Alternatively, shallow waters in warmer climates are routinely subject to isothermic conditions and to subsequent mixing. In both systems, during storm events particles which are already suspended in the water column are mixed with particles resuspended from the sediments. These resuspended particles and the contaminants associated with them are often modified by physicochemical and biological processes upon passage through the water column and upon their residence in the sediments. The net result upon mixing is a myriad of particle contaminant combinations, as well as interactions with new, newly altered, and preexisting components.

Conditions change in temperate lakes with the onset of summer when lakes stratify, separating the warm epilimnion from the colder hypolimnion. During this period, stratification limits mixing of waters and contaminants between the two phases. In late spring and early summer, phytoplankton blooms in the epilimnion followed by ensuing zooplankton expansions which redirect contaminants along alternative particle pools. These organisms which dominate the particle pool at this time not only are selective in the fraction of particles consumed (relative to size, organic carbon content, taste, and contaminant loads), but also perform biotransformational modification of contaminants and repackage as well as redistribute associated particles. These later activities, occurring largely as a result of feeding behavior and fecal deposition, change the character of particles to which contaminants are associated. As summer progresses and the nutrients driving the plankton bloom become rate limiting, dead and dying organisms and fecal material settles down through the trophogenic waters. The larger particles and associated contaminants traversing down through the thermocline are largely confined to the hypolimnion after passage until the lake destratifies in the fall. This provides opportunity for contaminant modification under near anaerobic tropholytic conditions by microbial and physiochemical processes. Numerous variations on this scheme are observed with different types of water bodies, community structures, climatic, and nutrient conditions.

8. Sediments — The Benthic Environment

Sediments are primary repositories for abiotic and biotic materials in the aquatic environment. By direct association with sediments or by association with settling particles many xenobiotic chemicals accumulate to significant levels in the sediments (Primuzic et al., 1982). These accumulations have been correlated to changes in benthic community structure (Nalepa and Landrum, 1988), elevated body burdens of contaminants in aquatic life (Malins et al., 1984; Wood et al., 1986; NOAA, 1987), and overt toxicity.

Sediment composition is highly variable within and between aquatic environments. Widely ranging organic carbon contents, particle size distribution, clay content, clay type, redox characteristics, cation exchange characteristics, and pH are observed under both natural and anthropogenically altered conditions. Natural processes such as wind, wave, and current action continually disperse and size fractionate particles. Such fractionation may result in clay suspended in the water column, sand or larger particles in hydraulically scoured areas, and silt in depositional areas. An extended example of this process, referred to as sediment focusing, has been demonstrated in the Great Lakes. In this process, wind in near shore areas resuspends sediments. The resuspended sediments are transported upon currents until particles size sort according to settling rates and the energetics of the aquatic environment. This process continues repetitively until, in deeper waters, seiches and currents transport fine grained organic rich sediments to basins where the local energy regime is insufficient to resuspend the sediments. The toxicological significance of this process is that contaminants may preferentially concentrate with particular sediment types or sizes. Early studies by Richardson and Epstein (1971) demonstrated that compounds such as 2-bis(*p*-chlorophenyl,1,1,1-trichloroethane) (DDT) and methoxychlor concentrate in fine clay particles.

Endosulfan on the other hand was preferentially associated with larger materials. The net result of these interactions is that contaminants may be nonuniformally distributed in the sediments of either lotic or lentic environments.

Organic carbon content of sediments is an important determinant in the interaction with contaminants (Weber et al., 1983). Several investigators have demonstrated that the sorption of neutral organic chemicals to sediments is directly related to the sediment's organic carbon content (K_{oc}) (Chiou et al., 1979; Means et al., 1979). In general, contaminants with high K_{ow}s (octanol/water partition coefficients-lipid solubility) are more strongly sorbed to organic carbon than those compounds with low K_{ow}s. The sorption process, while apparently resulting primarily from a partitioning of the contaminant between aqueous and organic phases, can also be influenced by pH, sediment surface area (Means et al., 1982), chemical structural parameters (Sabljic and Protic, 1982), and particle size. Studies have demonstrated that clay and other fine fractions may contribute significantly to sedimentary sorption independent of their organic carbon content. However, sorption differences within silt and clay fractions have been demonstrated to be largely the result of varying organic carbon content. Larger particles such as sand, when compared to the fine fractions on an equal organic carbon basis, exhibited much lower sorbent capabilities (Karickhoff et al., 1979).

Sediment sorbed contaminants are not equally available for desorption, hence equilibrium and bioavailability for aquatic organisms. High K_{ow} contaminants which readily bioconcentrate from water are slowly desorbed from sedimentary organic carbon resulting in low organism bioavailability (Landrum and Robbins, 1990). Conversely, low K_{ow} compounds may rapidly desorb from sediments providing for high environmental bioavailability; however, by their polar nature exhibit less bioaccumulative potential. Based on sediment desorption criteria, many sorbed contaminants appear to reside in reversible and resistant pools. The relative proportion of which depends on sorption duration (DiToro et al., 1982; Karickhoff and Morris, 1985).

Pollutants may be distributed on the sediments, in the water around sediment particles (interstitial or pore water), in the overlying water, and in the aquatic life inhabiting the benthic environment (Neff, 1984) (Figure 13.4). In most cases sedimentary sorbed contaminant concentrations will be in the process of or in equilibrium between the individual components. In most cases with nonpolar contaminants, the concentrations will be higher in the sediments than the interstitial water and higher in the interstitial water than the overlying water column. It has been postulated that the interstitial water is the primary vehicle for the accumulation of sediment associated contaminants by infaunal organisms (Oliver, 1987; Knezovitch and Harrison, 1988). Uptake of contaminants from this pool is limited by the desorption rate from sediments, the size of the interstitial water volume (Landrum and Robbins, 1990), and the binding of contaminants to dissolved and colloidal materials in the interstitial water.

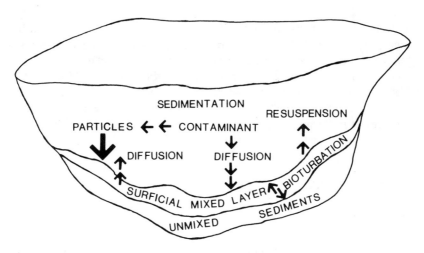

Figure 13.4.   Distribution of contaminants on and about the sediments.

Sedimentary desorption processes for many contaminants appear to be much slower than sorptive processes (Elzerman and Coats, 1987). These slow desorptive processes may require months to achieve true equilibrium between sediment particles and interstitial water (Karickhoff, 1980; Witkowski et al., 1988). Landrum (1989) has suggested that small desorption rate constants may explain the relatively slow rates of contaminant accumulation by benthic organisms.

Ingestion of pollutant-associated particles may provide another important route of contaminant assimilation in sediment feeding benthic organisms. A variety of infaunal organisms have been shown to feed selectively on fine organic rich particles in lieu of general sediment types (Eadie et al., 1985). This behavior appears to result in significant contaminant accumulation as often pollutants partition to similar particle types.

Sedimentation represents an important mechanism for the removal of hydrophobic organic compounds from the water column. Deposition onto the sediments and burial by subsequent uncontaminated particles, with time if taken alone, will limit availability to aquatic organisms. In most aquatic systems diffusional and advective processes as well as resuspension and bioturbation are operative in water-sediment interactions. Resuspension and bioturbation often result in a continual mixing of buried sediments with surficial sediments. The net result is a dynamic flux of contaminants from deeper sediments to surficial sediments and the water column. This mixing process increases the time in which contaminants contained in the sediments may be recycled back into the ecosystem.

A variety of organisms including clams and tubificid worms are involved in the recycling of contaminants from the sediments via the process referred to as bioturbation. Diverse organism life strategies result in differing contributions to the bioturbation process. Burrowing organisms like oligochaetes institute size selective feeding resulting in conveyor belt transport of sediments and contaminants from deeper strata to the surface (Karickhoff and Morris, 1985; Robbins, 1986). Other organisms such as surface active *Pontoporeia* are operative in mixing of only the surficial sediments. Bioturbation in all cases provides an important mechanism for reworking the sediment and increasing availability of buried contaminants.

C. TOXICANT UPTAKE

1. Waterborne

The gills of fish not only function in gas exchange, osmoregulation, ammonia excretion, and acid/base balance, but also are instrumental in the uptake of xenobiotic chemicals. In this role the gill is second only to the integument in surface area for diffusion and adsorption of exogenous materials (1 to 9 cm^2/g body weight). The large volume of water passed over the gills for normal physiological processes facilitates contaminant uptake by continually renewing toxicant concentrations at the gill, while countercurrent flow of blood and water enhances the exchange process. Passive diffusion, the predominate mechanism by which the gill extracts xenobiotics from water, can be limited by four factors: (1) the ventilation rate (the rate at which water flows over the gills), (2) the diffusion rate across the gill lamellae, (3) the rate of blood flow through the gills, and (4) the thickness of the stagnant water layer at the water/gill boundary (Hayton and Barron, 1990). Of these the diffusion gradient is controlled primarily by the ventilation rate and the rate of blood flow through the branchial basket (Erickson and McKim, 1990). For a number of contaminants a correlative relationship has been demonstrated between ventilation rate, xenobiotic uptake efficiency, and uptake efficiency for oxygen (McKim and Goeden, 1982).

Aquatic invertebrates use a variety of respiratory strategies. Uptake efficiencies (water clearance rates) for oxygen and xenobiotics, however, are not necessarily correlated (Landrum and Stubblefield, 1991). The uptake of certain polycyclic aromatic hydrocarbons in aquatic amphipods, for example, could not be solely accounted for by uptake across respiratory membranes (Landrum and Stubblefield, 1991). When the animal's surface area or surface area to body volume ratio is considered, the results appear to indicate that the animals may be extracting materials from the water phase via both respiratory and dermal routes.

The role of the dermis in the uptake of xenobiotics has not received extensive attention for aquatic vertebrates. Varanasi et al., (1978) produced suggestive evidence that naphthalene was absorbed across the dermis in rainbow trout. Similarly, Tovell et al. (1975) demonstrated that as much as 20% of the sodium lauryl sulfate burden found in goldfish was due to uptake across the dermis. From these and other studies it is apparent that the skin need not be a barrier to the external environment. The mucus layer above the skin and scales, important to the animal for environmental drag, osmoregulation, and disease/parasite protection, has also been shown to be affected by xeno–biotics and/or operative in the disposition of xenobiotics. Many metals such as lead, cadmium, zinc, and copper are known to induce increased mucus secretions (Heath, 1987). Additionally, other studies have demonstrated the significant uptake of selenium (Kleinow and Brooks, 1986) as well as other xenobiotics into the mucus layer of fish (Part and Lock, 1983). Although the toxicological implications of these processes are poorly understood, such data shows that the skin, scales, and mucus layer in fishes may be significant contributors to the whole body uptake, turnover, and accumulation of aquatic contaminants.

2. Food-Borne Contaminants

Ingestion may be an important route of contaminant uptake for aquatic organisms. This process is heavily dependent upon physiological determinants such as gastrointestinal pH, gastric emptying rate, intestinal motility, intestinal surface area, circulatory flow, enterohepatic recirculation, permeability of the mucosa, and completeness of digestion. Other features such as the nature of the food matrix, the amount of food ingested, and the characteristics of the ingested contaminants interplay with gross physiological as well as molecular transport considerations (diffusion, facilitated transport, pinocytosis, filtration) to determine the extent and nature of contaminant uptake from the diet.

Matrix composition (variations in food or sediment type) through differences in the strength of binding to the matrix, among other factors, can cause variations in assimilation. Assimilation coefficients as low as 3% have been demonstrated for phenolic compounds and related constituents in paper mill effluent (Niimi et al., 1990). More often, higher assimilations for lipophilic materials have been reported. Polychlorinated dibenzofurans, for example, exhibited assimilations of 41 to 44% (Muir et al., 1990), while other compounds have shown assimilations as high as 80 to 90% (Niimi and Cho, 1981).

Feeding, as a process, can directly impact the assimilation of xenobiotics. Studies with guppies have shown that an increase in xenobiotic concentration in a constant food matrix resulted in a linear increase in the assimilation of material from the GI tract (Clark and MacKay, 1991). Other investigations have demonstrated that the amount of food ingested (ration) may directly affect assimilations across gastric epithelia regardless of egestion rate (Dauble and

Curtis, 1990). In these studies, when a large ration was given, the stomach pH was buffered by the food. These conditions increased the amount of un-ionized compound, thereby facilitating the uptake from the stomach. Dietary preexposure can also impact the assimilation process. For example, when fathead minnows were pretreated with a dietary selenite exposure they exhibited a reduction in [75]Se[-] selenite retention upon subsequent ingestion (Kleinow and Brooks, 1986). Likewise, starvation can also produce significant interactions with the process of uptake by effecting processes such as biliary flow (Goodrich et al., 1991), intestinal integrity, and lipid metabolism (Boon and Duinker, 1985).

Aquatic invertebrates exhibit a wide diversity in digestive structures and function (Wetzel, 1975; Merrit and Cummins, 1978). Pelagic invertebrates such as the daphnids, copepods, and rotifers are primarily herbivores, filter feeding upon the seston which is comprised of both living and dead suspended particles. In any given season, up to 65% of the seston consists of dead algal cells with the remaining fraction composed largely of fecal material and allochthonous DOC, (Wetzel, 1975). Much of this material is considered dead or nonliving, yet the majority of these particles have a layer of bacteria, fungi, and/or protozoans utilizing the organic carbon fraction of the particle. The particles, although nonliving at their center, are alive on the outside and attractive food items for the pelagic grazers. Thus, the same high organic particles attract unicellular and zooplanktonic grazers, as well as lipophilic xenobiotics.

Most zooplankton grazers are not obligated to ingest the particles they have filtered. *Daphnia* are known to taste their food and some copepods have adapted chemical defenses that involve altering their taste to predators. There is little information on the selection or rejection of contaminated foodstuffs in zooplankton and very limited information in fishes. Such avoidance or attraction for contaminated foodstuffs may have significant impact on the trophic exchange of xenobiotics.

Food selection by benthic invertebrates as it relates to the bioaccumulation process has received more attention than in the pelagic invertebrates. Annelid worms (e.g., oligochaetes) are known to select sediment particles rich in organic content as do many of the larval insects (e.g., chironomids) (Klump et al., 1987) and sediment bedded clams (Boese et al., 1990). Experiments conducted to determine the bioaccumulation of xenobiotics (mostly chlorinated hydrocarbons and hydrophobic insecticides) have shown that in the selection of organic rich particles these animals also select particles with the highest levels of contamination (Fenchel et al., 1975). Conversely, there is some evidence that some of these creatures will avoid contaminated sediments. Both oligochaetes and chironomids have been shown to avoid sediments heavily contaminated with metals (Wentsel et al., 1977; McMurty, 1984).

**TABLE 13.1. Characteristics of Phase I and Phase II
Biotransformation Reactions**

Phase I	Phase II
Nonsynthetic	Synthetic
Reducing equivalents (NADPH) required	Requires energy and cofactors which act as high energy intermediates
Unmasks, adds, or alters functional groups	Uses endogenous molecules in conjugations

D. TOXICANT METABOLISM AND DEPOSITION

1. Biotransformation in Aquatic Organisms

Aquatic species have the ability to chemically alter xenobiotics to which they are exposed. This enzymatically mediated process referred to as biotransformation can result in the formation of products (metabolites) which may have different physicochemical properties from the parent compound. Biotransformation reactions can thus influence the distribution, accumulation, persistence, and toxicity of chemicals in both the environment as well as in the organisms.

Aquatic species are capable of biotransforming both endogenous and xenobiotic chemicals, much like their mammalian counterparts. While studies in aquatic species have qualitatively identified similar biotransformation pathways as in mammals, the relative contribution of a given pathway, reaction rates, and the products formed may differ. Biotransformation reactions can be classified as phase I or phase II according to the characteristics of the reaction as found in Table 13.1.

Phase I biotransformations are functionalization reactions adding, uncovering, or altering functional groups by oxidative, hydrolytic, or reductive pathways. Oxidative biotransformation reactions, the best characterized of these pathways in aquatic species, are largely catalyzed by a family of cytochromic enzymes referred to as cytochrome P450. Much of the current knowledge about cytochrome P450 in aquatic species has come from studies with fish. Successes in the purification of cytochrome P450 from rainbow trout (Williams and Buhler, 1982), scup (Klotz et al., 1983), and cod (Goksoyr, 1985) have confirmed earlier data, suggesting that like mammals, fish have multiple P450 forms. These isozymes, while offering polysubstrate capabilities, do exhibit some specificity in product formation and response to inducers. In general, when contrasted with mammalian P450, fish cytochrome P450 with associated reductases exhibit: (1) low specific activities, (2) ideal temperature compensation, (3) lower temperature optima, and (4) lack of response to phenobarbital-like inducers (Kleinow et al., 1990).

In addition to fish, P450 enzymes have been shown to be operative in aquatic plants, molluscs, crustacea (Foureman et al., 1978; Lee, 1981), in addition to fish, and other aquatic species. Of these groups, crustacean species exhibit one of the widest ranges in ability to monooxygenate chemicals. Benzo(a)pyrene, for example, is metabolized readily by the Florida spiny lobster (Little et al., 1985) but very slowly by the Maine lobster (Foureman et al., 1978). Molluscs, on the other hand, maintain as a group very low P450 dependent monooxygenase activities (Mix, 1984). This may account in part for high chemical burdens in these organisms.

Phase II reactions are responsible for the conjugation of endogenous compounds to polar functional groups such as COOH, NH_2, and OH. These functional groups, native on the parent molecule or as the result of phase I biotransformations, provide a functional handle for phase II reactions. Glycosylation, sulfation, acetylation, amino acid conjugation, and mercapturic acid formation are some of the more important phase II reactions reported in aquatic species. The pathways of conjugation utilized for any given compound and organism depend on the functional groups present, stearic considerations, and substrate concentration. Significant species differences exist in the relative contribution of each pathway for a given compound.

Glycosylation, an important phase II conjugation pathway, may occur for organic molecules containing carboxylic acid groups, aromatic or aliphatic hydroxyl groups, as well as thiol and nitrogen groups. The addition of glucuronic acid (glucuronidation) to xenobiotics appears as the major pathway of glycosylation in teleost fish whereas glucosidation has been demonstrated as an important pathway in invertebrates (James, 1987). Evidence in trout for inducibility, multiple forms, and substrate specificity (Andersson et al., 1985) exist for UDP-glucuronosyl transferase, the enzyme involved in glucuronidation.

Sulfation of aliphatic and aromatic hydroxyl groups has been demonstrated for a variety of fish species including guppy (Layiwola and Linnecar, 1981) and flounder (Gmur and Varanasi, 1982). Similarly, sulfate conjugates have been identified as major metabolites for a variety of invertebrates including shrimp (Sanborn and Malins, 1980), chiton (Landrum and Crosby, 1981), and lobster (Elmamlouk and Gessner, 1978). Significant variation in the extent of conjugation has been demonstrated to exist between aquatic species.

Acetylation of xenobiotics by vertebrate and invertebrate aquatic species has been demonstrated largely with compounds which are aryl or alkylamines (reviewed in James, 1987). Xenobiotics on this list include a variety of sulfonamides: sulfanilimide, sulfamethazole, and sulfadimethoxine; aminobenzoic acid derivatives including tricaine methane sulfonate, the widely used fish anesthetic, and nitroanisole. The acetylation of amino groups often produces a metabolite less polar than the parent compound.

Amino acid conjugation has been demonstrated in marine fish and crustacea. These reactions form a peptide linkage with carboxylic acid groups of xenobiotics.

The only rigorously identified metabolites of amino acid conjugates found so far in aquatic animals are the taurine conjugates (James, 1987).

Glutathione, a tripeptide consisting of cysteine, glutamine, and glycine is often conjugated with xenobiotics containing electrophilic centers. Once the conjugate is formed, other products may be formed from the cleavage of the cys-gly peptide bond and *N*-acetylation of the cysteinyl amino group. When all of the ensuing reactions occur, the end product is referred to as mercapturic acid. Glutathione, mercapturic acid, and other intermediate products of this enzymatic cascade have been found in bile, urine, and excreta of a wide variety of aquatic invertebrates and vertebrates (James, 1987). Glutathione-*S*-transferase, the enzyme responsible for the formation of the glutathione conjugate, has been purified from a variety of aquatic species (Sugiyama et al., 1981; Ramage and Nimmo, 1984; Foureman and Bend, 1984). In each of these species multiple forms of the enzyme were found. Glutathione conjugation serves a vital role in conjugating reactive compounds before they can bind to tissue macromolecules and initiate a variety of toxic reactions. The capacity of glutathione conjugation can be exceeded and depleted resulting in toxicity of reactive compounds.

In general, both phase I and phase II, but especially phase II reactions, increase water solubility and in many cases detoxify compounds. Exceptions, however, do exist to these generalizations with the nonpolar nature of many *N*-acetylated conjugates and the lethal synthesis of toxic metabolites as prominent examples. Conjugation reactions also result in the formation of organic anions. These organic anions, besides exhibiting increased water solubility, have been shown to be excreted into urine by facilitated renal tubular transport in a number of fish and invertebrates (Pritchard and James, 1982).

2. Bioconcentration and Bioaccumulation

In general, xenobiotic compounds which demonstrate the greatest accumulative potential are those which are recalcitrant to environmental degradation, poor substrates for biotransformation, and demonstrate lipid solubility. Uptake of chemicals into aquatic organisms may occur by a variety of portals including direct exposure via the water, surface contact with contaminated substrates, or ingestion of contaminated food and water. Technically, the term bioconcentration has been coined for the net assimilation of xenobiotics solely from water exposure. This is in contrast with bioaccumulation which generically refers to assimilation by all routes of possible exposure.

While the nature of the food chain and contaminant availability are important modulators of bioaccumulation, the physicochemical properties of the chemical are an overriding feature in both bioaccumulation and bioconcentration. Lipid solubility is paramount in this regard. Octanol/water partition coefficients (K_{ow}) are one laboratory means commonly employed to predict the ability of a chemical to accumulate in aquatic organisms (Veith et al., 1979). This coefficient, which essentially measures lipid solubility or nonpolar media solubility, is the ratio of the chemical soluble in the solvent octanol over that found in water (Chiou, 1981). In a similar regard, bioconcentration factors (BCFs) for very lipophilic compounds are in a true sense partition coefficients in that they represent the ratio of the compound between the highly lipid based internal environment of an organism and the water-based external environment. Bioconcentration factors determined by a variety of exposure and kinetic techniques have been produced for a wide array of chemicals in numerous organisms. Often these unitless coefficients are calculated from the steady state concentration of the xenobiotic within an organism divided by the environmental (e.g., water) concentration (Ernest, 1977). Direct correlations have been established between the octanol/water partition coefficient and an organism's BCF for the same material (Veith et al., 1980). Such relationships have been best illustrated by hydrophobic materials. The higher the chemical octanol/water partition coefficient, the higher the BCF. More recent work (McKim et al., 1985) has shown that this relationship between K_{ow} and BCF is functional only within a specific range of K_{ow} values. At K_{ow} below 1, the chemicals may not easily pass across membranes, due mainly to low lipid solubility; those with very high K_{ow} values may become bound to the membrane due to their very high affinity to lipids (Heath, 1987).

Besides lipid solubility, biotransformation has been shown to have a profound effect upon xenobiotic retention and bioaccumulation. The bioconcentration factors in *Gambusia affinis* for DDT and a closely related analog 2-bis(*p*-methylthiophenyl)l,l,l-trichloroethane have been shown to be 84,500 and 5.5, respectively (Kapoor et al., 1973). While these compounds are structurally related and demonstrate similar lipid solubilities, 2-bis-(*p*-methylthiophenyl)-1,1,1-trichloroethane was shown to be readily metabolized. Similarly, when experimental BCFs (35) for an alkylbenzene compound in bluegills were compared to predicted BCFs based on solubility considerations (6300), a large discrepancy existed (Werner and Kimerle, 1982). This could be attributed to metabolism.

Hand in hand with biotransformation is the environmental persistence of bioaccumulative toxicants. Perhaps the best known group of persistent environmental toxicants are the polychlorinated biphenyls (PCB). On examination, this group of over 209 congeners exhibit varying degrees of metabolism, induction potential, and environmental persistence. In large part, these characteristics are imparted by the degree and position of chlorination on the molecule (Safe, 1985). For example, the hexachlorinated biphenyl 2,4,5,-2',4',5' is recalcitrant to metabolism and very persistent in the environment. While with only two chlorines less than the hexachlorinated biphenyl, the tetrachlorinated biphenyl 3,4,-3',4' is metabolized and less persistent in the environment. The net effect is that, with time both in the environment and in organisms, the predominate PCB isomers available for and contributing to bioaccumulation are those which resist degradation.

3. Biomagnification

Biomagnification was a term coined in the 1960s for the observation that higher body burdens of certain contaminants such as DDT were found in successive levels in the food chain. It was inferred that these xenobiotics were conservatively transferred up trophic levels while relative community biomass followed a pyramid-like ascension. There has been considerable controversy about the existence or at least the importance of this process. A variety of studies with a number of compounds thought to biomagnify have demonstrated that tissue levels may result directly from water partitioning independent of dietary sources (Ellgehausen et al., 1980; Shaw and Connell, 1986). Likewise, laboratory studies employing short and single pathway food chains have demonstrated bioaccumulation (Bennett et al., 1986; Bertram and Brooks, 1986; Clark and MacKay, 1991).

Much of the emphasis on biomagnification today is on developing mathematical models based on either laboratory BCF/K_{ow} data and validation of the models by correlation to environmental data (Thomann, 1981; Rasmussen, et al., 1990; Environment Canada, 1991; Fordham and Reagan, 1991). Mechanistic models by Thomann (1989) and Thomann et al., (1986), considering exchange across the gills and via the food, suggest that for contaminants by which assimilation exceeds elimination biomagnification is a distinct possibility. Such models have suggested that 90 to 99% of contaminants in Lake Michigan lake trout represent concentration via the food chain. Additional studies by Oliver and Niimi (1988) have shown that bioconcentration factors for PCBs in Lake Ontario salmonids far exceed those expected from direct uptake from water (MacKay, 1982). Furthermore, these studies have shown that, as one progresses through the pelagic food chain, a greater divergence from the $BCF-K_{ow}$ line occurs with each trophic level independent of variations in lipid content. More recently, field studies examining lake trout from PCB contaminated lakes with varying trophic structures indicated that lake trout PCB concentrations increased with food chain length and lipid content (Rasmussen et al., 1990). A 3.5-fold biomagnification factor was correlated for each successive trophic level. There is still a great deal of uncertainty inherent in these models, for there are several confounding factors. Primary to the uncertainty is that the response of an ecosystem or community to xenobiotic contamination is not well understood (Cairns and Niederlehner, 1989). The effects of biotransformation and preexposure to other sublethal stressors (chemical or physiological) on the bioaccumulation process can cause difficulties in the interpretation of field data. Furthermore, these models are based on equilibrium equations while an ecosystem or the environment need not be at equilibrium. It is clear that additional studies will be necessary to further delineate the importance of biomagnification.

E. ENVIRONMENTAL TEMPERATURE

Environmental temperature has a pervasive influence on all aspects of aquatic organism physiology. From intestinal clearance time to muscle function to fatty acid profiles, temperature may significantly modify the physiology of poikilothermic organisms. Xenobiotic biotransformation, enzyme induction, residue retention, and toxicity are not immune to these effects. Perhaps the most well known and studied effects of temperature in aquatic toxicology are those upon toxicity and residue retention. Acute exposures at higher temperatures result in greater toxicity of some toxicants such as zinc, while chronic exposures result in a greater tolerance at high temperatures (Hodson and Sprague, 1975). Other contaminants exhibit little change in toxicity with temperature or even greater toxicity with cold temperatures (Brown et al., 1967). The mechanisms are often not definitively known for these temperature-related differences. For certain toxicants, however, decreased oxygen carrying capacity of warm water, as well as temperature effects on induction (James and Bend, 1980) and biotransformation (Koivusarri et al., 1981), have been shown to have a contributing effect. The complexity of temperature modulated interactions with processes dictating absorption, biotransformation, elimination, and toxicity make it difficult to predict temperature effects. Overall, there are no general statements which can be formulated regarding temperature and toxicity.

Xenobiotic residue retention can be dramatically influenced by acclimation temperature. For many xenobiotics in fish, an inverse relationship exists between environmental temperature and xenobiotic retention. While this phenomenon has been demonstrated with environmental contaminants such as naphthalene (Collier et al., 1978, Varanasi et al., 1981), the bulk of the supporting evidence has been developed with aquaculture drugs (Salte and Liestol, 1983; Kasuga et al., 1984). With these studies it is clear that significant differences exist between compounds in regards to residue retention. Few temperature-related residue studies are available for representative organisms of other trophic levels.

Several interesting studies have examined the influence of temperature on P450-mediated monooxygenase activity. When P450 activity was measured *in vitro* at the animals acclimation temperature, fish of disparate temperature profiles demonstrated similar activities. However, when *in vitro* assays were performed at an identical nonacclimation temperature, those animals acclimated to colder temperatures exhibited greater monooxygenase activities than those which were acclimated to warmer temperatures, (Koivusarri et al., 1981). Collectively, the studies in this area suggest that fish P450 activity responds to acclimation temperature in a compensatory manner. Studies examining the influence of temperature on phase II pathways, however, appears to indicate a stronger temperature dependency. Studies by Koivusarri (1983) show that if UDP-glucuronosyltransferase activity is measured at environmental temperatures, colder conditions resulted in lower activities.

Similar results have been demonstrated for the antimicrobial ormetoprim in catfish (personal communication; Kleinow, 1992). In these investigations warm-acclimated animals exhibited a selective formation of a metabolite which was much less prevalent on both an absolute and relative basis in cold-acclimated animals.

Temperature may also influence the time course of xenobiotic absorption and distribution. Several studies have suggested that higher tissue levels of xenobiotics were attained early on for warmer compared to cold temperatures. Similarly, reduced absorption rates have been reported for inducing agents with cold temperatures (James and Bend, 1980). This may in part be responsible for the increased time necessary for maximal monooxygenase induction at lower temperatures (James and Bend, 1980; Egaas and Varanasi, 1982; Andersson and Koivusarri, 1985).

F. CONCLUSIONS

This chapter provided a preliminary overview of aquatic toxicology. By necessity, many details, processes, points of interest, and important topics were omitted or lightly covered. It is the hope of the authors that the information presented will provide the contextual framework to which detail and complexity may be added.

The excitement and challenge of environmental aquatic toxicology is the complexity of interaction. As a discipline, aquatic toxicology includes diverse topics ranging from molecular biology to ecosystems, from geochemistry to statistics, from physical oceanography to limnology, invertebrate zoology to phycology, and so on. The point being, there are many active components of aquatic toxicology. What constitutes the field is a matter of semantics rather than reality. Toxicity, after all, is the endpoint. For that to be operative, many levels of interaction are encountered. Organisms react with chemicals on the molecular level and chemicals may affect organisms on the population level. Likewise, chemical interaction with the environment occurs on the particle and lake level. The toxicity of chemicals not only affects the biota, but through their toxicity, influence the normal processing events of the environment. Upon examination of aquatic toxicology as a field of study, area of research, or regulatory application, an underlying principle should be kept in mind: the toxicity of chemicals in the aquatic environment is intimately interactive and a collective event of biotic, abiotic, and physicochemical interactions.

REFERENCES

Andersson, T. and Koivusarri, U. Influence of environmental temperature on the induction of xenobiotic metabolism by β-naphthoflavone in rainbow trout, *Salmo gairdneri, Toxicol. Appl. Pharmacol.* 80:43–50, 1985.

Andersson, T., Personen, M., and Johansson, C. Differential induction of cytochrome P-450 dependent monooxygenase, epoxide hydrolase, glutathione transferase and UDP glucuronosyltransferase activities in the liver of the rainbow trout by β-naphthoflavone or clophen A 50, *Biochem. Pharmacol.* 34:3309–3314, 1985.

Bennett, W.N., Brooks, A.S., and Boraas, M.E. Selenium uptake and transfer in an aquatic food chain and its effects on fathead minnow larvae, *Arch. Environ. Contam. Toxicol.* 15:513–517, 1986.

Bertram, P.E. and Brooks, A.S. Kinetics of accumulation of selenium from food and water by fathead minnows, *Water Res.* 20:877–884, 1986.

Boese, B.L., Lee H., II, Specht, D.T., Randall, R.C., and Winsor, M.L. Comparison of aqueous and solid-phase uptake for hexachlorobenzene in the tellinid clam *Macoma Nasuta* (Conrad): A mass balance approach, *Environ. Toxicol. Chem.* 9:221–231, 1990.

Boon, J.P. and Duinker, J.C. Kinetics of polychlorinated biphenyl (PCB) components in juvenile sole *(Solea solea)* in relation to concentrations in water and to lipid metabolism under conditions of starvation, *Aquat. Toxicol.* 7:119–134, 1985.

Boyle, E.A., Sclater, F.R., and Edmond, J.M. On the marine geochemistry of cadmium, *Nature* (Lond). 263:42–44, 1976.

Brown, V.M., Jordan, D.H.M., and Tiller, B.A. The effect of temperature on the acute toxicity of phenol to rainbow trout in hard water, *Water Res.* 1:587–594, 1967.

Bruland, K.W., Knauer, G.A., and Martin, J.H. Zinc in northeast Pacific water, *Nature* (Lond). 271:741–743, 1978.

Cairns, J., Jr. and Niederlehner, B.R. Adaptation and resistance of ecosystems to stress: a major knowledge gap in understanding anthropogenic perturbations, *Specul. Sci. Technol.* 12:23–29, 1989.

Chiou, C.T. Partition coefficient and water solubility in environmental chemistry, in *Hazard Assessment of Chemicals: Current Developments.* Vol. 1. Academic Press., New York, 1981, 117–153.

Chiou, C.T., Peters, L.J., and Freed, V.H. A physical concept of soil-water equilibria for nontoxic organic compounds, *Science* 206:831–832, 1979.

Clark, K.E. and MacKay, D. Dietary uptake and biomagnification of four chlorinated hydrocarbons by guppies, *Environ. Toxicol. Chem.* 10:1205–1217, 1991.

Collier, T.K., Thomas, L.C., and Malins, D.C. Influence of environmental temperature on deposition of dietary naphthalene in coho salmon *(Oncorhynchus kisutch):* isolation and identification of individual metabolites, *Comp. Biochem. Physiol.* 61C:23–28, 1978.

Cooper, W.J. Sunlight-induced photochemistry of humic substances in natural water: major reactive species, in *Aquatic Humic Substance, Infuences on Fate and Treatment of Pollutants.* I.H. Suffett and P. MacCarthy, Eds. American Chemical Society, Washington, DC, 1989, 333–362.

Dauble, D.D. and Curtis, L.R. Influence of digestive processes on the absorption and fate of quinoline ingested by rainbow trout *(Oncorhynchus mykiss), Environ. Toxicol. Chem.* 9:505–512, 1990.

DiToro, D.M., Horzempa, L.M., Casey, M.M., and Richardson, W. Reversible and resistant components of PCB adsorption-desorption: adsorption concentration effects, *J. Great Lakes Res.* 8:336–349, 1982.

Eadie, B.J., Faust, W.R., Landrum, P.F., and Morehead, N.R. Factors affecting bioconcentration of PAH by the dominant benthic organisms of the Great Lakes, in *Polynuclear Aromatic Hydrocarbons: Eighth International Symposium on Mechanisms, Methods and Metabolism,* M.W. Cooke and A.J. Dennis, Eds. Battelle Press, Columbus, OH, 1985, 363–377.

Egaas, E. and Varanasi, U. Effects of polychlorinated biphenyls and environmental temperature on *in vitro* formation of benzo(a)pyrene metabolites by liver of trout *(Salmo gairdneri), Biochem. Pharmacol.* 31:561–566, 1982.

Eisenreich, S.J. The chemical limnology of nonpolar contaminants: polychlorinated biphenyls in Lake Superior, in *Sources and Fates of Aquatic Pollutants.* R.A. Hites and S.J. Eisenreich, Eds. American Chemical Society, Washington, DC, 1987, 393–469.

Ellgehausen, G., Guth, J.A., and Esser, H.O. Factors determining the bioaccumulation potential of pesticides in the individual compartments of aquatic food chains, *Ecotoxicol. Environ. Saf.* 4:134, 1980.

Elmamlouk, T.H. and Gessner, T. Carbohydrate and sulfate conjugations of *p*-nitrophenol by hepatopancreas of *Homarus americanus, Comp. Biochem. Physiol.* 61C:363–367, 1978.

Elzerman, A.W. and Coats, J.T. Hydrophobic organic compounds on sediments: equilibria and kinetics of sorption, in *Sources and Fates of Aquatic Pollutants.* Advances in Chemistry Series 216, American Chemical Society, Washington, DC, 1987, 263–318.

Environment Canada, Department of Fisheries and Oceans, and Health and Welfare Canada. *Synopsis, Toxic Chemicals in the Great Lakes and Associated Effects.* Minister Supply and Services Canada. EN 37-94/1990 E, 1991, 51.

Erickson, R.J. and McKim, J.M. A simple flow-limited model for exchange of organic chemicals at fish gills, *Environ. Toxicol. Chem.* 9:159–165, 1990.

Ernest, W. Determination of the bioconcentration potential of marine organisms, *Chemosphere* 11:731–740, 1977.

Fagerstrom, T. and Jernelov, A. Some aspects of the quantitative ecology of mercury, *Water Res.* 6:1193–1202, 1972.

Fenchel, T.M., Kofoed, L.H., and Lappalainen, A. Particle size — selection of two deposit feeders: the amphipod *Corophium volutaton* and the prosobranch *Hydrobia ulvae, Marine Biol.* 30:119–128, 1975.

Fordham, C.L. and Reagan, D.P. Pathways analysis for estimating water and sediment criteria at hazardous waste sites, *Environ. Toxicol. Chem.* 10:949–960, 1991.

Foureman, G.L. and Bend, J.R. The hepatic glutathione transferases of the male little skate, *Raja erinacea, Chem. Biol. Interact.* 49:89–103, 1984.

Foureman, G.L., Ben-Zvi, Z., Dostal, L., Fouts, J.R., and Bend, J.R. Distribution of ^{14}C-benzo(a)pyrene in the lobster, *Homarus americanus,* at various times after a single injection into the pericardial sinus, *Bull. Mt. Desert Isl. Biol. Lab.* 18:93–97, 1978.

Gass, T.E. To what extent is groundwater contaminated, *Water Well J.*, 34:26–27, 1980.

Gmur, G.J. and Varanasi, U. Characterization of benzo(a)pyrene metabolites isolated from muscle, liver, and bile of a juvenile flatfish, *Carcinogenesis* 3:1397–1403, 1982.

Goksoyr, A. Purification of hepatic microsomal cytochromes P450 from β-naphthoflavone-treated Atlantic cod *(Gadus morhua),* a marine teleost fish, *Biochem. Biophys. Acta.* 840:409–417, 1985.

Goodrich, M.S., Melancon, M., Davis, R., and Lech, J.J., The toxicity, metabolism and elimination of dioctyl sulfosuccinate (DSS) in rainbow trout *(Oncorhynchus mykiss), Water Res.* 25:119–124, 1991.

Griffin, R.A. and Chou, S.F.J. Attenuation of polybrominated biphenyls and hexachlorobenzene by earth materials. *Environ. Geol. Notes, Ill. State Geol. Surv.* Urbana, IL, 87:87–92, 1980.

Hassett, J.P. and Milicic, E. Determination of equilibrium and rate constants for binding of a polychlorinated biphenyl congener by dissolved humic substances, *Environ. Sci. Technol.* 19:638–643, 1985.

Hayton, W.L. and Barron, M.G. Rate-limiting barriers to xenobiotic uptake by the gill, *Environ. Toxicol. Chem.* 9:151–157, 1990.

Heath, A.G. *Water Pollution and Fish Physiology.* CRC Press, Boca Raton, FL, 1987, 245.

Hodson, P.V. and Sprague, J. B. Temperature-induced changes in acute toxicity of zinc to Atlantic salmon *(Salmo salar), J. Fish Res. Bd. Can.* 32:1–10, 1975.

James, M.O. Conjugation of organic pollutants in aquatic species, *Environ. Health Perspect.* 71:97–103, 1987.

James, M.O. and Bend, J.R. Polycyclic aromatic hydrocarbon induction of cytochrome P-450 dependent mixed function oxidases in marine fish, *Toxicol. Appl. Pharmacol.* 54:117–133, 1980.

Kapoor, I.P., Metcalf, R.L., Hirwe, A.S., Coats, J.R., and Khalsa, M.S. Structure activity correlations of biodegradability of DDT analogs, *J. Agric. Food Chem.* 21:310–315, 1973.

Karickhoff, S.W., Brown, D.S., and Scott, T.A. Sorption of hydrophobic pollutants on natural sediments, *Water Res.* 13:241–248, 1979.

Karickhoff, S.W. Sorption kinetics of hydrophobic pollutants in natural sediments, in *Contaminants and Sediments*, Vol. 2. R.A. Baker, Ed. Ann Arbor Science, Ann Arbor, MI, 1980, 193–206.

Karickhoff, S.W. and Morris, K.R. Impact of tubificid oligochaetes on pollutant transport in bottom sediment, *Environ. Sci. Technol.* 19:51–56, 1985.

Kasuga, Y., Sngitani, A., Yamada, F., Arai, M., and Morikawa, S. Oxolinic acid residues in tissues of cultured rainbow trout and ayu fish, *J. Food Hyg. Soc. Jpn.* 25:512–516, 1984.

Kleinow, K.M. and Brooks, A.S. Selenium compounds in the fathead minnow *(Pimephales promelas).* II. Qualitative approach to gastrointestinal absorption, routes of elimination and influence of dietary pretreatment, *Comp. Biochem. Physiol.* 83:71–76, 1986.

Kleinow, K.M., Haasch, M.L., Williams, D.E., and Lech, J.J. A comparison of hepatic P450 induction in rat and trout *(Oncorhynchus mykiss):* Delineation of the site of resistance of fish to phenobarbital-type inducers, *Comp. Biochem. Physiol.* 96C:259–270, 1990.

Kleinow, K.M. School of Veterinary Medicine, Louisiana State University, Baton Rouge, LA 70803 (504) 346-3256. Personal communication, 1992.

Klotz, A.V., Stegeman, J.J., and Walsh, C. An aryl hydrocarbon hydroxylating hepatic cytochrome P-450 from the marine fish *Stenotomus chrysops, Arch. Biochem. Biophys.* 226:578–592, 1983.

Klump, J.V., Krezoski, J.R., Smith, M.E., and Kaster, J.L. Dual tracer studies of the assimilation of an organic contaminate from sediments by deposit feeding oligochaetes, *J. Great Lake Res.* 10:267–272, 1987.

Knezovitch, J.P. and Harrison, F.L. The bioavailability of sediment-sorbed chlorobenzene to larvae of the midge, *Chironomus decorus, Ecotox. Environ. Safety* 15:226–241, 1988.

Koivusarri, U., Harri, H., and Hanninen, O. Seasonal variation of hepatic biotransformation in female and male rainbow trout*(Salmo gairdneri), Comp. Biochem. Physiol.* 70:149–157, 1981.

Koivusarri, U. Thermal acclimatization of hepatic polysubstrate monooxygenase and UDP-glucuronosyl-transferase of mature rainbow trout *(Salmo gairdneri), J. Exp. Zool.* 227:35–42, 1983.

Landrum, P.F. Bioavailability and toxicokinetics of polycyclic aromatic hydrocarbons sorbed to sediments for the amphipod, *Pontoporeia hoyi, Environ. Sci. Technol.* 23:588–595, 1989.

Landrum, P.F. and Crosby, D.G. Comparison of the disposition of several nitrogen-containing compounds in the sea urchin and other marine invertebrates, *Xenobiotica.* 11:351–361, 1981.

Landrum, P.F., Reinhold, M.D., Nihart, S.R., and Eadie, B.J. Predicting the bioavailability of organic xenobiotics to *Pontoporeia hoyi* in the presence of humic and fulvic materials and natural dissolved organic matter, *Environ. Toxicol. Chem.* 4:459–467, 1985.

Landrum, P.F. and Robbins, J.A. Bioavailability of sediment-associated contaminants to benthic invertebrates, in *Sediments: Chemistry and Toxicity of In Place Pollutants.* R. Baudo, J.P. Giesy, H. Muntau, Eds. Lewis Publishers, Ann Arbor, MI, 1990, 237–257.

Landrum, P.F. and Stubblefield, C.R. Role of respiration in the accumulation of organic xenobiotics by the amphipod *Diporeia* sp. *Environ. Toxicol. Chem.* 10:1019–1028, 1991.

Layiwola, P.J. and Linnecar, D.F.C. The biotransformation of [14C] phenol in some freshwater fish, *Xenobiotica* 11:167–171, 1981.

Lee, R.F. Mixed function oxygenases in marine invertebrates, *Marine Biol. Lett.* 2:87–105, 1981.

Leversee, G.J., Landrum, P.F., Giesy, J.P., and Fannin, T. Humic acids reduce bioaccumulation of some polycyclic aromatic hydrocarbons, *Can. J. Fish. Aquat. Sci.* 40 (Suppl. 2):63–69, 1983.

Little, P.J., James, M.O., Pritchard, J.B., and Bend, J.R. Temperature dependent disposition of [14C] benzo(a)pyrene in the spiny lobster *(Panulirus argus), Toxicol. Appl. Pharmacol.* 77:325–333, 1985.

MacKay, D. Correlation of bioconcentration factors, *Environ. Sci. Technol.* 16:274–278, 1982.

Malins, D.C., McCain, B.B., Brown, D.W., Chan, S.L., Myers, M.S., Landahl, J.T., Prohaska, P.G., Friedman, A.J., Rhodes, L.D., Burrows, D.G., Gronlund, W.D., and Hodgins, H.O. Chemical pollutants in sediments and diseases of bottom-dwelling fish in Puget Sound, Washington, *Environ. Sci. Technol.* 18:705–713, 1984.

McKim, J.M. and Goeden, H.M, A direct measure of the uptake efficiency of a xenobiotic across the gills of brook trout (*Salvelinus fontinales*) under normoxic and hyposic conditions, *Comp. Biochem. Physiol.* 72:65–74, 1982.

McKim, J., Schmieder, P., and Veith, G. Absorption dynamics of organic chemical transport across trout gills as related to octanol-water partition coefficient, *Toxicol. Appl. Pharmacol.* 77:1–10, 1985.

McMurty, J.M. Avoidance of sublethal doses of copper and zinc by tubificid oligochaetes, *J. Great Lakes Res.* 10:267–272, 1984.

Means, J.C., Hassett, J.J., Wood, S.G., and Banwart, W.L. Sorption properties of energy-related pollutants and sediments, in *Polynuclear Aromatic Hydrocarbons.* P.W. Jones and P. Leber, Eds. Ann Arbor Science Publ. Inc., Ann Arbor, MI, 1979, 327–340.

Means, J.C., Wood S.G., Hassett, J.J., and Banwart, W.L. Sorption of amino and carboxy-substituted polynuclear aromatic hydrocarbons by sediment and soil, *Environ. Sci. Technol.* 16:93–98, 1982.

Merrit, R.W. and Cummins, K.W. *An Introduction to the Aquatic Insects of North America.* Kendall/Hunt Publ., Dubuque, IA, 1978, 1–722.

Mix, M.C. Polycyclic aromatic hydrocarbons in the aquatic environment: occurrence and biological monitoring, in *Reviews in Environmental Toxicology* Vol. 5. E. Hodgson, Ed. Elsevier/North Holland, 1984, 51–102.

Muir, D.C.G, Yarechewski, A.L., Metner, D.A., Lockhart, W.L., Webster, G.R.B., and Friesen, K.J. Dietary accumulation and sustained hepatic mixed function oxidase enzyme induction by 2,3,4,7,8-pentachlorodibenzofuran in rainbow trout, *Environ. Toxicol. Chem.* 9:1463–1472, 1990.

Nalepa, T.F. and Landrum, P.F. Benthic invertebrates and contaminant levels in the Great Lakes: effects, fates and role in cycling, in *Toxic Contaminants and Ecosystem Health: A Great Lakes Focus.* M.S. Evans, Ed. John Wiley & Sons, New York, 1988, 77–102.

Neff, J. Bioaccumulation of organic micropollutants from sediments and suspended particulates by aquatic animals, *Fresenius Z. Anal. Chem.* 319:132–136, 1984.

Niimi, A.J. and Cho, C.Y. Uptake of hexachlorobenzene (HCB) by rainbow trout (*Salma gairdneri),* and an examination of its kinetics in Lake Ontario salmonids, *Can. J. Fish. Aquat. Sci.* 38:1350–1356, 1981.

Niimi, A.J., Lee, H.B., and Kisson, G.P. Kinetics of chloroguacols and other phenolic derivatives in rainbow trout (*Salmo gairdneri), Environ. Toxicol. Chem.* 9:649–653, 1990.

NOAA. A summary of selected data on chemical contaminants in tissues collected during 1984, 1985 and 1986, in *National Status and Trends Program for Marine Environmental Quality, Progress Report, Technical Memorandum,* NOS OMA 38, National Oceanic and Atmospheric Administration, Rockville, MD, 1987, 88.

Oliver, B.G. Biouptake of chlorinated hydrocarbons from laboratory — spiked and field sediments by oligochaete worms, *Environ. Sci. Technol.* 21:785–790, 1987.

Oliver, B.G. and Niimi, A.J. Trophodynamic analysis of polychlorinated biphenyl congeners and other chlorinated hydrocarbons in the Lake Ontario Ecosystem, *Environ. Sci. Technol.* 22:388–397, 1988.

OTA. Protecting the nation's groundwater from contamination. Vols. I and II. OTA-0-233 and OTA-0-276, Office of Technology Assessment, U.S. Congress, U.S. Government Printing Office, Washington, DC, 1984.

Part, P. and Lock, R.A.C.. Diffusion of calcium, cadmium and mercury in a mucous solution from rainbow trout, *Comp. Biochem. Physiol.* 76:259–263, 1983

Perdue, E.M. and Wolfe, N.L. Modification of pollutant hydrolysis kinetics in the presence of humic substances, *Environ. Sci. Technol.* 16:847–852, 1982.

Primuzic, E.T., Benkovitz, C.M., Gaffney, J.S., and Walsh, J.J. The nature and distribution of organic matter in the surface sediments of world oceans and seas, *Org. Geochem.* 4:63–77, 1982.

Pritchard, J.B. and James, M.O. Metabolism and urinary excretion, in *Metabolic Basis of Detoxication*. W.B. Jakoby, J.R. Bend, and J. Caldwell, Eds. Academic Press, New York, 1982, 339–357.

Ramage, P.I.N. and Nimmo, I.A. The substrate specificities and subunit compositions of the hepatic glutathione-*S*-transferase of rainbow trout *(Salmo garirdneri)*, *Comp. Biochem. Physiol.* 78B:189–194, 1984.

Rasmussen, J.B., Rowan, D.J., Lean, D.R.S., and Carey, J.H. Food chain structure in Ontario lakes determines PCB levels in lake trout *(Salvelinus namaychus)* and other pelagic fish, *Can. J. Fish. Aquat. Sci.* 47:2030–2038, 1990.

Richardson, E.M. and Epstein, E. Retention of three insecticides on different size soil particles suspended in water, *Soil Sci. Soc. Am. Proc.* 35:884–887, 1971.

Robbins, J.A. A model for particle-selective transport of tracers with conveyor belt deposit feeders, *J. Geophys. Res.* 91:8542–8558, 1986.

Sabljic, A. and Protic, M. Relationship betwen molecular connectivity indices and soil sorption coefficients of polycyclic aromatic hydrocarbons, *Bull. Environ. Contam. Toxicol.* 28:162–165, 1982.

Safe, S. Polychlorinated biphenyls (PCBs) and polybrominated biphenyls (PBBs): biochemistry, toxicology and mechanism of action, *CRC Crit. Rev. Toxicol.* 13:319–393, 1985.

Salte, R. and Liestol, K. Drug withdrawal from farmed fish. Depletion of oxytetracycline, sulfadiazine and trimethoprim from muscular tissue of rainbow trout *(Salmo gairdneri)*, *Acta Vet. Scand.* 24:418–430, 1983.

Sanborn, H.R. and Malins, D.C. The disposition of aromatic hydrocarbons in adult spot shrimp *(Pandalus platyceros)* and the formation of metabolites of naphthalene in adult and larval spot shrimp, *Xenobiotica* 10:193–200, 1980.

Shaw, G.R. and Connell, D.W. Factors controlling bioaccumulation in food chains, in *PCBs and the Environment*, Vol. I. CRC Press, Inc., Boca Raton, FL, 1986, 135–141.

Slonin, A.R. Acute toxicity of selected hydrazines to the common guppy, *Water Res.* 11:889–895, 1977.

Sugiyama, Y., Yamada, T., and Kaplowitz, N. Glutathione-*S*-transferase in elasmo-branch liver. Molecular heterogeneity, catalytic and binding properties and purification, *Biochem. J.* 199:749–756, 1981.

Thomann, R.V. Equilibrium model of fate of microcontaminants in diverse aquatic food chains, *Can. J. Fish. Aquat. Sci.* 38:280–296, 1981.

Thomann, R.V. Bioaccumulation model of organic chemical distribution in aquatic food chains, *Environ. Sci. Technol.* 23:699–707, 1989.

Thomann, R.V., Connolly, J.P., and Nelson, A.T. The Great Lakes ecosystem — modelling the fate of PCBs, in *PCBs and the Environment*, Vol. III. CRC Press, Inc., Boca Raton, FL, 1986, 153–180.

Tovell, P., Howes, D., and Newsome, C. Absorption, metabolism and excretion by goldfish of the anionic detergent sodium lauryl sulphate, *Toxicology* 4:17, 1975.

Tsuda, T., Aoki, S., Kojima, M., and Harada, H. The influence of pH on the accumulation of tri-*n*-butyltin chloride and triphenyltin chloride in carp, *Comp. Biochem. Physiol.* 95:151–154, 1990.

Varanasi, U., Uhler, M., and Stranahan, S. Uptake and release of napthalene and its metabolites in skin and epidermal mucus of salmonids, *Toxicol. Appl. Pharmacol.* 44:277–289, 1978.

Varanasi, U., Gmur, D.J., and Reicher, W.L. Effect of environmental temperature of naphthalene metabolism by juvenile starry flounder *(Platichtys stellatus)*, *Arch. Environ. Contam. Toxicol.* 10:203–214, 1981.

Veith, G.D., DeFoe, D., and Bergstedt, B. Measuring and estimating the bioconcentration factor of chemicals in fish, *J. Fish. Res. Bd. Can.* 36:1040–1048, 1979.

Veith, G.D., Macek, K., Petrocelli, S., and Carroll, J. An evaluation of using partition coefficients and water solubility to estimate bioconcentration factors for organic chemicals in fish, in *Aquatic Toxicology*. J. Eaton, P. Parrish, and A. Hendricks, Eds. ASTM STP707, American Society for Testing and Materials, Philadelphia, PA, 1980, 116–129.

Weber, W.J., Jr., Voice, T.C., Pirbazari, M., Hunt, G.E., and Ulanoff, D.M. Sorption of hydrophobic compounds by sediments, soils and suspended solids. II. Sorbent evaluation studies, *Water Res.* 17:1443–1452, 1983.

Wentsel, R., McIntosh, A.M., and McCafferty, W.P. Avoidance response of midge larvae *(Chironomus tentans)* to sediments containing heavy metals, *Hydrobiologia* 55:171–176, 1977.

Werner, A.F. and Kimerle, R.A. Uptake and distribution of C_{12} alkylbenzene in bluegill *(Lepomis macrochirus)*, *Environ. Toxicol. Chem.* 1:143–146, 1982.

Wershaw, R.L., Burcar, P.J., and Goldberg, M.C. Interaction of pesticides with natural organic material, *Environ. Sci. Technol.* 3:271–273, 1969.

Wetzel, R.G. *Limnology*. Saunders College Publishing, Philadelphia, PA., 1975, 743.

Williams, D.E. and Buhler, D.R. Purification of cytochromes P-448 from β-naphthoflavone treated rainbow trout, *Biochem. Biophys. Acta* 717:398–404, 1982.

Witkowski, P.J., Jaffe, P.R., and Ferrara, R.S. Sorption and desorption dynamics of Aroclor 1242 to natural sediment, *J. Contaminant Hydrol.* 2:249–269, 1988.

Wood, L.W., Jones, P.A., and Richards, A. Possible sediment scavenging of chlordane and contamination of aquatic biota in Belmont Lake, New York. *Bull. Environ. Contam. Toxicol.* 36:159–167, 1986.

Zitko, V. and Carson, W.G. A mechanism of the effects of water hardness on the lethality of heavy metals to fish, *Chemosphere* 5:299–303, 1976.

Chapter

14

Impacts of Xenobiotics on Estuarine Ecosystems

Judith S. Weis and Peddrick Weis

A. ESTUARINE SYSTEMS AND SALT MARSHES

An estuary is defined as a "semi-enclosed coastal body of water which has a free connection with the open sea and within which sea water is measurably diluted with fresh water derived from land drainage" (Cameron and Pritchard, 1963). Thus an estuary includes brackish seas, river mouths, lagoons, and tidal marshes. The most notable characteristic of such environments is the variation in salinity due to the mixing of fresh and salt water and thus the need for organisms to be able to cope with those changes. The tendency of an estuary to stratify into waters of different density (salinity) with fresh water on top of the salt water depends on several factors, including its depth and width, and the degree of disturbance by tides and winds. River-dominated estuaries normally have a large volume of fresh water input, a relatively deep channel with fresh water at the surface, and a salt water wedge underneath. Marine-dominated estuaries have a smaller input of fresh water and are more homogeneous due to vertical mixing.

Estuaries undergo temporal variation with the diurnal tidal cycle, the neap and spring tidal cycles, and seasonal changes in fresh water inputs. The unusual stresses of the fluctuating estuarine system including temperature, light, oxygen, turbidity, desiccation, and salinity stress results in a restricted number of species being able to thrive in this environment. Turbidity varies seasonally; it is greatest during the rainy season when erosion from land increases the sediment load. Dissolved oxygen is usually near saturation in surface waters due to diffusion from the air and production by aquatic plants. However,

0-8493-8851-1/94/$0.00+$.50
© 1994 by CRC Press, Inc.

oxygen concentration decreases with depth due to bacterial oxidation of organic matter and respiratory activities of organisms.

Estuaries are highly productive, due in part to the input of nutrients from fresh water and the ability of the estuaries to trap and release nutrients. Primary producers include the microscopic phytoplankton eaten by zooplankton such as larvae of many benthic invertebrates. Much of the primary productivity is contributed by benthic algae and benthic subtidal and intertidal higher plants including marsh grasses such as *Spartina*. In salt marshes these plants can act as pumps to draw nutrients out of the substrate and into the water column (Kennish, 1986). Due to the high rate of primary productivity in an estuary, additional nutrients are available for passage up the food web resulting in the support of large populations of benthic animals such as mussels and clams, as well as abundant juvenile fish and birds. Little of the primary production of marsh grasses is consumed by grazers. Instead, most is transferred to higher trophic levels via decomposers. The detritus resulting from the decomposition of the marsh grasses is an important link to the higher trophic levels. There is seasonal nutrient cycling between salt marshes and adjacent mud flats; most nutrients are retained and used in the system rather than being exported out to sea.

Humans use estuaries as a source of food and for transport, recreation, and disposal of wastes. These activities have, in many cases, altered their physical, chemical, and biological characteristics and often conflicted with one another. As more of the human population lives nearer the coast, pressures on the limited resources of estuaries are increased. In the U.S. over half the population lives within 50 miles of a coast. Public awareness of pollution in estuaries has increased considerably. Toxic contamination has led to closures of several fisheries, including the striped bass fishery in New York. The Hudson-Raritan Estuary, for example, has been subjected to serious industrial pollution problems for more than a century resulting in a reduction in diversity and abundance of living marine organisms. Several commercial and recreational fisheries have closed due to chemical contamination (Panel on Water Quality of the Hudson-Raritan Estuary, 1987). Consequently, there is considerable concern over toxics and their long-term effects on the estuarine biota and on human health.

B. INPUTS OF TOXIC SUBSTANCES INTO ESTUARIES

Estuaries receive inputs of contaminants from a variety of sources. The most direct inputs are from point sources, that is, pipe discharges or dumping of wastes directly into the water body. Among the wastes discharged in this manner is municipal waste water, which constitutes a large fraction of the freshwater input in urban areas. This waste water contains a wide variety of inorganic and organic pollutants. In New York Harbor, 13% of the inflow is treated municipal waste water (Mueller et al., 1976). Nonpoint sources, primarily urban and agricultural runoff, are important sources of contaminants discharged into estuaries and rivers that later empty into estuaries. Agricultural pesticides are used intensively during the spring and summer, which coincides with the spawning and sensitive early life stages of many estuarine species as well as marine species that spawn in the estuary. Increased runoff following spring thaw and spring rains wash sediments with elevated concentrations of toxics into the estuary at this critical time. Runoff after storms can cause great fluctuations in water quality and have critical impacts on estuarine organisms (Weis and Weis, 1984; Trefry et al., 1986).

Additional contaminants in an estuary can arise from oil spills and leaching of antifouling paint and wood preservatives from structures placed into the water. While spectacular oil spills such as the 1989 catastrophe at Port Valdez, Alaska receive much attention from the public because they are newsworthy, result in bird and mammal deaths, and impair recreation and fishing, most urban point and nonpoint source discharges also contain petroleum hydrocarbons. These wastes and runoff account for a greater amount of hydrocarbon input than the tanker accidents. Leaching of antifouling paints from boats and wood preservatives such as pentachlorophenol, creosote, and chromated copper arsenate (CCA) can also affect the estuarine environment. Discharges from power plants which use ambient water for cooling can stress estuarine organisms through temperature changes, entrainment, and impingement, the toxicity of the biocide chlorine.

Additional inputs come from the river water entering the estuary and from atmospheric deposition. The atmosphere can be a significant source of polycyclic aromatic hydrocarbons (PAHs), polychlorinated biphenyls (PCBs), and metals such as lead, mercury, cadmium, and zinc. Atmospheric inputs are found in some cases to be in the same order of magnitude as inputs from rivers and discharges. Wet deposition results from uptake of gases and particles into cloud droplets followed by precipitation. Dry deposition involves the direct uptake of pollutants by the sea surface. Both metals and organic compounds are deposited into estuaries this way. Most organic compounds and metals do not dissolve readily in sea water but are absorbed on particles. These enter the sediment or are incorporated into organic complexes at the sea surface and concentrated in the surface microlayer.

C. ROLE OF SEDIMENTS AND THE SURFACE MICROLAYER

The sediment plays a major role in the dynamics and fate of many contaminants in an estuary since it acts as a sink for metals and many organic contaminants. Fine grain particles such as silts and clays provide a large surface area for adsorption of chemicals from sea water. Changes in salinity can alter the flux of pollutants in and out of the sediment and, therefore, alter the availability of toxics to biota. Suspended sediments in rivers tend to remain in suspension due to the velocity of the water. As the river widens into the estuary, however, the water energy decreases and sediments, with adsorbed contaminants, settle in estuarine areas with little tidal or wave movement. The sedimentation flux depends on the settling velocity of the particles, which depends on the particle size distribution. In estuaries, flocculation also occurs where freshwater and saltwater meet and mix. At this juncture, the suspended matter forms larger flocculates which are heavy and settle out. This area of high turbidity is known as the turbidity maximum. The extent of this area depends on the concentrations of suspended material in the river and the sea water, as well as the estuarine circulation pattern. This zone plays an important role in remobilization and availability of toxics adsorbed onto the sediment particles. Deposition of toxics to the benthos is assisted by biological processes, such as production of feces by animals.

Particles in the sediment may also be resuspended. Resuspension is a function of the sediment structure itself, activities of burrowing organisms, and the flow velocity of the water above the sediment. This velocity is determined by such factors as river flow rate, tides, and winds. Resuspension of contaminated sediments can release chemicals into the water column from which they can be accumulated by organisms. (For this reason, dredging in urban harbors has become an issue.) Since contaminants are concentrated on fine grain sediment particles, they will not be uniformly distributed in estuaries.

The air/sea interface, comprising the upper 1 mm, is another site for the concentrating of anthropogenic materials. Concentrations of heavy metals, petroleum hydrocarbons, pesticides, and PCBs which are 10^2 to 10^4 greater than those in the water column have been found (Hardy, 1982). This surface microlayer is also an important habitat for plankton (termed neuston) including larvae of many commercially important species.

D. FATE — TRANSPORT, DEGRADATION, AND ABSORPTION

Chemical contaminants entering the estuary are partitioned into the water column, suspended matter, sediments, and organisms. Most organic contaminants that enter the estuary will ultimately be transformed and degraded. These processes include physicochemical degradation such as photodegradation and biodegradation pathways which occur primarily in microorganisms but also in animals and plants. These processes remove the parent compound from the environment but may replace it with one of equal or greater toxicity. The rate of these changes varies with environmental factors such as temperature, salinity, and the nature of the microbiota in the area. The precise pathways of degradation have not been determined for many pollutants but are known for certain components of oil and organochlorine compounds. Petroleum hydrocarbons can be broken down by marine bacteria and by photochemical oxidation at the sea surface. In general, the lighter fractions of oil, which evaporate rapidly after a spill, are more toxic. The heavier fractions accumulate in sediments and may persist for a long time in salt marshes. Bioavailability depends on solubility, particle size, and organic content of the sediments.

Organochlorines and PCBs are persistent lipophilic chemicals that enter estuaries through agricultural runoff and industrial discharges. They are not readily metabolized by microbiota and concentrate in organisms. They are passed up the food chain through biomagnification and reach their highest concentrations in the top carnivores, which include marine birds and mammals. Before being banned, dichlorodiphenyltrichloroethane (DDT) accumulation caused severe population declines in fish-eating birds. Microbial activity in sediments slowly transforms trichloroethane (TCE) and PCBs through reductive dechlorination (Brown et al., 1987).

Heavy metals are natural components of the estuarine environment. In some cases, metals are required for the normal functioning of an organism such as those found in blood pigments (Fe, Cu) or as cofactors of enzymes (Zn). Other metals that are not required can be toxic when their levels are elevated. This occurs in some estuarine systems used by industry for waste disposal. "Enrichment factors" represent a heavy metal load of sediments in relation to the natural background levels. Metals are also distributed into the water column and sediments, where they can be absorbed by organisms. They bioaccumulate in organisms, although most do not biomagnify through the food chain (Mance, 1987). The environmental conditions, physical state, and chemical form of the metal determine its fate in the estuary. In water and sediment, metals undergo chemical transformation or "speciation" allowing them to exist in different forms such as a free ion or in a methylated form. The bioavailability and toxicity of a metal are highly dependent on this speciation. The ionic forms of cadmium and copper in sea water are the most bioavailable (Ray and McLeese, 1987). Plankton are highly sensitive to ionic copper, but in sea water most copper is complexed with hydroxide or carbonates or to organic materials. Such ligands are more concentrated in coastal rather than oceanic waters, thus reducing the toxicity of copper in estuaries (Sunda and Gillespie, 1979).

Organometals, in general, are more toxic than the inorganic form because their lipophilic properties enhance their passage through cell membranes. Sediment microbes can convert inorganic mercury into methylmercury, which is absorbed by biota more rapidly than the inorganic form allowing it to biomagnify (Jensen and Jernelov, 1969). Mercury preferentially accumulates in the nervous system, which is lipid-rich.

Sediment particles make up the final sink for metals, and their availability depends on partitioning. Metals tend to be more available in oxidized sediments than reduced ones. The turbidity maximum in an estuary, the major area of sedimentation, results in the greatest mobilization and flux of metals between the sediments and the water column. Sediment-bound metals can also be taken up by marsh grasses and later excreted or released from plant detritus.

E. TOXICITY STUDIES

There have been many laboratory experiments investigating effects of contaminants on particular estuarine species. Initial studies focused on determining lethal levels and the LC_{50}, the concentration at which 50% of the test organisms are killed within a given time, usually 96 h. Recent studies have focused more on sublethal effects at the biochemical, cellular, and organismal levels to determine the effects on physiology, immunology, growth, reproduction, behavior, and development (Figure 14.1). We must understand the effects and compensatory mechanisms at each level of biological organization and the links to the next level of organization (Capuzzo et al., 1988). While most studies have exposed organisms to the contaminant in water, sediment bioassays are being developed as well. These are more environmentally realistic, with benthic animals exposed to water and sediment. It is not possible to discuss the wide variety of laboratory bioassays in this chapter (see Chapter 19). In any laboratory study, problems exist with extrapolation of data to field situations due to the many different environmental and ecological variables. Also, the role of diet in relation to water as a source of toxicants is not well understood.

F. MONITORING CONTAMINANT LEVELS AND BIOACCUMULATION

The simplest approach to determine environmental pollution is to analyze the amount of pollutant present in the sediment and/or water and in the organisms. Though water concentrations have been the focus of much attention, and governments set "water quality criteria" for levels of chemicals, water concentrations are extremely variable in time and space, and most chemicals of concern do not reach high levels in the water column, even in highly contaminated estuaries. Instead, they accumulate in the sediments, which should thus be the major focus of environmental monitoring for pollution. We must know what the "normal" background level of a contaminant should be. For some anthropogenic contaminants such as PCBs the normal background level is zero, while for heavy metals certain levels are always present. The "normal" metal content for an organism can vary greatly with species, age, sex, and environmental variables.

The concentration of a chemical in the organism compared to that in the sediment in which it lives is the "concentration factor", an expression of the

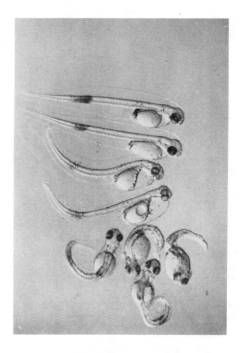

Figure 14.1. Fish embryos are very responsive to a variety of environmental contaminants. These newly-hatched winter flounder *(Pseudopleuronectes americanus)* embryos were exposed to 10% treated municipal waste water from the New York-New Jersey Harbor estuary. (Photo by P. Weis.)

accumulating capacity of the organism for the particular chemical. The liver is usually the organ in which the greatest accumulation occurs, although muscle levels are more relevant for human consumption (Figure 14.2). Epidemiological studies on humans who consumed fish contaminated with PCBs have, in fact, revealed that deleterious health effects may ensue (Swain, 1988).

Many studies are designed to relate the concentration of contaminants in sediments with levels in organisms. Correlations have been found in most cases. In general, urbanized estuaries have elevated levels of metals, chlorinated hydrocarbons, petroleum hydrocarbons, and aromatic hydrocarbons. However, levels of contaminants in sediments can have extensive spatial and temporal variability due to seasonal differences in runoff.

An important factor in the uptake of a contaminant is its bioavailability. Bioavailability depends on the chemical structure of the compound, its ability to bind to particles, the organic content of the particles, and the biology of the organism. Studies have been undertaken in which "clean" organisms have been transplanted into polluted sites for specific periods. Pesch (1979) studied accumulation of copper in a deposit-feeding polychaete and found that animals placed in mud accumulated metals to a greater degree than animals placed in sediments containing larger particles. The uptake of metals varies considerably

Figure 14.2. Recreational fishing takes place in urbanized estuaries in which seafood may be contaminated with a variety of chemicals. (Photo by P. Weis.)

depending on environmental and other factors. The presence of certain trace metals can have a major impact on the bioavailability of zinc and lead to bivalve mollusks (Luoma and Bryan, 1979). A weak correlation between Zn concentrations in *Scrobicularia plana* with sediment concentrations was noted, but this correlation did not apply to *Macoma balthica*. The levels of iron and manganese were important in determining the availability of the Zn. Increasing concentrations of Fe led to increasing bioavailability of Zn to *M. balthica*. For lead and arsenic, increased Fe led to decreased bioavailability to *S. plana*. The only consistent effect was that of organic matter, which always coincided with decreased bioavailability of metals.

The temperature, salinity, and oxygen content of the water as well as season, height in the water, and biological factors such as size, growth rate, age, sex, reproductive condition, feeding habits, physiology, genetics, and fat content of the organism can all alter the uptake and metabolic turnover of contaminants. Organic compounds that have a low water solubility will concentrate in lipids including energy reserves and gonadal reserves, as well as membrane lipids and cellular macromolecules. The levels accumulated will depend on the lipid content of the animal (Neely et al., 1974).

Pharmacokinetic models can be used to study uptake from food and water, metabolic conversions, bioaccumulation, and excretion. Excretion can be via urine, bile, gills, and the release of eggs in spawning. Bioaccumulation is best explained through the use of multicompartment models in which distribution among different body compartments (tissues) is considered. Bioconcentration is dependent on relative rates of uptake and exchange of toxics between different body compartments (O'Connor and Pizza, 1987).

Although monitoring the levels of chemicals in the environment or in organisms is a common procedure for studying pollution, it is highly variable, and it is often difficult to distinguish natural changes from those caused by humans. For that reason, long term records can be extremely valuable for documenting changes and for generating testable hypotheses (Wolfe et al., 1987). Such long term data sets have shown, for example, that copper pollution in the Fal Estuary (southwest England) has declined slowly but that zinc pollution has not (Bryan et al., 1987). Similarly, DDT contamination has decreased in U.S. estuaries (Mearns and van Ness, 1987). Monitoring data by itself has another major drawback in that it does not give insight into the biological effects of the contaminants.

G. ABNORMALITIES IN FIELD-COLLECTED ORGANISMS

Observations of abnormalities in biota can be a good initial indicator of adverse effects from pollution. Fish diseases such as fin erosion, skeletal deformities, and tumors have been noted in several contaminated ecosystems (Sindermann, 1983). Fin rot (Figure 14.3) is a condition involving progressive death of tissue and erosion of fins that is prevalent in benthic fish, mainly flatfish. Fin rot may be caused by exposure to contaminated sediments and can also increase mortality. This condition has been observed in flatfish from highly polluted areas such as the New York Bight (Murchelano and Ziskowski, 1976). Fin rot has also been noted in Southern California fish in which elevated levels of aromatic hydrocarbons and DDT derivatives were found in their stomach contents and liver (Malins et al., 1988).

Increased incidence of liver tumors and other hepatic pathological changes in fish from contaminated areas have been noted. High incidences of hepatic tumors were observed by Malins et al., (1984) in English sole from Puget Sound (Elliott Bay), where greatly elevated levels of polynuclear aromatic hydrocarbons were found in the sediments. Elevated levels of PAH metabolites were also found in the bile of these fish. Tumor incidence was correlated with free radicals in the liver microsomes. Laboratory studies indicated that high molecular weight hydrocarbons were carcinogenic (Malins et al., 1988). Liver neoplasms in 8% of winter flounder from Boston Harbor were correlated with age and concentrations of PAHs and PCBs (Murchelano and Wolke, 1985). Cells from these tumors contained activated oncogenes (McMahon et al., 1988). Observations that tumors in field-caught specimens correlated with environmental contamination were supported by laboratory studies in which several chemicals were found to be carcinogenic to fish (Couch and Harshbarger,

Figure 14.3. A winter flounder *(P. americanus)* with fin rot disease. (Photo by R. A. Murchelano.)

1985). Cancerous diseases also occur in mollusks; however, Mix (1986) conducted a thorough review and concluded there was no clear evidence that neoplasms in bivalves are pollution related.

High incidences of shell disease, a brownish black erosion of the exoskeleton, have been reported in benthic crustaceans from polluted sites including the New York Bight. At this site, elevated concentrations of mercury, cadmium, PCBs, DDT and metabolites, and petroleum hydrocarbons were found (Murchelano, 1982).

Although scientists have correlated the incidence of disease with accumulation of certain chemicals in an organism, it is often difficult to establish a clear causal relationship. One case in which a relationship was substantiated by laboratory studies is that of deformities in mollusks caused by the antifouling paint additive tributyltin (TBT). Antifouling agents are used in boat paint to prevent colonization by barnacles and other fouling organisms on the hull of boats. Aggregations of these organisms impair boat speed and efficiency. In the 1970s boat paints containing TBT became more popular than the traditional copper-based paints because of their longer effectiveness. Alzieu and Heral (1984) noted that in the vicinity of marinas in France abnormalities of the oyster *Crassostrea gigas* were common. The shells were unusually thick, and within the thickened shell the animal was stunted and thus unmarketable. The deformed oysters contained high levels of tin, suggesting that this metal rather than other chemicals related to boating activity was responsible for the malformations. Similar malformations were noted in the vicinity of marinas in England and on the U.S. West Coast. Laboratory experiments (Waldock and Thain, 1983) confirmed that TBT was causing the abnormality. The thickening of the shell could be produced in the laboratory with concentrations as low as 150 parts per trillion (ppt) of TBT.

In another field observation of an abnormality followed by laboratory confirmation, Bryan et al. (1986) found that TBT caused anatomical malformations in female snails *Nucella lapillus*. A masculinization phenomenon called "imposex" was noted in snails in areas with extensive boating activity. This condition was considered responsible for the decline of the species off the coast of England. Imposex could be induced in the laboratory following exposure to 20 ppt of TBT. Thus, in this case, field observations of an unusual malformations could be verified by laboratory experimentation.

H. INDICES

Pollution effects can be monitored by examining sublethal indices in organisms inhabiting polluted vs. clean estuaries. No single pollution index, however, can provide predictive capability for evaluating population changes, thus multiple responses should be measured and considered simultaneously (Capuzzo and Kester, 1987).

Body condition indices can be derived by comparing weight/length relationships in a variety of organisms. They have been used in mollusks (modified as flesh/shell relationships), where a decrease in index was recorded as the level of the pollutant (copper) increased (Wilson and McMahon, 1981). The condition index, however, can change with the season, reproductive state, and age of the organism, so samples must be chosen carefully to avoid these confounding effects.

A group of rapid responses were tested in mussels *(Mytilus edulis)* that had been caged at sites with varying levels of pollution. The sessile lifestyle of this species, its extensive distribution and abundance, makes it a useful "sentinel" organism for such studies. Mollusks in general, and mussels in particular, have become important in global monitoring programs (Bayne et al., 1985). Among the indices examined were scope for growth (discussed below), lysosomal stability, and metal-binding proteins, all of which were altered in the highly contaminated areas (Johnson and Lack, 1985). Lysosomes are organelles in which cellular debris including metals and organic chemicals can accumulate. Cytochemical techniques, including the measurement of lysosomal membrane stability and enzyme activity, have been successfully used in pollutant impact assessment (Moore, 1985). A quantitative relationship has been noted between effects on lysosomal membrane, organelle enlargement, enzyme changes, and the scope for growth.

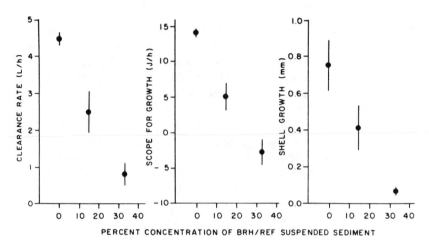

Figure 14.4. Clearance rate, scope for growth, and actual growth increment in mussels exposed to various proportions of contaminated (BRH) and clean (REF) sediment. (From Nelson, W.A. *Pollution Physiology of Estuarine Organisms.* University of South Carolina Press, Columbia, SC, 1987.)

The scope for growth is an index of physiological fitness based on the energy budget of the organism (Bayne et al., 1985). It represents production, which depends on the amount of food available, the efficiency with which the organism can extract energy from the food, and the demands of routine metabolism and excretion (P = A − [R + U], where P = production, A = assimilation, R = respiration, and U = energy lost in excretion). The scope for growth provides an assessment of the energy status of the animal, as well as insight into the individual components such as respiration, filtration rate, etc. This widely used indicator is depressed in animals from polluted sites and correlates well with direct measurements of actual growth (Widdows, 1985). There was a good correlation between the scope for growth in mussels exposed in the laboratory and the field to contaminated sediments, with a reduced scope for growth value being associated with a reduced clearance rate and shell growth (Figure 14.4) (Nelson, 1987).

Another indicator that relates to the energy budget is the O:N ratio (ratio of oxygen consumed to nitrogen excreted), which provides an index of relative utilization of protein to carbohydrate and lipid consumed in energy metabolism. The catabolism of protein increases under stress, thus decreasing the O:N ratio (Widdows, 1985). Adenylate energy charge is an index of metabolic energy available from the adenylate pool. It is calculated from tissue concentrations of ATP, ADP, and AMP (AEC = [ATP + 1/2 ADP]/[ATP + ADP + AMP]). It can be an indicator of stress from pollution and other causes (Ivanovici, 1980).

Another biological index of effects that can be monitored is the cytochrome P_{450} monooxygenase system activity, or mixed function oxidase (MFO) system, which participates in the detoxification of foreign organic compounds.

This system is inducible, and its presence can be interpreted as an indicator of exposure to organic pollutants (Bayne et al., 1985). The presence of this enzyme system may have deleterious consequences in that it may convert some compounds to mutagenic and carcinogenic derivatives (Stegeman and Kloepper-Sams, 1987).

Metallothioneins are low molecular weight, cysteine-rich proteins with a high binding affinity for certain metals. Increased synthesis of these proteins occurs following exposure to metals, and binding of the metals to the protein reduces their toxic effects. Elevated levels of metallothioneins are found in organisms from metal-polluted sites, thus the levels can be used as a monitoring tool (Engel and Roesijadi, 1987). Toxicity by the metal is believed to occur only after the binding capacity of metallothioneins is exceeded and metals "spill over" into the metal-sensitive enzyme pool. Metallothioneins are protective mechanisms that allow organisms to be exposed to or accumulate greater concentrations of a metal without toxic effects.

Another measurement of toxic effects in organisms from contaminated areas is chromosomal damage. Increased incidence of chromosomal aberrations and sister chromatid exchange were noted in pelagic fish eggs from polluted areas (Longwell and Hughes, 1980). Exposure of these eggs may have resulted from the contaminated surface microlayer or during passage from the ovary of the adult. Impaired immune function of macrophages from fish from contaminated regions has also been noted (Weeks and Warriner, 1984). Activity returned to control values when fish were placed in clean water.

Impaired reproduction has been noted in fish from pollution impacted areas. Fish from contaminated areas showed decreased hatching success of embryos artificially inseminated and reared in the laboratory (Dethlefsen, 1988). McCain et al., (1988) noted impaired reproductive success in flatfish from a contaminated area within Puget Sound. Impaired reproduction may be due in part to interference with the activity of reproductive hormones. A decrease in circulating steroid hormones has been documented in many fish exposed to toxicants (Thomas, 1988). Impaired reproduction has been shown in top carnivores such as marine mammals and birds following exposure to high levels of toxic substances.

I. POPULATION EFFECTS

There are reports of slower growth of individual organisms in populations of bivalves living in polluted areas (Elliott and McLusky, 1985). However, there may be considerable variation within an area, and confounding factors including the structure of the population can interfere with definitive conclusions of pollution effects on an individual growth. An extreme change in population size is needed before the effect can be differentiated from natural variability. Thus, it is very difficult to detect changes in this parameter in a population due to pollution.

Populations of organisms that have been chronically exposed to pollutants may develop increased tolerance or resistance to these toxicants. Temporary tolerance may result from short term physiological acclimation due to mechanisms such as synthesis of metallothioneins and activation of the MFO system. Tolerance may also arise from genetic adaptation due to selection in a polluted environment, resulting in an increase in organisms with resistant genotypes. Genetically based tolerance persists when the organisms are placed in clean water and will be passed on to later generations bred in the laboratory. Genetically based tolerance is found frequently in phytoplankton from contaminated areas. Studies of metal tolerance in populations of estuarine invertebrates were reviewed by Bryan (1976). Populations of the polychaete *Nereis diversicolor* from sediments rich in a particular metal were often resistant to that metal. The increased resistance was not altered by exposing resistant worms to uncontaminated sediments, so the tolerance was considered genetic. Resistance to lead and copper in isopods *(Asellus meridianus)* from polluted environments persisted in later generations bred in the laboratory (Brown, 1976). More direct evidence for genetic selection was found by Nevo et al. (1981), who studied allozymic variation of phosphoglucomutase (PGM) genotypes in the shrimp *Palaemon elegans* following exposure to mercury. They found differential tolerance of certain genotypes to mercury, suggesting that they are adaptive. Nevo et al. (1984), found that the genotype of the gastropod *Monodonta turbinata* that was more resistant to mercury in the laboratory was also more prevalent in polluted sites, indicating that selection for mercury tolerance occurred at the polluted site.

In fresh water fish exposed to metals there are reports of stress, including earlier maturation, increased fecundity, and decreased longevity as opposed to increased tolerance (McFarlane and Franzin, 1978). Weis and Weis (1989) reviewed extensive literature on the mummichog *(Fundulus heteroclitus),* an estuarine fish that persists in highly polluted areas. Resistance of embryos from clean areas to the teratogenic effects of methylmercury (CH_3Hg) is highly variable and is dependent on the female producing the eggs. A population from a polluted environment showed much greater and more uniform embryonic tolerance. This tolerance was also seen in eggs and sperm before fertilization. Interestingly, embryos and gametes did not exhibit increased tolerance to inorganic mercury, but instead, were more susceptible. Eggs were also far less able to cope with salinity changes than eggs from the clean population. After hatching, larvae and juveniles from the polluted population did not exhibit enhanced tolerance to CH_3Hg but were as susceptible as larvae from the clean population. Mechanisms of tolerance by the embryos were related to decreased chorionic permeability to CH_3Hg and faster development through sensitive embryonic stages, mechanisms not relevant to larvae. Adults from the polluted site had a decreased resistance, became reproductive at a smaller size and younger age, and did not grow as well or live as long as fish from the clean population. Thus, increased tolerance to a toxicant at early life stages (gametes and embryos) was accompanied by decreased tolerance to other environmental variables and evidence of stress in the adults.

J. COMMUNITIES

Community changes can be observed in response to pollution. Most studies have used macrobenthos which are diverse, numerous, relatively stationary, and easily collected. In field studies, the community composition pre- and post-exposure can be assessed. In many cases, however, knowledge of the community before pollution occurred is not available or is incomplete. Estuarine communities are highly variable spatially and temporally. Long term records of biological data are therefore extremely valuable for documenting community changes and distinguishing natural changes (due to climatic factors, cyclic phenomena, etc.) from those caused by human activities. A time series of benthic faunal data in certain lochs in Scotland demonstrated impacts of effluent discharges from a paper mill between 1966 and 1980, which increased carbon inputs and affected population levels of various species (Pearson, 1975).

In general, after severe pollution benthic community changes can be observed. Some species are no longer represented, some species show great population decreases, and other species undergo strong fluctuations. These may include population booms of certain "opportunistic" species shortly after a catastrophic event such as a major oil spill which initially causes mass mortality. One opportunistic organism appearing as an indicator of marine pollution is the polychaete *Capitella capitata*, a complex of several sibling species within which is a succession from the most opportunistic to the least opportunistic following the discharge of a pollutant (Grassle and Grassle, 1974). Results from a long term study following an oil spill showed mass mortality followed by dominance by *Capitella* at the heavily oiled areas for a year (Sanders et al., 1980). In these areas fauna underwent significant fluctuations and recovery took almost 10 years. Degradation of oil in the sediments of the marsh was also a slow process. Physiological and behavioral disorders persisted for many years in biota at the highly polluted sites.

Livingston (1984) has conducted long term, multidisciplinary, integrated field and laboratory studies covering climate, nutrients, water quality, productivity, and biological trophic components. These components include phytoplankton, zooplankton, benthic macrophytes, microbiota, benthic infauna, epibenthic invertebrates, and fishes (Figure 14.5). The data indicated that different organisms responded in different ways. This data was used to develop models of projected population and community distribution and their responses to physical factors, predator/prey interactions, and pollution stress.

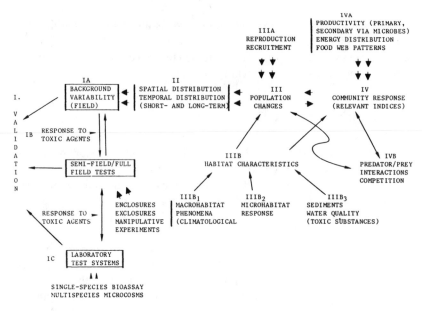

Figure 14.5. Research approach (laboratory/field) to impact analysis in studies of Livingston (1984).

Another approach to investigating community level responses to pollution is to examine the richness or heterogeneity of the species using such indices as the Shannon-Wiener index, which can quantify the degree of change in a community (Gray, 1976). This index is widely used in fresh water pollution assessment in which a low value indicates a polluted environment. Low indices, however, are not always representative of polluted situations in estuaries, suggesting that this is not a very reliable tool in the study of estuarine pollution. Even without the input of pollutants, the dynamic, fluctuating estuarine environment can be stressful to its inhabitants, and thus it is difficult to separate natural from pollution related stresses in studies of estuarine community diversity.

K. ECOSYSTEMS — MESOCOSMS

It is often difficult to prove that elevated concentrations of xenobiotic chemicals cause effects on the estuarine ecosystem. On one hand, field data show wide variations and it is difficult to prove cause/effect relationships. On the other hand, laboratory studies performed on individuals of a single species are difficult to extrapolate to the "real world" with its many environmental variables and species interactions. For example, temperature and salinity can affect the responses to pollutants. In general estuarine animals are more susceptible to heavy metals at lower salinities and higher temperatures (McLusky et al., 1986).

To better approach the "real world", experimental ecosystems have been developed to study the fate and effects of pollutants. Microcosms are laboratory simulations of a portion of the ecosystem. They may be derived from natural communities or synthesized from laboratory cultures. They are often limited to primary, secondary, and detrital trophic levels. While microcosms are usually established in the lab, mesocosms are larger and are set up outdoors where natural temperature and light are present. Water is frequently taken from an adjacent ecosystem. These large systems help to bridge the gap between the artificiality of laboratory single-species studies and the complexity of conditions that exists in the real world. Mesocosms permit experimentation with whole ecosystems and are often more sensitive tests than single species laboratory tests. Menzel and Steele (1978) reviewed the use of mesocosms (large plastic enclosures) in the study of pollutants on natural plankton communities. In most studies, the addition of a single chemical was made and the enclosed experimental ecosystem was compared with untreated controls for phytoplankton, zooplankton, and bacteria. Physico-chemical factors relating to the fate of the added pollutant were monitored. When 5 µg/l Hg was added (Kuiper et al., 1983) primary production was initially inhibited but then rose to the levels of controls. Examining the fate of the mercury in the mesocosms revealed that ionic mercury was transformed to volatile metallic mercury which then disappeared from the system by evaporation.

Micro- and mesocosms are useful in studies of the biodegradation of organic pollutants in estuaries. They are also useful in studying indirect effects due to biological interactions such as competition or predation. For example, addition of 1 to 50 µg/l cadmium (Cd) caused decreased densities of the comb jelly *Pleurobrachia pileus*, resulting in increased numbers of copepods on which the comb jellies prey (Kuiper, 1981).

Differences in plankton species composition due to interactions in an ecosystem can also be found. Phytoplankton assemblages in dialysis bags placed in an estuary responded to the addition of PCBs by decreasing in size, indicating that the large diatoms were more sensitive than the small ones. Such changes could increase the number of trophic levels needed to reach harvestable fish and could divert energy flow toward gelatinous predators (O'Connors et al., 1978). The existence of such a shift, however, has not been verified in natural systems. When contaminants caused decreased density of copepods, the phytoplankton bloomed and became dominated by larger species. This was probably due to selective grazing by copepods on the larger species of phytoplankton (Steele and Gamble, 1982).

More realistic estuarine mesocosms have been developed that contain sediment as well as water. The Model Tidal Flat ecosystem (MOTIF), containing macrofauna, meiofauna, phytobenthos, bacteria, phytoplankton, and zooplankton, has been used to study the fate and effects of oil spills and dispersants on a model of the Wadden Sea (Kuiper et al., 1984). The Marine Ecosystem Research Laboratory (MERL) mesocosms, consisting of 5 m-deep flow-through tanks with 13.1 m^3 water and 0.76 m^3 sediment simulating the benthic and pelagic system of Narragansett Bay, have been used intensively for ecosystem studies (Grice and Reeve, 1982). Treatment with number 2 fuel oil over several months resulted in uptake by the benthos as well as changes in its composition (Wade and Quinn, 1980). Recovery from oil was also studied.

Mesocosms lack much of the variation inherent in field studies and can contribute greatly to the basic understanding of the functioning of estuarine ecosystems and enhance our knowledge of contaminant effects on these systems. In mesocosms, one can simultaneously study the fate of the chemical and its effects on biota under nearly natural conditions. They can provide important information for more accurate risk analysis and the setting of standards for contaminants. A drawback to these systems is the inability to contain larger predators that are important in natural environments. Another problem is that, as with single species tests, only one chemical at a time is studied, and the complex interactions of contaminants which occur in the real world are not investigated. Polluted estuaries almost always have a variety of toxic contaminants, and interactions among these may occur. Additivity occurs when the toxicity of a mixture of chemicals is approximately equal to the sum of the individual toxicities. Synergism occurs when the toxicity of two substances in combination is greater than the sum of their individual toxicities, and antagonism occurs when the overall toxicity is reduced. The interactions of any two chemicals can be different in different organisms and can be altered by the presence of additional contaminants. Furthermore, other environmental variables such as salinity can alter the way in which two chemicals interact with each other. Pollutants can also alter tolerance to these natural variables. For example, metals can impair osmoregulation making natural salinity changes more stressful — potentially a major problem in an estuary (Calabrese et al., 1982).

L. THREAT OF GLOBAL WARMING

It is predicted by many atmospheric scientists that the increase in natural greenhouse gases such as carbon dioxide, methane, and chlorofluorocarbons in the atmosphere will cause changes in the global climate and a rise in the sea level. If these projected climatic changes do occur, estuaries will be subjected to even greater stresses than they have been in the past. Some economically important species are quite sensitive to increases in water temperature. Increased temperature will also enhance the toxicity of pollutants. Altered rainfall patterns and rising sea level will alter the salinity structure and stratification, and hence the circulation patterns within estuaries and the oceans. Thus, ecosystem structure and function will be altered. Furthermore, if the rate of sea level rise exceeds the rate of accretion due to sedimentation, existing tidal marshes will be inundated. While the intertidal zone will be diverted inland, marshes may be unable to migrate inland because of roads and development. There is a possibility that vast areas of estuaries and tidal marshes will be wiped out altogether.

M. CONCLUSION

The goals of the Clean Water Act were to ensure that U.S. waterways were both "fishable and swimmable". Obviously, many of our estuaries are far from this condition. In response to the passage of the Clean Water Act, improved sewage systems were built which significantly reduced inputs of conventional pollutants, which improved water quality. The problem of toxics has proven to be more intractable, since many of these chemicals are persistent and remain in the sediments for extended periods. They pose a problem for estuarine organisms and for the health of those who consume seafood from contaminated estuaries.

Although the oceans have a high capacity for absorbing waste ("assimilative capacity"), the bulk of our waste is not discharged into the vast deep ocean, but into shallow estuaries with far less assimilative capacity. The competing demands on estuaries (i.e., industry vs. conservation) are often unreconcilable. Good management is therefore needed to optimize the uses of the estuary. Technology has been developed to treat most point sources, but high costs are associated with the processes. It is much more difficult to manage nonpoint sources. Waste reduction is preferable to treatment.

Scientists need to develop predictive ecological hazard assessments to determine what effects a contaminant will have on a given resource. Such assessments will play an important role in environmental decisions. A hazard assessment is an objective, but probabilistic, calculation requiring data on environmental concentration of pollutants, levels of exposure to the organism, and biological effects. Holistic strategies need to focus on contaminant loading, determination of physical transport, biogeochemical fate, and ecological effects. Mathematical models need to be developed for the behavior of contaminants in the ecosystems they enter. Since estuarine organisms are under considerable stress from temperature, salinity, oxygen, and turbidity, good ecosystem management is needed to protect them from further toxic stresses.

REFERENCES

Alzieu, C. and Heral, M. Ecotoxicological effects of organotin compounds on oyster culture, in *Ecotoxicological Testing for the Marine Environment*. Vol 1. G. Persoone, E. Jaspers, and C. Claus, Eds. State University of Ghent, Belgium, 1984, 187–196.

Bayne, B., Brown, D., Burns, K., Dixon, D., Ivanovici, A., Livingstone, D., Lowe, D., Moore, M., Stebbing, A., and Widdows, J. *The Effects of Stress and Pollution on Marine Animals*. Praeger, New York, 1985, 384.

Brown, B. Observations on the tolerance of the isopod *Asellus meridianus* rac to copper and lead, *Water Res.* 10:555–559, 1976.

Brown, J., Wagner, R., Feng, H., Bedard, D., Brennan, M., Carnahan, J., and May, R. Environmental dechlorination of PCBs, *Environ. Toxicol. Chem.* 6:579–593, 1987.

Bryan, G. Some aspects of heavy metal tolerance in aquatic organisms, in *Effects of Pollutants on Aquatic Organisms*. A. Lockwood, Ed. Cambridge University Press, Cambridge, 1976, 7–34.

Bryan, G., Gibbs, P., Hummerstone, L., and Burt, G. The decline of the gastropod *Nucella lapillus* around southwest England: evidence for the effect of tributyltin from antifouling paints, *J. Mar. Biol. Assoc. U.K.* 66:611–640, 1986.

Bryan, G., Gibbs, P., Hummerstone, L., and Burt, G. Copper, zinc and organotin as long-term factors governing the distribution of organisms in the Fal Estuary in southwest England, *Estuaries* 10:208–219, 1987.

Calabrese, A., Gould, E., and Thurberg, F. Effects of toxic metals in marine animals of the N.Y. bight: some laboratory observations, in *Ecological Stress and the N.Y. Bight: Science and Management*. G. Mayer, Ed. Estuarine Research Federation, Columbia, SC, 1982, 281–297.

Cameron, W. and Pritchard, D. Estuaries, in *The Sea Vol. 2*. M.N. Hill, Ed. John Wiley & Sons, New York, 1963, 306–324.

Capuzzo, J. and Kester, D. Biological effects of waste disposal: experimental results and predictive assessments, in *Biological Processes and Wastes in the Ocean*. J.M. Capuzzo and D.R. Kester, Eds. Krieger Publishers, Malabar, FL, 1987, 1–15.

Capuzzo, J., Moore, M., and Widdows, J. Effects of toxic chemicals in the marine environment: predictions of impacts from laboratory studies, *Aquat. Toxicol.* 11:303–311, 1988.

Couch, J. and Harshbarger, J. Effects of carcinogenic agents on aquatic animals: an environmental and experimental overview, *J. Environ. Sci. Health Part C Environ.* 3:63–105, 1985.

Dethlefsen, V. Status report on aquatic pollution problems in Europe, *Aquat. Toxicol.* 11:259–286, 1988.

Elliott, M. and McLusky, D. Invertebrate production ecology in relation to estuarine management, in *Estuarine Management and Quality Assessment*. J. Wilson and W. Halcrow, Eds. Plenum Press, London, 1985, 85–104.

Engel, D. and Roesijadi, G. Metallothioneins: a monitoring tool, in *Pollution Physiology of Estuarine Organisms*. W. Vernberg, A. Calabrese, F. Thurberg, and F.J. Vernberg, Eds. University of South Carolina Press, Columbia, SC, 1987, 421–438.

Grassle, J.F. and Grassle, J.P. Opportunistic life histories and genetic systems in marine benthic polychaetes, *J. Mar. Res.* 32:253–284, 1974.

Gray, J.S. The fauna of the polluted river Tees estuary, *Estuar. Coast. Mar. Sci.* 4:653–676, 1976.

Grice, G.D. and Reeve, M.R. Eds. *Marine Mesocosms*. Springer Verlag, New York, 1982, 430.

Hardy, J. The sea surface microlayer: biology, chemistry, and anthropogenic enrichment, *Prog. Oceanog.* 11:307–328. 1982.

Ivanovici, A. Application of adenylate energy charge to problems of environmental impact assessment in aquatic organisms, *Helgol. Meeresunter.* 33:556–565, 1980.

Jensen, S. and Jernelov, A. Biological methylation of mercury in aquatic organisms, *Nature* 223:753–754, 1969.

Johnson, D. and Lack, T. Some responses of transplanted *Mytilus edulis* to metal-enriched sediments and sewage sludge, *Mar. Env. Res.* 17:277–280, l985.

Kennish, M.J. *Ecology of Estuaries, Vol 1. Physical and Chemical Aspects*. CRC Press, Inc., Boca Raton, FL, 1986, 254.

Kuiper, J. Fate and effects of cadmium in marine plankton communities in experimental enclosures, *Mar. Ecol. Prog. Ser.* 6:161–174. 1981.

Kuiper, J., Brockman, U., Groenewoud, H. van Het, Hoornesman, G., and Roele, P. Effects of mercury on enclosed plankton communities in the Rosfjord during POSER, *Mar. Ecol. Prog. Ser.* 14:93–105, 1983.

Kuiper, J., de Wilde, P., and Wolff, W. Effects of an oil spill in outdoor model tidal flat ecosystems, *Mar. Poll. Bull.* 5:102–106, 1984.

Livingston, R. Aquatic field monitoring and meaningful measures of stress, in *Concepts in Marine Pollution Measurements*. H. White, Ed. University of Maryland Press, College Park, MD, 1984, 681–692.

Longwell, A. and Hughes, J. Cytologic, cytogenetic, and developmental state of Atlantic mackerel eggs from sea surface waters of the N.Y. Bight and prospects for biological effects monitoring with ichthyoplankton, *Rapp. P.-V. Reun. Cons. Int. Explor. Mer.* 179:275–291, 1980.

Luoma, S. and Bryan, G. Trace metal availability: modeling chemical and biological interactions of sediment-bound zinc, in *Chemical Modeling in Aqueous Systems*. E. Jenne, Ed. *Am. Chem. Soc. Symp. Ser.* 93:577–609, 1979.

Malins D., McCain, B., Brown, D., Chan, S.-L., Myers, M., Landahl, J., Prohaska, P., Friedman, A., Rhodes, L., Burrows, D., Gronlund, W., and Hodgins, H. Chemical pollutants in sediments and diseases in bottom-dwelling fish in Puget Sound, Washington, *Environ. Sci. Technol.* 18:705–713, 1984.

Malins, D., McCain, B., Landahl, J., Myers, M., Krahn, M., Brown, D., Chan, S.-L., and Roubal, W. Neoplastic and other diseases in fish in relation to toxic chemicals: an overview, *Aquat. Toxicol.* 11:43–67, 1988.

Mance, G. *Pollution Threat of Heavy Metals in the Aquatic Environment*. Elsevier Applied Science, London, 1987, 372.

McCain, B., Brown, D., Krahn, M., Myers, M., Clark, M., Jr., Chan, S.-L., and Malins, D. Marine pollution problems, North American West Coast, *Aquat. Toxicol.* 11:143–162, 1988.

McFarlane, G. and Franzin, W. Elevated heavy metals: a stress on a population of white suckers, *Catostomus commersoni* in Hamell Lake, Saskatchewan, *J. Fish. Res. Bd. Can.* 35:963–970, 1978.

McLusky D., Bryant, V., and Campbell, R. The effects of temperature and salinity on the toxicity of heavy metals to marine and estuarine invertebrates, *Ocean. Mar. Biol. Ann. Rev.* 24:481–520, 1986.

McMahon, G., Huber, L., Stegeman, J., and Wogan, G. Identification of a c-ki-ras oncogene in a neoplasm isolated from winter flounder, *Mar. Environ. Res.* 24:345–350, 1988.

Mearns, A. and van Ness, K. Order from chaos? History of chlorinated pesticide contamination of U.S. coastal fauna., in *Proc. Oceans '87 Conf. Coastal and Estuarine Pollution.* Marine Tech. Society, 1987, 1538–1543.

Menzel, D. and Steele, J. The application of plastic enclosures to the study of pelagic marine biota, *Rapp. P.-V. Reun. Cons. Int. Explor. Mer.* 173:7–12, 1978.

Mix, M. Cancerous diseases in aquatic animals and their association with environmental pollutants: a critical literature review, *Mar. Env. Res.* 20:1–141, 1986

Moore, M. Cellular responses to pollutants, *Mar Poll. Bull.* 16:134–139, 1985.

Mueller, J., Jeris, J., Anderson, M., and Hughes, C. Contaminant inputs to the N.Y. bight, *U.S. Department of Commerce NOAA Tech. Memo ERL MESA-6.* Boulder CO, 1976, 347.

Murchelano, R. Some pollution-associated diseases and abnormalities of marine fishes and shellfishes: a perspective for the New York bight, in *Ecological Stress and the New York Bight: Science and Management.* G. Mayer, Ed. Estuarine Research Fed. Columbia, SC, 1982, 327–346.

Murchelano, R. and Wolke, R. Epizootic carcinoma in the winter flounder *Pseudopleuronectes americanus, Science* 228:587–589, 1985.

Murchelano, R. and Ziskowski, J. Fin rot disease studies in the New York bight, *Am. Soc. Limnol. Oceanog. Spec. Symp.* 2:329–336, 1976.

Neely, W., Branson, D., and Blau, G. The use of the partition coefficient to measure the bioconcentration potential of organic chemicals in fish, *Environ. Sci. Tech.* 8:1113–1115, 1974.

Nelson, W. A Comparison of the physiological condition of the blue mussel, *Mytilus edulis* after laboratory and field exposure to a dredged material, in *Pollution Physiology of Estuarine Organisms.* W.B. Vernberg, A. Calabrese, F.P. Thurberg, and F.J. Vernberg, Eds. University of South Carolina Press, Columbia, SC, 1987, 185–205.

Nevo E., Berl, T., Beiles, A., and Wool, D. Mercury selection of allozyme genotypes in shrimps, *Experientia* 37:1152–1154, 1981.

Nevo, E., Ben-Shlomo, R., and Lavie, B. Mercury selection of allozymes in marine organisms: prediction and verification in nature, *Proc. Nat. Acad. Sci.* 81:1258–1259, 1984.

O'Connor, J. and Pizza, J. Dynamic polychlorinated biphenyls in striped bass from the Hudson River. III. Tissue disposition and routes for elimination, *Estuaries* 10:68–77, 1987.

O'Connors, H., Wurster, C., Powers, C., Biggs, D. and Rowland, R. Polychlorinated biphenyls may alter marine trophic pathways by reducing phytoplankton size and production, *Science* 201:737–739, 1978.

Panel on Water Quality of the Hudson-Raritan Estuary, *The Hudson-Raritan: State of the Estuary.* New Jersey Marine Sciences Consortium, Fort Hancock, NJ, 1987, 36.

Pearson, T. The benthic ecology of Loch Linnhe and Loch Eil, a sea loch system on the west coast of Scotland. IV. Changes in the benthic fauna attributable to organic enrichment, *J. Exp. Mar. Biol. Ecol.* 20:1–41, 1975.

Pesch, C. Influence of three sediment types on copper toxicity to the polychaete *Neanthes arenaceodentata, Mar. Biol.* 52:237–245, 1979.

Ray, S. and McLeese, D.W. Biological cycling of cadmium in marine environments, in *Cadmium in the Aquatic Environment.* J. Nriagu and J. Sprague, Eds. John Wiley & Sons, New York, 1987, 199–221.

Sanders, H., Grassle, J., Hampson, G., Morse, L., Garner-Price, S., and Jones, C. Anatomy of an oil spill: long term effects from the grounding of the barge *Florida* off West Falmouth, MA, *J. Mar. Res.* 38:265–380, 1980.

Sindermann, C. An examination of some relationships between pollution and disease, *Rapp. P.-V. Reun. Cons. Int. Explor. Mer.* 182:37–43, 1983.

Steele, J. and Gamble J. Predator control in enclosures, in *Marine Mesocosms.* D. Grice and M. Reeve, Eds. Springer Verlag, New York, 1982, 227–238.

Stegeman, J. and Kloepper-Sams, P. Cytochrome P-450 isozymes and monooxygenase activity in aquatic animals, *Environ. Health Persp.* 71:87–95, 1987.

Sunda, W. and Gillespie, P. The response of a marine bacterium to cupric ion and its use to estimate cupric ion activity in sea water, *J. Mar. Res.* 37:761–777, 1979.

Swain, W. Human health consequences of consumption of fish contaminated with organochlorine compounds, *Aquat Toxicol.* 11:357–377, 1988.

Thomas, P. Reproductive endocrine function in female atlantic croaker exposed to pollutants, *Mar. Env. Res.* 24:179–183, 1988.

Trefry, J., Nelson, T., Trocine, R., Metz, S., and Vetter, T. Trace metal fluxes through the Mississippi River Delta system, *Rapp P.-V. Reun. Cons. Int. Explor. Mer.* 186:277–288, 1986.

Wade, T. and Quinn, J. Incorporation, distribution and fate of saturated petroleum hydrocarbons in sediments from a controlled marine ecosystem, *Mar. Env. Res.* 3:15–33, 1980.

Waldock, M. and Thain, J. Shell thickening in *Crassostrea gigas*: organotin antifouling or sediment induced?, *Mar. Poll. Bull.* 14:411–415, 1983.

Weeks, B. and Warriner, J. Effects of toxic chemicals on macrophage phagocytosis in two estuarine fishes, *Mar. Env. Res.* 14:327–335, 1984.

Weis, J.S. and Weis, P. A rapid change in methylmercury tolerance in a population of killifish, *Fundulus heteroclitus* from a golf course pond, *Mar. Environ. Res.* 13:231–245, 1984.

Weis J.S. and Weis, P. Tolerance and stress in a polluted environment: the case of the mummichog, *BioScience* 39:89–95, 1989.

Widdows, J. Physiological responses to pollution, *Mar. Poll. Bull.* 16:129–134, 1985.

Wilson, J. and McMahon, R. Effects of high environmental copper concentration on the oxygen consumption, condition and shell morphology of natural popoulations of *Mytilus edulis* and *Littorina rudis, Comp Biochem. Physiol.* 70C:139–147, 1981.

Wolfe, D., Champ, M., Flemer, D., and Mearns, A. Long-term biological data sets: their role in research, monitoring, and management of estuarine and coastal marine systems, *Estuaries* 10:181–193, 1987.

Chapter

15

Wildlife Toxicology

Gregory J. Smith and Russell J. Hall

A. INTRODUCTION

The field of wildlife toxicology involves the assessment of chemical hazards in the environment to wildlife populations. Research is conducted to determine if either (1) an apparent population decline, die-off, or other problem observed in wild species is the result of exposure to an environmental toxin or (2) a measured or predicted environmental concentration of a chemical is likely to result in exposure to and adverse biological effects in certain wildlife species. Moreover, the individual species or groups of organisms potentially affected and the pathways of exposure are of considerable interest. From this perspective, wildlife toxicology is a subdiscipline of environmental toxicology that seeks to determine if exposure of wildlife to a chemical has either resulted in a problem or is likely to result in a problem if the chemical is released into the environment. There are two major driving forces in wildlife toxicology: assessing causes and mechanisms of problems in the field and predicting potential adverse effects through experiments and testing. Both of these goals ultimately require a combination of field and laboratory studies to describe a cause and effect relationship with any certainty. Combining the results of laboratory and field investigations is a critical process in understanding toxicological hazards to wildlife populations. Moreover, in addition to biological field and laboratory data required to assess or predict toxic hazards of chemicals to wildlife, analytical chemistry is almost always required at some point in the comprehensive investigation. Chemical analysis often serves to identify those contaminants or metabolites present, and their respective concentrations, and may often provide for some diagnosis of an observed wildlife problem.

0-8493-8851-1/94/$0.00+$.50
© 1994 by CRC Press, Inc.

However, in many instances chemical residue data obtained from field samples has prompted further laboratory and field investigations to determine if any undetected biological effects are occurring.

In this chapter, we address some of the major features of wildlife toxicology, including its inception, complexity, approaches and methods, and relation to other subdisciplines of environmental toxicology. These features will be illustrated through examples of problems in which chemicals in the environment have been demonstrated or implicated in threatening populations of wild animals. This work should not be viewed as a comprehensive summary of the discipline or of the important literature produced in this field. Original papers cited in this chapter will contain more references and will discuss in more detail specific aspects of wildlife toxicology.

B. FUNDAMENTAL PRINCIPLES OF WILDLIFE TOXICOLOGY

The field of wildlife toxicology is a subdiscipline of environmental toxicology. It represents a heterogeneous mixture of wildlife ecology, toxicology, population biology, chemistry, botany, biometry, and other biological and biomedical sciences. The principles that guide experimental design and interpretation are likewise derived from these and other fields of study. There are a few fundamental principles that form the basis for wildlife investigations. Moreover, these principles are not unique to wildlife toxicology but are derived from other biological sciences and applied to wildlife problems.

One fundamental concept of wildlife toxicology, especially when field research is conducted, is exposure vs. effects. Demonstrating exposure to a chemical may or may not be relatively straight forward. In the laboratory, exposure is one of the variables manipulated to achieve an effect. In the environment, showing exposure usually involves collecting of animal tissues or organs and performing chemical analysis to determine if exposure has occurred. While this approach may have several limitations, such as the inability to detect the specific chemical with a given analytical method, the inability to detect the absence of a chemical in a tissue due to accumulation/depuration kinetics, or even the inability to definitively collect the correct tissue that represents a site of accumulation, demonstrating exposure is almost always easier than showing resulting effects from that exposure. A good example is that of polychlorinated biphenyls (PCBs) in wildlife tissue. Tissues could be collected from millions of apparently healthy and normal wildlife species that would contain some quantifiable concentration of PCBs. However, an effect resulting from this exposure may not have occurred in one individual if concentrations were below those that produce some biological effect. The cause/effect or dose/response relationship that is central to laboratory studies is rare, if not nonexistent, in the study of environmental contaminants and their effects on wild populations. This demonstrates the need for integrated field and laboratory approaches to solving complex environmental problems of contaminants in wildlife populations.

Another important concept is the dose/response or concentration/effect relationship that is central to classical toxicology. This applies not only to laboratory and controlled field experiments, but also to the assessment of contaminant problems in the environment. Studies of the effects of pollutants from point sources have used this concept most effectively by attempting to establish the occurrence of a graded response in wildlife with changes in distance (exposure) from the source.

Treatment-related effect is related to dose responsiveness in the sense that the wildlife toxicologist must, at some point, ascribe an adverse biological response observed in a wildlife population to a specific "treatment", that is, exposure to a contaminant of concern. Ultimately, the question is asked about the impact of any adverse effect on the population. This is perhaps the greatest challenge to the field of wildlife toxicology today, determining population effects of contaminants.

C. COMPLEXITY OF WILDLIFE TOXICOLOGY MODELS

Wildlife species are often considered important natural resources as well as indicators of the environmental health of an ecosystem. The term "wildlife" has no precise definition and when interpreted by some would mean primarily wild mammals and birds. Others might include reptiles, amphibians, fish, and even wild plants and lower food chain organisms. Even if the term wildlife is restricted to the 2166 species of amphibians, reptiles, birds, and mammals that inhabit North America (Banks et al., 1987), the differences in the life history, physiology, ecology, and potential sensitivity to pollutants among these groups of species are enormous. It is not unusual to have the acute oral toxicity of a compound differ by orders of magnitude for two different species within the same genus (Tucker and Crabtree, 1970).

Biological and ecological factors that contribute to differences among wildlife species in their response to a contaminant that occurs in the environment can be summarized as follows:

1. Differences contributing to differential contaminant exposure of species
   - Geographic range, including migration corridors
   - Specific habitat requirements (e.g., wetlands and forests)
   - Diurnal activity patterns (e.g., use of farm fields in relation to pesticide application times)
   - Food habits and potential for biomagnification to occur in lower food chain organisms

2. Differences contributing to differential responses to contaminant exposure
   - Biochemical differences in detoxification mechanisms
   - Differences in absorption via skin and/or gastrointestinal pathway
   - Differences in depuration mechanisms (e.g., female birds can eliminate some contaminants in their eggs)
   - Different life stage sensitivities
   - Reproductive system differences
   - Target organ differences (certain groups of wildlife species do not even have the same internal organs)

The differences among wildlife species are large and contribute to the complexity of wildlife toxicology investigations. Most studies focus on single species problems and try to determine which contaminant(s) are responsible for an observed problem. The complexity of these investigations is further increased when multiple contaminant exposure is likely to occur. Simultaneous or sequential exposure of an animal to more than one contaminant can have a dramatic impact on the outcome of their exposure. Moreover, chemicals may interact to produce antagonistic, additive, or synergistic effects.

One final but important consideration in evaluating the complexity of contaminant effects on wildlife is the great diversity of xenobiotics that wildlife species may be exposed to. Classic studies in wildlife toxicology dealt with chlorinated hydrocarbon pesticides such as dichlorodiphenyltrichloroethane (DDT). However, it has become clear that an actual application of a chemical to the environment is not necessary to produce a severe toxic hazard to wildlife populations. Management of agricultural drainwater in the western U.S. provides a good example of how land management can produce hazards to wildlife without a single spray plane leaving the ground.

1. History of Wildlife Toxicology and Its Role in Ecology

The field of wildlife ecotoxicology began as an outgrowth of wildlife ecology and wildlife management, and recruitment of the skills of professional toxicologists to the task occurred only as the field developed. These roots have had a continuing impact in that, even today, wildlife management agencies have a greater investment in the field than do those agencies directly involved with pollution control or chemical regulation. In fact, wildlife ecotoxicology arose from wildlife ecology rather than from the biomedical professions, as did other branches of toxicology, adding dimensions to the broader field of toxicology that otherwise would have developed more slowly.

The specific origins of wildlife ecotoxicology as an organized field of study are something of a fluke. The Great Depression of the 1930s had both economic and environmental impacts, one of which was that poor agricultural practices severely depleted wildlife populations. The U.S. Fish and Wildlife Service, working with the Department of Agriculture, sought to develop and evaluate new farming methods, searching for those that would be economically sustainable and would best conserve soil, water, and wildlife populations. Among the innovations evaluated beginning in 1943 was the miracle insecticide DDT. Had the toxicology of DDT been simpler it is likely that Fish and Wildlife Service scientists would have quickly gone on to other problems. By the time the research, begun in 1943, had come full circle, 30 years had elapsed and the profession of wildlife toxicology had become well established. Hall (1987) gives a more complete account of the origins of ecotoxicology within the U.S. Fish and Wildlife Service.

One indirect contribution of the early studies in wildlife ecotoxicology was that their results provided much of the basis for the publication of *Silent Spring* by Rachel Carson (1962). This work, a huge popular success, was instrumental in alerting the nation and the world to actual and potential hazards of synthetic chemicals in the environment.

The development of wildlife ecotoxicology preceded most of the legislative and regulatory actions that govern environmental chemicals, and most existing regulatory guidelines for assessing chemical hazards to wildlife grew out of methods developed in early studies. The primary federal legislative authority underlying early efforts was the Migratory Bird Treaty Act of 1918, and if there has been one consistent focus on chemical contamination, it has been an emphasis on chemical pollution that results in the "taking" of migratory birds. Because wildlife ecotoxicology did not originate from chemical regulation, the scope of its investigations has never been limited to chemicals subject to intense regulatory scrutiny. Also, significant foci have ranged from agricultural chemicals to spent lead shot, to industrial wastes, to acid precipitation, to the inadvertent byproducts of irrigation.

Within the discipline of ecology, environmental toxicology belongs under the study of limiting factors (Odum, 1959). Concepts concerning the action of environmental factors in limiting the distribution and abundance of free-living populations have been expressed as various rules and corollaries, some of the more important of which follow:

- Of factors that may limit populations, that which is present in the least favorable amount will tend to be limiting (i.e., only one factor at a time may limit a population); though a chemical in the environment might be potentially limiting, it will effectively do so only if some other environmental factor is not more limiting.
- The effects of limiting factors may vary seasonally and geographically.
- Different factors may be limiting at different stages of the life cycle, with the period of reproduction often being most sensitive.

- One limiting factor may alter the tolerance of a population for other limiting factors.
- Species with the greatest tolerance for all limiting factors tend to be most widely distributed geographically.

2. Studies of DDT

Important findings based on studies of effects of DDT on wildlife include (1) the frequency and duration of exposure of animals to certain chemicals may be more important than the magnitude of exposure, (2) the physiological condition of animals at the time of exposure to a chemical may affect lethality, particularly as it affects storage of chemicals in fat, (3) metabolic transformation of pesticides may produce compounds that are more toxic or more persistent in tissues than the parent compounds, (4) compounds with relatively low overt toxicity to adults may depress populations by impacting reproduction, and (5) even closely related wildlife species may vary greatly in their sensitivity to certain pesticides.

The first large scale efforts to determine the impacts of DDT on wildlife were not wholly successful. The shortcomings of these attempts had two results. First, the hazards of DDT were underestimated and were not fully known for many years. Second, the failure of the first assessments led to the application of new methods and capabilities which significantly increased the ability to address other problems in the future. Early field experiments (Hotchkiss and Pough, 1946; Mitchell, 1946; Stewart et al., 1946; Stickel, 1946) were based on monitoring the wildlife populations of sprayed and unsprayed areas in comparable habitats. They provided a few indications that heavy (5 lb/acre) applications of DDT in forests impact wildlife, but they did not find serious effects. Elsewhere, significant mortality of wildlife from DDT applications was observed in the field (Benton, 1951; Springer, 1961; Wurster et al., 1965). Though well designed and executed, the first studies were conceived without any knowledge of the environmental dynamics, physiological kinetics, or reproductive toxicity of the compound. Later field research developed methods to overcome many of the failings of the first studies.

Experimental studies seeking to explain the complex findings from the field (Stickel, 1973) showed the important roles of chronic intake, accumulation in lipids, and mobilization to the brain of stored residues in lethal toxicity, not only for DDT but also for many of the other organochlorines in common use. Investigations of heptachlor, aldrin, dieldrin, and other pesticides (Stickel, 1973; Stickel, 1975) likewise used the methods of experimental biology and analytical chemistry to better understand mechanisms and hazards of these chemicals.

These investigations that explained phenomena observed in the field and could predict hazard under certain environmental conditions were made possible only by advances in analytical chemistry that permitted the detection of pesticide residues in minute quantities in small amounts of tissue. Overcoming initial difficulties in separating chemical residues from the complex array of organic compounds occurring naturally in animal tissues and the significant discovery that polychlorinated biphenyl (PCB) residues had confounded accurate estimation of DDT residues were important advances in analytical chemistry that enabled continued progress in both field and laboratory research (Bagley et al., 1970; Mulhern et al., 1970).

Declines in productivity at sublethal exposures to DDT were observed in the field (Stickel and Heath, 1965; Stickel et al., 1966), and several experimental investigations involving breeding populations of wildlife species were begun. The mechanism of reduced productivity was not understood until observations in the field suggested that reduced eggshell strength was involved (Ratcliffe, 1967; Hickey and Anderson, 1968). Reanalysis of experiments in progress (Porter and Wiemeyer, 1969) and later experiments (Heath et al., 1969; Longcore et al., 1971) confirmed that DDT in the diet reduces eggshell thickness and strength and that the mechanism can reduce reproductive success in birds. They further indicated that the effect is much stronger in some species than in others; the kestrels studied by Porter and Wiemeyer (1969) are less sensitive than certain of their close relatives (Hickey and Anderson, 1968), but more sensitive than waterfowl (Longcore et al., 1971). Some other species of birds are highly resistant to the same effect (Stickel, 1975).

3. Cholinesterase Inhibitors

The cholinesterase-inhibiting pesticides are a large group of chemicals that grew out of developmental work on nerve gasses conducted largely during the Second World War. Their use increased markedly after the banning or restriction of DDT and other organochlorines, and organophosphates and carbamates, collectively the cholinesterase inhibitors, are now the class of insecticides in greatest use. Early studies of effects on wildlife used evaluations of acute (Tucker and Crabtree, 1970; Hudson et al., 1984) or subacute (Heath et al., 1972; Hill et al., 1975) toxicity tests to identify chemicals of particular hazard to birds and selected other wildlife. Acute testing involved the administration of chemicals in gelatin capsules, 14 d of observation following dosage, and calculation of the LD_{50}, the mean amount lethal to 50% of the experimental population. In subacute testing, chemicals were mixed into the diet to mimic likely exposure in the wild, and animals were fed treated diets for 5 d followed by an additional 3 d of observation. The statistic reported was the LC_{50}, the concentration of the chemical in the diet that is lethal to half the experimental population over the 8-d period of the test. Chemicals that are toxic based on chronic exposure should appear much more toxic when given subacutely than acutely, whereas those with low chronic toxicity should have reasonably close values for LD_{50} and LC_{50} statistics when amounts of chemical consumed are similar. Most cholinesterase inhibitors are relatively less toxic when consumed subacutely than when dosed acutely. Indeed, some acutely toxic chemicals are essentially nontoxic when given subacutely because they cause food avoidance and are not consumed by birds (Grue, 1982).

The mechanism of toxicity is inhibition of cholinesterase, the enzyme that permits transmission of impulses between neurons by breaking down the neurotransmitter acetylcholine and permitting its recycling. Measurement of cholinesterase activity in the brain to indicate exposure to or death from cholinesterase inhibitors was used in medicine, and the method employed there (Ellman et al., 1961) was adapted by Ludke et al. (1975) and Hill and Fleming (1982) for use on birds. Additional work (Hall and Clark, 1982) suggested that relationships developed for birds may also apply to reptiles.

Acute and subacute toxicity led to a ranking of cholinesterase inhibiting pesticides based on the sensitivity of birds to the chemicals. Sensitivity alone, however, is not a totally reliable predictor of risk; some highly toxic compounds are rarely implicated in die-offs of free-living wildlife, and compounds not among the most toxic may commonly cause mortality in the field (Smith, 1987).

Many attempts to discover effects of long term exposure to cholinesterase inhibitors or effects on reproduction comparable to those caused by some organochlorines have shown that such effects are unlikely. Powell and Gray (1980), Grue et al. (1982), and Grue and Shipley (1984), respectively, dosed nestling starlings daily with an organophosphate, dosed foraging adults, and searched for important modifications of behavior in experimentally sprayed areas, but they found that the most significant effect was a slight and ephemeral decrease in growth of young. Powell (1984) extended similar detailed investigations to field applications of pesticides. Stromborg (1981) incorporated graded concentrations of diazinon into the diets of breeding quail and noted reduced reproductive success. These effects, however, seemed attributable to sickening of the adults at the higher pesticide concentrations rather than to some other mechanism. Haseltine and Hensler (1981) exposed young from a breeding colony of American black ducks *(Anas rubripes)* to phosphamidon in the feed, but found only minor reductions in growth. Studies of the effects of temephos on mallard *(Anas playtrynchos)* reproduction indicated some apparent reduction of duckling survival (Franson et al., 1983), but later research (Fleming et al., 1985) indicated effects on duckling survival only when exposure was at the highest level and when ambient temperatures were cold. Research (Rattner et al., 1982; Rattner and Franson, 1984), on the interaction of cholinesterase inhibitors and extremes of temperature, indicated increased toxicity at cold temperatures and reduced cold tolerance with sublethal exposures, but these effects occurred soon after exposure while other signs such as cholinesterase depression persisted. Effects of cholinesterase inhibitors on salt gland function (Eastin et al., 1982; Rattner et al., 1983) were measurable, but did not affect osmoregulation.

Experiments examining the potential for secondary poisoning by cholinesterase inhibitors indicated that poisoned birds (Hill and Mendenhall, 1980) and exposed amphibians (Hall and Kolbe, 1980; Fleming et al., 1982; Powell et al., 1982) can produce mortality in animals eating them. Field observations showed that both adult gulls and nestlings may be killed by consuming poisoned insects (White et al., 1979; White and Kolbe, 1985) and that both magpies and their predators can be poisoned by consuming hair pulled from cattle treated with acaricides (Henny et al., 1985).

Pesticides in greatest demand are those that are selectively toxic to pest species but relatively nontoxic to nontarget species, particularly humans. The organophosphates and carbamates tend to be selective in their action, and the mechanisms of selectivity have been the subject of research (Potter and O'Brien, 1963, 1964; Murphy et al., 1968; Benke et al., 1974; Andersen et al., 1977; Wang and Murphy, 1982). Owing to the selectivity of the cholinesterase inhibitors, it may not be safe to predict the responses of one species based on information gained from another. Nevertheless, there have been suggestions that such predictions are possible (Kenaga, 1979). Experimental investigations that challenged amphibians (Hall and Kolbe, 1980; Fleming et al., 1982; Hall, 1990), reptiles (Hall and Clark, 1982), songbirds (Fleming and Grue, 1981; Schafer et al., 1983), raptors (Rattner and Franson, 1984), herons (Smith et al., 1986), and wild rodents (Graf et al., 1976; Rattner and Michael, 1985) in addition to the domestic rodent, waterfowl, and gallinaceous bird species usually tested sought preliminary information on variability of responses among vertebrate groups. Research on amphibians confirmed early reports (Edery and Schatzberg-Porath, 1960) that they are highly resistant to most cholinesterase inhibitors. That this resistance may be an adaptation to life in freshwater ponds is suggested by the recent discovery (Cook et al., 1989) that toxic freshwater algae produce cholinesterase inhibitors.

4. Agricultural Drainwater Contaminants and Wildlife

A combination of natural circumstances and anthropogenic activities in the western U.S. has resulted in perhaps the most noticeable wildlife contamination problem since the widespread use of DDT. This situation has resulted from irrigation of agricultural land, drainage of the fields to prevent salt and water build-up, and deposition of the waste drainwater containing salts and toxic elements into wetlands used by wildlife.

In 1982, concentrations of selenium in mosquitofish at the Kesterson National Wildlife Refuge and Reservoir in the San Joaquin Valley of California were about 100 times higher than determined for fish collected at a nearby area that did not receive agricultural drainwater (Presser and Ohlendorf, 1987). In

1983, field studies at Kesterson determined that reproductive success of birds nesting there was severely impacted and embryonic deaths and deformities occurred at alarming rates in certain species (Ohlendorf et al. 1986). Later investigations that included hydrology, geochemistry, ecological field studies, and laboratory toxicological investigations showed that several naturally occurring elements were mobilized and transported in agricultural drainwater. Although selenium was initially the principal element of concern, elements such as boron, arsenic, molybdenum, and others were present in agricultural drainwater systems and considered to be potentially harmful to wildlife species.

The result of the highly visible effects found in the drainwater-contaminated ponds of the Kesterson Wildlife Refuge was a multiyear research effort involving more than 100 agencies and coordinated by the San Joaquin Drainage Program (San Joaquin Valley Drainage Program, 1990). Over the course of several years, wildlife investigations were conducted to determine (1) the occurrence and distribution of various agricultural drainwater contaminants in wildlife and wildlife food items, (2) the effects of these contaminants on reproduction, survival, physiology, and behavior in selected wildlife species, and (3) the possible interactions of drainwater contaminants to produce adverse effects in the species exposed. A large portion of the wildlife toxicology studies conducted by the federal government were done by the Patuxent Wildlife Research Center, while the fisheries investigations were done at the National Fisheries Contaminant Research Center also of the U.S. Fish and Wildlife Service.

Field studies were conducted throughout the San Joaquin Valley to determine embryotoxic effects of drainwater contaminants on birds (Ohlendorf et al., 1986) and mammals (Clark, 1987; Clark et al., 1989) and to determine the extent of contamination of wildlife food organisms (Hothem and Ohlendorf, 1989). Experimental reproduction studies were conducted to determine the effects of specific concentrations of selenium on reproduction in mallards (Heinz et al., 1987) and black-crowned night-herons (Smith et al., 1988). The chemical form of selenium was determined to be important in the interpretation of its potential biological effects (Heinz et al., 1989). The uptake and depuration kinetics were studied in the mallard to assess how long migrating birds may need to be exposed to selenium before effects would result (Heinz et al., 1990), and the routes of exposure were determined for wildlife habitats (Lemly and Smith, 1987). Sublethal physiological (Hoffman et al., 1989) and behavioral effects of selenium (Heinz and Gold, 1987) were determined using controlled experiments applying biomedical clinical techniques and those of ethology, respectively. The effects of boron, another important drainwater contaminant, was evaluated for mallard reproduction (Smith and Anders, 1989) and duckling behavior (Whitworth et al., 1991).

Ultimately, these progressions of studies in the field and laboratory using a variety of techniques and approaches provided a comprehensive picture of the extent and effects of the hazards of agricultural drainwater contaminants. Selenium, boron, and other drainwater contaminants are widespread in the western U.S. where soils have formed from former marine sediments, and irrigation and drainage are now essential for agriculture to exist. The land/water management practice of irrigating and draining creates contaminated water that allows for the accumulation of toxic elements in wildlife food chains. If those food chain concentrations reach high enough levels, wildlife species may experience death by direct acute toxicity, teratogenesis in embryos, and severe reproductive impairment sufficient to impact wildlife populations, at least on a local scale.

D. CONCLUSIONS

Wildlife toxicology attempts to explain actual or potential effects of chemicals in the environment on free-living biota. The problem is complex not only because of the enormous number of chemicals released into the environment but also because of the great variety of species potentially affected, the many potential mechanisms for exposure, and the limited ability to measure and predict effects on wild populations. Owing to the inherent complexity of its subjects, wildlife toxicology has always been a hybrid discipline, applying the techniques of wildlife ecology, toxicology, environmental chemistry, and other fields in coordinated approaches to problems. Most professionals in wildlife toxicology come from backgrounds in biology, ecology, chemistry, biochemistry, and biometry. Consequently, wildlife toxicology has been a fertile ground for multidisciplinary research, and the approaches developed to address research problems have extended far beyond those practices in the traditional disciplines on which it rests.

Analyses of the problems presented by the selected examples in this chapter illustrate the range of problems met by the wildlife toxicologist and the variety of approaches taken to understand and address the hazards they pose. With DDT, chronic toxicity proved ultimately more important than acute toxicity, and target organs and processes were not recognized until there had been much misdirected research. In addition, hazard resulted primarily from metabolites rather than the parent compound, and accumulation in the environment and magnification by biological processes were not fully understood until advances were made in analytical chemistry. Differences in species sensitivity also confounded early attempts to understand actions and hazards of the chlorinated hydrocarbon pesticides. With cholinesterase inhibitors, the primary hazard to wildlife ultimately was from direct toxicity of these chemicals or their metabolites. The greatest concerns for wildlife were unpredictable results from selective toxicity and unusual routes of exposure. Agricultural drainwater contamination of wildlife illustrated the accumulation of natural environmental elements to toxic levels, interaction of toxic substances, modification of toxicity by chemical form, and a variety of sublethal and lethal toxic effects at different levels of exposure.

Each environmental problem presented to the wildlife toxicologist will have its own set of physical, chemical, and biological components that require the application of multidisciplinary approaches to resolve. While assessment and understanding of observed wildlife population impacts remains to be a major challenge in the field of wildlife toxicology, predictions of environmental hazards to wildlife will continue to increase in importance to resource managers and require new and innovative approaches.

REFERENCES

Andersen, R.A., Aaraas, I., Gaare, G., and Fonum, F. Inhibition of acetylcholinesterase from different species by organophosphorus compounds, carbamates and methylsulphonylfluoride, *Gen. Pharmac.* 8:331–334, 1977.

Bagley, G.E., Reichel, W.L., and Cromartie, E. Identification of polychlorinated biphenyls in two bald eagles by combined gas chromatography-mass spectrometry, *J. Assoc. Offic. Agr. Chem.* 53:251–261, 1970.

Banks, R.C., McDiarmid, R.W., and Gardner, A.L. Checklist of vertebrates of the United States, the U.S. Territories, and Canada, U.S. Fish Wildl. Serv. Resour. Publ. 166:79, 1987.

Benke, G.M., Cheever, K.L., Mirer, K.L., and Murphy, S.D. Comparative toxicity, anticholinesterase action and metabolism of methyl parathion and parathion in sunfish and mice, *Toxicol. Appl. Pharmacol.* 28:97–109, 1974.

Benton, A.H. Effects on wildlife of DDT used for control of Dutch elm disease, *J. Wildl. Manage.* 15:20–27, 1951.

Carson, R. *Silent Spring.* Houghton Mifflin, Boston, 1962, 368.

Clark, D.R., Jr. Selenium accumulation in mammals exposed to contaminated California irrigation drainwater, *Sci. Total Environ.* 66:147–168, 1987.

Clark, D.R., Jr., Ogasawara, P.A., Smith, G.J., and Ohlendorf, H.M. Selenium accumulation by raccoons exposed to irrigation drainwater at Kesterson National Wildlife Refuge, California, 1986, *Arch. Environ. Contam. Toxicol.* 18:787–794, 1989.

Cook, W.O., Beasley, V.R., Lovell, R.A., Dahlem, A.M., Hooker, S.R., Mahmood, N.A., and Carmichael, W.W. Consistent inhibition of peripheral cholinesterases by neurotoxins from the freshwater cyanobacterium *Anabaena flosaquae*: studies of ducks, swine, mice and a steer, *Environ. Toxicol. Chem.* 8:915–922, 1989.

Eastin, W.C., Fleming, W.J., and Murray, H.C. Organophosphate inhibition of avian salt gland Na, K-ATPase activity, *Comp. Biochem. Physiol.* 73C:101–107, 1982.

Edery, H. and Schatzberg-Porath, G. Studies on the effects of organophosphorus pesticides on amphibians, *Arch. Int. Pharmacodyn.* 124:212–224, 1960.

Ellman, G.L., Courtney, K.D., Andres, V., Jr., and Featherstone, R.M. A new and rapid colorimetric determination of acetylcholinesterase activity, *Biochem. Pharmacol.* 7:88–95, 1961.

Fleming, W.J. and Grue, C.E. Recovery of cholinesterase activity in five avian species exposed to dicrotophos, an organophosphorus pesticide, *Pestic. Biochem. Physiol.* 16:129–135, 1981.

Fleming, W.J., de Chacin, H., Pattee, O.H., and Lamont, T.G. Parathion accumulation in cricket frogs and its effect on American kestrels, *J. Toxicol. Environ. Health* 10:921–927, 1982.

Fleming, W.J., Heinz, G.H., Franson, J.C., and Rattner, B.A. Toxicity of abate 4E (temephos) in mallard ducklings and the influence of cold, *Environ. Toxicol. Chem.* 4:193–199, 1985.

Graf, G., Guttman, S., and Barrett, G. The effects of an insecticide stress on genetic composition and population dynamics of a population of feral *Mus musculus*, *Comp. Biochem. Physiol. C. Comp. Pharmacol.* 55:103–110, 1976.

Grue, C.E. Response of common grackles to dietary concentrations of four organophosphate pesticides, *Arch. Environ. Contam. Toxicol.* 11:617–626, 1982.

Grue, C.E. and Shipley, B.K. Sensitivity of nestling and adult starlings to dicrotophos, an organophosphate insecticide, *Environ. Res.* 35:454–465, 1984.

Grue, C.E., Powell, G.V.N., and McChesney, M.J. Care of nestlings by wild female starlings exposed to an organophosphate pesticide, *J. Appl. Ecol.* 19:327–335, 1982.

Hall, R.J. Impact of pesticides on bird populations, in *Silent Spring Revisited*. G.J. Marco, R.M. Hollingworth, and W. Durham, Eds. American Chemical Society, Washington, DC, 1987, 85–111.

Hall, R.J. Accumulation, metabolism and toxicity of parathion in tadpoles, *Bull. Environ. Contam. Toxicol.* 44:629–635, 1990.

Hall, R.J. and Clark, D.R., Jr. Responses of the iguanid lizard *Anolis carolinensis* to four organophosphorus pesticides, *Environ. Pollut.* (Ser. A) 28:45–52, 1982.

Hall, R.J. and Kolbe, E. Bioconcentration of organophosphorus pesticides to hazardous levels by amphibians, *J. Toxicol. Environ. Health* 6:853–860, 1980.

Haseltine, S. and Hensler, G. Growth of mallards fed phosphamidon for 13-day periods during three different developmental stages, *Environ. Pollut.* (Ser. A) 25:139–147, 1981.

Heath, R.G., Spann, J.W., and Kreitzer, J.F. Marked DDE impairment of mallard reproduction in controlled studies, *Nature* 224:47–48, 1969.

Heath, R.G., Spann, J.W., Hill, E.F., and Kreitzer, J.F. Comparative dietary toxicities of pesticides to birds, U.S. Fish and Wildlife Service Special Science Report — Wildlife No. 152. 57, 1972.

Heinz, G.H., Hoffman, D.J., Krynitsky, A.J., and Weller, D.M.G. Reproduction in mallards fed selenium, *Environ. Toxicol. Chem.* 6:423–433, 1987.

Heinz, G.H., Hoffman, D.J., and Gold, L.G. Impaired reproduction of mallards fed an organic form of selenium, *J. Wildl. Manage.* 53:418–428, 1989.

Heinz, G.H. and Gold, L.G. Behavior of Mallard ducklings from adults exposed to selenium, *Environ. Toxicol. Chem.* 6:863–865, 1987.

Heinz, G.H., Pendleton, G.W., Krynitsky, A.J., and Gold, L.G. Selenium accumulation and elimination in mallards, *Arch. Environ. Contam. Toxicol.* 19:374–379, 1990.

Henny, C.J., Blus, L.J., Kolbe, E.J., and Fitzner, R.J. Organophosphate insecticide (Famphur) topically applied to cattle kills magpies and hawks, *J. Wildl. Manage.* 49:648–658, 1985.

Hickey, J.J. and Anderson, D.W. Chlorinated hydrocarbons and eggshell changes in raptorial and fish-eating birds, *Science* 162:271–273, 1968.

Hill, E.F. and Fleming, W.J. Anticholinesterase poisoning of birds: field monitoring and diagnosis of acute poisoning, *Environ. Toxicol. Chem.* 1:27–38, 1982.

Hill, E.F. and Mendenhall, V.M. Secondary poisoning of barn owls with Famphur, an organophosphate insecticide, *J. Wildl. Manage.* 44:676–681, 1980.

Hill, E.F., Heath, R.G., Spann, J.W., and Williams, J.D. Lethal dietary toxicities of environmental pollutants to birds. U.S. Fish and Wildlife Service Special Science Report — Wildlife No. 191, 61, 1975.

Hoffman, D.J., Heinz, G.H., and Krynitsky, A.J. Hepatic glutathione metabolism and lipid peroxidation in response to excess dietary selenomethionine and selenite in mallard ducklings, *J. Toxicol. Environ. Health* 27:263–271, 1989.

Hotchkiss, N. and Pough, R.H. Effects on forest birds of DDT used for gypsy moth control in Pennsylvania, 1945, *J. Wildl. Manage.* 10:202–207, 1946.

Hothem, R.L. and Ohlendorf, H.M. Contaminants of foods of aquatic birds at Kesterson Reservoir, California, 1985, *Arch. Environ. Contam. Toxicol.* 18:773–786, 1989.

Hudson, R.H., Tucker, R.K., and Haegele, M.A. *Handbook of Toxicity of Pesticides to Wildlife,* 2nd ed. U.S. Fish and Wildlife Service Resource Publication 153, 1984, 90.

Kenaga, E.E. Acute and chronic toxicity of 75 pesticides to various animal species, *Down to Earth* 35:25–31, 1979.

Lemly, A.D. and Smith, G.J. Aquatic cycling of selenium: implications for fish and wildlife. Fish and Wildlife Leaflet No. 12, U.S. Fish and Wildlife Service, Washington, DC, 1987,10.

Longcore, J.R., Sampson, F.B., and Whittendale, T.W., Jr. DDE thins eggshells and lowers reproductive success of captive black ducks, *Bull. Environ. Contam. Toxicol.* 6:485–490, 1971.

Ludke, J.L., Hill, E.F., and Dieter, M.P. Cholinesterase (ChE) response and related mortality among birds fed ChE inhibitors, *Arch. Environ. Contam. Toxicol.* 3:1–21, 1975.

Mitchell, R.T. Effects of DDT spray on eggs and nestlings of birds, *J. Wildl. Manage.* 10:192–194, 1946.

Mulhern, B.M., Reichel, W.L., Locke, L.N., Lamont, T.G., Belisle, A., Cromartie, E., Bagley, G.F., and Proutly, R.M. Organochlorine residues and autopsy data from bald eagles 1966–68, *Pestic. Monit. J.* 4:141–144, 1970.

Murphy, S.D., Lauwreys, R.R., and Cheever, K.L. Comparative anticholinesterase action of organophosphorus insecticides in vertebrates, *Toxicol. Appl. Pharmacol.* 12:22–35, 1968.

Odum, E.P. *Fundamentals of Ecology.* W.B. Saunders, Philadelphia, 1959, 546.

Ohlendorf, H.M., Hoffman, D.J., Saiki, M.K., and Aldrich, T.W. Embryonic mortality and abnormalities of aquatic birds: apparent impacts of selenium from irrigation drainwater, *Sci. Total Environ.* 52:49–63, 1986

Porter, R.D. and Wiemeyer, S.N. Dieldrin and DDT: effects on sparrow hawk eggshells and reproduction, *Science* 165:199–200, 1969.

Potter, J.L. and O'Brien, R.D. The relation between toxicity and metabolism of paraoxon in the frog, mouse and cockroach, *Ent. Exp. Appl.* 6:319–325, 1963.

Potter, J.L. and O'Brien, R.D. Parathion activation by livers of aquatic and terrestrial vertebrates, *Science* 144:55–57, 1964.

Powell, G.V.N. Reproduction of redwinged blackbirds in fields treated with the organophosphate insecticide, Famphur, *J. Appl. Ecol.* 21:83–95, 1984.

Powell, G.V.N. and Gray, D.C. Dosed free-living nestling starlings with an organophosphate pesticide, Famphur, *J. Wildl. Manage.* 44:918–921, 1980.

Powell, G.V.N., DeWeese, L.R., and Lamont, T.G. A field evaluation of frogs as a potential source of secondary organophosphorus insecticide poisoning, *Can. J. Zool.* 60:2233–2235, 1982.

Presser, T.S. and Ohlendorf, H.M. Biogeochemical cycling of selenium in the San Joaquin Valley, California, USA, *Environ. Manage.* 11:805–821, 1987.

Ratcliffe, D.A. Decrease of eggshell weight in certain birds of prey, *Nature* 215:208–210, 1967.

Rattner, B.A. and Franson, J.C. Methyl parathion and fenvalerate toxicity in American kestrels; acute physiological responses and effects of cold, *Can. J. Physiol. Pharmacol.* 62:787–792, 1984.

Rattner, B.A. and Michael, S.D. Organophosphorus insecticide induced decrease in plasma luteinizing hormone concentration in white-footed mice, *Toxicol. Lett.* 24:65–69, 1985.

Rattner, B.A., Sileo, L., and Scanes, C.G. Hormonal responses and tolerance to cold of female quail following parathion ingestion, *Pestic. Biochem. Physiol.* 18:132–138, 1982.

Rattner, B.A., Fleming, W.J., and Murray, H.C. Osmoregulatory function in ducks following ingestion of the organophosphorus insecticide fenthion, *Pestic. Biochem. Physiol.* 20:246–255, 1983.

Schafer, E.W., Jr., Bowles, W.A., Jr., and Hurlbut, J. The acute oral toxicity, repellency and hazard potential of 998 chemicals to one or more species of wild and domestic birds, *Arch. Environ. Contam. Toxicol.* 12:355–382, 1983.

Smith, G.J. Pesticide use and toxicology in relation to wildlife: organophosphorus and carbamate compounds, U.S. Fish and Wildlife Service Resource Publication No. 170, 171, 1987.

Smith, G.J. and Anders, V.P. Toxic effects of boron on mallard reproduction, *Environ. Toxicol. Chem.* 8:943–950, 1989.

Smith, G.J., Spann, J.W., and Hill, E.F. Cholinesterase activity in black-crowned night-herons exposed to fenthion-treated water, *Arch. Environ. Contam. Toxicol.* 15:83–86, 1986.

Smith, G.J., Heinz, G.H., Hoffman, D.J., Spann, J.W., and Krynitsky, A.J. Reproduction in black-crowned night-herons fed selenium, *Lake Reserv. Manage.* 4:175–180, 1988.

Springer, P.F. Relationship of mosquito control to conservation, *Proc. Pap. 29th Ann. Conf. California Mosquito Control Assoc.* 1961, 83–85.

Stewart, R.E., Cope, J.B., Robbins, C.S., and Brainerd, J.W. Effects of DDT on birds at the Patuxent Research Refuge, *J. Wildl. Manage.* 10:195–201, 1946.

Stickel, L.F. Field studies of a *Peromyscus* population in an area treated with DDT, *J. Wildl. Manage.* 10:216–218, 1946.

Stickel, L.F. Pesticide residues in birds and mammals, in *Environmental Pollution by Pesticides.* C.A. Edwards, Ed. Plenum Press, London, 1973, 254–312.

Stickel, L.F. and Heath, R.G. Wildlife studies, Patuxent Wildlife Research Center, in *Effects of Pesticides on Fish and Wildlife.* U.S. Fish and Wildlife Service Circular No. 226, 1965, 3–30.

Stickel, L.F., Chura, N.J., Stewart, P.A., Menzie, C.M., Prouty, R.M., and Reichel, W.L. Bald eagle pesticide relations, *Trans. N. Amer. Wildl. Nat. Res. Conf.* 31:190–200, 1966.

Stickel, W.H. Some effects of pollutants in terrestrial ecosystems, in *Ecological Toxicology Research.* A.D. McIntyre and C.F. Mills, Eds. Plenum Press, New York, 1975, 25–74.

Stromborg, K.L. Reproductive tests of diazinon on bobwhite quail, in *Avian and Mammalian Wildlife Toxicity: Second Conference.* D.W. Lamb and E.E. Kenaga, Eds. Am. Soc. Test. Mat., Spec. Tech. Publ. No. 757, 1981, 19–30.

Tucker, R.K. and Crabtree, D.G. Handbook of toxicity of pesticides to wildlife, U.S. Bur. Sport Fish. Wildl., Resour. Publ. No. 84:131, 1970.

Wang, C. and Murphy, S.D. Kinetic analysis of species difference in sensitivity to organophosphate insecticides, *Toxicol. Appl. Pharmacol.* 66:409–419, 1982.

White, D.H. and Kolbe, E.J. Secondary poisoning of Franklin's gulls in Texas by monocrotophos, *J. Wildl. Dis.* 21:76–78, 1985.

White, D.H., King, K.A., Mitchell, C.A., Hill, E.F., and Lamont, T.G. Parathion causes secondary poisoning in a laughing gull breeding colony, *Bull. Environ. Contam. Toxicol.* 23:281–284, 1979.

Whitworth, M.R., Pendleton, G.W., Hoffman, D.J., and Camardese, M.B. Effects of dietary boron and arsenic on the behavior of mallard ducklings, *Environ. Toxicol. Chem.* 10:911–916, 1991.

Wurster, C.F., Jr., Wurster, D.H., and Strickland, W.N. Bird mortality after spraying for Dutch elm disease with DDT, *Science* 148:90–91, 1965.

Section IV

Chapter

16

Environmental Health

Camille J. George and William J. George

A. GENERAL CONSIDERATIONS

Environmental health is a difficult subject to address because the term embraces the health of all individuals causally affected by components of the environment. For this chapter, environmental health encompasses the discipline which centers around disease-causing chemicals or pathogens that occur naturally in our environment and those introduced by man. As such, environmental health is central to the solution of problems resulting from expanding populations which are depleting our natural resources, including our forests and water supply. Equally important are environmental problems resulting from present day living standards which depend upon tens of thousands of man-made chemicals.

B. MAN AND HIS ENVIRONMENT

The high standard of living enjoyed by members of industrialized societies is in large part due to the availability of naturally occurring and synthetic chemicals used for production of food, treatment of disease, manufacture of consumer goods, etc. Today there are about 3.5 million organic compounds that have been introduced into our environment. Although these chemicals benefit man, there are nonetheless many adverse effects on man and the environment as a result of their direct or indirect toxicity.

0-8493-8851-1/94/$0.00+$.50
© 1994 by CRC Press, Inc.

C. PRINCIPLES OF CHEMICAL TOXICITY

Toxicity is the ability of a chemical to cause adverse effects. In 1567, Paracelsus stated, "What is it that is not poison? All things are poison and nothing is without poison. It is the dose only that makes a thing not poison." Dose, along with time of exposure, defines the type of toxicity. Acute toxicity is the ability of a substance to cause damage as a result of a relatively short, one-time exposure. LD_{50} expresses the acute toxicity of chemicals and indicates that 50% of the animals died at a given dose. Chronic toxicity results from low level exposure over a long period. There is less known about chronic toxicity than acute toxicity because the symptoms are more subtle and complex. The threshold of a chemical is the minimum quantity that results in toxicity. This varies with species, with individuals within species, and even with time in the same individual; therefore, it is impossible to determine the threshold precisely by experimentation. The dose/response of a substance illustrates that as the dose increases, the intensity of the effects increases until the maximal effect is reached (Table 16.1 and Figure 16.1). Frequent exposure to relatively large amounts of a chemical, subacute toxicity, can produce both chronic and acute symptoms (Ottoboni, 1984).

Also influencing toxicity is the pathway by which the chemical enters the body. There are three main routes of exposure: dermal (through the skin), inhalation (through the lungs), and oral (through the gastrointestinal tract). The most common route for chemicals is dermal. Most chemicals are not equally toxic by all three routes. For example, vitamin D is highly toxic when ingested but not acutely or chronically toxic by skin. Two reasons why toxicity varies with the route of exposure are the amount of chemical that enters the body and the path the chemical follows through the body (Ottoboni, 1984).

Another factor that influences the intensity of toxicity is age. Newborn or infant animals are more sensitive than adults to ill effects of chemicals but less sensitive than adults to others. This is due to differences in the activity of enzymes that metabolize foreign substances. For example, newborns are incapable of detoxifying phenacetin, hexobarbital, and amido- and antipyrene and are unable to form glucuronides. Similarly, alcohol dehydrogenase activity in the neonate has been shown to be lower than that found in adults (Raina et al., 1967; Smith et al., 1971; Pikkarainen and Raina, 1967). However, the assumption that infants are always more sensitive than adults to drugs based on metabolic differences is not always valid. The problem of detoxification is compounded by lower renal function in infants. Renal excretion is as important in the disposal of foreign substances within the body as is hepatic biotransformation.

The sex of an individual may also play a factor in the toxicity of chemicals. In general, women are more susceptible to drugs than men. Some striking examples are hexobarbital, phenacetin, acetylsalicylic acid, benzene, nicotine,

**TABLE 16.1. Relationship Between Blood Alcohol Levels and
Selected Effects**

Blood Concentration	Effects
20–30 mg%	Most would probably not show much of an effect threshold
50 mg%	Stimulation in terms of behavior and an increase in confidence
100 mg%	Coordination is impaired, personality is well expanded, and judgment is impaired, legal cutoff for DWI
150 mg%	In addition to effects at 100 mg%, mood swings and related emotional eruptions occur (half of the people at this level are clearly intoxicated)
200 mg%	All faculties are effected and the individual is unmistakenly drunk
300 mg%	Stuporous type individual who is effectively under anesthesia
400 mg%	Begin to present in coma
>400 mg%	Death due to respiratory arrest

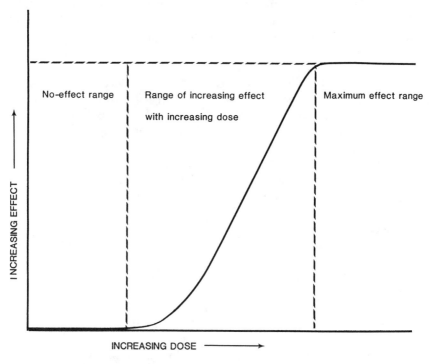

Figure 16.1. Typical dose response curve indicating increased effect with increasing dose or exposure.

warfarin, and strychnine (Doull, 1980). However, the exceptions far outnumber the rule; the sexes, in general, respond equally to drugs (Gerald, 1974).

Dietary factors can influence toxicity by producing changes in body composition, physiologic and biochemical functions, and nutritional status of the individual. Diets of reduced caloric intake increase the toxicity of caffeine and DDT in rats, and low protein diets have been shown to increase the toxicity of a variety of pesticides and other agents. On the other hand, low protein diets have been found to protect rats against the hepatotoxicity and lethality of exposure to carbon tetrachloride and dimethylnitrosamine but not chloroform and aflatoxin. Chloroform toxicity can be enhanced by high-fat diets or by the addition of a microsomal inducer such as DDT to the diet (Doull, 1980).

An animal's state of health also influences its response to toxins. The individual's response to a substance is modified by disease, and some toxicants only exert their effect in the presence of disease. Pentobarbital prolongs sleep in people with liver disease due to the inability of the tissue to biotransform and inactivate the drug. Diarrhea speeds the passage of substances through the gastrointestinal tract and less drug is absorbed; therefore, a larger dose is necessary to produce the desired effect. Body weight also influences the concentration of a drug in the blood and at the site of action. Abnormally thin people have less drug tolerance than extremely obese individuals (Gerald, 1974).

Individual susceptibility to chemicals occurs in all species resulting in a different response in individuals exposed to the same concentration of chemical (Ottoboni, 1984). Unusual reactions to drugs may be inherited without the hereditary trait being directly associated with the metabolism or disposition of the drug. There are several inherited red blood cell enzymatic deficiencies which result in unusual adverse effects if certain drugs are given. The best known example is the variety of deficiencies of erythrocyte glucose-6-phosphate dehydrogenase (G6PD). The variants with <30% of the normal activity of the enzyme develop hemolytic reactions to primaquine and other drugs. The reduced level of G6PD results in a decreased production of NADPH, the reduced cofactor for glutathione reductase synthesis. An adequate supply of glutathione is critical for maintaining the integrity of the erythrocyte membrane. A variety of chemicals causes hemolysis in individuals deficient in G6PD. Males are more likely to show the deficiency and drug sensitivity than females since the gene responsible for G6PD is carried on the X chromosome (LaDu et al., 1971).

The presence of other chemicals can also modify the toxicity of substances. Because of their complex actions on the body, combinations of chemicals may have effects which might not have been expected if only the primary action of an individual drug was considered (James et al., 1978). Interactions between compounds can occur in several ways. The substances can interact with each other chemically, resulting in a decreased response. Chemical interaction can also change the rate of absorption, the degree of protein binding, and the rate of metabolism or excretion of one or both of the interacting chemicals. The responses of an organism to combinations of chemicals may be increased or decreased due to toxicologic responses at the receptor sites. The most common response is an additive effect which occurs when the combined effects of two chemicals is equal to the sum of their individual effects. For example, when two organophosphate pesticides are given together to an animal, the resulting cholinesterase inhibition is usually additive.

Synergism occurs when the combined effect of two chemicals is greater than the sum of their individual effects. Both carbon tetrachloride and ethanol are hepatotoxic, but when given together a more extensive and severe injury occurs than the mathematical sum of their individual effects. Potentiation occurs when one chemical which does not have a toxic effect on a particular organ is given with another chemical so it enhances the toxic response. Isopropanol, for example, is not hepatotoxic, but when given with carbon tetrachloride, the hepatotoxicity of carbon tetrachloride is much greater than carbon tetrachloride alone.

Antagonism occurs when two chemicals interfere with each other's actions or when one chemical interferes with the action of another chemical. During severe barbiturate intoxication, blood pressure falls dramatically. This response can be effectively overcome by intravenous administration of a vasopressor agent such as norepinephrine (Klaassen and Doull, 1980).

Adaptation to substances also influences toxicity. Adaptation is the process by which individuals exposed to subtoxic levels become tolerant to later doses of the chemical in amounts that would normally be harmful (Ottoboni, 1984). There is some evidence that chronic exposure to low doses of cadmium and lead results in the increase of proteins that bind the metals (Hammond and Beliles, 1980). The blood level at which cattle develop severe encephalopathy following exposure to lead is <80 mcg/100 ml. However, when cattle received 5 to 6 mg of lead per kg body weight per day orally, the concentration of lead in the blood exceeded 100 mcg/100 ml within 2 to 4 months, and with continuous exposure lead lasted for 4 years without any apparent harm to the animals. Pretreatment with lead reduces the sensitivity of cattle and sheep to acutely toxic amounts. The intranuclear inclusion bodies that develop during lead exposure sequester the lead in the kidney, thus making it less toxic. However, the intranuclear inclusion bodies are present only for a relatively short time in workers exposed to lead. Thus, their protection functions during a limited period of exposure.

A more practical example is alcohol, which has the distinction of being the only potent pharmacological agent by which self-intoxication is socially acceptable. In the U.S., two thirds of all adults use alcohol occasionally. It is by far the most serious substance abuse problem in America when measured by accidents, lost productivity, crime, broken homes, and health effects. The

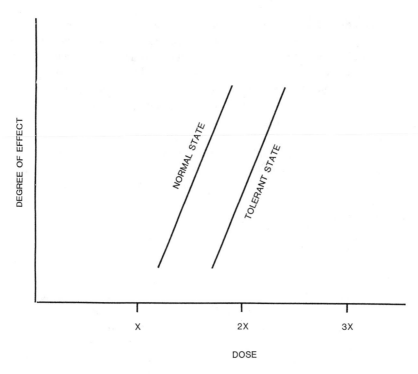

Figure 16.2. Dose response curve indicating different sensitivity in normal and tolerant individuals.

effect of alcohol is directly related to the concentration of alcohol in the bloodstream, and this in turn is affected by absorption, distribution, metabolism, and excretion of the alcohol. The direct effects of alcohol are the result of primary and continuous depression of the central nervous system. In comparing blood alcohol levels with the effects on the nervous system, there are considerable data indicating threshold plasma concentration, and increasing results leading from modest depression to coma can occur. Some degree of tolerance to alcohol occurs with chronic use. Such tolerance may be due to metabolic adaptation, functional adaptation, apparent tolerance, and learned behavior. Such adaptation is a result of enzyme induction and increased rate of metabolism of alcohol as well as a pharmacodynamic tolerance in which an increased amount of alcohol is needed to produce a specific effect (Figure 16.2). Figure 16.2 shows the adaptive behavior of the brain to increasing doses of alcohol. Taking into consideration the effects of alcohol in individuals who have consumed varying quantities of alcohol, it is accepted that significant impairment of judgment occurs in individuals who have a blood level >100 mg%. While the threshold may be 30 mg%, some individuals do not show significant effects at this concentration due to the presence of adaptation. Alcohol intoxication can be differentiated from other drug intoxication with similar symptoms by determining the blood alcohol concentration. This is done by several methods, such as gas chromatography and UV spectrophotometry.

D. SPECIFIC CHEMICALS AND THEIR IMPACT ON HEALTH

When exposure to toxic chemicals occurs in the workplace, the above mentioned phenomena and effects must be considered. Some chemicals to which workers are exposed occupationally are carbon tetrachloride, chlorobenzene, cresols, cyanide, lead, polychlorinated biphenyls, dioxins, and carbon monoxide.

1. Carbon Tetrachloride

Carbon tetrachloride is a clear nonflammable liquid with a characteristic odor. Exposure is by all three major routes, although absorption through the skin is not as significant as oral or inhalation routes. Children and adolescents are more susceptible to the toxic effects of carbon tetrachloride on the liver. The oral LD_{50} for rats is 1.76 ml/kg, and the estimated oral lethal dose for humans is 0.05 ml/kg.

Acute poisoning with carbon tetrachloride is usually more severe than chronic toxicity, until chronic exposure injures internal organs beyond repair. On contact with skin, carbon tetrachloride produces dermatitis characterized by reddening, chapping, and sensitization. Secondary infections sometimes occur due to the irritation of the skin. Ingestion of carbon tetrachloride produces gastrointestinal, renal, hepatic, and cardiac injury, whereas, inhalation produces predominantly nervous system injury accompanied by injury of the GI, renal, hepatic, and cardiovascular systems. Death from acute carbon tetrachloride toxicity is usually due to respiratory failure or ventricular fibrillation. The mortality rate is 90% for severe exposures, and death usually occurs 8 h to 10 d if no treatment is sought.

Chronic exposure to carbon tetrachloride produces the same symptoms as acute exposure. The difference is that, with chronic exposures, the level of carbon tetrachloride that produces toxicity is not as high as that which causes acute toxicity. Chronic exposure of laboratory animals to carbon tetrachloride results in an increased incidence of carcinomas. The absorption of carbon tetrachloride in the GI tract is enhanced by fats, phenobarbital, and organic solvents such as acetone, *n*-butyl alcohol, methanol, and methyl ethyl ketone. These compounds are promoters of carbon tetrachloride toxicity. A water concentration of 4 mcg/l increases the lifetime cancer risk by 1 in 100,000. Recovery from chronic carbon tetrachloride poisoning is possible (George et al., 1985).

2. Chlorobenzene

Chlorobenzene, a colorless, flammable liquid with a mild odor, is absorbed by the skin, lungs, and GI tract. The recommended threshold limit value is 75 ppm and, as with carbon tetrachloride, can be measured by gas chromatography methods. Acute exposure to chlorobenzene causes irritation of the skin, eyes, and nasal mucosa; the central nervous system is also affected. Drowsiness, incoherence, and depression have been reported. Exposures to laboratory animals have also produced kidney damage. The symptoms of acute poisoning of a 2-year-old have been reported. After ingesting 5 ml, the child was pale and unconscious for 2.5 h, the pulse was weak, the facial muscles twitched, and reflexes were negative. After 3 h consciousness returned, the pulse improved, and recovery was complete.

Chronic poisoning by chlorobenzene results in damage to bone marrow and kidneys. The nervous system, vision, and GI tract may also be disturbed. Recovery depends on the extent of damage to the internal organs, but the symptoms subside once the exposure is terminated. No carcinogenic effects have been recorded for chlorobenzene (George et al., 1985).

3. Cresols

Cresols are brownish-yellow or pinkish liquids which are very corrosive. They can be absorbed by the skin, lungs, and GI tract and are potentially lethal. The recommended threshold limit value is 5 ppm in air, at which concentration an odor is detectable.

Acute poisoning from cresols occurs most often as a result of absorption through the skin. Contact causes a burning sensation, redness, itching, ulceration, and brownish discoloration. Symptoms occur 20 to 30 min after absorption. Headaches, dizziness, mental confusion, rapid respiration, muscle weakness, nausea, and dimmed vision all occur. Ingestion and inhalation of cresols produces white burns on mucous membranes. Kidney and liver damage may also occur with severe exposure resulting in death. Oral ingestion of at least 8 g produces circulatory collapse and rapid death.

Chronic exposure sometimes results from absorption through the skin. Chronic toxicity causes inflammation of hair follicles, spotty pigmentation, and disorders of the central nervous and GI systems. Lesions may be seen in the liver, kidney, heart, and brain. Most symptoms subside when exposure is terminated.

One of the cresols, *para*-cresol, has been implicated in the production of malignant scrotal tumors in coal tar workers. Cresols are considered promoters in carcinogenesis (George et al., 1985). Analysis is by gas chromatography/mass spectroscopy (GC/MS).

4. Cyanide

Cyanide is present in rodent poisons, photographic solutions, and certain seeds such as apple and apricot. Cyanide can enter the body by the three main routes. Alkaline cyanide salts cause toxicity after ingestion. The ingestion of sodium cyanide or potassium cyanide, even in small amounts, may result in death within minutes to hours depending on the route and duration of exposure. Rapid death results from cyanide inhalation (Rumack and Lovejoy, 1986).

Only a very low level of cyanide is necessary to produce toxic effects; symptoms of toxicity appear at a blood levels of 0.2 mcg/ml. A blood level greater than 1 mcg/ml is usually lethal. The first sign of cyanide toxicity is a bitter almond odor from the victim's breath. Other symptoms are salivation, nausea, anxiety, paralysis, disorientation, cardiac arrhythmias, and coma (Waxler et al., 1947). Cyanide has also been shown to produce damage to several areas of the brain (Levine, 1967; Bass, 1968).

Cyanide inhibits cytochrome oxidase, an important enzyme in respiration. Treatment with sodium nitrite increases methemoglobin (Stewart, 1974) which has a higher affinity for cyanide than does cytochrome oxidase. Treatment with sodium thiosulfate enhances the excretion of the cyanide as thiocyanate (Chen and Rose, 1952; Stewart, 1974). Both the thiocyanate metabolite and cyanide itself can be measured by gas chromatography.

5. Polychlorinated Biphenyls (PCBs)

Polychlorinated biphenyls (PCBs) are a class of compounds that for over 60 years, have been used as insulating fluids, hydraulic and lubricating fluids, heat exchange fluids, and additives in adhesive inks and paints. The very properties that made PCBs attractive to industry, such as resistance to fire and persistence in the environment, are the same properties that have resulted in their toxicological problems. Mixtures of these persistent compounds which are contaminated with other agents have been shown to produce adverse organ and system effects on a variety of animal models. Since the mid 1970s production and use of PCBs have been curtailed due to their chronic toxicity and concern about their environmental persistence. Due to the lipophilic and hydrophobic nature of these compounds, they tend to bioaccumulate in living tissues and the food chain (George et al., 1988).

There are four major routes by which PCBs enter the environment: (1) industrial accidents, (2) incomplete destruction of PCB-containing products, (3) weathering of PCB-containing products, and (4) leaking from landfills (Clayton and Clayton, 1981). Most PCBs enter the environment from pumps leaking into the cooling water of capacitors and transformers (Durfee, 1975). PCBs can enter municipal sewage systems and can spread throughout the environment via diffusion, rain, snow, and dust (Clayton and Clayton, 1981).

PCBs are chlorinated aromatic hydrocarbons with the basic structure $C_{12}H_{10-x}Cl_x$. The number of chlorine substituents on the biphenyl molecule can range from 1 to 10, thereby producing 10 forms of PCBs. There are two hundred and nine theoretical PCB isomers with molecular weights ranging from 154 to 494 Da. The number of chlorine atoms in each isomer determines the classification and nomenclature (Durfee, 1975).

PCBs are commercially produced by chlorinating biphenyls with anhydrous chlorine in the presence of a catalyst such as ferric chloride. This reaction occurs at very high temperatures and results in a crude product that requires further purification using an alkali wash or distillation (Durfee, 1975). The final commercial product is a mixture of PCBs with varying chlorine concentrations and various impurities. These impurities include related chemical compounds such as polychlorinated dibenzofuran (PCDF) and polychlorinated quaterphenyls (PCQs) that are formed during the production of PCBs (Coulston and Pocchiari, 1983).

Many PCB mixtures are marketed around the world. The most common trade names for PCB mixtures in the U.S. are Aroclor®, Chloretol®, DyKanol®, Inerteen®, and Pyranol®. In 1929, the Monsanto Company began production of PCBs. PCBs were used in many capacities until the regulation of their use in 1971 (Hurkat, 1977). PCB usage is classified into (1) closed uses, (2) nominally closed uses, and (3) open end uses. Closed uses include electrical insulation for wires, cables, and condensers, as well as a coolant/dielectric in transformers and capacitors. Nominally closed uses of PCBs include hydraulic fluids, heat transfer fluids, and high pressure lubricants. These closed uses do not totally isolate PCBs from the environment, but for the most part they do not allow direct contact. The open ended uses include addition of PCBs to paints, ink dyes, plasticizers, protective coating for woods, dedusting agents, adhesives, pesticide extenders, and microencapsulation of dyes for carbonless duplicating paper (Committee on the Assessment of Polychlorinated Biphenyls in the Environment, 1979). This application of PCBs results in their distribution directly into the environment. However, since 1971 PCB applications have been limited to closed end systems (U.S. EPA, 1976; Hurkat, 1977).

PCBs are valuable in the above-mentioned uses in industry because of their properties of stability and fire resistance. Pure PCBs are solids at room temperature (25°C), and their melting points range from 54°C to 310°C, depending upon their configuration and makeup. In general, melting points increase with chlorination. PCBs are fat soluble, water insoluble, and very stable. They are very resistant to degradation, oxidation, and other agents such as acids or bases. They can withstand temperatures up to 1600°F (870°C) (Clayton and Clayton, 1981).

There is no evidence that PCBs naturally occur; therefore, all PCBs present in the environment must have originated from human dissemination. Since PCBs have a tendency to bioaccumulate, concern has been expressed about PCB incorporation into the food chain and their subsequent bioaccumulation. As a result, the uppermost trophic levels have the highest concentration of PCBs in the environment.

Aquatic organisms such as fish accumulate the greatest concentrations of PCBs. This finding was widely publicized in the early 1970s as a result of contamination of fish caught in Lake Michigan. Lake trout accumulated PCBs to concentrations about 3 million times that of the environment (Metcalf et al., 1975). PCBs are distributed primarily in adipose tissue, explaining why fish like the catfish, with 10% body fat, accumulate more PCBs than fish like perch, with 4% body fat (Kleinert, 1976).

Although PCBs accumulate to a greater extent in adipose tissue, the liver must be considered an important target organ because of its pivotal role in the excretion and metabolism of many toxins from the body (Prestt et al., 1970). The hepatic microsomal enzymes metabolize certain PCBs into less lipophilic compounds which facilitate their excretion (Schnellman et al., 1983, 1984). The nature of the PCB is important because the rate of metabolism and the lipophilicity of a particular PCB dictate its rate of excretion. Both the chlorine content and location on the biphenyl ring affect its disposition, but in general, excretion is inversely proportional to the degree of chlorination (Schnellman et al., 1983; George et al., 1988).

PCBs are metabolized by dechlorination and arene oxide formation mediated by mixed function oxidases (MPO). These processes result in more polar and water soluble metabolites. The MPO system can be induced following exposure of the animal to PCBs and PCDFs (Schuetz et al., 1986). Induction of arylhydrocarbon hydroxylase (AHHase) occurs following PCB exposure, and toxicity of PCBs may be associated with induction potency (Kimbrough, 1987). It has also been suggested that toxicity of PCBs may be related to increased intermediary metabolism. Genetic variations result in differences in arylhydrocarbon hydroxylase responsiveness. Individuals with high AHHase inducibility may be more sensitive to toxic effects of PCBs (Poland and Glover, 1977). Hydroxylation, the major pathway of metabolism for PCBs, may involve arene oxide intermediates. Arene oxide intermediates may elicit carcinogenic, cytotoxic, or mutagenic effects. Increased production of arene oxides may be due to the additive activity of PCBs and polybrominated biphenyls (PBBs) (Ingle, 1965). Additionally, hydroxylated chlorobiphenyls are more hazardous than the parent PCBs (George et al., 1987). Laboratory experiments with rats indicate that the toxicity of other chemicals, including carbon tetrachloride, chloroform, and bromobenzene, increases when the rats are given PCBs (Ingle, 1952; Hyde and Falkenburg, 1976).

Experiments with pregnant rats indicate placental transfer of PCBs during pregnancy. However, higher levels of PCBs were noted in mother's milk due, no doubt, to the large percentage of fat in the milk, and the lipophilic nature of PCBs (Jacobson et al., 1984).

Some of the toxic effects of PCBs such as hepatic injury appear to result from electrophilic intermediates produced during metabolism (Shimado and Sato, 1978). PCB mixtures also contain several other chlorinated compounds, such as polychlorinated naphthalenes and polychlorinated dibenzofurans (PCDFs) (Sittig, 1985). The presence of these PCDFs in PCB mixtures may arise from the distillation process during purification (George et al., 1987).

The effects of PCDFs have been widely studied due to their association with PCBs and their toxicity. PCDFs inhibit food consumption and, to a lesser degree, suppress appetites in laboratory rats. PCBs also exhibit such effects, but to a lesser degree. Low levels of PCDFs decrease hemoglobin concentration, hematocrit, and mean corpuscular volume. Concentrations of PCDFs greater than 10 ppm may result in erythrocyte count reductions as well (Wright et al., 1972). Serum glutamic-pyruvic transaminase activity and testosterone concentrations may also decrease following PCDF exposure. Conversely, an increase in serum glutamic oxaloacetic acid transaminase was noted in rats following PCDF treatments. When mixtures of PCBs and PCDFs are given experimentally, cholesterol and cholinesterase activity are both elevated, while triglycerides and aminopeptidase activity decrease (Wastler et al., 1975).

Due to their lipophilicity, PCBs inhibit ATPase activity in organisms. Understandably, this inhibitory effect on ATPase activity decreases as the water solubility of PCBs increases. It is postulated that the lipophilic portion of the ATPase molecule readily associates with PCBs. Allosteric changes in the enzyme complex secondary to this lipophilic aggregation are probably the mechanism for the inhibitory action of PCBs on ATPase activity (Radcleff et al., 1955).

Chloracne is the most common symptom of PCB poisoning in humans. However, chloracne is not unique to PCB exposure as it results from exposure to many chlorine-containing compounds. This condition appears as severe acne and may even produce facial scarring. The acneiform eruptions of chloracne, which form both closed and open comedones, appear to result from follicular excretion of PCBs with sebum. The excretion continually stimulates the skin to produce acneiform eruptions which are characteristic of PCB poisoned patients (Urabe and Asahi, 1984). Other skin manifestations of PCB poisoning include hyperpigmentation, alopecia, and porphyria (Kimbrough, 1987).

Many different animal models have been used to assess PCB toxicity. Rhesus monkeys displayed chloracne, loss of eyelashes, and subcutaneous edema when given diets of 300 ppm of Aroclor 1248® (Norback and Allen, 1975). Administration of a 65% chlorinated biphenyl mixture to laboratory animals resulted in atrophy and lesions of the liver (Wolfe et al., 1963). Male dogs given 100 mcg/g diets of PCB had reduced growth rates and liver enlargement (Keplinger et al., 1971). Topical PCB exposure in guinea pigs, rats, and rabbits resulted in epidermal thickening at the application site, fatty degeneration, and central atrophy of the liver (Ottolenghi et al., 1974). High concentrations (1000 ppm) of Kaneclor® have resulted in cholangiofibrosis in laboratory rats (Huff and Gerstner, 1978). PCB toxicity in chickens reduces egg production and fertility (Platonou and Reinhart, 1973). Additionally, doves have demonstrated reduced egg hatching rates (Buck et al., 1975).

One of the primary concerns about PCBs toxicity is the effect on mammalian reproduction and the reproductive system. PCBs produced menstrual cycle irregularities and increased abortion rates in monkeys (IARC, 1979). Kihlstrom et al. (1975) demonstrated decreased implantation capability in mice when both mother and father nursed from PCB-treated mothers. Burke and Fitzhugh (1970) exposed rats to various concentrations of PCBs and noted changes in mating behavior. At dosages of 100 ppm of Aroclor 1242® the animals displayed decreased mating indices in the second generation (Kimbrough et al., 1972). Similar dosages of Aroclor 1254® resulted in a decreased survival rate and an increased stillborn rate (Burke and Fitzhugh, 1970).

The lack of enough data on human exposure to PCBs is the limiting factor in determining PCB toxicity in humans. Most data on human exposure result from rare accidental PCB exposures. The first of these accidental exposures occurred in 1968 when thousands of people were exposed to very high concentrations of PCBs and PCDFs in Fukuoka, Japan. This exposure resulted from contamination of rice oil which was later used for cooking. The second major exposure of humans to PCBs occurred a decade later in 1978, in Taiwan. Again, high concentrations of PCBs were ingested by humans through the use of contaminated rice oil. Most data concerning PCB toxicity in humans was obtained from these two main incidents and several other minor incidents.

PCB ingestion by humans results in symptoms such as weakness, nausea, headaches, impotence, insomnia, anorexia, loss of weight, and abdominal pain. Muscular spasms and pain may also be associated with PCB exposure. PCB toxicity to the eye is characterized by inflammation, burning, edema, and cysts of the tarsal glands and conjunctiva (Kimbrough, 1987). The Meibomian gland hypertrophies and produces a cheese-like discharge (Urabe and Asahi, 1984).

Skin changes that result from PCB exposure include chloracne, hyperpigmentation, alopecia, nail deformity, xerosis, and follicular hyperkeratosis (Kikuchi, 1984). These symptoms improved dramatically or returned to normal within 2 to 5 years following exposure in the Fukuoka incident. Prolonged exposure to PCBs can elicit hepatic damage, but the severity of such damage is not certain (Urabe and Asahi, 1984).

The extent of the effects of PCB exposure on the nervous system are, at most, vague. Decreases in both motor and sensory nerve conduction velocities have been claimed following PCB exposures (Chia and Chu, 1984). In 1980, a neurological study was performed on the cohorts of the people exposed to PCBs in the 1978 accident in Taiwan. Dull, nonpulsating headaches, dizziness, or light headaches as well as nausea and vomiting were reported by about one half of the participants in the study. Neither vertigo nor tinnitus was noted in these patients. There was no correlation between these effects and serum PCB concentrations (Chia and Chu, 1984).

Large dosages of highly chlorinated compounds induce tumor growth in both rats and mice, but the role of moderate to high levels of PCBs in promotion of tumor growth is unclear (George et al., 1987). Oxidation of PCBs may produce arene oxide intermediates which are potential tumor promoting agents (IARC, 1974). Hepato-cellular carcinomas and nodular hyperplasia have been found in mice exposed to high levels of PCBs (Nagasaki et al., 1972; Ito et al., 1973).

There have been extensive data collected on the carcinogenic effects of PCBs in laboratory animals but few involving humans (Furukawa et al., 1983). Limited accidental and prolonged occupational exposures are the only source of these data. Of the thousands of individuals exposed in Japan in 1968, only four autopsies were performed to determine the absence or presence of possible PCB-related effects. Three of these autopsies were performed on adults and the fourth on a stillborn infant (Kikuchi et al., 1969). High levels of PCB were found in skin and fatty tissue (Kikuchi et al., 1969, 1971). Hepatic damage was noted in only one of the four autopsies. This individual had multilobular hepatic cirrhosis accompanied by many hepatic carcinoma nodules (Kikuchi, 1972). The stillborn child was heavily pigmented (often referred to as a "brown baby"). This pigmentation may have been due to the passage of PCBs across the placenta to the developing fetus (Kikuchi, et al., 1969).

A study of over 2500 plant employees at a capacitor factory was conducted to investigate an increased death rate which was ascribed to liver cancer in this population (Brown and Jones, 1981). However, the incidence of cancer was found to be related inversely to the duration and latency of exposure to PCBs (Griffin and Chian, 1979). This finding suggests that occupational exposure to PCBs was not the causative agent.

PCBs can be degraded by several methods. Hydrolysis is possible, but the stability of PCBs makes this method impractical (Hutzinger et al., 1974; Nisbet, 1976). Even though PCBs are very stable, chemical degradation can occur under controlled chemical reactions including oxidation, reduction, nitration, isomerization, and nucleophilic reactions (Hutzinger et al., 1974; Leifer et al., 1983). Another method for the elimination of PCBs from the environment is photodegradation. Ultraviolet radiation from sunlight can dechlorinate and polymerize PCBs (Lehman, 1954). A solvent such as hexane is necessary for this reaction. Biodegradation is another natural process by which PCBs may be degraded. The degree and rate of biodegradation depend upon position and degree of chlorination of the PCB involved (Onishi and Trench, 1981; Brown et al., 1985). Some bacteria can readily transform PCBs of low chlorination but encounter difficulty in degrading more highly chlorinated PCBs. These PCBs are usually oxidized to their corresponding chlorobenzoic acids, but other hydroxy and meta-cleavage compounds may also be formed (Goldstein, 1983). Bacteria that biodegrade PCBs include *Acinetobacter, Alcaligenes, Arthrobacter, Achromobacter, Nocardia,* and *Pseudomonas* (Furukawa et al., 1983). Finally, PCBs can be degraded by incineration. This process requires heating the PCBs to at least 950°C. When PCBs are incinerated at 950°C a residue of 100 mg hexachlorobenzene/kg is formed. Heating between 500 and 600°C results in the formation of degradative products, such as PCDFs (Buser, 1985).

Many researchers believe that significant adverse effects are elicited by PCBs when they are given to animals under laboratory conditions. However, except for the development of chloracne, there is little evidence to suggest that PCBs are toxic to man. Carcinogenic effects resulting from chronic exposure to PCBs in the workplace or environment has not been proven.

6. Polychlorinated Dibenzo-*p*-dioxins (PCDDs, Dioxins)

Polychlorinated dibenzo-*p*-dioxins (PCDDs), also known as dioxins, are produced as by-products of several chemical reactions. Photodegradation of dioxins does occur; however, they may become resistant to photodegradation as well as microbial degradation. These compounds accumulate in body fat, and the half-life in humans is about 6 to 10 years. The most studied dioxin is 2,3,7,8-tetrachlorodibenzo-*p*-dioxin (TCDD). TCDD has proven lethal to some laboratory animals such as the guinea pig. However, acute exposure of humans is much less severe. Symptoms include chloracne, irritability, fatigue, headache, immunological changes, behavioral changes, decreased sex drive, and blurred vision. Some epidemiological studies have suggested that exposure to dioxins may lead to birth defects. No relationship has been established between dioxin exposure and an increased incidence of cancer.

Very little is known about the chronic toxicity of dioxins in humans. Most data concerning chronic exposures are derived from epidemiological studies that are often contradictory and thus inconclusive. The long term effects of dioxin exposure on humans and the dose eliciting toxicity are unclear (Wilson et al., 1988).

7. Carbon Monoxide

Carbon monoxide is an asphyxiant which elicits its adverse effect by combining with hemoglobin to form carboxyhemoglobin, which cannot carry oxygen. This reaction decreases the oxygen-carrying capacity of blood (Amdur, 1986). Fainting may result when the oxygen level of blood drops below normal, and death may result (Smith, 1986).

Small amounts of carboxyhemoglobin are present in everyone's blood. Although smokers normally have levels of about 5%, nonsmokers have only about 0.5% carboxyhemoglobin. Symptoms of carbon monoxide poisoning develop in the naive or previously unexposed individual when carboxyhemoglobin levels reach 2%. An exposure of 50 ppm carbon monoxide for 90 min results in a level of 2.5%, at which time interval discrimination is impaired. At 5% psychomotor faculties are impaired, and at higher levels impairment of the heart occurs in individuals previously unexposed to carbon monoxide (Amdur, 1986).

Chronic exposure to carbon monoxide may result in adaptation, which includes an increased hematocrit, increased hemoglobin concentration, and increased blood volume (Amdur, 1986). These compensations reestablish the oxygen content of blood to normal levels. As with ethanol, exposure to carbon monoxide can be verified by UV spectrophotometry or gas chromatography.

8. Lead

Lead is a naturally occurring bluish or silvery grey soft metal found in the earth's crust and in all compartments of the biosphere (ATSDR, 1986). Lead sulfide, a naturally occurring form of lead, has been mined for over 4000 years, and its production, which has increased dramatically since the industrial revolution, has resulted in increased environmental pollution (Sawyer et al., 1985). Lead has not been found to have any biological function, and its presence in the body has always been considered a sign of environmental pollution (Fischbein, 1983).

The U.S. consumes 1.3 million tons of lead and releases an estimated 100,000 tons into the environment annually (ATSDR, 1988). This has raised concerns about the disposal of lead waste, the possible entry of lead compounds into the water table, and the uptake by animals of lead from aqueous solutions. Accumulation of lead in the body depends upon concentration and duration of exposure. Urban residents have the highest blood lead levels (20 to 25 µg/dl); suburban (15 to 20 µg/dl), and rural populations (10 to 14 µg/dl) have lower levels (Mahaffey et al., 1982).

In recent years attention has turned to the potential risks in the general population, but especially children. Subclinical effects of lead have been identified at relatively low levels of exposure. Exposure of the general population to lead and its compounds results from breathing air, drinking water, and eating many foods that contain lead.

The concentration of lead in surface water is extremely variable depending upon sources of contamination, lead content of sediments, and various environmental factors (such as pH, temperature, etc.) (ATSDR, 1988). Although lead levels as high as 890 ppb have been detected in surface waters, the typical range throughout the U.S. is from 5 to 30 ppb (ATSDR, 1988).

According to one report (U.S. EPA, 1986), lead concentrations in tap water in most households varied from 7 to 11 ppb with the higher levels found in the "first draw" samples of water from the tap as a result of the dissolution of lead from solder joints.

The primary routes of exposure are ingestion and inhalation (Harrison and Laxen, 1981; Lippmann, 1990). In the U.S. the total daily intake of lead for an adult varies from <0.1 mg/d to >2 mg/d (Kehoe, 1961). After absorption, lead is transported by the blood to the soft tissues and bone. Lead concentrations in blood differ with age (Mahaffey et al., 1982). Children less than 7 years old have significantly higher blood lead levels than older children. Blood lead levels decrease during adolescence, apparently due to bone development and concomitant deposition of lead in the bones. Gastrointestinal absorption ranges from about 8% in adults to 40% in infants (Ziegler et al., 1978).

Historically, lead is the most identifiable chemical toxin (Waldron, 1985). Lead causes toxicity by associating with sulfhydryl groups of enzymes (Shane, 1991). The primary target organs are the brain, peripheral nervous system, kidney, bone marrow, liver, as well as gastrointestinal and reproductive systems. For health and performance, the central nervous system effects of lead are the most significant (Needleman, 1980; Rutter and Jones, 1983).

E. POPULATION GROWTH AND EFFECTS ON THE ENVIRONMENT

Our ever increasing population has been central to many environmental issues, especially within the last two decades. By the year 2000, it is estimated that over 5 billion people will inhabit our earth. Feeding these people will be increasingly difficult as 500 million of the present inhabitants are malnourished or starving. In fact, if the present rate of population growth continues, over 1 billion billion (1×10^{18}) people are projected to populate the earth in the year 3000 A.D. This is obviously absurd since this would require that over 1700 people would inhabit every square yard of the earth's surface (including oceans, deserts, and mountain tops) (Nadakavukaren, 1986). There must, therefore, be a significant limitation to human population growth.

The maximum growth potential of an organism is termed as its biotic potential. The biotic potential is an ideal unlimited growth rate for a given organism. However, in reality, growth rates do not increase without limit. Instead, food shortages, overcrowding, disease, predation, and toxic waste accumulation limit the extent of the growth rate. Such limitations on population growth is referred to as environmental resistance (Nadakavukaren, 1986).

There are two types of population growth curves in which the ordinate represents the number of organisms and the abscissa represents time. These growth curves are referred to as S-curves and J-curves. The former was developed by G. F. Gause in 1932 with growth experiments using *Paramecium caudatum* in a tank. The growth rate of the *Paramecium* population went first through a lag phase where the population grew slowly. With time, the growth rate and number of organisms increased exponentially. Eventually, the population exceeded the limits of the food supply and overcrowding occurred. At this point, the rate of growth decreased. This population limit in the population growth curve is termed the carrying capacity of the environment.

The J-curve is very similar to the S-curve except the growth rate does not decrease as the population approaches the carrying capacity of the environment. Instead, the population continues to grow at the same rate until the population greatly exceeds the carrying capacity and the environmental resistance causes the population to "crash". This type of growth curve is more dynamic than the S-curve, but more "population crashes" occur. This type of growth is seen in organisms such as lemmings and algae (Nadakavukaren, 1986).

Some view human population growth as being represented by the J-curve and believe that the human population will continue to grow until the environmental resistance causes a "population crash". The doubling time for the earth's population has been steadily decreasing. In 1650, the doubling time was 200 years. This time decreased to 80 years in 1850 and to 41 years in 1985 (Nadakavukaren, 1986). As the doubling time decreases, the population growth rate will continue to escalate.

Thomas Robert Malthus (1766–1834), an economist-clergyman, predicted a dreary future for the earth in "An Essay on the Principle of Population as it Affects the Future Improvement of Society". He saw that the growth of the population is geometrical while the agricultural growth remains arithmetical. As a result, the population will always outgrow the supply of food. The land will support only a certain number of people. The human population will always remain "in check" by natural disasters such as floods, hurricanes, droughts, earthquakes, famine, war, and disease. Whether these natural disasters will actually maintain the earth's population at a manageable level is debatable (Nadakavukaren, 1986).

This warning by Malthus was ignored by most people at that time. After all, he introduced this theory when agricultural advances and cultivation of vast American prairies produced what seemed to be an endless supply of food. The western world was quite capable of producing enough food to keep ahead of the rapidly growing population. It was not until recently, with the advent of large scale starvation in the third world countries, that Malthus's theory has gained credence.

Technological innovations in agriculture continue to increase crop yields per acre of land, but not without some adverse effects. Crop rotation maintains the productivity of land that would otherwise have lain fallow. Nitrogen producing microbes and nitrogen containing fertilizers are being used in ever growing quantities. However, replacement of organic matter in the soil cannot keep pace with the ever growing demands on the soil. Decomposition is not rapid enough in some cases to maintain organic components of the soil. Furthermore, poorly organized farmlands must contend with an increasing amount of soil erosion. This erosion not only leaches minerals and important organic matter out of usable soil, but also washes top soil away quicker than new soil can be generated from rock.

Increasing agricultural demand and overgrazing of grasslands has dramatically affected the amount of functional land available today. Much of the once functional grasslands and farmlands have now become desert in a process called desertification, the cumulative result of extensive grazing, trampling of foliage by herds, erratic rainfall, and elimination of forests by people for firewood. The Sahara Desert is a prime example of increased desertification. The size of this desert has been steadily encroaching on the once fertile lands of Northern Africa. If this process continues, much of the currently productive farmland and grasslands will be irreversibly damaged.

The increased demand for farmland has resulted in the draining of many marshes and wetlands for agricultural lands. Additionally, people in the past viewed these wetlands as inconveniences that breed insects such as mosquitoes. Only recently has the utility of these lands has been appreciated. The wetlands not only provide an environment for many diverse forms of wildlife, but also serve as natural filtering systems for water. The swamps and marshes maintain water in one place long enough to ensure that it is clean when it filters into underground springs. Additionally, the vast wetlands of the southern Mississippi River Valley serve as tremendous buffers than can retain large quantities of water and prevent flooding of dry land. Clearly, the loss of such wetlands would greatly affect humans and wildlife alike.

The combination of the above-mentioned stressors on population dynamics are synergistic. With the degradation of the available farmland, loss of wetlands, desertification of once fertile lands, and the rapidly increasing number of mouths to feed, farmers are attempting to produce more food with fewer resources. These farmlands cannot tolerate repeated abuse without one day completely failing to produce food, thus causing worldwide famine. In light of the delicate balance of our growing population, one must rationally hope that Malthus was correct.

It is interesting that developing nations have a high population growth rate (birth rate minus death rate). Since death rates are high in these countries, the high growth rate is a direct reflection of the fertility rate (number of births per 1000 women between 15 and 44 years old) and birth rate (number of births per 1000 people). High infant mortality and shortened life expectancy in these countries require high fertility rates. As a result, the developing nations of the world have relatively young populations. Large portions of these populations are below the age of 30. The younger populations contribute to higher fertility and higher growth rates. These result in the need for more housing and jobs.

Unfortunately, younger workers tend to be more inexperienced and less productive. As a result, the country over all has a decreased gross national product. With small gross national products, less money is available for education in these countries. Additionally, larger numbers of school aged children live in these countries, compounding the limitation of money available for education. Thus, it is not surprising that countries with lower gross national products have higher fertility rates and shorter life expectancies.

F. NOISE POLLUTION AND EFFECTS ON ENVIRONMENTAL HEALTH

In an age of toxic waste and hazardous occupational chemical exposures, noise is seldom thought of as a pollutant. Noise is an omnipresent component of all of our lives. Sounds permeate all aspects of our lives. We are exposed to noises as soft as the innocent chirping of avian life in the rural setting to the blaring of automobiles in the densely populated metropolitan areas. When considering noise pollution, busy airports, and neighbors playing annoyingly loud music initially come to mind. However, noise pollution encompasses much more.

Motorcycles, automobiles, and trucks produce a large percentage of urban noise pollution. This source of noise pollution can be recognized by anyone living near a major highway. Traffic noise also contributes to that constant "din" that is heard some distance from the major highways at night. With the ever increasing population, more vehicles will travel our roads and contribute to noise pollution in the years to come. Not only will the number of vehicles increase, but the number of automobiles with smaller 4-cylinder engines is also increasing. These smaller engines work harder to accomplish the same result and consequently, produce more sound (Nadakavukaren, 1986).

The projected population increase will also require construction of more homes and large buildings. In urban areas, this form of noise pollution is already of major concern. In the years to come, urban areas will continue to grow and expose more people to this form of noise pollution. The growing population and advancing technology have also contributed to increases in air travel. Additional high decibel noise has resulted from this heavier air traffic and construction of additional airports in areas that were once not affected by this form of noise pollution.

Many people have their own opinions as to what constitutes noise pollution. However, most people will agree that annoying noises bother them. Different people perceive noises differently, so noise pollution varies from person to person. For example, the noise of loud music and people at a party is not annoying to the party participants, but may be very distracting to neighbors trying to sleep. In general, people tolerate noise better if they are responsible for its production, note some utility for it, or know its origin (Bugliarello et al., 1976).

Sound is wavelike in nature. Sound is produced when a source vibrates to create successive compression and expansion waves through the air. These waves travel outward from the source in a longitudinal manner at different amplitudes and frequencies. Amplitudes and frequencies are responsible for various characteristics of sound. Amplitude is a term used to express the amount of negative and positive displacement a wave exhibits. The amount of energy used to produce the sound wave is directly proportional to the amplitude and also the loudness. The intensity of a sound is determined largely by the amplitude. The decibel (dB) is a number on a relative scale based upon the sound detection limits of the human ear. The threshold of human hearing is the reference point 0 with 194 representing the maximal threshold of hearing in the human. The scale is logarithmic, so small changes in the decibels indicate a large change in loudness. For example, a 30 dB increase describes a sound 1000 times louder (Nadakavukaren, 1986). To put this scale in more practical terms, a broadcast studio has sound of about 20 dB. Most conversations occur in the range of 40 to 50 dB (U.S. EPA, 1977).

Frequency is a term used to indicate the number of wave cycles that are produced per unit time. The standard unit used to express the number of wave cycles per unit time is the hertz (Hz). A hertz is equivalent to 1 wave cycle per second. The human ear can detect frequencies ranging from 20 to 20,000 Hz but hears best between 1000 and 3000 Hz. Human speech is produced in the range of 300 to 4000 Hz (Nadakavukaren, 1986).

Noise pollution affects human health in many ways that are usually ignored. Most adverse effects of noise pollution occur on the chronic level. Prolonged exposure to certain levels of noise can contribute to hearing loss, stress, teratogenic effects, impaired learning, decreased productivity, and impaired sleep (Kavaler, 1975; Terry, 1979).

Hearing loss occurs more frequently than most people believe. Many people have subtle hearing losses and never realize their loss. Temporary deafness or tinnitus may result from sudden loud noises and may last several hours before resolution. This type of temporary deafness is referred to as temporary threshold shift. Sudden noises occur too quickly to allow the tensor tympani muscle to contract and stabilize the ossicles of the middle ear, thus the sound is conducted without attenuation. Such noises include a close gunshot or loud rock music through headphones (U.S. EPA, 1978).

Hearing loss results in damage (as in temporary hearing loss) or loss of receptor hair cells in the Organ of Corti. These hair cells are located throughout the cochlea of the inner ear and detect vibrations in the fluid of the inner ear. Loud noises can damage or even destroy these hair cells. Repeated exposure to noise levels between 70 and 85 dB has been shown to produce hearing loss (U.S. EPA, 1978). This type of hearing loss associated with extensive noise exposure is referred to as sociocusis. Natural death of the hair cells occurs with aging. This is termed presbycusis and usually begins with the loss of the higher frequency receiving hair cells located nearest the round window.

Since hearing loss usually begins with the loss of higher frequency detection, most hearing screening tests use 4000 Hz sounds to test. Hearing loss in this range is noticed by difficulty in hearing the tick of a clock or the ringing of a telephone. These people generally have difficulty distinguishing spoken sounds such as th, ch, sh, s, f, p, or t (Nadakavukaren, 1986).

Noise can be stressful for many people. This function of noise was clearly used for survival by our primitive ancestors. Loud noises warned of the approach of hostile tribes or dangerous animals and initiated the "fight or flight" response. This immediate response from the sympathetic nervous system results in an increase in heart and respiration rates, blood pressure, and muscle contraction. Chronic exposure to noise can result in headaches, ulcers, colitis, and even elevated serum cholesterol (Nadakavukaren, 1986). The necessity for such an "alert" system is evident in a primitive setting; however, this survival reaction has become a nuisance in modern man. The noises that trigger this system may not be associated with impending danger. The noises are now man-made and often chronic in nature. Such exposure results in a chronic stimulation of the sympathetic nervous system and the production of stress which affect millions of people in society today. Factory workers in noisy environments have more medical problems, especially respiratory infections, than fellow workers in quiet settings. Rhesus monkeys exposed to noisy environments similar to that of a factory setting, have displayed a 30% increase in blood pressure. This hypertension continued for some time after the noise was reduced. The study suggests that noise can result in adverse medical conditions that may not be readily relieved by removal of the source (Terry, 1979).

Noise can also affect the sleep cycle and impair learning. Not only can noise make it difficult to fall asleep and remain in a restful sleep, but it can also alter the stages of the sleep cycle. Sleep patterns may be altered so less time is spent in the deeper stages of sleep. The results of decreasing particular stages of sleep can be as detrimental to physical and mental health as decreasing sleep in general. Sleep is essential for rebuilding the body, assimilating memories, and resting the conscious functions of the mind. Stage IV sleep is necessary for anabolic repairs to the physical body. Decreasing this stage of sleep can, therefore, seriously compromise physical health as lack of sleep may affect mental health. Different people require different amounts of sleep and are therefore affected differently by noise pollution. In general, ill people, older people, and women are more sensitive to the effects of noise on sleep (Bugliarello et al., 1976).

For years, people have noted that high levels of noise in a learning environment, be it at home or school, adversely affected learning in students. Distracting noise can make concentration very difficult for children trying to develop good study habits. Children exposed to high levels of noise during their education have displayed not only impaired reading ability but also impaired language development as well. Sounds such as "B" and "V" are difficult to discriminate in settings with high noise levels (Nadakavukaren, 1986). As a result of sound discrimination difficulties, children may develop distorted speech patterns.

Reading skills have been evaluated over the years using standardized reading examinations for most grade school and high school students. Some of the data from such tests have shown impaired reading skills in noisy surroundings. For example, students in noisy areas scored considerably lower on reading tests than their counterparts in the same building in quieter areas. High levels of noise can also adversely affect job performance in factories. Tasks requiring high levels of accuracy are affected most by noise. The type of noise is also important in affecting workers. Loud, sporadic noises produce the worst effects on workers. These sounds disturb concentration and work rhythm, distort perception of time, and increase variability in work performance (U.S. EPA, 1978). In general, quality of work is affected to a greater degree than quantity of work.

The repercussions of a noisy work environment on workers can extend into the home setting. These workers tend to be more tense, irritable, easily frustrated, and have difficulty relaxing. Such conditions are intensified if the home setting is also noisy (U.S. EPA, 1978).

Many people have become particularly concerned over the possible teratogenic effects of noise. Women have believed for centuries that noise could penetrate the uterus and elicit a reaction, such as kicking, in the fetus. Studies indicate that noise may increase the fetal heart rate (Kavaler, 1975). However, noise-induced maternal stress may affect the fetus even more. Stress of this nature may result in altered production of maternal hormones necessary for fetal growth and development. Studies have also shown in these mothers that uterine vessels can undergo constriction and reduce the blood flow to the uterus and fetus. Other studies of mothers living near airports have cited increases in the incidence of harelip, cleft palate, and spina bifida (U.S. EPA, 1978).

G. CONCLUSION

There are some common misconceptions about toxins. One is that the effects of toxins are always harmful. Often there are beneficial effects from trace exposures to foreign chemicals. Another is the exact level which is safe following chronic exposures. Knowledge of the threshold of a chemical that affects an individual resulting in a subtle ill effect in unknown. Some scientists believe that certain chemicals that enter the human body can be stored permanently resulting in the accumulation of high concentrations. Once the chemical storage sites become saturated, a steady state results so the amount of compound exiting the body is equal to the amount entering the body (Ottoboni, 1984).

These issues are often contested in court. To prove the cause of adverse health effects as a result of chemical exposure, one must prove that there is a direct link between exposure and the resultant injury. Such causation issues rely heavily on statistical, epidemiologic, and toxicologic evidence for resolution. The difficulty lies in that the cause/effect relationship being sought is often obscured by a multitude of extraneous probabilistic evidence. There are many possible causes for any one event and the identification of the dominant factor which caused the injury must be determined by careful assessment of scientific data. It is, therefore, the responsibility of the scientific community to be clear and precise about the evidence offered and of the legal community to accept the limitations of causal chain reasoning (Troyen and Brennan, 1987).

The number of possible sources of exposure to hazardous chemicals in the workplace is extremely high and includes metals (lead), aliphatic hydrocarbons, aromatic hydrocarbons (benzene), alcohols (ethanol), and halogenated carbon compounds (carbon tetrachloride). The relationship between chemical exposure and adverse effects has been established for many compounds. Primary to the toxicologic mechanisms involved are adaptation, absorption, distribution, metabolism, excretion, and related pharmacogenetic factors. Clinical findings and laboratory techniques often provide the final piece of evidence linking a chemical to an injurious effect.

REFERENCES

Agency for Toxic Substances and Disease Registry. Toxicological profile for lead. Agency for Toxic Substances and Disease Registry, Atlanta, GA, 1986, 1–207.

Agency for Toxic Substances and Disease Registry. The nature and extent of lead poisoning in children in the United States Agency for Toxic Substances and Disease Registry, Atlanta, GA, 1988, I.1–G.5.

Amdur, M.O.Air pollutants, in *Casarett and Doull's Toxicology. The Basic Science of Poisons*, 3rd ed. C.D. Klaassen, M.O. Amdur, and J. Doull, Eds. Macmillan Publishing Company, New York, 1986, 801–824.

Bass, N.H. Pathogenesis of myelin lesions in experimental cyanide encephalopathy, *Neurology* 18:167–177, 1968.

Brown, D.P. and Jones, M. Mortality and industrial hygiene study of workers exposed to polychlorinated biphenyls, *Arch. Environ. Health* 36:120–129, 1981.

Brown, J.F., Wagner, R.E., Dedard, D.L., Brennan, M.J., Carnaban, J.C., May, R.J., and Tofflemire, T.J. PCB dechlorination in Upper Hudson sediments, *Proc. Am. Chem. Soc.*, Miami Beach, FL, ENVR-0012, 1985.

Buck, W.B., Osweiler, D.G., and Van Gelder, G.A. Clinical and diagnostic veterinary toxicology.*Farm. Chem.* 138(13):36, 1975.

Bugliarello, G., Alexandre, A., Barnes, J., and Wakstein, C. *The Impact of Noise Pollution*, Pergamon Press, New York, 1976,1–461.

Burke, J. and Fitzhugh, O.G. Status report of chemistry and toxicology of PCBs, Suppl. I, December 1, 1970, U.S. Food and Drug Administration, Washington, DC, 1970, 1–30.

Buser, H.R. Formation, occurance and analysis of polychlorinated dibenzofurans (PCDs), dioxins and related compounds, *Environ. Health Perspect.* 60:259–267, 1985.

Chen, K.K. and Rose, C.L. Nitrite and thiosulfate therapy in cyanide poisoning, *JAMA* 149:113–119, 1952.

Chia, L.G. and Chu, F.L. Neurological studies on polychlorinated biphenyl (PCB) poisoned patients, *Am. J. Ind. Med.* 5(1–2):117–126, 1984.

Clayton, G. and Clayton, F. Halogenated cyclic hydrocarbons, in *Patty's Industrial Hygiene and Toxicology*, Vol. 2B, New York, 1981, 3645–3669.

Coulston, F. and Pocchiari, F. *Accidental Exposure to Dioxins — Human Health Aspects.* Academic Press, New York, 1983, 1–294.

Doull, J. Factors influencing toxicology, in *Casarett and Doull's Toxicology. The Basic Science of Poisons*, 2nd ed. J. Doull, C.D. Klaassen, and M.O. Amdur, Eds. Macmillan Publishing Company, New York, 1980, 70–83.

Durfee, R. Production and usage of PCBs in the United States, in Proc. of the National Conf. on Polychlorinated Biphenyls, Chicago, 1975. EPA-560/6-75-004. U.S. Environmental Protection Agency, Washigton, DC, 1975, 103–107.

Fischbein, A. Environmental and occupational lead exposure, in *Environmental and Occupational Medicine.* W.N. Rom, Ed. Little, Brown and Co., Boston, 1983, 433–447.

Furukawa, K., Tomizuka, N., and Kamibayashi, A. Metabolic breakdown of kaneclors (polychlorobiphenyls) and their products by *Acinetobacter* sp., *Appl. Environ. Microbiol.* 46(1):140–145, 1983.

George, C.J., Bennett, G.F., Simoneaux, D.K., and George, W.J. Polychlorinated biphenyls. Environmental Institute for Waste Management Studies, University of Alabama Press, Tuscaloosa, AL, 1987, 1–48.

George, C.J., Bennett, G.F., Simoneaux, D.K., and George, W.J. Polychlorinated biphenyls: a toxicological review, *J. Hazardous Mater.* 18:113–144, 1988.

George, W.J., Martin, L.A., and White, L.E. *Toxicity Profiles of Selected Organic Solvents.* University of Alabama Press, Tuscaloosa, AL, 1985, 4:1–101.

Gerald, M.C. *Pharmacology: An Introduction to Drugs.* Prentice-Hall, Inc., Englewood Cliffs, NJ, 1974, 1–524.

Goldstein, B. Toxic substances in the atmospheric environment, *J. Air Pollut. Control Assoc.* 33(5):454–467, 1983.

Griffin, R.A. and Chian, E.S.K. Attenuation of water soluble polychlorinated biphenyls by earth materials, *Environ. Geol. Notes* 86:1–99, 1979.

Hammond, P.B. and Beliles, R.P. Metals, in *Casarett and Doull's Toxicology. The Basic Science of Poisons*, 2nd ed. J. Doull, C.D. Klaassen, and M.O. Amdur, Eds. Macmillan Publishing Company, New York, 1980, 409–467.

Harrison, R.M. and Laxen, D.P.H. *Lead Pollution: Causes and Control.* Chapman and Hall, New York, 1981, 5, 29, 133–158.

Huff, J.E. and Gerstner, H.B. Kepone — a literature summary, *J. Environ. Pathol. Toxicol.* 1(4):377–395, 1978.

Hurkat, P.C. Polychlorinated biphenyls, *Indian J. Anim. Sci.* 47(10):671, 1977.

Hutzinger, O., Safe, S., and Zitko, V. *The Chemistry of PCBs.* CRC Press, Cleveland, OH, 1974, 1–269.

Hyde, D.M. and Falkenburg, R.L. Neuroelectrical disturbance as an indicator of chronic chlordane toxicity, *Toxicol. Appl. Pharmacol.* 37(3):499–515, 1976.

Ingle, L. Chronic oral toxicity of chlordane to rats, *Arch. Ind. Hyg. Occup. Med.* 6:354–367, 1952.

Ingle, L. *A Monograph of Chlordane: Toxicological and Pharmacological Properties.* University of Illinois, Urbana, IL, 1965, 1–67.

International Agency for Research on Cancer. IARC Monographs on the Carcinogenic Risks of Chemicals to Humans, Lyon, France. Suppl. 1974, 7:261–281.

International Agency for Research on Cancer. IARC Monographs on the Carcinogenic Risks of Chemicals to Humans, Lyon, France. Suppl. 1, 1979, 1–41.

Ito, N., Nagasaki, H., Arai, M., Maklura, S., Sugihara, S., and Hirao, K. Histopathologic studies on liver tumorigenesis induced in mice by technical polychlorinated biphenyls and its promoting effect on liver tumors induced by benzene hexachloride, *Gann.* 51:1637–1664, 1973.

Jacobson, J.L., Fein, G.G., Jacobson, S.W., Schwartz, P.M., and Dowler, J.R. The transfer of polychlorinated biphenyls (PCBs) and polybrominated biphenyls (PBBs) across the human placenta and into maternal milk, *Am. J. Pub. Hlth.* 74(4):378–379, 1984.

James, J.D., Braunstein, M.L., Karig, A.W., and Hartshorn, E.A. *A Guide to Drug Interactions.* McGraw-Hill, New York, 1978, 1–368.

Kavaler, L. *Noise, the New Menace.* The John Day Company, New York, 1975, 1–206.

Kehoe, R.A. The metabolism of lead in health and disease. The Harben Lectures, *J. R. Inst. Pub. Hlth. Hyg.* 24:1–81, 1961.

Keplinger, M.L., Fancher, O.E., and Calandria, J.C. Toxicologic studies with polychlorinated biphenyls, *Toxicol. Appl. Pharmacol.* 19:402–403, 1971.

Kihlstrom, J.E., Lundberg, C., and Orberg, J. Sexual functions of mice neonatally exposed to DDT or PCB, *Environ. Physiol. Biochem.* 5(1):54–57, 1975.

Kikuchi, M. An autopsy case of PCB poisoning with liver cirrhosis and liver cell carcinoma, *Fukuoka Acta Med.* 63:387–391, 1972.

Kikuchi, M. Autopsy of patients with yusho, *Am. J. Ind. Med.* 5(1–2):19–30, 1984.

Kikuchi, M., Hashimoto, M., Hozumi, M., Kaga, K., Oyoshi, S. and Nagakawa, M. An autopsy case of stillborn of chlorobiphenyls poisoning, *Fukuoka Acta Med.* 60:489–495, 1969.

Kikuchi, M., Mikagi, Y., Hashimoto, M., and Kojima, T. Two autopsy cases of chronic chlorobiphenyls poisoning, *Fukuoka Acta Med.* 62:89–103, 1971.

Kimbrough, R.D. Human health effects of polychlorinated biphenyls (PCBs) and polybrominated biphenyls (PBBs), *Ann. Rev. Pharmacol. Toxicol.* 27:87–111, 1987.

Kimbrough, R.D., Linder, R.E., and Gaines, T. Morphological changes in liver of rats fed polychlorinated biphenyls, *Arch. Environ. Hlth.* 25:3540–3640, 1972.

Klaassen, C.D. and Doull, J. Evaluation of safety: toxicologic evaluations, in *Casarett and Doull's Toxicology. The Basic Science of Poisons*, 2nd ed. J. Doull, C.D. Klaassen, and M.O. Amdur, Eds. Macmillan Publishing Company, New York, 1980, 11–27.

Kleinert, S.J. The PCB problem in Wisconsin. Report for the Joint Hearing of the Assembly of Environmental Quality Committee and Senate and Assembly of Natural Resources Committees on HR212 Administrative Rules, Madison, WI, September 21, 1976, 124–126.

LaDu, B.N., Mandel, H.G., and Way, E.L. *Fundamentals of Drug Metabolism and Drug Disposition*, The Williams & Wilkins Company, Baltimore, 1971, 308–327.

Lehman, A.J. Assoc. Food Drug Office, *U.S. Q. Bull.* 16:3, 1952.

Lehman, A.J. Assoc. Food Drug Office, *U.S. Q. Bull.* 18:3, 1954.

Leifer, A., Brink, R.H., Thom, G.C., and Partymiller, K.G. Environmental transport and transformation of polychlorinated biphenyls. EPA-560/5-83-025. Office of Pesticides and Toxic Substances, U.S. Environmental Protection Agency, Washington, DC, 1983, 14–25.

Levine, S. Experimental cyanide encephalopathy. *J. Neuropathol. Exp. Neurol.* 26:214–222, 1967.

Lippmann, M. Lead and human health: background and recent finding, *Environ. Res.* 51:1–24, 1990.

Mahaffey, K.R., Annest, J.L., and Roberts, J. National estimates of blood lead levels: United States, 1976–1980, *N. Engl. J. Med.* 307:573–580, 1982.

Metcalf, R.L., Sanborn, J.R., Po-Yang, L., and Nye, D. Laboratory model ecosystem studies of the degradation and fate of radiolabeled tri-, tetra-, and penta-chlorobiphenyl compared with DDE, *Arch. Environ. Contam. Toxicol.* 3:151–165, 1975.

Nadakavukaren, A. *Man and Environment,* 2nd ed., Waveland Press, Prospect Heights, IL, 1986, 39–128, 291–286.

Nagasaki, H., Tomii, S., Mega, T., Masugami, M., and Ito, N. Hepatocarcinogenicity of polychlorinated biphenyls in mice, *Gann* 63:805, 1972.

Needleman, H. *Low Level Lead Exposure, the Clinical Implications of Current Research*, Raven Press, New York, 1980, 1–322.

Nisbet, I.C. Criteria document for PCBs. Report No. PB-225 397, U.S. Department of Commerce, Washington, DC, 1976, 1–135.

Norback, D.H. and Allen, J.R. Pathobiological responses of primates to polychlorinated biphenyl exposures. Presented at the National Conference on Polychlorinated Biphenyls, Chicago, IL, November 19–21, 1975.

Onishi, H.A. and Trench, W.C. PCBs in perspective, *Ind. Wastes* Sept/Oct:30–35, 1981.

Ottoboni, M.A. *The Dose Makes the Poison*, Vincente Books, Berkeley, CA, 1984, 74–90.

Ottolenghi, A.D., Haseman, J.K., and Suggs, F. Teratogenic effects of aldrin, dieldrin, and endrin in hamsters and mice, *Teratology* 9:11–16, 1974.

Pikkarainen, P.H. and Raina, N.C. Development of alcohol dehydrogenase activity in the human liver, *Pediat. Res.* 1:165–168, 1967.

Platonou, N.S. and Reinhart, B.S. The effects of polychlorinated biphenyls (Aroclor 1254) on chicken egg production, fertility and hatchability, *Can. J. Comp. Med.* 37:341–346, 1973.

Poland, A. and Glover, E. Chlorinated biphenyl induction of aryl hydrocarbon hydroxylase activity: a study of the structure-activity relationship, *Mol. Pharmacol.* 13(5):924–938, 1977.

Prestt, I., Jeffries, D.J., and Moore, N.W. Polychlorinated biphenyls in wild birds in Britain and their avian toxicity, *Environ. Pollut.* 1:3, 1970.

Radcleff, R.D., Voodward, G.T., Nickerson, W.J., and Bushland, R.C. Polychlorinated biphenyls and ATPase activity. *U.S. Dep. Agric. Tech. Bull.* 7:1122, 1955.

Raina, N.C., Koskinen, M., and Pikkarainen, P. Developmental changes in alcohol-dehydrogenase activity in rat and guinea pig liver, *Biochem. J.* 103:623–626, 1967.

Rumack, B.H. and Lovejoy, F.H., Jr. Clinical toxicology, in *Casarett and Doull's Toxicology. The Basic Science of Poisons,* 3rd ed. C.D. Klaassen, M.O. Amdur, and J. Doull, Eds. Macmillan Publishing Company, New York, 1986, 879–901.

Rutter, M. and Jones, R.R. *Lead Versus Health; Sources and Effects of Low Level Lead Exposure,* John Wiley & Sons, New York, 1983, 1–379.

Sawyer, M., Kerny, T., and Spector, S. Lead intoxication in children. Interdepartmental Conference, University of California, San Diego, *West J. Med.* 143:357–364, 1985.

Schnellman, R.G., Putnam, C.W., and Sipes, I.B. Metabolism of 2,2′,4,4′,5,5′-hexachlorobiphenyl by human hepatic microsomes, *Biochem. Pharmacol.* 32(21):3333–3339, 1983.

Schnellman, R.G., Volp, R.F., Putnam, C.W., and Sipes, I.B. The hydroxylation, dechlorination, and glucuronidation of 4,4′-dichlorobiphenyl (4-DCB) by human hepatic microsomes, *Biochem. Pharmacol.* 32(21):3503–3509, 1984.

Schuetz, E., Wrighton, S., Safe, S., and Guzelian, P. Regulation of cytochrome P-450$_p$ by phenobarbital and phenobarbital-like inducers in adult rat hepatocytes in primary monolayer culture and *in vivo, Biochemistry* 25(5):1124–1133, 1986.

Shane, B.S. Unpublished results, 1991.

Shimado, T. and Sato, R. Covalent binding *in vitro* of polychlorinated biphenyls to microsomal macromolecules. involvement of metabolic activation by a cytochrome P-450-linked monooxygenase system, *Biochem. Pharmacol.* 27(4):585–590, 1978.

Sittig, M. *Handbook of Toxic and Hazardous Chemicals and Carcinogens,* 2nd ed., Noyes Publications, Park Ridge, NJ, 1985, 737–739.

Smith, M., Hopkinson, D.A., and Harris, H. Developmental changes and polymorphism in human alcohol dehydrogenase, *Ann. Hum. Genet.* 34:251–271, 1971.

Smith, R.P. Toxic responses of the blood, in *Casarett and Doull's Toxicology. The Basic Science of Poisons,* 3rd ed. C.D. Klaassen, M.O. Amdur, and J. Doull, Eds. Macmillan Publishing Company, New York, 1986, 223–244.

Stewart, R. Cyanide poisoning, *Clin. Toxicol.* 7:561–569, 1974.

Terry, L.L. *Health and Noise,* EPA Journal, Vol. 5, Number 9, Office of Public Awareness, Environmental Protection Agency, Washington, DC, 1979, 20–41.

Troyen, A. and Brennan, M.D. Untangling causation issues in law and medicine: hazardous substance litigation, *Ann. Intern. Med.* 107(5):741–747, 1987.

Urabe, H. and Asahi, M. Past and current dermatological status of yusho patients, *Am. J. Ind. Med.* 5(1–2):5–11, 1984.

U.S. Environmental Protection Agency. PCBs in the United States: industrial use and environmental distribution. PB-252-012. U.S. Environmental Protection Agency, Washington, DC, 1976, 34–334.

U.S. Environmental Protection Agency. *The Urban Noise Survey,* U.S. Environmental Protection Agency, Washington, DC, 1977.

U.S. Environmental Protection Agency. *Noise: A Health Problem.* Office of Noise Abatement and Control, U.S. Environmental Protection Agency, Washington, DC, 1978.

Waldron, H.A. Chasing the lead, *Br. Med. J.* 291:366–367, 1985.

Wastler, T.A., Offutt, C.K., Fitzsimmons, C.K., and DesRosiers, P.E. Effects of polychlorinated hydrocarbons on serum lipoprotein levels. National Technical Information Service, PB-253, 1975, 979.

Waxler, J. Whittenberger, J.L., and Dumke, P.R. The effect of cyanide on the electrocardiogram of man, *Am. Heart J.* 34:163–173, 1947.

Wilson, J.M., Zimmerman, R., Sengupta, A.K., Bennett, G.F., and George, W.J. The toxicity of dioxins, a review of described biologic effects and proposed mechanisms. Environmental Institute for Waste Management Studies, University of Alabama Press, Tuscaloosa, AL, 1988, 1–191.

Wolfe, H., Durham, W.F., and Armstrong, J.F. Health hazards of the pesticides endrin and dieldrin, *Arch. Environ. Hlth.* 6:458–464, 1963.

Wright, A.S., Potter, D., Wooder, M.F., and Donniger, C. The effects of dieldrin on the subcellular structure and function of mammalian liver cells, *Fd. Cosmet. Toxicol.* 10:311–332, 1972.

Ziegler, E.E., Edwards, B.B., Jensen, R.L., Mahaffey, K.R., and Fomon, S.J. Absorption and retention of lead by infants, *Ped. Res.* 12:29–34, 1978.

Chapter

17

Occupational Toxicology

Norbert P. Page

A. INTRODUCTION

Early in the 20th century, with the introduction of the automobile, an expanding petrochemical industry, and production line methods, the workplace environment changed from one of mainly small, isolated shops to one dominated by large factories employing many workers. In many cases, workers were crowded into buildings with poor ventilation and minimal safety controls, and the hazards of chemicals became more evident. Petroleum and mined materials, e.g., asbestos and metals, could now be removed from the earth, transported great distances, and introduced into a plethora of manufactured products. The nature of the workplace today is quite varied, from workers on large production lines in manufacturing plants to workers in small groups or individuals working alone such as in the application of pesticides. Regardless of the nature of employment, workers in virtually all places are potentially exposed to chemicals. In fact, the Occupational Safety and Health Administration (OSHA) estimated in 1987 that 32 million workers were potentially exposed to chemical hazards (OSHA, 1987).

As the industrial and chemical revolution progressed, knowledge of the harmful effects of chemicals primarily resulted from exposures occurring in the workplace and the effects that were observed in worker populations. It was not until the 1930s and 1940s, that safety testing with laboratory animals began, and only then by some of the larger companies to assess toxicity for high volume chemicals. The concerns were for immediate acute effects such as death or skin and eye damage with little regard for long term effects or the potential for birth defects. Competing corporations ardently protected their

0-8493-8851-1/94/$0.00+$.50
© 1994 by CRC Press, Inc.

data with minimal exchange and publication of the health effects data. (To some extent this problem still exists today, although government regulations now require that adverse health effects data be made available to the public.)

B. GOVERNMENT REGULATIONS FOR INDUSTRIAL CHEMICALS

Before the establishment of the American Conference of Governmental Industrial Hygienists (ACGIH) in 1938, there was no meaningful governmental surveillance, so the protection of workers was haphazard. Since that time numerous laws (federal, state, and local governments) have been implemented that require the careful adherence by industrial hygienists and occupational health officials.

The ACGIH is a professional society established in 1938 by employees of federal or state organizations and educational institutions engaged in occupational health activities. It was formed to meet this obvious need to protect workers from unhealthy workplace environs. The primary goal was to provide a forum for information exchange and to promote actions in the common interest of worker safety (Lippman, 1983). The two major activities of the ACGIH are (1) to determine safe or "acceptable exposure levels" for workplace chemicals by development of threshold limit values (TLVs), and (2) to promote the development of better methods to measure exposure levels, e.g., air sampling instruments.

The TLVs are recommendations for airborne concentrations of substances which represent conditions under which ACGIH believes that nearly all workers may be repeatedly exposed daily without adverse health effects. These are not official regulatory standards but rather recommended maximum exposure limits. They are not absolute safety levels and a small percentage of workers may experience effects at the TLV level or below due to the wide variability in individual susceptibility (ACGIH, 1990).

There are three forms of TLVs that consider the length of exposure, i.e., TLV-TWA (8-h time weighted average), TLV-STEL (short term exposure limit), and TLV-C (ceiling). The TLV-TWA is an average concentration for a normal 8-h workday and a 40-h workweek that workers may be exposed to day after day. The TLV-STEL is supplementary to the TLV-TWA and consists of a concentration that workers can be exposed to for no more than a short period (15 min). The TWA-STELs protect from the acute effects of substances whose primary toxic effects are of a chronic nature. They are recommended only when acute toxic effects have been reported from high, short term exposures of animals or humans. The TLV-C is a concentration that should not be exceeded at any time. In some cases, such as irritant gases, the TLV-C may be the only TLV category.

For those substances that are absorbed as the result of skin, mucous, or eye exposure, a large body dose may result; in those cases, the TLVs are also provided with a "skin notation". This alerts employers to the need to protect workers from skin or mucous membrane absorption even if respiratory protection has been provided. Those substances for which there has been a determination of carcinogenicity are also identified with a "confirmed or suspected" carcinogen tag.

As of 1990, ACGIH has developed TLV-TWAs for approximately 700 chemical substances with TLV-STELs for about 20% of these (about 20% have skin notations). Approximately 100 of the substances were also designated as A1 (confirmed) or A2 (suspected) carcinogens. TLVs have also been developed for several physical agents, including ultrasonic, ultraviolet, laser, radiofrequency/microwave, upper- and ultrasonic acoustic radiations, audible noise, static magnetic fields, heat stress, and hand-arm vibration. The recommendations of the National Council for Radiation Protection and Measurements have been adopted by the ACGIH for ionizing radiation exposure.

1. OSHA Act of 1970

The concern of the public and Congress over the escalating number of serious accidents due to uncontrolled workplace hazards led to the passage of the Occupational Safety and Health Act (OSHAct) of 1970. The OSHAct was to "assure, so far as possible, that every working man and woman in the Nation had a safe and healthful working environment." The OSHAct created two federal agencies, the Occupational Safety and Health Administration (OSHA) and the National Institute for Occupational Safety and Health (NIOSH).

OSHA was assigned the responsibility for promulgating legally enforceable standards which dictate practices appropriate to protect workers, while NIOSH was established to conduct research on occupational safety and health. The initial thrust of OSHA's Standards Program was to adopt the 1968 TLVs as OSHA federal standards for permissible exposure limits (PELs). The 1968 TLVs were based primarily on acute effects such as poisoning, irritation of the eyes or respiratory tract, and skin rashes.

The 1968 TLVs did not consider carcinogenic, teratogenic, or mutagenic effects. Since then, additional PELs have been promulgated, most related to carcinogenic effects. The NIOSH Pocket Guide to Chemical Hazards (DHHS, 1987) provides information on the PELs. NIOSH has evaluated many industrial substances or processes and has recommended occupational safety and health standards for many of them (DHHS, 1988). OSHA has accepted many of these recommendations, rejected a few, and is considering several others. A complete listing of OSHA PELs has recently been published (OSHA, 1989).

TABLE 17.1. OSHA Regulated Carcinogens

Substance	Date of Final Standard
2-Acetylaminoflourene	1/29/74
Acrylonitrile	1/17/78
4-Aminobiphenyl	1/29/74
Asbestos	6/20/86
Benzene	9/11/87
Benzidine	1/29/74
Beryllium	10/17/75
Bis(chloromethyl)ether	1/29/74
Coal pitch volatiles	1/21/83
Coke oven emissions	10/11/76
1,2-Dibromethane [EDB]	(Proposed)
1,2-Dibromo-3-chloropropane [DBCP]	3/17/78
3,3'-Dichlorobenzidine	1/29/74
4-Dimethylaminoazobenzene	1/29/74
4-Dimethylazobenzene	1/29/74
Ethyleneimine	1/29/74
Ethylene oxide	6/22/84
Formaldehyde	12/04/87
Inorganic arsenic	5/05/78
4,4-Methylenebis-(2-chloroaniline) [MOCA]	(Vacated)
Methyl chloromethyl ether	1/29/74
α-Naphthylamine	1/29/74
β-Naphthylamine	1/29/74
4-Nitrobiphenyl	1/29/74
η-Nitrosodimethylamine	1/29/74
β-Propriolactone	1/29/74
Trichloroethylene	(Proposed)
Vinyl chloride	10/04/74

2. OSHA Regulation of Carcinogens

OSHA has now published proposed regulations that identify 28 workplace carcinogens and proposed permissible exposure limits, established work practices, and medical surveillance requirements for them. The first substance to be regulated was asbestos in 1972. This was followed in 1973 by a proposed rule covering 14 substances, with final rules promulgated in 1974. Since that time, proposed rules were issued for 13 other substances. Of the 28 substances proposed as carcinogens, final rules were issued for 26 of them. However, nine of the final standards were challenged in court and two, MOCA and benzene, were struck down. A reproposed benzene rule has since been finalized. Thus, as of this writing, regulations are in effect for 25 carcinogenic substances as listed in Table 17.1.

Realizing the enormous task of promulgating individual standards for the multitude of potential carcinogenic industrial chemicals, OSHA chose to develop a generic policy to streamline the standards rulemaking process.

3. OSHA Cancer Policy

Since information was mounting that many workers were at high risk to develop cancer, OSHA was compelled to find an efficient means to determine the potential cancer hazards and introduce reasonable control measures to reduce the exposures. Toward this end, OSHA (1980a) promulgated a generic cancer policy.

The proposed cancer policy stipulated that OSHA would periodically publish a candidate list of potential occupational carcinogens. They were to be listed in two categories: category I was for those substances of the highest concern, based on evidence of carcinogenicity in humans, long term animal bioassays, and results that were in concordance with other data, e.g., short term tests; category II was for those which met the criteria for category I, but for which the evidence was only "suggestive".

OSHA thought that a generic carcinogen policy would speed the regulation of carcinogens by limiting the debate over generic issues in future regulatory proceedings for individual carcinogens. An important (and controversial) feature of the proposed OSHA policy was that quantitative risk assessments would not be employed to set exposure standards but rather only to set priorities for regulation. It was OSHA's intent that once a substance was declared an occupational carcinogen, exposure standards would be based only on feasibility for control measures. The first "candidate list" was released in 1980 and consisted of 107 chemicals (OSHA, 1980b). However, chemicals were not classified into category I or II at that time since OSHA intended to publish an updated candidate list on an annual basis and would then list them by categories.

Soon after OSHA issued its cancer policy, the Supreme Court ruled against the 1978 OSHA final benzene standard and stipulated that OSHA could only regulate exposures posing a "significant" risk to the health of workers and only then if the regulation could be demonstrated to reduce the risk significantly. This prompted OSHA to amend its carcinogen policy in 1981 to allow consideration of estimated risk along with feasibility in setting health standards for occupational carcinogens (OSHA, 1981). This eviscerated the thrust of the proposed cancer policy. In 1982, OSHA suspended the publication of the candidate lists (OSHA, 1982) and has not published one since. As of this writing the OSHA Cancer Policy remains dormant. (Note: a final rule was issued for benzene in 1987, based on additional data and the conduct of a quantitative risk assessment that demonstrated significant risk at the previous PEL.)

4. OSHA Hazard Communication Standard

In 1983, OSHA promulgated the Hazard Communication Standard (HCS) directing employers in the manufacturing industry to inform their employees of the hazards with which they work (OSHA, 1983). This was modified in 1987 to expand coverage to employees in nonmanufacturing industries (OSHA, 1987). The basic provisions of the HCS consist of hazard determination, labeling, preparation of material safety data sheets (MSDSs), and employee training.

Employers are required to review the available scientific data relative to the hazardous chemicals that they manufacture or use and to report the information to their employees and to those who purchase or use their products.

OSHA defined "hazardous chemicals" as

- Toxic and hazardous substances for which OSHA has issued a permissible exposure standard (29 CFR 1910, subpart Z)
- Those for which ACGIH has prepared a TLV
- Suspect or confirmed carcinogens as reported by the National Toxicology Program or International Agency for Research on Cancer (IARC) or those that OSHA has regulated as carcinogens

Labeling is required on all storage containers of hazardous chemicals in the workplace and those leaving the workplace. The labels must identify the chemicals, their hazards, and the name and address of the manufacturer or responsible party. Employers are responsible for obtaining or developing a MSDS for each hazardous chemical produced or used in their workplace. The MSDS must contain

- The identification of the chemical
- Physical and chemical characteristics of the hazardous chemical
- Known acute and chronic health effects and related health information
- Exposure limits
- Whether the chemical is considered to be a carcinogen by NTP, IARC, or OSHA
- Precautionary measures
- Emergency and first aid procedures
- The identification of the organization responsible for preparing the sheet

The MSDSs are to be readily accessible to employees in the work areas and for manufactured materials, and they must be provided to those purchasing the substances. Excellent reviews by Marsick and Byrd (1990) and Marsick and Adkins (1990) provide further information required of the MSDSs and recommended sources of information for their preparation. Figure 17.1 is the MSDS format recommended by OSHA.

In addition to labeling and MSDSs, the employers are also required to establish a training and information program for employees exposed to

MATERIAL SAFETY DATA SHEET

(Company Name and Address)

IDENTITY (As Used on Label and Use)

Note: Blank spaces are not permitted. If any item is not applicable, or no information is available, the space must be marked to indicate that.

Section I. Identification

Manufacturer's Name

Emergency Telephone Number

Address (Number, Street, City, State, and ZIP Code)

Telephone Number for Information

Date Prepared

Signature of Preparer (optional)

Section II. Hazardous Ingredients/Identity Information

Hazardous Components	OSHA PEL	ACGIH TLV	Other Limits Recommended	% (optional)

Section III. Physical/Chemical Characteristics

Boiling Point	Specific Gravity (H₂O = 1)
Vapor Pressure (mm Hg.)	Melting Point
Vapor Density (AIR = 1)	Evaporation Rate (Butyl Acetate = 1)
Solubility in Water	
Appearance and Odor	

Section IV. Fire and Explosion Hazard Data

Flash Point (Method Used)	Flammable Limits	LEL	UEL

Extinguishing Media

Special Fire Fighting Procedures

Unusual Fire and Explosion Hazards

Section V. Reactivity Data

Stability	Unstable		Conditions to Avoid
	Stable		

Incompatibility (Materials to Avoid)

Hazardous Decomposition or Byproducts

Hazardous Polymerization	May Occur		Conditions to Avoid
	Will Not Occur		

Section VI. Health Hazard Data

Route(s) of Entry:	Inhalation?	Skin?	Ingestion?

Health Hazards (Acute or Chronic)

Carcinogenicity	NTP?	IARC Monographs?	OSHA Regulated?

Signs and Symptoms of Exposure

Medical Conditions Generally Aggravated by Exposure

Emergency and First Aid Procedures

Section VII. Precautions for Safe Handling and Use

Steps to be Taken in Case Material is Released or Spilled

Waste Disposal Method

Precautions to Be Taken in Handling and Storing

Other Precautions

Section VIII. Control Measures

Respiratory Protection (Specify Type)

Ventilation	Local Exhaust	Special
	Mechanical (General)	Other

Protective Gloves	Eye Protection

Other Protective Clothing or Equipment

Work/Hygienic Practices

Figure 17.1. Recommended format for material safety data sheet (OSHA).

hazardous chemicals in their work area at the time of initial assignment and whenever a new hazard is introduced into their work area. A training plan is required which will train employees in the reading and interpretation of labels and MSDSs and instruct employees in obtaining and using the available hazard information.

5. Emergency Planning and Community Right-to-Know Act

Regulations have also been promulgated extending the dissemination of information on chemicals used in industry to the public residing nearby the industrial facilities. Title III of the CERCLA Superfund Amendments and Reauthorization Act of 1986 (SARA) requires that industry inform the public regarding these hazardous chemicals (U.S. EPA, 1988).

The responsible industry must

- Provide a representative to a required Local Emergency Planning Committee (LEPC)
- Notify the LEPC in the event of a release to the environment of a listed hazardous substance that exceeds the reportable quantity (RQ) for that substance
- Submit copies of MSDSs or a list of MSDS chemicals to the LEPC, State Emergency Response Commission, and local fire departments
- Report to the LEPC any hazardous chemicals (for which OSHA MSDSs have been prepared) that were present in the facility at any time during the year at a level above the specified thresholds
- Report annually to the EPA and state officials an inventory of chemicals that have been released to the environment in excess of specified threshold quantities

In addition to the Federal OSHA and SARA requirements, many states have "Right-to-Know" laws. The occupational health specialists must be cognizant of state and local regulations and assure compliance with those provisions as well as the federal requirements.

C. MEASUREMENT OF OCCUPATIONAL EXPOSURES

To assure compliance with the TLVs or PELs, knowledge of individual worker exposures are needed. Assessment of the exposure to chemical agents primarily involves three approaches: (1) examination of exposure records, (2) environmental monitoring, and (3) biological monitoring. Records that can be used as a source of exposure data are industrial records, information from suppliers (such as MSDSs), data pertaining to environmental discharges, and data provided by the company, its insurance carrier, and government agencies. Data from such sources are rather qualitative in nature but can be valuable in directing more definitive efforts for environmental monitoring or biological monitoring.

1. Environmental Monitoring

The two main types of environmental monitors consist of general area sampling and personal sampling (Hermann and Peterson, 1988). In general area sampling, the measurements are not specific for individuals but rather for fixed stations or areas, e.g., near machines and equipment. This type of sampling can detect sources of exposure or points of release, e.g., welding arcs. Area sampling, coupled with knowledge of a worker's activity (time present in various areas where the exposures have been documented), can often be sufficient to estimate a person's exposure profile. Extrapolation of area sampling to estimating individual exposure is rarely a simple matter. Most exposures have multiple release points with complex spatial and temporal variations. Workers usually move through the workplace in a nonpredictable manner, so detailed recordings of such movements are needed (Lynch, 1979). Area sampling can also detect "fugitive emissions", leaks or inadvertent and random loss of containment or control (e.g., due to plugged drain, inoperative fan, loss of temperature control of a vat, etc.). Constant monitoring is desirable to detect the fugitive emission at the earliest time and prevent extensive injury. Leak detection by fixed sensors in critical locations which are connected to alarms or remote indicators are now commonly used for some of the more toxic chemicals, e.g., hydrogen sulfide, carbon monoxide, hydrogen cyanide, nitrogen dioxide, sulfur dioxide, chlorine, and other highly toxic agents. Extensive and continuing general area sampling must record temporal fluctuations of concentrations at specific locations to relate to worker movements and accurately determine the individual exposure profile.

The advantages of general area sampling are that the main sources of exposure can be pinpointed and the monitoring equipment can be rigged to collect very large samples and can be coupled with automatic recording and analytic devices. They are adequate when precise worker exposure is not necessary. Many monitors are capable of issuing alerts when exposures are excessive. On the other hand, some area monitors are quite complex and require well trained technicians for proper use and maintenance. Keep in mind that to relate general area monitoring results to an individual's exposure the individual's work pattern and specific areas worked must be recorded to integrate their exposures. A variation of area monitoring at fixed locations is the use of "grab samples". Such area sampling is done on an ad hoc basis and can collect larger volumes at specific locations as the need warrants.

Personal monitoring involves recording actual exposures on an individual basis and overcomes the requirement for recording workplace activity. The earliest personal samplers consisted of film badges and pocket chambers used to record exposures to ionizing radiations. Similar devices are available as passive chemical dosimeters in which gases can diffuse onto solid solvents (like film badges) for direct reading or later analysis. Colorimetric detector tubes (change in color indicates exposure to the chemical) are attached to the worker's clothing and absorb gas or vapors. They are most valid for the determination of peak concentrations rather than the integration of total exposure over the workday. Colorimetric tubes are available for several hundred atmospheric contaminants.

More sophisticated exposure measurements can be obtained by using battery powered pumps continuously drawing workplace air through the sampling device. These sampling devices are worn on the person and travel with the worker. They often consist of a pump worn on the worker's belt or in a pocket with the sampler (air intake) clipped to the lapel near the breathing zone. Particulates may be gathered on filters while vapors and gases are absorbed onto activated charcoal for future analysis. Newer devices can measure several chemicals simultaneously, e.g., H_2S, CO, SO_2, $\%O_2$, and chlorine. Some have top-mounted LCD displays and will emit audible or visual alarms when PELs, TLVs, STELs, or ceilings are exceeded. Many of these newer samplers have multiple function control switches so the worker can obtain a reading for gas concentration, peak value, average value, and STEL (average for last 15 min).

The personal monitor has the advantage of providing a good indication of the individual worker's exposure regardless of their activity (work pattern). The main disadvantage is that the personal monitors may not reveal the most important source(s) of exposure or the time sequence of the exposures. The personal sampler must be lightweight, portable, and not affected by motion. They also can be somewhat of a nuisance to the worker, so the cooperation of the worker must be obtained ahead of time.

In planning for the exposure analysis, several decisions are required including what to sample, where to sample (which work areas), how to sample (area, grab, or personal), how long to sample (may depend on sensitivity of equipment and ambient level), and period for sampling (full workday or 15-min exposure) (Soule, 1979).

2. Biological Monitoring

For some substances, biological monitoring can provide a more useful measurement of harmful exposure than area or personal sampling of volatile materials in the workplace. For example, noninhalation exposure (mainly skin) of phenol, fat-soluble hydrocarbons, and solvents and substances with low vapor pressure, e.g., benzidine, may represent a large hazard (Lynch, 1979). On the other hand, although some skin absorption of benzene occurs, by far the most important exposure is via inhalation. Biological monitoring directly assesses the amount (or dose) of the substance that has been absorbed by the worker, regardless of the route(s) of exposure. Substances that do not enter the body are not detectable by biological monitoring and, except for irritation, are of little concern. As discussed by Bernard and Lauwerys (1989), biological monitoring integrates the absorption of the substance by all routes and from all possible sources (not just workplace). It also considers factors that influence absorption and distribution of the substance such as age, sex, nutritional and health status, and physical activity. While there is great interest (and much research) in improving biological monitoring, the knowledge of the parameters measured by the biological monitoring and associated toxic effects (and the time relationships of exposure) is limited to only a small portion of the industrial chemicals.

The main biological materials used for biological monitoring are urine, blood, and exhaled air (breath analysis). Other biological materials (e.g., fat, hair, nails, saliva, milk, and placenta) are collected and analyzed in special situations. Either spot or 24-h urine sampling can be used for inorganic chemicals and those organic chemicals which are rapidly biotransformed to water-soluble metabolites. Blood is the more appropriate biological fluid for substances poorly biotransformed. Breath analysis is used for exhaled volatile materials. Hair and nail clippings are often used to estimate exposures to heavy metals. However, several problems exist in the use of hair and nail clippings. At best, they reflect historical exposure due to the slow growth rates. Accuracy in detection depends on the portion of the hair or nail sampled. Milk samples may contain fat-soluble substances that have been absorbed by the female worker (Waritz, 1979). Body fat and bone marrow samples may reveal fat-soluble substances; however, invasive procedures are required to obtain these biological samples and they are thus only selectively used.

Biological monitoring also provides information about substances in the workplace in which a major portion of the absorbed body dose resulted from skin exposure and penetration. From that viewpoint, it might alert the industrial hygienist to a hazard that might go undetected by general area and personal samplers (that sample only workplace air).

Bernard and Lauwerys (1989) have categorized biological monitoring methods as either selective or nonselective tests. Selective tests consist of measurements of the unchanged chemical in biological media (blood serum, RBCs, hair, urine, etc.) or specific metabolites excreted in the urine. They are now the most commonly used tests. Use of urine has higher worker acceptance (than blood sampling) and often is preferable since it is less influenced by very recent exposure than is the unchanged chemical in expired air or blood. Examples of nonselective urine tests are the presence of diazopositive metabolites (produced by many aromatic amines), thioethers (produced by many electrophilic materials), creatinine (induced by many substances), and mutagenic activity (also produced by numerous materials).

The ACGIH (1990) has adopted biological exposure indices (BEIs) for 22 substances and has announced the intention of adopting BEIs for five more substances. Most BEIs pertain to the presence of the parent substance or its metabolite in the exhaled air, blood, or urine. For several, however, the BEIs are bioeffects markers such as creatinine in urine and reduced RBC cholinesterase.

Considerable effort is being expended to develop new and improved methods for biological monitoring. Two recent reviews of biological monitoring have been published by the National Research Council (NRC, 1989a, b). At this writing, NRC subcommittee reports have been published for pulmonary toxicology, reproductive toxicology, environmental neurotoxicology, and immunotoxicology. The NRC has categorized biomarkers into three categories: biomarkers of exposure, biomarkers of effect, and biomarkers of susceptibility.

Most of the prior discussion has pertained to biomarkers of exposure and to some extent biomarkers of effect. Examples of biological markers according to the NRC categorization are (1) biomarkers of exposure — parent compound and its metabolites in blood, urine, saliva, hair, and milk, (2) biomarkers of effects — respiratory function (e.g., airway resistance, clearance of particles), reproductive performance (e.g., sperm motility), enzyme levels (e.g., RBC cholinesterase), urinary proteins (e.g., metallothionein, creatinine), and DNA adduct formation, and (3) biomarkers of susceptibility — inborn differences in metabolism, variations in immunoglobulin concentrations, and conditions of target tissue. While there are many promising leads, much research is needed to define the relationships between biomarkers of exposure and effect and toxicity.

An excellent article reviewing the use of biological markers of toxicology has recently been published by Henderson et al. (1989). As pointed out by those authors, for biological markers to gain their potential, information on the toxicokinetics and degree of specificity for toxic effects must be generated for each chemical for which biological markers of exposure are developed.

D. ASSESSMENT OF OCCUPATIONAL HAZARDS

While great strides have been made in reducing chemical and radiation exposures in the workplace, recent events have dramatically demonstrated that accidents can and will happen. The not-too-infrequent deaths that occur in the application of neurotoxic pesticides (e.g., parathion) and the deaths of over 3000 persons due to the accidental release of methyl isocyanate at Bhopal, India serve as stark reminders of the potential acute hazards of chemicals. The increased risk of bladder cancer in the dye industry, liver cancer in vinyl chloride-exposed workers, lung cancer from employment as asbestos workers or in uranium mines, and bone cancer in radium dial painters are examples that illustrate the chronic/carcinogenic hazards that can occur within occupations where there are chemical and radiation exposures.

In assessing the potential hazards or risks from chemical and radiation exposures, not only the inate toxicity of the substance, but the nature of the occupational exposure must be considered. The most toxic of chemicals is of little risk if it is fully contained and exposure never occurs. (Keep in mind that methyl isocyanate was contained but accidental releases can and do happen.) In contrast, an exposure to a minimally toxic material can represent an unacceptable hazard if exposure levels are high.

Hazard and risk have been used interchangeably by some scientists, and in the past the terminology has created considerable confusion. However, hazard is considered as the potential for the substance to cause a particular health effect, while risk is the probability that the hazard will be manifest under specific exposure situations.

The National Research Council (NRC, 1983) standardized the terminology and approaches in the risk assessment process, organizing the process into four-steps:

- *Hazard identification* — determines the inate ability of an agent to cause adverse effects, e.g., cancer or birth defects.
- *Dose/response assessment* — characterizes the relation between dose of an agent administered or received and the incidence of an adverse effect in exposed human populations.
- *Exposure assessment* — measures or estimates the intensity, frequency, and duration of human exposures to an agent in the environment.
- *Risk characterization* — estimates the incidence of a health effect under the various conditions of human exposure described in the exposure assessment.

Since risk assessment is covered in detail in another chapter, this discussion will be about the hazard identification and dose/response assessment with emphasis on occupational hazards. The hazard identification step relies heavily on four types of data: epidemiology data, animal bioassay data, results of short term testing, and comparison of structural similarity to known toxic agents. The dose response assessment utilizes the epidemiologic and animal bioassay data to predict effects at various levels of exposure and from different routes.

Epidemiology data are the most desirable for assessing risk; however, they often are least available or are complicated by uncertainties in exposure estimates and other uncontrollable factors. Gamble and Battigelli (1979) classify epidemiology studies as cohort-type, case control, or cross-sectional studies. The cohort study defines the independent variable and follows a group of individuals to examine the incidence of a condition. In contrast, the case control study defines a disease or condition in a group of individuals contrasted to a similar group of nonaffected individuals (controls) to trace back the significant causative factors. The cross-sectional study defines a group by both disease (dependent variable) and characteristics (independent variables).

Epidemiologic studies can be either retrospective or prospective in nature. Retrospective studies examine the relationships between a population exposed to a specific chemical or condition in the past (perhaps several years or decades) and effects that are currently manifest. Prospective studies consist of present and future studies that follow a defined population exposed to specific chemicals or other agents and examine the incidence of effects that may develop in the future. Prospective studies define with greater precision the time relationship and magnitude of risk associated with the specific exposure.

Case reports may be suggestive of an effect, but without sufficient numbers of cases no cause and effect relationship can be established. With case control studies, cases of a specific disease or condition are collectively analyzed to determine if a relationship to a causative agent can be identified. Usually the existing epidemiology data are ill-defined as regards the exposure, or the exposures are at very high levels such that estimation of effects at lower levels, more typical of the usual occupational levels, must be extrapolated. For further details of epidemiologic study design and conduct, the reader is referred to another chapter in this text and the chapter on epidemiology by Gamble and Battigelli (1979).

Animal bioassay data are normally the most available and potentially useful data for hazard identification. It is assumed that results from animal experimentation are applicable to human hazard evaluation unless there are data to refute this position. Information from more than one species and in repetitive studies enhances the predictability confidence in the animal data to human hazard. Animal toxicity tests have been reasonably standardized and can evaluate a wide variety of health endpoints.

Considerable effort has gone into the development of short term tests, primarily employing cell systems. The most useful tests are cell systems such as the Salmonella bacterial test (Ames assay) that reveal mutagenicity which can be used as a predictor of carcinogenic potential. Since the cell systems cannot account for the pharmacokinetics of chemicals in the body and cell/organ interactions, the short term tests are limited to predicting mechanisms for chemical interactions with the cellular or test system. At best, they can be viewed only as weakly predictive of effects that might be found in whole animal testing or human exposure. Nevertheless they contribute to the hazard evaluation database that is to be considered.

Without epidemiologic and animal bioassay data, the toxicologist often resorts to a comparison of the chemical under evaluation with chemicals of similar structure and chemical/physical properties. Toxic effects are often associated with specific molecular moieties and can be predicted with reasonable confidence. A word of caution is needed since, in many cases, slight changes in configuration of a chemical can eliminate or change the inate toxic property. Structural activity relationships are best used to prioritize for animal bioassay testing.

The more typical situation is to conduct dose response assessments with animal bioassay data. In this case, two extrapolations are necessary: animal to human and high dose to low dose. Both are fraught with major uncertainties as discussed in the chapter on risk assessment in this text.

Exposures in the workplace usually involve inhalation of vapors or small particulate materials, or skin and eye contact. Ingestion is of less concern in the well managed workplace. Rarely are exposures well known in the workplace since area and personal monitoring are often neglected except in unusual situations. To gain knowledge about exposures to specific substances, in 1974 NIOSH conducted the National Occupational Hazard Survey (NOHS). Recently, an extension of this survey has been performed known as the National Occupational Exposure Survey (NOES). However, those surveys do not identify individual exposure levels of value in specific risk assessments.

The final estimate of risk should include the uncertainties of the data and the dose response and exposure assessments. Based on the risk characterization, the regulatory agencies can decide on a number of options, e.g., to take no action, impose a total ban from production or use, apply specific hazard labels, restrict exposures, or postpone a decision pending the collection of information.

Animal bioassay toxicity tests using experimental animals provide the data used for hazard evaluation and dose response assessment. Animal testing procedures have evolved over the past 50 years so reasonably standardized procedures are now in use. While OSHA and NIOSH have not developed specific toxicity test guidelines, they accept data developed under guidelines issued by other government agencies (e.g., EPA TSCA Section 4 health testing guidelines) and those of international learned bodies such as WHO and the Organization for Economic Cooperation and Development (OECD) (Page, 1985).

The following are terms often encountered in risk assessments:

Acute toxicity — the toxic effects resulting from a brief exposure of a few minutes to a few hours (the effects may not, however, be manifest for several days)

Dermal/eye irritation — the production of reversible changes to the skin or eye after exposure

Dermal/eye corrosion — the production of irreversible tissue damage to the skin or eye after exposure

Cumulative toxicity — the adverse effects of repeated doses occurring as a result of prolonged action on, or increased concentration, of the administered substance or its metabolites in susceptible tissue

Subchronic toxicity - adverse effects which follow repeated daily exposure for part of a lifespan (approximately 10%)

Chronic toxicity — adverse effects which follow long term exposure or have a long latency period for development

Oncogenicity — the property of being able to cause tumor formation (benign or malignant)

Carcinogenicity — the property of being able to cause malignant tumors, especially carcinomas

Exposure — the concentration of material in the air, food, or on the skin to which the animal is exposed (this is not the dose received by the animal)

Dosage — term comprising dose, frequency, and duration of dosing

Dose — the amount of test substance that enters the body, usually expressed as weight of test substance per unit body weight (e.g., mg/kg or mg/kg/d)

Dose response — the relationship between dose and magnitude of biologic effect

No observed adverse effect level (NOAEL) — the maximum dose which did not produce an observed adverse effect (expressed as mg chemical/kg body weight/d)

Lowest observed adverse effect level (LOAEL) — the lowest dose used in a test that produced an observed adverse effect (usually expressed as mg/kg/d)

Since another chapter in this text provides an in-depth discussion of animal bioassay methodology, only a brief review of pertinent studies important in occupational toxicology are reviewed here. Acute toxicity tests are usually the first studies to be performed. They determine the toxic effects (usually lethality) resulting from a brief single exposure. They establish the toxicity of the test substance relative to other chemical substances and may provide information on the mode of toxic action, median lethal dose (LD_{50}), and differences in toxic response as related to sex and route of exposure. Groups of animals are administered single doses (at least three dose level groups are used) and are observed for 14 d.

Accidental exposures of the skin and eye are common in industrial operations. Hazards may be related to the physical form, with a liquid or particulate having the greatest potential for entering via the body surface. Physical properties, notably the pH (highly acidic or basic), can indicate a potential for producing extensive tissue damage or corrosion. Allergic sensitization, such as that which can occur following exposure by the dermal or inhalation route, presents problems to a significant number of workers. It is an immunological-mediated cutaneous reaction to a substance, characterized by pruritus, erythema, edema, papules, or vesicles. Allergic reactions involve at least one exposure to initiate sensitization and a later exposure to illicit the immune reaction.

While acute toxicity deals with the adverse effects of single doses usually at fairly high dose levels, human exposure to chemical substances is more often in the form of repeated doses at levels which do not produce immediate toxic effects. Delayed effects may occur due to two primary mechanisms: (1) accumulation of tissue damage which can exceed the threshold for repair or compensation, and (2) accumulation of the chemical in tissues leading to a long term body exposure even after cessation of the ambient exposure.

Subchronic toxicity refers to toxic effects produced as the result of repeated exposure to a chemical over a part of the average lifespan. The division between subchronic and chronic dosing regimes is usually taken as 10% of the test animal's lifespan. Such studies provide information on toxic effects, target organs, reversibility of effects, and an indication of a "no observed adverse effect level (NOAEL)." The NOAEL is used in the risk assessment to generate acceptable exposure levels (see chapter on risk assessment in this text). The chronic toxicity study determines the effects of a test substance following long term repeated exposure. This test should demonstrate effects which require either cumulative injury or those effects that require a long latent period for them to become clinically manifest (e.g., cancer).

Animal test procedures have become rather standardized (Page, 1985; U.S. EPA, 1982). In the acute dermal test, the test substance is applied to the skin in graduated doses and left in place for 24 h, and the animals observed for 14 d. Adult rats, rabbits, or guinea pigs are used. In the acute inhalation test, the preferred species is the rat with exposure for at least 4 h.

In the eye irritation test the substance is applied to one eye at a dose of 0.1 ml with the untreated eye serving as a control. The material is left in the eye for 24 h without washing. The degree of irritation/corrosion is evaluated and scored for up to 21 d, unless positive results are evident earlier. In the dermal irritation test, 0.5 ml of liquid or 0.5 g of solid or semisolid substance is applied as a single dose to the shaved skin for 4 h, with observations for up to 14 d unless positive effects are observed before that. The rabbit is the preferred test animal for eye and dermal irritation studies. For the dermal sensitization tests, the test substance is applied to the shaven skin of young adult guinea pigs. In this test, animals are initially exposed to a sensitizing dose and after a period of not less than one week they are re-exposed to a challenge exposure to determine whether they have become hypersensitized by the test chemical.

Neurotoxicants act on several diverse targets within the nervous system and by a wide variety of mechanisms. A class of chemicals of great concern are those that inhibit cholinesterase. It is known that some organophosphorus cholinesterase inhibitors can induce delayed neurotoxicity in addition to acute effects. An acute delayed neurotoxicity test is available to detect this potential for delayed neurotoxicity. In this test, the material is administered orally to adult domestic hens (chickens) that have been protected from acute cholinesterase inhibition with atropine. The hens are observed for at least 3 weeks to detect delayed neurotoxic effects. The hen is a sensitive test species that has been shown to predict delayed neurotoxicity reliably in mammals.

The subchronic toxicity study is designed to illucidate the no observed effect level and toxic effects associated with continuous or repeated exposure to a substance for 90 d. The test is not capable of determining those effects that have a long latency period for development (e.g., carcinogenicity and life shortening). As a dose response relationship is desired, at least three dose levels and a control are used with the animals treated for 90 d. Rats are employed for oral and inhalation studies with rabbits preferred for subchronic dermal studies. Dogs or primates may be used in situations where a nonrodent test species is desired.

In the chronic and carcinogenicity studies, animals are observed for a major portion of their lifespan (at least 18 months) for the development of late-occurring toxicity and neoplastic lesions. A compound of unknown activity is usually tested in two mammalian species due to species variability. For carcinogenicity studies, rats and mice are normally the species of choice with hamsters sometimes used as the third choice. The exposure route usually mimics the human exposure although properties of the test chemical may require a modification (e.g., gavage instead of feed mixture).

No experimental animal serves as a generic surrogate for humans, and the choice of test species is heavily influenced by availability, cost to house and feed, and experience in their use. As discussed previously, rodents and other small laboratory animals are used most extensively in toxicological studies. Of the nonrodents, dogs and primates have been selectively utilized. The advantage of large animals is that they facilitate the performance of clinical and biochemical examinations that are difficult with rodents and they represent species more phenotypically similar to humans.

REFERENCES

ACGIH. 1990–1991 Threshold limit values for chemical substances and physical agents and biological exposure indices. American Conference of Governmental Industrial Hygienists, Cincinnati, OH, 122, 1990.

Bernard, A. and Lauwerys, R. Biological monitoring of exposure to industrial chemicals, in *Occupational Health Practice*, H. Waldron, Ed. Butterworths, London, 1989, 203–216.

DHHS. NIOSH Pocket guide to chemical hazards. DHHS (NIOSH) Publication No. 85-114, second printing. U.S. Department of Health and Human Services, Washington DC, 1987

DHHS. NIOSH Recommendations for occupational safety and health standards 1988. National Institute for Occupational Safety and Health, Centers for Disease Control, U.S. Department of Health and Human Services, Washington DC, Centers for Disease Control Morbidity and Mortality Weekly Report Vol. 37, No. S-7. Supplement, August 26, 1988. HHS Pub (CDC) 88-8017.

Gamble, J. and Battigelli, M. Epidemiology, in *Patty's Industrial Hygiene and Toxicology*, Vol. I, 3rd ed. G. Clayton and F. Clayton, Eds. John Wiley & Sons, New York, 1979, 113–134.

Henderson, R., Bechtold, W., Bond, J., and Sun, J. The use of biological markers in toxicology, *Crit.Rev. Toxicol.* 20(2):65–82, 1989.

Hermann, E. and Peterson, J. Evaluation, in *Fundamentals of Industrial Hygiene*, 3rd ed. B. Plog, Ed. National Safety Council, Washington, DC, 1988, 387–396.

Lippman, M. ACGIH: its background, membership, and activities, *Ann. Am. Conf. Gov. Ind. Hyg.* 5:5–11, 1983.

Lynch, J. Measurement of worker exposure, in *Patty's Industrial Hygiene and Toxicology*, Vol. III, 3rd ed. G. Clayton, and F. Clayton, Eds. John Wiley & Sons, New York, 1979, 217–255.

Marsick, D. and Adkins, J. Chemical regulation and computer resources, *Chemtech.* January, 1990.

Marsick, D. and Byrd, D., III. Resources for material safety data sheets (MSDS) preparation. Toxicological information series, III. *Fundam. Appl. Toxicol.* 15:1–5, 1990.

NRC. Risk Assessment in the federal government: managing the process. National Research Council. National Academy Press, Washington, DC, 191, 1983.

NRC. Biologic markers in pulmonary toxicology. National Research Council. National Academy Press, Washington, DC, 179, 1989a.

NRC. Biologic markers in reproductive toxicology. National Research Council. National Academy Press, Washington, DC, 395, 1989b.

NRC. Biological markers in immunotoxicology. National Research Council, National Academy Press, Washington, DC, 206, 1992.

NRC. Environmental neurotoxicity. National Research Council, National Academy Press, Washington, DC, 154, 1992.

OSHA. Identification, classification and regulation of potential occupational carcinogens. U.S. Department of Labor, Washington, DC, Federal Register 45, 5002, January 22, 1980.

OSHA. List of substances which may be candidates for further scientific review and possible identification, classification, and regulation as potential occupational carcinogens. U.S. Department of Labor, Washington, DC, Federal Register 45, 53672, August 12, 1980.

OSHA. Identification, classification and regulation of potential occupational carcinogens: conforming deletion. U.S. Department of Labor, Washington, DC, Federal Register 46, 4889, January 19, 1981.

OSHA. Identification, classification and regulation of potential occupational carcinogens. U.S. Department of Labor, Washington, DC, Federal Register 47, 187, January 5, 1982.

OSHA. Hazard communication, final rule. U.S. Department of Labor, Washington, DC, Federal Register 48, 53280, November 25, 1983.

OSHA. Hazard communication, final rule. U.S. Department of Labor, Washington, DC, Federal Register 52, 31852, August 24, 1987.

OSHA. Air contaminants — permissible exposure limits (Title 29 CFR Part 1910.1000). OSHA 3112. U.S. Department of Labor, Washington, DC, 85, 1989.

Page, N. International harmonization of toxicity testing, in *Evaluation of Drugs and Chemicals*, W. Lloyd, Ed. Hemisphere Publishing Corporation, New York, 1985, 453–465.

Soule, R. Industrial hygiene sampling and analysis, in *Patty's Industrial Hygiene and Toxicology*, Vol. I, 3rd ed., G. Clayton, and F. Clayton, Eds. John Wiley & Sons, New York, 1979, 707–769.

U.S. Environmental Protection Agency. Health effects test guidelines (EPA 560/6-82-001). Office of Pesticides and Toxic Substances, U.S. Environmental Protection Agency, Washington, DC, 1982.

U.S. Environmental Protection Agency. Title III fact sheet: emergency planning and community right-to-know. U.S. Environmental Protection Agency, Washington DC, 8, 1988.

Waritz, R. Biological indicators of chemical dosage and burden, in *Patty's Industrial Hygiene and Toxicology*, Vol. III, 3rd. ed., G. Clayton and F. Clayton, Eds. John Wiley & Sons, New York, 1979, 257–318.

Chapter

18

Environmental Epidemiology

Jerry D. Rench

A. INTRODUCTION

Epidemiology is often defined as the study of the distributions and determinants of illnesses and injuries in human populations (Mausner and Bahn, 1974). The roots of epidemiological investigation are at least 100 years old and center around the causes and preventative measures of infectious disease epidemics such as cholera and plague which required new methods of investigation. Concerns about the origin of noninfectious diseases, however, occurred as early as the fourth century, B.C., when Hippocrates recognized the problem of lead toxicity in the mining industry (Clayton, 1973).

Environmental epidemiology, that aspect of the field dealing with noninfectious disease agents in the environment, has become a prominent discipline in understanding the consequences of human exposure to chemicals and radiation. This field of environmental health is an observational science of human populations. The environmental epidemiologist observes populations to determine the causes and effects of environmental contamination. This approach is unlike that taken in classic toxicological studies in which exposures of animals to chemicals are controlled by the investigator. However, there are exceptions to this approach in epidemiology such as clinical investigations, in which patients are intentionally "exposed" to a substance or procedure to evaluate the efficacy of a new therapeutic agent. For the most part, environmental epidemiology is limited to studies of unintentional exposures to toxic substances. In these studies, the epidemiologist has to deal with whatever

0-8493-8851-1/94/$0.00+$.50
© 1994 by CRC Press, Inc.

exposure or disease conditions exist. The strength of epidemiological over toxicological investigation is that human populations are the basis for study.

The purpose of this chapter is to provide an overview of the basic concepts and methods of study used in environmental epidemiology. This chapter is not intended to be an epidemiology primer. Several texts provide basic instruction and guidance on epidemiological concepts and study methods (Mausner and Bahn, 1974; Lilienfeld and Lilienfeld, 1980; Checkoway et al., 1989). A student who is unfamiliar with this field will find this chapter useful in explaining some of the basic concepts needed to read and understand epidemiological information.

Fortunately, diseases do not occur randomly. Epidemiologists recognize that illnesses occur in patterns that reflect the underlying causes. For example, the question can be raised about whether lung cancer occurs randomly in people or in persons with certain characteristics which make them more prone to this disease. A common goal in environmental epidemiology is to identify and understand the risk factors and etiological factors (causes) associated with an illness. By definition, "risk factors" are those characteristics such as a person's age or sex that cannot be modified or controlled but enhance an individuals likelihood of developing a disease. "Etiological factors" are a subcategory of risk factors which can be controlled and include physical, chemical, or biological agents that are responsible for initiating or promoting disease.

B. THE EPIDEMIOLOGICAL APPROACH TO UNDERSTANDING DISEASE

One approach that epidemiologists take to identify factors that cause or influence disease occurrence is to evaluate time, person, and place characteristics. Disease patterns can vary considerably over time, a fact that has provided information about etiologic and risk factors. For example, lung cancer mortality in the U.S. has risen dramatically since the early 1960s, but stomach cancer mortality has been decreasing steadily for at least the past five decades (Page and Asire, 1985). The reason for the increase in lung cancer mortality has been attributed largely to an expanding population of cigarette smokers, and the decline in stomach cancer mortality may be related in part to refrigeration and less reliance on certain food preservatives. As an example of the importance of personal factors in understanding disease, it is well known that asbestos workers have a considerably higher risk of succumbing to mesothelioma and lung cancer than persons who have not worked with these fibers (U.S. EPA, 1986a). A prominent finding regarding exposure of asbestos workers is the synergistic effect between asbestos exposure and a personal characteristic, cigarette smoking. The occupation of an asbestos worker is also considered as an important personal factor relating to the etiology of the disease.

Investigating the areas in which a disease occurs is another important factor. For example, Mason et al. (1975) compiled mortality rates for cancer sites by county in the U.S. The results of this effort were colorful maps of the U.S. depicting counties with elevated and depressed cancer rates compared to the national average for the period 1950 to 1960. For melanoma of the skin, counties with elevated mortality rates were predominantly in the South where exposure to the sun is higher and possibly related to the etiology of the disease.

C. SOURCES OF HEALTH STATISTICS

Nearly any environmental contaminant can cause a range of effects depending upon the dose and exposure conditions. These effects may range from death following severe exposure such as carbon monoxide poisoning to irreversible effects such as congenital malformations or cancer or reversible conditions such as a dermatitis from contact with an organic solvent. One of the problems in studying diseases is obtaining reliable information of the disease entity that is sufficiently complete for scientific study. Data collection systems exist in many countries to record information of certain health events. For example, all deaths in the U.S. recorded in death certificates can be assimilated into a data base for further study. These certificates also include a description of the underlying cause and other diseases contributing to death. Environmental epidemiology studies are frequently based on mortality because of the certainty that a death certificate will be filed if a death occurs. Also, diseases which result in mortality are of considerable interest for public health assessment because of the suffering and cost associated with death, particularly when it occurs at a young age. The drawback of studying mortality as an indicator of a chemically related disease is that not all persons will necessarily die from that exposure, and death certificates sometimes have erroneous or incomplete disease information. Confirmation of death is also difficult in some cases as only about 11% of the causes of death are confirmed by autopsy. Morbidity data (for nonfatal illnesses) are often preferred as a measure of these diseases, but data collection systems for many nonfatal illnesses do not exist or have limited coverage. Frequently, illnesses due to chemical exposure are often undiagnosed, lead poisoning in children being a well known example. Some of the sources of health information used in epidemiological studies are identified in Table 18.1.

D. SOURCES OF EXPOSURE INFORMATION

Measurement of exposure is a component of epidemiologic studies in which the objective is to assess a relationship between an environmental

TABLE 18.1. Sources of Information on Diseases for Environmental Epidemiology Studies

Vital statistics offices
  Death certificates (causes of death)
  Birth certificates (congenital malformations, birth weight)
National Death Index (vital status)
Social Security Administration (vital status)
Disease registries
  Population-based cancer registries (e.g., Connecticut)
  Birth defects registries (e.g., Birth Defects Monitoring
    Program)
Medical records (disease and poisoning cases)
  Companies
  Hospitals
  Clinics
  Insurance
  Medicaid
  Health Maintenance Organizations
Unions, professional organizations (cause of death and disease
  cases)
Special studies
  Screenings
  Personal interviews

contaminant and a disease. The processes by which xenobiotic substances are released from their sources and enter the environment and a living organism is complex and varies considerably from one chemical to another. There are a multitude of ways in which humans can be exposed, with personal factors determining the dose that is absorbed. For example, the uptake of an ambient air pollutant such as ozone is a function of a person's size, their ventilation rates during exposure, and amount of time spent in the contaminated area. It is now also understood that the exposure concentration often does not correlate well with the concentration of the contaminant (biologically effective dose) in the target organ. Considerable interindividual variability exists in the rate in which xenobiotics are absorbed, distributed, metabolized, and excreted by humans. Thus, the traditional method of estimating dosage by monitoring the concentration of contaminants to which humans are exposed by ingestion or inhalation are no longer in themselves the ideal approach to estimating the exposure. In epidemiologic studies, the effect of not being able to assign doses accurately is referred to as misclassification. People in study populations can be easily misclassified into dose groups to which they do not belong.

For many populations there is very little information about exposure to contaminants. Some chemicals have been monitored by federal and state agencies who have had regulatory responsibilities for ensuring the quality of drinking water, air, and food. However, in general the availability of reliable and complete monitoring data for populations large enough to consider in

TABLE 18.2. Selected Sources of Chemical Exposure Data for Epidemiological Studies

Data Source	Description
Company and military records	Data collected by companies and services including industrial hygiene monitoring, biological monitoring, personnel and payroll records (job title, length of employment/exposure), process, engineering or production records
Various databases STORET (Storage and Retrieval of Water Quality Information)	U.S. EPA database containing toxic pollutant concentrations from water quality monitoring sites throughout U.S.
SAROAD (Storage and Retrieval of Aerometic Data)	U.S. EPA database containing monitoring results of air quality monitoring stations in U.S.
NHAT (National Human Adipose Tissue Data)	U.S. EPA database containing chemical concentrations measured in human adipose tissue
National Priority List of Superfund Sites	List of hazardous waste sites designated for clean-up by U.S. EPA. Information is available on contaminants at waste sites, their location, and exposed populations
U.S. Census of Manufacturing	Data from U.S. counties on proportion of population employed in various industries
Personal interviews	Study subjects recall activities and exposures to substances (e.g., smoking habits, use of solvent or pesticides)
Chemical Exposure Registries	Specialty registries that list individuals who have been exposed to a contaminant (e.g., beryllium)

epidemiological studies is scarce. Table 18.2 lists some of the data bases and sources of exposure information used in epidemiology studies.

E. MEASURES OF DISEASE AND RELATIVE RISK

It would be very difficult to study the consequences of exposure to environmental contamination in humans if disease could not be measured. The occurrence of a disease or an illness in a population can be expressed as a rate or as a ratio of the number of cases or deaths in a population compared to the total number of individuals in the population. If one compares the occurrence of disease in two populations that differ with regard to exposure to a chemical, it probably would not be meaningful to make the comparison on the basis of the number of cases unless the populations were identical in size. Rather, a comparison of the rates of disease would be more appropriate, because this approach considers varying population sizes and can provide probability estimates for assessing risk factors and extrapolating the results to other populations.

1. Disease Rates

A crude disease or death rate (R) can be expressed as

$$\frac{\text{number of cases or deaths}}{\text{size of population}}$$

In order for the measure of disease to be considered a rate, the individuals counted in the numerator must come from the population at risk in the denominator. If this is not the case, the measure is usually referred to as a ratio, which can also be a useful measure in epidemiology. An example of a ratio is the number of mesothelioma cases in a group of asbestos workers relative to the number of lung cancer cases.

One of the problems encountered in comparing crude rates between populations is that these measures may not be directly comparable because of differences in risk factors that influence the occurrence of the disease. For example, many diseases resulting in morbidity and mortality are strongly related to age. As age increases, the probability of developing a disease such as cancer or heart disease also increases, and therefore age must be considered in the analysis of the study results. Age adjustment is a statistical procedure that is employed to remove the effects of aging when comparing rates for populations that may have differing age distributions. The procedure can also be used for other demographic characteristics such as sex or race if these factors influence the risk of a disease.

Adjusted rates (also called standardized rates) are fictitious summary statistics, because they are weighted fractions in which the values are based on distributions from another population. In the direct method of age adjustment, a standard population (with a known age, sex, or race structure) is used to weight the age-specific rates of each population or exposure group being studied. In the indirect method of adjustment, the age-, sex-, or race-specific disease rates of a large and unexposed population such as that from a state or the U.S. in general serve as the standard and are used to estimate expected numbers of disease cases or deaths in the study population. The expected numbers are then compared with the observed numbers. The indirect method of adjustment is typically used when the exposed population is too small to calculate stable disease rates.

In studying nonfatal (morbidity) diseases, two types of rates can be determined: incidence and prevalence. Incidence refers to number of new cases of disease (numerator) that has occurred within some specified time, such as a year. Prevalence rates depend on the number of existing cases in a population at a particular time. Prevalence is related to incidence in so far as prevalence (P) varies as incidence (I) and duration (t) vary ($P \approx I \times t$). For some types of illnesses, it is difficult to determine when a disease is manifest, therefore the best way to measure the disease is by the prevalence rate. For example, some chemicals in the work place affect the respiratory system. Studies of decrements in respiratory function attempt to identify persons with deficits because without continued surveillance it is not possible to know when the changes in respiratory function may have occurred. However, incidence or mortality are usually the preferred measures of disease in situations where etiological analyses or risk assessments are needed.

2. Relative Risk

Certain types of epidemiologic studies are designed to measure the increase in disease, above background levels, associated with an activity or exposure to an agent. This excess risk may be measured in several ways, but one of the common methods is to express risk as relative risk. Relative risk (RR) is the ratio of the rate of disease in an exposed population to that in an unexposed population. For example, Hirayama (1981) studied the risk of lung cancer mortality in the nonsmoking wives of men who smoked. The adjusted mortality rate for lung cancer was 15.5 per 100,000 for nonsmoking wives of smokers and 8.7 deaths per 100,000 for nonsmoking wives of nonsmokers. By dividing the rate of the exposed group by the rate of the unexposed group, a relative risk of 180 (expressed as a percentage) was obtained, indicating an 80% increase in risk. The larger the value of relative risk, the more convincing the evidence that an etiological relationship exists between exposure and the disease.

3. Standardized Mortality Ratio

Another measure of disease occurrence is the standardized mortality ratio (SMR), a measure of risk relative to that of another population called the standard or reference population. In many studies the standard population is the control or unexposed group, while the study population is one that has experienced exposure to some agent. For example, using death rates of U.S. white males as a standard, Rinsky et al. (1981) found a SMR of 560 (p < 0.001) for leukemia in workers exposed to benzene in the rubber industry. A value of 100 would have indicated there was no excess risk of cancer compared to the standard population. The SMR considers differing age structures between the study and standard populations and the use of this method is known as the indirect adjustment method. The SMR method of analysis is frequently used in occupational cohort studies which are described in more detail in the following sections. The SMR is a measure of relative risk for a particular cause of death.

In a hypothetical example, if 50 bladder cancer deaths were observed in a group of workers involved in a particular industrial process and only 10 deaths were expected, then the SMR for this cause of death would be 500 (50/10 × 100). An interpretation of this example is that if the workers were dying at the same rate as the standard or unexposed population, one would have expected only 10 deaths, but in this example there were 50 deaths or 5 times more than expected.

If a statistically significant increase in a SMR is reported in a study and the confidence interval (CI), set at 95% by many investigators, for the SMR does not overlap with 100, there is a "statistical association" between the process and the cause of death. Such an association does not necessarily constitute an etiological relationship as will be discussed in greater detail later. In instances in which the health event under study is morbidity rather than mortality, the measure is known as the standardized morbidity ratio or the standardized incidence ratio (SIR).

4. Proportionate Mortality Ratio

The proportionate mortality ratio (PMR) is useful in determining the importance of a specific cause of death in relation to all deaths in a particular population. The PMR method of analysis is based on the premise that the distribution of cause-specific deaths in a large population can be used as a baseline to determine if there is a disproportionate number of deaths due to a specific cause in a population with exposure to a specific agent. The PMR is a ratio rather than a rate because the denominator consists of deaths rather than a living population at risk. The PMR can be described by the following examples. Suppose that 5% of all deaths in a certain state can be attributed to lung cancer, but in a particular community 10% of all deaths are attributed to lung cancer. The ratio between the proportion of lung cancers in the state and that in the community might indicate that one or more factors may be influencing the occurrence of death due to this disease. In a second example Greene et al. (1979) conducted a study on printing plant workers from the U.S. Government Printing Office (GPO) because previous surveys of cancer in these workers had been inconclusive and two workers had died of liver cancer, a rare tumor. Using the distribution of cause-specific deaths in Washington, DC as a standard, the investigators found a PMR of 144 (95% CI: 110 to 190) for lymphatic and hematopoietic cancers. The PMR for multiple myeloma was 220 (95% CI: 120 to 400) among these workers. Many of the excess deaths occurred in personnel working in the composing room. The criticism of the PMR-type of analyses is that distribution of the number of deaths in a population (used as the denominator) fluctuates from one group to another, and an excess number of deaths in one group means a deficit in another. Thus, an elevated or depressed PMR may not necessarily reflect the true risk in a population under study.

5. Odds Ratio

The odds ratio (OR) is an estimate of relative risk derived from case-control studies, described in more detail earlier in Section E. This measure of risk is the ratio of odds or proportions, that is, the proportion of a group experiencing an exposure to the proportion not experiencing the same exposure. For example, Alavanja et al. (1989) conducted a case-control study of non-Hodgkin's lymphoma deaths and matched control deaths, all of which were identified through a file of U.S. Department of Agriculture employees dying between 1970 and 1979. They reported an odds ratio of 2.6 (p < 0.05) for soil conservationists after 1960, indicating that this group of USDA workers were disproportionately represented among the cases compared to controls. It is hypothesized that the positive results seen for non-Hodgkin's lymphoma and certain other cancers in workers employed in agriculture-related occupations may be related to herbicide or insecticide exposure.

F. TYPES OF EPIDEMIOLOGIC STUDIES

One of the best ways to appreciate the epidemiological approach to studying disease is to review the types of studies employed which have evaluated environmental contaminants. The most commonly conducted studies are ecological, cross-sectional, case-control, and cohort.

1. Ecological Studies

For many environmental pollutants, we know that human exposures can vary considerably depending upon place of residence or work. Much of the health information assembled in this country is done on the basis of where a disease is diagnosed, which typically coincides with a person's place of residence. An ecologic study design is one in which the disease experience of a population from one geographic area is compared to that of another. In these studies, which may focus on counties or other small political units, comparisons or correlations are made between areas with differing levels of exposure. An example of an ecological study is the one by Griffith et al. (1989) where the cancer mortality in U.S. counties with hazardous waste sites was compared with those counties without waste sites. Table 18.3 shows the types of cancers that were studied and some of the results. The odds ratio reported in this study measured the extent to which there was a disproportionate excess of cancer mortality in those counties with waste sites compared to the remaining U.S. counties without waste sites. The investigators found that in counties with waste sites significant elevations in cancer mortality for lung, bladder, stomach, large intestine, and rectum was diagnosed. However, it should not be concluded that this study provides sufficient evidence that

TABLE 18.3. Odds Ratios for Selected Cancer Sites
Comparing 339 Counties with Hazardous Waste Sites to
2726 Counties Without Waste Sites for Significantly
Elevated Rates of Cancer Deaths

Cancer Sites	Males	Females
Pancreas	1.4	1.4
Lung	2.0[a]	5.2[a]
Prostate	1.8	—
Kidney	1.3	0.4
Bladder	5.8[a]	2.1[a]
Soft tissue	1.3	0.6
Leukemia	0.7	0.5
Esophagus	4.7[a]	1.8
Stomach	3.6[a]	2.7[a]
Large intestine	5.9[a]	4.3[a]
Rectum	9.4[a]	3.7[a]
Liver	1.1	1.6
Breast	—	6.5[a]

[a] $p > 0.002$.

Source: Griffith et al., 1989.

substances released from the waste sites cause cancer. It is possible that other
characteristics of the counties such as poor air quality, employment in certain
industries, or life style characteristics of the people (e.g., excessive smoking)
could have contributed to the elevated cancer mortality rates. Ecologic studies are subject to "ecologic fallacy", in which factors other than the one under
study could be etiologically associated with the disease of interest. The
principal characteristic of an ecologic study is that exposure is estimated for
the population and not on an individual basis. Exposure patterns can vary
considerably from one individual to the next within the boundary of the study
area. For these reasons the ecologic study design is frequently used to
identify trends or to generate hypotheses about disease risk factors that merit
further study with more analytical approaches, such as cohort or case-control
study designs.

2. Cross-Sectional Studies

The cross-sectional study evaluates the prevalence of a disease or illness in
a population. Typically, these investigations examine health endpoints such as
symptoms or clinical factors that are indicative of existing or imminent disease.
In a cross-sectional study the prevalence of a disease or clinical parameter is
compared among one or more exposure groups. For example, Goldsmith and
Shy (1988) tested the hypothesis that employment in a hardwood furniture

plant was associated with the prevalence of respiratory symptoms and impairment of pulmonary function. They found that certain symptoms such as frequent sneezing and eye irritation were significantly correlated with dusty jobs and that the prevalence odds ratio for each symptom was 4.0 (95% CI 1.1 to 14.7 and 1.0 to 16.6, respectively).

3. Case Control Studies

As the name of this type of study design implies, case-control studies are based on a comparison of exposure factors in cases with a disease to control subjects without the disease. The study by Woods et al. (1987) is an example of a case-control investigation in which occupational exposure to phenoxyacetic acid herbicides and chlorinated phenols was evaluated among persons who had developed soft tissue sarcoma (STS) or non-Hodgkin's lymphoma (NHL) and those who had not developed these cancers. There is concern that these pesticides may be carcinogenic. Interest in these compounds also comes from the fact that a contaminant, tetrachlorodibenzo-p-dioxin (TCDD), has been introduced during product formulation in the past. TCDD was a contaminant of the Agent Orange herbicide used in Vietnam. In the study of STS and NHL, cancer cases were identified through a cancer surveillance system in the state of Washington and were matched with controls living in the same general area as the cases. In evaluating subjects with any exposure to phenoxyacetic acid, there was no evidence of an association between exposure and either of the two types of cancer, as indicated by the odds ratio of 0.8 (95% CI: 0.5 to 1.2) for STS and 1.07 (95% CI: 0.8 to 1.4) for NHL. However, the odds ratios for NHL was significantly elevated among men who had been farmers (OR: 1.3, 95% CI:1.0 to 1.7) and forestry herbicide applicators (OR: 4.8, 95% CI:1.2 to 19.4).

Case-control studies are sometimes referred to as retrospective studies because the investigation looks back in time to determine exposure status after the case and control subjects have been identified. The advantage of this study design is that a population of subjects does not have to be enumerated. However, if the cases and controls are not representative of other persons with and without the disease, it may be difficult to draw conclusions.

4. Cohort Studies

The cohort study design is a frequently used method in environmental epidemiology. Cohort investigations are commonly used to assess possible adverse health effects of workers exposed to a certain type of contaminant. Occupational studies are undertaken because of the ease in identifying persons that are known to have been exposed to a substance and because workers often have had higher exposure levels than other segments of the population. The cohort study design involves identification of a group of individuals with exposure to a chemical and a group without exposure during a specified period. At the beginning of the study the individuals are without disease and then they are followed over time to estimate and compare disease occurrence.

If groups are identified according to current exposure conditions and followed into the future, the study is referred to as a prospective cohort design. If exposure is defined according to historical exposure conditions and follow-up proceeds forward in time, perhaps to the present date, then the study is taking a retrospective cohort approach. Occupational studies of work place contaminants are frequently retrospective cohort investigations in which a study population is defined as those individuals working in a particular plant, company, or industry during a specified period and their disease experience is tracked over time.

As an example of this type of study, consider the investigation by Simonato et al. (1986), who pooled mortality data from workers employed in factories across Europe to assess the health effects of man-made mineral fibers (MMMF). These fibers include rock wool, slag wool, glass wool, and continuous filament fibers. Some of these fibers are being considered as substitutes for asbestos. Table 18.4 shows the number of deaths of European workers involved in the production of MMMF. These epidemiologists reported a statistically significant increase in risk for all causes of death, all malignant neoplasms (SMR = 111), lung cancer (SMR = 125), and accidents, poisonings, and violence (SMR = 153). The elevated mortality risk for lung cancer is noteworthy, because similar risks have been seen in MMMF workers in the U.S. (Enterline et al., 1983). The SMR of 111 for all causes of death in this study was also unusually high. In other occupational cohort studies, SMRs for all causes of death are typically less than 100, often near 80, because workers have better health during their working years compared to the general population, as will be discussed later.

G. INTERPRETATION OF EPIDEMIOLOGIC STUDIES

Epidemiologic studies can be very useful in understanding the effects of environmental contaminants on people, but like any other discipline there can be pitfalls in interpreting epidemiologic data. To analyze properly or design epidemiologic studies one must be aware of the impacts of precision and bias.

1. Precision

Precision refers to the potential for random error (or chance) to affect study results. Epidemiologists attempt to control this type of error by using study and control populations that are as large as possible, but there are usually limitations to the size of groups that are available for study. For example, in occupational

TABLE 18.4. Analysis of Selected Causes of Death Among Workers Exposed to Man-Made Mineral Fibers From European Factories

Cause of Death	Observed Number of Deaths	Expected Number of Deaths[a]	SMR[b]	95% Confidence Interval
All causes	2719	2457.0	111	107–115
All malignant neoplasms	661	597.7	111	102–119
Stomach	64	68.5	93	72–119
Intestine, except rectum	37	40.5	91	64–126
Lung	189	151.2	125	108–144
Bladder	24	17.9	134	86–199
Leukemia	17	21.3	80	47–128
Diseases of respiratory system	165	164.9	100	85–117
Diseases of digestive system	123	107.8	114	95–136
Accidents, poisonings, violence	406	264.8	153	139–169

[a] Expected number of deaths based on national age-, sex-, cause-, and calendar-specific death rates.
[b] Standardized mortality ratio.

Source: Simonato et al., 1986.

cohort studies the investigator has no control over the historical circumstances of the factory and the number of workers that have been exposed to a particular chemical in a process. For this reason, it is common practice to conduct cohort studies on workers from more than one plant or company. In general, the larger the study and control populations, the more likely the investigator can improve precision, which in effect means that the confidence intervals surrounding a risk estimate are more narrow.

The issue of precision often arises in studies with negative results, that is, in investigations such as cohort studies in which no statistically significant difference is seen in risk between exposed and unexposed groups. The question that needs to be asked in those instances is whether the study sample was large enough to allow for a reasonable probability of detecting an elevated risk if it existed. This characteristic of a study is called power and it depends upon several factors. In a cohort study, these factors include the level of statistical significance that is being employed (e.g., $p < 0.05$), the disease rates in the exposed and unexposed populations, the number of study subjects, and the relative size of the two study groups (Checkoway et al., 1989). Power calculations are made during the design phase of a study to determine whether the study population is sufficiently large to detect an increase in risk of a particular

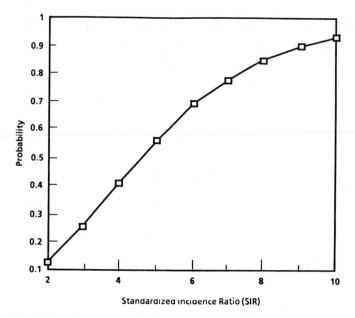

Figure 18.1. Relationship between probability of rejecting null hypothesis and SIR value for bladder cancer for an occupational cohort investigation with 40,000 person-years of observation and p = 0.03.

magnitude. If a study design has only limited power, the investigation may be abandoned or there may be incentive to marshall the resources to conduct an industrywide study to increase the size of the study groups.

Figure 18.1 provides an example of how power was considered in a study and shows the relationship between the standardized incidence ratio for bladder cancer and the probability of rejecting the null hypothesis of no elevated cancer risk in an occupational cohort involved in the production of a particular chemical (Unger, 1990). This cohort consisted of approximately 40,000 person-years of observation, and the statistical test was designed to achieve a level of significance of p = 0.03. This study provided relatively little power to detect an elevated risk, since the SIR would have to be a value of almost 800 to achieve an 80% probability of rejecting the null hypothesis (epidemiologists frequently use levels of 80% or greater). A SIR of this magnitude would suggest a strong carcinogenic effect in the cohort. To detect a smaller increase in risk, a considerably larger cohort would have to be assembled. The implication of a negative finding in this example (e.g., a nonstatistically significant SIR of 150 at p = 0.03) is that one would not be able to conclude with any degree of certainty that work in the occupational environment was not associated with bladder cancer, because the study could have detected only a very large level of risk.

2. Bias

Epidemiologists attempt to control bias (or systematic error) in the design or the analysis of a study. Bias is a key issue to assess in any study regardless of whether or not it is showing positive or negative results. For example, one would want to ask whether or not an association found between exposure to a toxic substance and a disease is real, once chance alone has been ruled out. It is possible that a bias may have existed in the way that the study was conducted, which could explain the positive (or negative) results rather than exposure to the toxic substance being studied. The types of bias that can be found in epidemiologic studies include selection bias, information bias, and confounding factors.

Selection bias occurs when the study groups are not representative of the populations from which they came. It is usually not feasible to study an entire population, and therefore a subset of the groups that are theoretically available must be selected.

One of the best known types of selection bias is the healthy worker effect, which occurs in occupational cohort studies. The study by Austin and Schnatter (1983) of the mortality experience of petrochemical workers from a Texas refinery provides an example of this effect. In that study, the expected number of deaths were based on the mortality experience of the U.S. population. The investigators observed 765 deaths due to all causes and expected 923.8, if the workers mortality experience had been the same as that of the U.S. The SMR of 83 (765/923.8 $\times$ 100) for all causes of death in this cohort of workers is typical of cohort investigations. The lower mortality in workers is attributed to the fact that compared to the general population, persons who are healthy are more likely to be employed and to stay healthy and employed. In many studies selection bias is controlled by selecting a control group that minimizes the potential for this type of bias. One approach in cohort investigations is to use standard populations that more closely approximate those of the study population. For example, rather than using U.S. mortality data as a standard, some investigators use a control population consisting of unexposed workers, if reliable mortality rates are available.

Information bias is another consideration when epidemiologic studies are being evaluated. This type of bias occurs when study subjects are misclassified in their disease or exposure status. Examples are the instances in which a rare form of a disease is misclassified as another disease or when the exposure status of an individual is incorrectly assigned. Determining the levels of exposure for an individual in a cohort or case-control study can be difficult in studies of diseases with long latency periods, because the exposures of interest may have occurred decades earlier and historical exposure information may be scanty or poorly documented. Information bias is an important factor to consider in negative studies. If misclassification is likely to affect the study and control subjects equally, then the bias is referred to as nondifferential and its effect is to drive results toward the null hypothesis (i.e., no association). If misclassification is more prominent in one of the study groups (exposed or unexposed subjects), then the effect estimate (e.g., the odds ratio or the SMR) can be moved toward or away from the null value (Checkoway et al., 1989). An example of this type of bias, called differential bias, can be envisioned in a case-control study of a particular type of birth defect in which the investigators interview mothers to assess exposures to chemicals in the household. Compared to the control group, the mothers of children with a birth defect may be more likely to recall use of or exposure to a particular chemical.

Another type of bias that can easily occur in an epidemiologic study is referred to as confounding. Bias of this type occurs when the study populations are different with respect to the presence of risk factors which influence the occurrence of the disease. Demographic factors such as age, sex, or race can be confounding factors, if not properly controlled. Another common confounder in studies that evaluate potential lung carcinogens is smoking. The question that is asked after an investigator finds an association between exposure and a respiratory disease in which smoking has been associated with the disease in the past is the extent to which the positive results can be explained by smoking rather than to the exposure factors under consideration.

Confounding can be controlled in the design or the analysis of the study results. In some case-control studies in which study subjects or their survivors are interviewed, the opportunity exists to collect data and other information on risk factors and to control the bias in the analyses. In historical cohort studies, which may not involve contact with each study subject, it can be difficult to collect the personal information needed to assess some potential confounding factors such as smoking history.

H. ASSESSING ETIOLOGICAL RELATIONSHIPS

One purpose of conducting environmental epidemiologic studies is the need to determine the cause of illnesses. A second purpose of epidemiologic studies is to identify those exposure levels that pose minimal or no health risks. The process of understanding the potential consequences of human activities is health risk assessment, discussed in Chapter 21. A strength of epidemiology is that it is an observational science, but the results of studies can be subject to bias or chance alone. A single epidemiologic investigation reporting a positive finding is rarely regarded as conclusive evidence of an etiological relationship. The problem of assessing causality is that direct or obvious connections between exposure and disease are hard to recognize. For example, does every person who smokes cigarettes develop lung cancer? The answer is no because the occurrence of disease depends on so many factors including dose, exposure to other agents, latency, susceptibility, genetic predisposition, and competing illnesses (someone may not have a chance of developing lung cancer because they die of heart disease first). Therefore, determining whether or not exposure to an environmental contaminant causes a chronic illness requires a weight of evidence approach in which several lines of evidence are examined.

There are at least five criteria that epidemiologists recognize as essential for assessing the etiological significance of statistical associations. These criteria along with others were originally proposed by Hill (1965) and are strength of association, consistency of association, temporality, dose-response relationship, and biological plausibility.

Causal relationships are sometimes established without each of these criteria being met or satisfied. Generally, there is evidence in two or more of these criteria when an etiological relationship is considered to be present. To this list we also need to add a criterion that considers the quality of the studies. Investigations which have minimal bias or few confounding factors are more valid in assessing causal relationships.

1. Strength of Association

Epidemiologic studies that test hypotheses report risk ratios that indicate how strongly a factor is associated with a disease. For example, consider the study by Dement et al. (1983) in which a SMR of 315 was found for cancer of the respiratory organs among white male asbestos workers. This threefold increase in risk is substantially more convincing that an etiological relationship with chrysotile asbestos exposure exists than if the investigators had found a SMR of 120 indicating a 20% increase in risk. Assuming that an investigator reports a statistically significant positive association and that bias is ruled out as an obvious explanation of the result, then one becomes more convinced of an etiological relationship if the measure of risk is large. That is, the larger the relative risk estimate the more likely that the association is etiological. In the many studies that have been conducted with workers that have been exposed to asbestos fibers, the statistically significant SMRs for lung cancer from the various studies have ranged from 157 to 875 for groups predominantly exposed to crocidolite asbestos, from 196 to 380 for amosite asbestos, and 136 to 824 for predominantly chrysotile asbestos (U.S. EPA, 1986a). The U.S. EPA has concluded that an etiological relationship exists between exposure to each type of asbestos fiber and lung cancer and mesothelioma. The asbestos industry has argued that the association between cancer and the chrysotile form of asbestos is not sufficient to be considered causal.

2. Consistency of Association

Establishing that an environmental exposure is etiologically related to a disease can be likened to developing a prosecutor's case in a trial. The more witnesses to an event the more likely it is that a jury or judge will be convinced that the event took place as described. A similar analogy can be drawn in causal inference in epidemiology. Consistency of association refers to the extent to which a positive association has been seen in more than one exposed population studied by different investigators using different methods. Significantly elevated lung cancer risks associated with crocidolite asbestos have been reported in at least four different studies involving mining or manufacturing workers from Canada, Australia, and the U.K. (U.S. EPA, 1986a). Even the best designed study can be subject to bias. Therefore, when different investigators using different study groups and methods report an elevated relative risk, then a causal basis for the relationship becomes more convincing. Consistency and strength of association are considered to be two of the more important criteria in establishing causality.

3. Temporality

It is obvious that if exposure to an etiological factor has not preceded the occurrence of disease an etiological relationship is not likely. In prospective studies in which populations are defined according to whether or not they have been exposed to a factor, it is considerably easier to establish the temporal aspects of disease history. However, in a case-control study of cancer in which the existence or absence of disease is first established and then the occurrence of etiological factors are compared among cases and controls, it may be difficult to establish with confidence that exposures occurred before onset of disease, particularly when one considers the long latency of cancer. Ecological studies that correlate disease rates and exposure factors in counties or other geographical areas usually are unable to establish that exposure has preceded disease because the study design does not permit this type of observation.

4. Dose-Response Relationship

Some epidemiological studies provide an opportunity to categorize study participants into exposure dose groups. When disease risks increase as exposure levels increase, a causal relationship becomes more credible because it is unlikely that study bias would occur in a trend-like fashion. Table 18.5 shows the dose-response relationship observed by Dement et al. (1983) in which there was a dramatic increase in lung cancer SMRs as exposure levels increased. The lack of a dose-response relationship in a study does not necessarily negate an etiological relationship, because establishing doses for individuals in a study can be difficult to ascertain accurately, particularly when this information is being constructed from historical records. Also, it is considerably difficult to define biologically effective doses which often do not correlate well with work place or environmental levels of a contaminant.

TABLE 18.5. Lung Cancer Mortality Risks According to Cumulative Exposure Level to Chrysotile Asbestos in White Males with 15 or More Years of Latency

Cumulative Exposure (fiber/cc × days)	Observed Cases	Expected Cases	SMR[a]
<1,000	5	3.58	140
1,000–10,000	9	3.23	279[b]
10,000–40,000	7	1.99	352[b]
40,000–100,000	10	0.91	1099[b]
>100,000	2	0.11	1818[b]
Total	33	9.82	336[b]

[a] Standardized mortality ratio.
[b] $p < 0.05$.

Source: Dement et al., 1983.

5. Biological Plausibility

Epidemiological information should not be evaluated in a vacuum but with other health and biological information, such as the results of toxicological, pharmacokinetic, and other studies. For example, asbestos fibers deposited and retained in the lung supports the conclusion that these fibers are capable of causing lung cancer. One of the unconfirmed issues about the carcinogenicity of asbestos fibers is the role that fiber dimensions play in the etiology of carcinogenicity. Toxicological studies have shown that long and thin fibers have greater carcinogenic potency than short and thin fibers. There has been very little epidemiological evidence to support this hypothesis and therefore it has been difficult to establish a causal relationship between fiber size and degree of carcinogenicity. To establish that the health risks of short fibers pose a lower hazard than long fibers could influence regulatory policy.

6. Application of Causality

After an agent is causally related to a disease, prevention programs can be implemented to reduce exposures and to monitor periodically persons at a higher risk of developing an illness, such as a former asbestos worker. Another application of causal epidemiological data is in health risk assessments or the development of new regulatory programs. For example, one of the charges of the National Toxicology Program is to update periodically and publish the list of carcinogens that can be used by risk assessors. The list is divided into substances that are known to be carcinogens and those that are reasonably anticipated to be carcinogens (NIEHS, 1989). The definition of a known carcinogen is one in which evidence from human studies indicates a causal relationship between exposure and human cancer, while those that are reasonably anticipated to be carcinogens have sufficient evidence of carcinogenicity in animals but limited evidence in humans. Limited evidence refers to those instances in which a causal interpretation is credible but alternative explanations such as chance, bias, or confounding factors cannot be adequately ruled out.

I.   EPIDEMIOLOGY IN RISK ASSESSMENT

Risk assessments are conducted for a variety of purposes. Many are driven by public or government interest in determining whether or not an activity or situation poses an unreasonable health risk to workers or the public. For example, among the many current concerns are the cancer health risks associated with high radon levels in homes (U.S. EPA, 1986b). Health risk assessments are frequently conducted without the benefit of epidemiological data because there are none. In these instances, the assessments are based on toxicological studies with laboratory animals and extrapolation of the experimental results to humans. Occasionally, epidemiological data may be used as a basis for developing exposure guidelines or identifying health risks. However, epidemiological studies are often relevant to a particular population or set of exposure conditions, and a larger and different population needs to be considered in the assessment. For example, many epidemiological studies of environmental chemicals are undertaken on occupational groups which usually experience higher exposure levels than those occurring in the general population. Furthermore, workers may be exposed and followed for a limited portion of their lifetime as opposed to the general population which can be exposed and at risk to disease for an entire lifetime. For these reasons and others, risk assessments usually involve biological models to portray exposure and disease relationships and to allow assessments of disease risks over a range of exposure conditions. Table 18.6 presents a few examples of risk assessments conducted with epidemiological data as a basis for the risk estimates.

The risk assessment approach depends in part on the assumptions that are made about a chemical and how it initiates or promotes disease. Some chemicals produce toxic effects only after a threshold dose is achieved. Many exposure guidelines are based on results of human and animal studies that show no adverse health effects when exposure levels do not exceed certain levels. The classic approach toward establishing acceptable exposure guidelines in these instances is to first identify the human dosage from the knee of the dose response curve at which effects start to occur if dosages increase. With that

TABLE 18.6. Examples of Risk Assessments Conducted with Carcinogens

Risk Assessment	Approach	Findings
Mesothelioma and lung cancer from asbestos (U.S. EPA, 1986a)	Linear model for each tumor type: absolute risk model for mesothelioma with minimum induction, nonthreshold relative risk model for lung cancer	Continuous exposure through inhalation to 0.0001 fibers/ml will cause lifetime risks of 1.9 mesothelioma deaths and 1.7 excess lung cancer deaths per 100,000 individuals
Lung cancer mortality from ^{222}Rn released from uranium mill tailings (U.S. EPA, 1986c)	Linear relative risk model (risk calculations assume minimum induction period, range of risk coefficients, decay product equilibrium fractions, condition of mine tailings and other factors)	There are 1 to 6 lung cancer deaths per year in the U.S. among persons residing more than 80 km from deposits of uranium mill tailings
Leukemia from benzene released by coke by-product recovery plants (U.S. EPA, 1984)	Linear nonthreshold model	For those persons residing within 20 km of a by-products plant emitting benzene at preregulation levels, the lifetime risk from continuous exposure was 6.4×10^{-3} or 2.2 cases per year

dosage level, extrapolations may need to be made from the study conditions to a broader situation such as converting 8-hour exposures in a work place setting to 24-hour exposures possibly occurring in a home. The threshold dose derived from a study may be further modified by imposing a margin of safety to account for interindividual variability and uncertainty in the scientific data. The ambient air quality standards for sulfur oxide and particulates were derived using this approach.

When a nonthreshold mechanism of action is assumed, as is frequently the case for carcinogenic effects, models are used to determine potential disease risks. These models may be simple, in which it is assumed that the dose-response curve is linear and passes through the origin, or they may be more sophisticated, such as the Armitage-Doll model based on the assumption of multiple stages for the carcinogenesis process. For more detailed descriptions of the methods and approaches used in risk assessments of epidemiological data, resources such as Checkoway et al. (1989), should be consulted.

REFERENCES

Alavanja, M.C.R., Blair, A., Merkle, S., Teske, J., Eaton, B., and Reed, B. Mortality among forest and soil conservationists, *Arch. Environ. Health* 44:94–101, 1989.

Austin, S.G. and Schnatter, A.R. A cohort mortality study of petrochemical workers, *J. Occ. Med.* 45:304–312, 1983.

Checkoway, H., Pearce, N.E., and Crawford-Brown, D.J. *Research Methods in Occupational Epidemiology*. Oxford University Press, New York, 1989, 344.

Clayton, G.D. Chapter 1, *Introduction in the Industrial Environment — Its Evaluation and Control*. National Institute for Occupational Safety and Health, U.S. Department of Health, Education and Welfare, Washington, DC, 1973, 719.

Dement, J.M., Harris, R.L., Symons, M.J., and Shy, C.M. Exposures and mortality among chrysotile asbestos workers. Part II: mortality, *Am. J. Ind. Med.* 4:421–433, 1983.

Enterline, P.E., Marsh, G.M., and Esmen, N.A. Respiratory disease among workers exposed to man-made mineral fibers, *Am. Rev. Resp. Dis.* 128:1–7, 1983.

Goldsmith, D.F. and Shy, C.M. An epidemiologic study of respiratory health effects in a group of North Carolina furniture workers, *J. Occ. Med.* 30:959–965, 1988.

Greene, M.H., Hoover, R.N., Eck, R.L., and Fraumeni, J.F., Jr. Cancer mortality among printing plant workers, *Environ. Res.* 20:66–73, 1979.

Griffith, J., Duncan, R.C., Riggan, W.B., and Pellom, A.C. Cancer mortality in U.S. counties with hazardous waste sites and ground water pollution, *Arch. Environ. Health* 44:69–74, 1989.

Hill, A.B. The environment and disease: association or causation?, *Proc. R. Soc. Med.* 58:295–300, 1965.

Hirayama, T. Non-smoking wives of heavy smokers have a higher risk of lung cancer: a study from Japan, *Br. Med. J.* 282:183–185, 1981.

Lilienfeld, A.M. and Lilienfeld, D.E. *Foundations of Epidemiology*. Oxford University Press, New York, 1980, 375.

Mason, T.J., McKay, F.W., Hoover, R., Blot, W.J., and Fraumeni, J.F., Jr. Atlas of cancer mortality for U.S. counties: 1950–1969. DHEW Publication Number (NIH) 75-780. National Cancer Institute, National Institutes of Health, Bethesda, MD, 1975, 103.

Mausner, J.S. and Bahn, A.K. *Epidemiology, An Introductory Text*. W.B. Saunders Company, Philadelphia, 1974, 377.

National Institute of Environmental Health Sciences. Fifth annual report on carcinogens — summary 1989. NTP 89-239. National Institute of Environmental Health Sciences, Research Triangle Park, NC, 1989, 340.

Page, H.S. and Asire, A.J. *Cancer Rates and Risks*. 3rd ed. U.S. Department of Health and Human Services, Washington, DC, 1985, 136.

Rinsky, R.A., Young, R.J., and Smith, A.B. Leukemia in benzene workers, *Am. J. Ind. Med.* 2:217–245, 1981.

Rothman, K.J. *Modern Epidemiology*. Little, Brown and Company, Boston, 1986, 358.

Simonato, L., Fletcher, A.C., Cherrie, J., Anderson, A., Bertazzi, P.A., Charnay, N., Claude, J., Dodgson, J., Esteve, J., Frentzel-Beyme, R., Gardner, M.J., Jensen, O.M., Olsen, J.H., Saracci, R., Teppo, L., Winkelmann, R., Westerholm, P., Winter, P.D., and Zocchetti, C. The man-made mineral fiber European historical cohort study: extension of the follow-up, *Scand. J. Work Environ. Health* 12 (Suppl. 1):34–47, 1986.

U.S. Environmental Protection Agency. National emissions standards for air pollutants; proposed standards for benzene emissions from coke oven by-product recovery plants. 49 FR 23522-23527. U.S. Environmental Protection Agency, Washington, DC, 1984.

U.S. Environmental Protection Agency. Airborne asbestos health assessment update. EPA/600/8-84/003F. Environmental Criteria and Assessment Office, U.S. Environmental Protection Agency, Washington, DC, 1986a, 198.

U.S. Environmental Protection Agency. A citizen's guide to radon — what it is and what to do about it. OPA-86-004. U.S. Environmental Protection Agency, Washington, DC, 1986b, 14.

U.S. Environmental Protection Agency. Final rule of radon-222 emissions from licensed uranium mill tailings. EPA 520/1-86-009. Office of Radiation Programs, U.S. Environmental Protection Agency, Washington, DC, 1986c, 214.

Unger, A. Personal communication. Battelle. Arlington, VA, 1990.

Woods, J.S., Polissar, L., Severson, R.K., Heuser, L.S., and Kulander, B.G. Soft tissue sarcoma and non-Hodgkin's lymphoma in relation to phenoxyherbicide and chlorinated phenol exposure in western Washington, *JNCI* 78:899–909, 1987.

Chapter

19

Detection — Analytical

Christine A. Purser and Arthur S. Hume

A. INTRODUCTION

The federal government has made it necessary for environmental pollutants to be identified and quantified to determine that compliance to regulations are within acceptable tolerance limits.

Since the industrial revolution began, man has been making or dumping chemicals into the environment, contaminating the air we breathe, the water we drink, and the soil on which we live. By nature or man's design, many of these chemicals such as pesticides are intended for the destruction of undesirable animal and plant life, but can in certain concentrations present a hazard to human life.

Fortunately, many persons finally became concerned and unified enough to act about the pollution of their environment. In 1969 the National Environmental Policy Act (NEAP) was passed. In 1970, it was followed by the formation of the U.S. Environmental Protection Agency (EPA). This was the beginning of a proliferation of federal and state legislation to be passed for the benefit of our environment (Cram, 1989; Ryan, 1989).

The EPA is divided into four program offices: (1) Water, Air, and Radiation, (2) Solid Waste and Emergency Response, (3) Pesticides, and (4) Toxic Substances; these offices are responsible for developing and issuing environmental regulations (Friedman, 1990). Each office is also responsible for methods used to determine regulatory compliance of its protocols and regulations. Therefore, the EPA has no single methods book for analysis of pollutants. Upon request, the EPA will supply a list of over 400 methods of analysis from the National Technical Information Services (U.S. EPA, 1989). Another agency interested in the environment, for health reasons, is the National Institute for

0-8493-8851-1/94/$0.00+$.50
© 1994 by CRC Press, Inc.

Occupational Safety and Health (NIOSH). NIOSH has methods for monitoring about 500 airborne contaminants in the ambient air in the workplace. These methods describe collection and analysis of vaporous or particulate contaminants as organic solvents, inorganic acids, pesticides, and metals.

These methods are continually revised and updated as improved collection techniques and analyses are developed.

The EPA has been trying to integrate its approval of methods procedures to facilitate the acceptance of newer and better methods and instruments. According to Cram (1989), it may soon take 10 years or longer to bring a technique from conception (published in the literature) to the market as a commercial product. Another 10 years are required for a given technique or technology to be validated and approved as an EPA method. The development and application of capillary technology is an excellent example of this time lag disparity. Capillary gas chromatography was first introduced in the literature in 1957, the first commercial capillary column was introduced in 1977, and the first EPA method using a capillary column was approved in 1984. Cram (1989) states that most laboratories doing EPA work use EPA methods at least 60% of the time. However, up to 25% of the time, modified EPA or non-EPA methods are used. The latter methods are usually employed either to solve unique or proprietary problems or in situations where the EPA methods are nonexistent or nonapplicable to the sample matrix.

Over the years technology has lowered the limits of detection in instrumentation by developing newer, more sensitive instruments. New methods of retrieving pollutants from complex matrices have also evolved from this technological explosion. Without this new technology and improved level of sensitivity, the presence of many chemicals would not be recognized. There have been significant decreases in detection limits of most methods over the past 30 years. The more sensitive measurements of chemicals have permitted improved studies of pollutants for health and risk assessment purposes. Some previously established "safe levels" are not as risk free as once thought. The sensitivity of methods has progressed from a parts-per-thousand (ppt) range in the 1960s to a parts-per-billion (ppb) range in the 1980s. It is predicted that a parts-per-trillion (ppt) range will be achieved in the 1990s. In fact, today many laboratories successfully use methods for determining concentrations at the ppt level. Also, methods have been developed for dioxin analysis at the 1 to 2 parts-per-quadrillion (ppq) level in waste water (Cram, 1989). The key to attaining higher sensitivities in environmental analytical measurements for industry lies not only in new or enhanced detection technologies, but also in more emphasis on sample preparation, matrix clean-up techniques, and preconcentration techniques.

According to Cram (1989) "the productivity of modern environmental-analytical instrumentation is highly dependent upon the number of tasks which can be performed in parallel, that is, with multi-user, multi-instrument, and multi-tasking capabilities." Highly interactive computer equipment (data systems) is the key to productivity in the laboratory. The interface of computer equipment to laboratory instruments makes it possible for instrumental data to be directly routed into data bases or information management systems. Software programs are now becoming available to automate the process of checking data for quality, adherence to protocol, and completeness, while also allowing for invoicing, project management, and generation of departmental reports (Wright, 1989). Productivity is being enhanced by the capabilities of some software to use the result of one experiment to design the next. There is a continued interest in the development of computer systems which decrease total analysis and reporting time, thereby lowering the cost per sample. Data systems are allowing laboratory personnel to focus on data interpretation rather than "baby sitting" the instrument. The computer management of chemical information in a multivendor laboratory requires processing, reviewing, reporting, and archiving of data. Typically, samples are divided into several aliquots and analyzed by various techniques, for example, gas chromatography (GC), high performance liquid chromatography (HPLC), gas chromatography/mass spectrometry (GC/MS), atomic absorption (AA), and noninstrumental techniques. The job of the chemical management information system is to pull together this data, both raw and reduced, into a single processing environment. Decision makers can then best use the information in sample reports while improving the overall productivity of their laboratory.

A combination of techniques, such as a separation technique with an identifying technique, is providing new analytical power, allowing the analytical chemist to examine nature and our ecology in ways never before possible (Finnigan and Poppiti, 1989). GC/MS was one of the first of these combination techniques. Others coming into their own include liquid chromatography/mass spectrometry (LC/MS), multistage mass spectrometry (MS/MS), supercritical fluid chromatography/mass spectrometry (SCF/MS), and gas chromatography/Fourier transform infrared (GC/FTIR).

Mobile laboratories are also being developed. They can be taken to field sites where the analysis is performed on the spot, thereby eliminating the delays and costs associated with shipping, storage, and laboratory backlogs (Cram, 1989; Grupp et al., 1989; Spartz et al., 1989).

Today, quality in analytical measurements is the foundation of the environmental laboratory operation. To insure quality, it is urgent for good laboratory practice to include a quality assurance (QA) program (EPA Publication No. 600/9-76-005). Such a program should include all phases of environmental analysis: collection and storage of samples, chemical and physical analyses, interpretation of data, and the publication of results. Two concepts used in a QA program are quality control (QC), a mechanism used for controlling errors, and quality assessment, a system used to verify that the entire analytical system is working within acceptable statistical limits (Gautier and Gladney, 1987).

Quality assurance planning in the field and laboratory can improve the confidence level of environmental data. Some important QA parameters include detection limits (typically at 99% confidence), precision of measurements, and accuracy. Different analyses have their own control limits (CL) which comprise acceptable precision and accuracy ranges. The client must relay to the laboratory the CLs that are needed for their samples. If the data is to be used for screening an environmental site, the CLs need not be nearly so rigid. However, if the site is involved in litigation, the CLs need to be rigid. In any case, when CLs are not met the laboratory must have follow up procedures by either rechecking results, repeating analyses, or providing explanations of difficulties encountered.

Some controls used to reduce errors in the collection of samples are (1) travel blinds and rinses of field equipment which monitor introduction of contamination during transportation or sampling, respectively and (2) duplicates to provide a measure of analytical precision. Duplicate samples collected in the field are taken from the same source. Once the samples are in the laboratory, duplicates are homogenized and two samples are taken from the homogenate to eliminate any effects of the sampling procedure (Bryden and Smith, 1989a).

The determination of accuracy is more involved than that of precision. Accuracy denotes the nearness of a measurement to its accepted value. The basic difference between this term and precision is that accuracy involves a comparison with a true or accepted value. In contrast, precision compares a result with other measurements made the same way (Skoog and West, 1975). To measure accuracy, the following steps must be taken: (1) calibrations must be established using high quality commercial or laboratory prepared standards, (2) calibrations must be checked against standards from a second source, and (3) periodic analyses of reference materials must be made. To confirm accuracy on a batch-to-batch basis, sample duplicates must be spiked with known quantities of compounds under investigation. Laboratories should be able to provide spike recovery limits for different matrix types (Bryden and Smith, 1989a). The basic success of the environmental laboratory in the future will require increased quality, sensitivity, and productivity.

The quality and integrity of laboratory data will improve through the development of more rigorous methods and automatic error checking. However, the greatest concerns of the environmental analysis industry are the adequate staffing with trained personnel in analytical and environmental chemistry, instrument operation and maintenance, data handling, and interpretation skills. This is, and will continue to be, one of the basic limitations to quality and productivity.

Both the National Science Foundation (NSF) and the EPA are addressing the need for joint environmental analytical science and engineering programs in the U.S. by funding training and research programs (Cram, 1989).

Our intention in this chapter is to describe some of the methods used by the EPA and others to analyze the various air, water, and soil samples for environmental pollutants such as the analyses of metals, petroleum, solvents, pesticides, halogenated aromatic compounds, inorganic substances, and herbicides.

B. SAMPLE COLLECTION

In determining the environmental quality of a sample, we must keep in mind that analysis is no better than the collection of the sample. To obtain meaningful data from the analyses, a plan of attack must be devised. First, we must know what tests on which samples are needed. Equally important is the necessity of determining which regulatory jurisdictions apply to the proposed investigation. For example, one should know pertinent ordinances of the cities, sewage districts, treatment and landfill facilities, water quality boards, health departments, as well as EPA requirements. Water and air investigations are the easiest to address, since all states are subject to the requirements of the Safe Drinking Water Act. Also, all businesses are subject to air standards set by the Occupational Safety and Health Administration, OSHA (Supelco, Inc., 1985). Contaminated soil or hazardous waste samples are more difficult to handle than others since definitions as to what pollutants and levels are considered safe or hazardous are not yet standardized.

The most widely followed list of contaminants is the EPA's "Priority Pollutants" list which includes over 100 organic compounds of varying types along with asbestos, cyanide, and 13 metals. Cities, counties, states, and various federal agencies and departments may have additional requirements which may apply when more than one agency claims jurisdiction.

After the regulations are understood, a "sampling plan" can be worked out to determine what information is needed (Keith, 1988). A consolidation of the types of most frequently used analyses, their methodologies, sampling containers, sample preservatives, and allowed holding times is published by Bryden and Smith in the July 1989 issue of *American Laboratory*.

OSHA has set standards and exposure limits for airborne contaminants in the workplace. A compilation of OSHA standards and sampling procedures is presented in GC/HPLC Bulletin 769D (Supelco, Inc., 1985). Included in this report are the names, molecular weights, densities, OSHA Standards (TWA and ceiling level), sample volumes (liters), sample rates (ml/min), sample times, desorption solutions (eluent), GC or HPLC columns, and collection devices of airborne contaminants.

When samples have been properly collected, they need to be taken to a certified laboratory where the environmental analyses can be performed. The laboratory should supply clients with all pertinent information about sample collection, such as the volume of the sample and any preservatives required for analyses, as well as appropriate containers. Portions of individual samples should be saved if further investigation is needed.

C. SAMPLING METHODOLOGIES

1. Collection for Volatile Organic Compounds (Chlorinated and Aromatic Hydrocarbons) in Aqueous Solutions

The important factor in the collection of a volatile organic compound is to prevent analyte loss during collection and transportation. Sampling methods call for adjusting the pH of the water sample to 2 or less with an acid, such as hydrochloric (Bryden and Smith, 1989b). If samples contain residual chlorine, sodium thiosulfate is added to avoid further formation of halogenated organic compounds. To reduce volatile organic loss, when filling the sample vial caution should be exercised to avoid turbulence which would aerate the sample. Samples should be collected in duplicate. For QA purposes, a pair of travel blanks prepared from organic free water should accompany each batch of sample vials to the field and back. Duplicates and spikes are also recommended when sampling for volatiles. For soil samples, 250 ml wide mouth glass jars are usually used with adhesive-free PTFE-lined caps. As with water, we must keep aeration of the sample to an absolute minimum.

To prevent contamination, samples should be isolated by placing the jars in individual plastic bags. They should then be transported immediately to the laboratory in a cooled container (4°C). Freezing samples by packing them in dry ice is not recommended. Jars may break and samples may absorb carbon dioxide.

2. Collecting Other Organic Compounds

To extract and concentrate enough of the analyte for detection, large samples of semivolatile organics must be collected for analysis, for example, a liter of water is required for analysis. However, aeration during collection of the sample is not critical since the organic compound is nonvolatile. Rinsing the container is necessary to recover any adsorbed analyte that may have adhered to the glass surface. Duplicate samples need to be collected.

3. Collecting Inorganic Compounds

For the analysis of most dissolved metals, water samples are collected in 500 or 1000 ml plastic bottles containing nitric acid (pH <2). Nitric acid and laboratory digestion prevent precipitation of metals (Bryden and Smith, 1989b); however, exposure to air can cause precipitation and oxidation of metals. In the analysis for hexavalent chromium, or any metal in which speciation is important, the sample should be neither acidified nor exposed to air.

Analysis of dissolved metals requires that the water sample be passed through a 0.45 μm filter. Some devices which are used in sampling to limit exposure of water to air are a filter in line with the pump discharge (when using a pump), inert gas pressure (to filter the sample), or vacuum or "zero headspace" devices. For QA purposes, when using filters we must remember to also collect a filter blank (Bryden and Smith, 1989b).

After collection, samples should be cooled to 4°C and transported to the environmental laboratory without delay. Caution should be taken when corrosive preservatives or those which generate toxic gases are used (especially if ions such as sulfide or cyanide are present).

4. Collecting Liquid Samples

Sampling for volatile and semivolatile organic compounds require special treatment to limit introduction of air into the water sample. There are other pollutants which are concerns for drinking water. These concerns include water quality standards, the presence of coliform bacteria, and lead from pipes. The quality of drinking water samples is influenced by the water's residence time in pipes or by system accessories. The person taking the sample must know what information is needed from the analysis so he can determine where to take the sample. For example, if the main concern is the consumption of the water from the tap, then the sample should be collected from the faucet. If the concern is with the softener, then the sample should be taken from the outlet on the softener.

At waste sites and underground tank facilities, monitoring wells are used to obtain water samples. To get reliable samples when the monitoring well water becomes stagnant, the well may be purged of about three volumes of water until conductance, pH, and temperature have stabilized (Bryden and Smith, 1989b). There are various well pumping and sampling devices available for retrieving underground water. Some available pumps are surface or bladder pumps for shallow, wide diameter wells or deep, narrow diameter wells, or submersible, air-lift or bladder pumps for deep, wide diameter wells. Surface water sampling usually means directly filling the container with water, being careful not to lose the preservative.

Sometimes, it may be necessary to collect a sample over an extended period (i.e., monitoring the effluent from an industrial waste discharge pipe). In such a case, an autosampler is desirable. Such an apparatus could collect a sample every 15 to 20 min and even pool samples on an hourly basis. Instruments which can sample the effluent and analyze it for contents of pollutants are being manufactured. These devices are being used in remote areas where the test is performed on site and the results are sent back to the main laboratory by computer/phone line interface. If a problem exists, the computer can shut down the operation until the necessary adjustments or repairs are made. Now, automatic samplers are not recommended for the sampling of volatile compounds because evaporation of the analyte and aeration of the sample can present a problem.

Waste water samples must be kept cool, since they are usually rich in bacteria that can rapidly alter its composition.

5. Collecting Solid Samples

The collection of solid samples in large areas can be approached in two ways: (1) the probablistic approach, which involves laying out a grid over the area and sampling at several randomly chosen coordinates or (2) the deterministic approach, which uses site histories and random sampling to select areas in which samples are to be collected (Bryden and Smith, 1989b). Although the latter method is not as thorough and has the possibility of biased results, sampling and analytical costs are almost always lower. One clean (background) sample must be collected for reference. Samples from a large area are usually batched to save on cost; however, this is inappropriate for the analysis of volatile compounds which could be lost in the transfer.

Some soil sampling devices are the trowel, shovel, or hand corer for shallow soil samples (to 1 ft.). A split spoon sampler with metal liners, adapted for hand use, is often employed for sampling at depths of less than 3 ft. Other devices useful in collecting samples of depths of less than 3 ft include the power auger and the corer. For depths below 3 ft, a drill rig is necessary. Many drilling systems use a split spoon or a split barrel sampler. These devices are driven by a weight through a hollow stem auger.

Volatile compounds are better retained in soil samples when the samples are tightly packed into metal sleeves.

Much of the above discussion on sampling soil can also be used when sampling marine environment sediments. Various dredge or "clam shell" samples can be used when sampling shallow sediments.

6. Collecting Hazardous Waste

Due to the dangers of collecting samples from sources such as drums, transformers, and tanks, OSHA considers it necessary that these samples be collected by trained personnel. OSHA requires at least 40 h of training for these workers. The training includes the recognition of hazards in the opening of drums, entering confined spaces, and working at heights.

7. Collecting Air Samples

The collection of air samples is considerably different from that of soil or water. Some type of trap for the airborne contaminants (for vapor or particles) is needed. The absorbent is exposed to the air in question for some period, sealed, and transported to the laboratory for analysis.

Supelco, Inc. has an entire line of absorbent traps for collecting airborne contaminants listed by OSHA. After a specified amount of air is drawn through the tube, the ends are sealed and the tube is refrigerated until analysis.

One popular way of collecting radon samples has been to place an open canister (EPA approved) filled with activated charcoal in the area to be sampled and leave it for 1 to 6 d. Correction factors for humidity and exposure time must be taken into consideration (EG and G Ortec, 1989).

8. Collecting Biological Samples

The methods of detection of environmental pollutants have been applied to biological materials to collect essential toxicokinetic data and to address the problem of determining if and to what extent exposure of persons to such chemicals has occurred.

The biological specimen might include human tissue, blood or urine, or animal tissue, that is, wildlife, domestic animals, produce, and meats. The isolation of most organic pollutants from biological tissue, particularly fat, is a formidable task since these chemicals themselves are highly fat soluble. Thus, extraction of these chemicals with solvents can be very difficult and troublesome as the chemical constituents of biological tissue are also soluble in the extracting solvent.

Elkins (1967) has proposed that biological threshold limits (BEL) used in biological monitoring be established at least for the chemicals more commonly used in industry. With biological monitoring comes the necessity of selection of types of specimens. Discussion of this problem was presented by Waritz (1979). As a specimen for biological monitoring, urine has distinct advantages over blood, such as ease of collection, absence of some analytical problems, and concentration of analytes. However, the detectable form of the chemical in urine may be a metabolite, so the metabolic pathway of each chemical to be measured must be available.

Specimens of blood must be collected by venipuncture, which can be somewhat of a procedural problem. Chemical analysis of blood presents some problems such as presence of protein and other endogenous substances, but the level of the chemical in the circulating blood is the best indicator of the extent of exposure to that chemical and of related toxicity.

Chemical breath analysis is done in some cases, if the chemical to be tested is volatile and enough is exhaled to be analyzed by available methods. There are few methods available for this type of procedure, but the one most commonly used is breath analysis for ethanol in drinking and driving situations.

The lipophilic characteristics of some chemicals result in their deposition in adipose tissue. The quantitative chemical analysis of this tissue could be an

TABLE 19.1. Chemical Metabolite Detection

Chemical	Metabolic Analyte	Reference
Benzene	Phenol	Hunter and Blair, 1972
Cyclohexane	Cyclohexanol	Perbelinni and Brugnone, 1980
p-Dichlorobenzene	p-Dichlorophenol	Fletcher and Barthel, 1970
Ethylbenzene	Mandelic acid	Bardodej and Bardodejova, 1970
Ethylene glycol		
monomethyl ether	Methyoxyacetic acid	Groeseneken et al., 1986
Formaldehyde	Formic acid	Gottschling et al., 1984

indication of past exposure, and also could provide information about the severity of the exposure.

The method for analysis of urine, blood, or breath depends on the chemical or chemicals to be tested for. The analysis of many environmental chemicals uses many of the same procedures already presented in this chapter. In some incidences, the detection of a metabolite or metabolites of the chemical in urine is evidence of exposure. A few examples are listed in Table 19.1.

In general, analyses of biological specimens are essentially the same as for other specimens of aqueous matrix. The biological specimens contain various biochemicals which can interfere with some analyses, particularly in the search for chemicals present in low concentrations. Consequently, methods must include additional "clean-up" procedures. As a result, methods for the chemical analysis of biological specimens usually are not as sensitive as methods for analysis of water.

D. SAMPLE CONTROL

Strict chain-of-custody documentation should be maintained for each sample collected. Every reputable laboratory has a chain-of-custody form which serves as a legal record of collection and later transfers. A field sample identification label enables the sampler and the data user to relate all pertinent data to the source of the sample. For example, "MW-2-071190-10" indicates a water sample from a monitoring well at "location two" sampled on July 11, 1990, at a depth of 10 ft. Sample labels are also critical to the identity of the sample and integrity of the analysis. Pertinent information on the label should include client name, preservatives used, type of analysis needed, and remarks.

All samples should be transported in cooled containers (4°C), with suitable padding to prevent breakage. To prevent moisture from contaminating samples, ice or an ice substitute used in packing should be in sealed plastic bags or jars. Hazardous materials must be labeled appropriately.

E. PURIFICATION TECHNIQUES

Once samples are collected, the analytes of interest must first be isolated from their matrix. The extent to which an analyte must be manipulated before it can be analyzed depends upon (1) how dirty the sample is, (2) how much analyte is present, and (3) what analysis is to be performed. For example, a sample of sludge containing trace quantities of a pesticide requires much more sample preparation than a sample of drinking water containing toxic levels of a pesticide. In the past, liquid/solvent or solid/solvent extractions have been popular for the extraction of organic molecules from water and soil samples. However, today, technology has given us more innovative means of analyte extraction. Greater analyte recoveries from cleaner preparations are being attained. For instance, the use of solid phase extraction tubes has become a popular means of retrieving organic compounds from liquids. After the addition of a buffering agent to the liquid, the sample is applied to a conditioned column. The column contains a packing material suitable for trapping the analyte of interest (polar, nonpolar, or ionic). After the impurities are washed from the trapped analyte, the analyte is eluted using a suitable solvent (Baker, 1982). Unless the analyte must be chemically altered (derivatized) for detection, it is ready for analysis. This column method has several advantages over more conventional methods in that sample preparation time and the use and exposure to solvents are decreased which lead to a lower cost of the analysis.

One new development in column technology that has improved the analysis of transition metals is the use of a chelation resin column for recovering transition metals from high salt matrices of estuaries and from soils containing high levels of alkali and alkaline earth metals. A chelation resin column is often used to treat the sample before analysis by ion chromatography. The sample is passed over the chelation column where the transition metals are selectively retained. The alkali and alkaline earth metals or common ions are passed through. The transition metals are then eluted with the proper solvent, and the eluent is concentrated by evaporation and analyzed (Joyce and Schein, 1989).

Another use of solid phase extraction is the impregnation of filter disks with solid phase material. These disks have recently been marketed by Analytichem International for the recovery of phthalates and pesticides from drinking water and ground water, respectively. Five hundred to one thousand milliliters of sample are filtered through an Empore Extraction Disk® impregnated with a C_8 (an eight carbon molecule) sorbent filters 500 to 1000 ml of sample into apparatus. The analytes are then eluted from the filter with the proper solvent. The solvent is concentrated and the analyte is then ready for analysis by HPLC or GC.

Extraction of analytes from solid materials can be accomplished by using hot solvent and a Soxhlet extractor. The solvent is vaporized and condensed on top of the solid material (in the extractor), and as it passes through the solid it extracts soluble compounds. When the condensed liquid fills the extractor, it automatically is siphoned back into the solvent flask. This process continues as more solvent is vaporized and condensed. When the extraction is finished, the solvent, with solute, is removed to a clean beaker or test tube and concentrated by evaporating the solvent. Care should be taken to ensure that the solute does not evaporate with the solvent (Shugar et al., 1973).

An ultrasonic method is another means of removing analytes from solid materials. The solid is added to the proper solvent, placed in an ultrasonicator, and sonicated for a specific length of time. The solvent containing the analyte is then removed and processed further.

In both of these methods, it is sometimes necessary to further purify the extracts by another method such as gel permeation chromatography (GPC). GPC removes high molecular weight materials, that is, humic acids (Marsden, 1989).

Two other related extraction techniques are distillation and the purge-and-trap system for GC and GC/MS. One example of a distillation procedure is the isolation of sulfites from water or solids. The sample is placed in a distillation apparatus with an inlet for nitrogen and one for acid. Solid samples are first emulsified in water then treated with acid, forming ionic sulfite which is released as sulfur dioxide (SO_2). The evolved SO_2 can then be swept by a stream of nitrogen into another solution for detection (Greyson and Zeller, 1987). In the purge-and-trap technique (used with GC and GC/MS), the sample is placed in a purge vessel where purge gas is bubbled through the water sample. A solid sample is immersed in water before being placed in the purge vessel. The purge gas extracts volatile organic compounds (VOCs) and carries them into an adsorbent trap. Adsorbent traps are composed of porous polymers containing functional groups designed to attract and trap a specific analyte. After purging is complete, the trap is quickly heated to a predesignated temperature. The contents of the trap are backwashed onto the GC column by the carrier gas. Separation of the VOCs occurs in this column.

F. SEPARATION

After the purification step, some samples are ready for direct analyte detection using atomic absorption (AA) for metals, gamma spectrometry for radon, Fourier-transform infrared spectrometry (FT-IR) for on-site VOCs, or several other analytical apparatuses. However, for most samples it is nearly impossible to extract only one analyte which will respond to the detector. Normally, several common analytes will be extracted with some impurities which also respond to the detector. Therefore, some other separation technique must be used before presentation of analytes to the detector. Chromatography has played a major role in separating mixtures of analytes and contaminants. Gas chromatography (GC) has been instrumental in transferring purer analytes to the detector.

According to McNair (McNair and Bonelli, 1969) the underlying principle of GC technique is the separation of volatile substances by percolating a gas stream (carrier gas) over a stationary phase. The sample mixture is partitioned between the carrier gas and the nonvolatile solvent (stationary phase) supported on an inert size-graded solid (solid support). As the sample passes over the stationary phase, the sample components are selectively retarded according to their distribution coefficients; thus, they form separate bands in the carrier gas. These components leave the column in the gas stream and enter the detector where their responses are recorded as a function of time or retention time (RT). The degree of separation of components in a mixture is dependent on several characteristics of the GC system and the samples themselves.

Selecting the proper stationary phase for the analyte(s) of interest is critical to separation. Nonpolar components require nonpolar stationary phases and polar components require polar stationary phases. The length and diameter of the column also have considerable effect on separation. Originally, GC analysis was performed using packed columns. These columns were usually metal or glass columns 6 ft long × 1/8 or 1/4 in. in diameter and were packed with a stationary phase which was coated onto a solid support. Today, fused silica capillary columns 30 m long × .25 to .75 mm in diameter are used, offering better efficiency and resolution of components. In these columns, the stationary phase is coated onto the inside of the column. Many approved EPA methods for GC use packed column technology. However, many of these methods are being updated to include the use of capillary columns (Supelco, Inc., 1988). Another advantage of capillary GC is that two different capillary columns can be connected to the same injection port with a two-hole ferrule. In this manner, when a sample is injected, one column can be used as a primary analytical column while the other column provides conformational analysis. For example, in the analysis of chlorinated pesticides EPA Method 608 uses the SPB-608 capillary column by Supelco, Inc. which separates 16 of Method 608 pesticides in less than 30 min. The assurance that these compounds have been correctly identified is high when used with the PTE-5 capillary column by Supelco, Inc. This column also separates these compounds within 30 min but shifts the elution order of nine of the pesticides and the relative retention times (RRT), compared to aldrin, of all 16 pesticides (Supelco. Inc., 1989).

Another technique used in GC, which aids in the separation of components, is temperature programming. Temperature programming is the controlled change of column temperature during an analysis. Temperature programming allows for the proper selection of temperature (above the boiling points of compounds) which results in well-resolved, nicely shaped peaks in a shorter analysis time than with an isothermal operation. If the range of boiling points of compounds of interest is 100°C or more programming is advisable (McNair and Bonelli, 1969).

Chemicals which can be analyzed by GC have the following characteristics: they must be volatile (gases) or made volatile and be thermally stable (no decomposition during vaporization or analysis on column). Some compounds which do not have these characteristics can be chemically altered by derivatization. This procedure usually involves the addition of a methyl, acetyl, or trimethylsilyl group to the problem functional group. In the GC analysis of phenoxyacid herbicides, the free acid portion of the molecule is derivatized with diazomethane before analysis producing the desired volatile compound (Marsden, 1989). For other compounds it might be necessary to use another separation technique where derivatization is not needed.

There are many nonvolatile, polar, thermally unstable organic compounds such as pesticides (Marsden, 1989) which do not lend themselves to analysis by GC. Anderson (1989) reports that about 80 to 85% of existing organic compounds lie outside the capabilities of GC. He states that "molecules which have polar or ionic substituents, or exceed molecular weight range of 600–800 atomic mass units (amu) usually have insufficient volatility to reach the detector in measurable concentrations." Therefore, the use of high performance liquid chromatography (HPLC) is becoming increasingly more popular to separate analytes before detection. Procedures for direct injection of the sample into the HPLC are being developed and would end the time consuming extraction procedures.

Analyses of biological specimens by HPLC have been applied to many chemicals such as phthlates (Draviam et al., 1980), polynuclear aromatic hydrocarbons (Obana et al., 1981), chlorophenols (Wright et al., 1981), and oxalic acid (Hughes et al., 1982).

The principle behind HPLC is a modification of the technique originally developed for GC (Lindsay, 1987). In liquid chromatography the eluent is a liquid called the mobile phase. It is passed over an adsorbent stationary phase such as silica that has been packed into a stainless steel column. The mixture to be separated is introduced at the top of the column and is washed over the stationary phase with a mobile phase. A weakly adsorbed analyte will travel down the column faster than one which is more strongly adsorbed by the solid. This method is called adsorption chromatography or liquid/solid chromatography (LSC).

Another sorption method useful for separating analytes is liquid/liquid chromatography (LLC), which uses a liquid phase coated onto a finely divided and uniformly inert solid support. Separation of analytes is due to differences of distribution coefficients between the mobile phase and the liquid (stationary) phase. In normal phase chromatography, the stationary phase is relatively polar and the mobile phase is relatively nonpolar, but the opposite is true in reverse phase LLC.

In recent years, ion chromatography has become a "powerful analytical technique for environmental laboratories" (Joyce and Schein, 1989) in the analysis of inorganic anions in water and wet depositions and in particulate matter in air such as Cl^-, NO_3^-, SO_4^{-2}, and cations from acid rain. Ion chromatography has also been used to monitor toxic by-products from improperly applied drinking water disinfectants (ClO_2, O_3, $O_3 + H_2O_2$) at levels of <100 ppb (Joyce and Schein, 1989). In this technique, the stationary phase is an ion-exchange material, usually a resin. Separations occur according to the strength of interactions between the solute ions and the exchange sites on the resin.

Another separation technique, based on the differences in the sizes and shapes of the solutes, is called exclusion chromatography (EC). In EC, the largest molecules travel most rapidly through the system. The introduction of chiral columns and mobile phases has solved the problem faced by environmental analysts in separating analytes which have optical isomers. This type of column is now used in pyrethroid pesticide analysis (Frost, 1989).

Sample preparation may be the limiting step in any analysis where the analyte must first be separated from a complex matrix such as sewage, industrial effluent, hazardous waste, or soil. Preparative HPLC and gel permeation chromatography are excellent tools for removal of co-extractives before injection in GC or HPLC analytical columns.

G. ANALYSIS OF SAMPLES

After it has been determined which environmental contaminant is suspected and the proper samples have been collected, a method for analysis is selected. The quality of the laboratory analysis cannot be stressed enough as results of the sample analysis mean nothing if the analysis has not been performed properly. Qualified personnel, acceptable controls, standards, calibrations, and instrument checks are of utmost importance for high quality analysis.

After the sample is collected and delivered to the laboratory, it is extracted from its matrix, introduced into the instrument where components can be further separated (usually chromatographically), and then introduced into a detector specifically selected for that analyte. Normally, the data acquired from the detector is enhanced by use of a multiplying device, for example, an electron or photo multiplier. Today, most pieces of equipment use a computer to collect the data from the instrument, allowing the operator to perform various manipulations of the data.

Tests performed on water, soil, and air samples vary extensively in their degree of difficulty. Analyses can range from simple wet chemistry techniques to the use of computerized equipment costing several hundreds of thousands of dollars such as highly sophisticated mass spectrometers (MS). Up to now, MS has been the most definitive means of analysis of organic substances. Their interface to gas chromatographs (GC) has made them vital pieces of equipment in the environmental laboratory. New and improved interfaces for coupling the MS to other separating devices such as HPLC, SCF, and some high performance thin layer chromatography will certainly enhance the versatility of the MS for future applications.

H. DETECTORS

The selection of the proper detector for the analyte is very important and depends on several factors such as the chemical characteristics of the analyte (selectivity) and the amount of analyte present (sensitivity).

1. GC Detectors

According to McNair and Bonelli (1969), the chromatographic detector is a device which indicates the presence and measures the amount of separated components of the injected sample in the carrier gas. Some GC detectors (ionization detectors) operate on the principle that the electrical conductivity of a gas is directly proportional to the concentration of charged particles within the gas. The ionization detectors have a constant current called background (noise) which is caused by column phase bleeding and impurities in the carrier gas.

In a flame ionization detector (FID) the effluent is mixed with hydrogen and burned in air or oxygen, producing ions and electrons that enter the electrode gap. Consequently, the gap resistance decreases, creating a flow of current in the external circuit (McNair and Bonelli, 1969). This current is measured in analog signals that are converted to digital signal for use in integrators or computers for qualitation and quantitation. FID has a much greater response to organic compounds than to fixed gases or water (McNair and Bonelli, 1969). To obtain high FID sensitivity, proper proportions of gas flow rates of air, hydrogen, and carrier gas must be maintained. Some uses for FID in environmental analysis are the detection of airborne carbon disulfide and trace amounts of solvents in water and air (Supelco, Inc., 1984, 1985).

Another version of the FID is the nitrogen/phosphorous detector (NPD), also called the thermionic sensitivity detector (TSD), thermionic emission detector (TED), alkali thermionic detector, or alkali flame detector. This detector operates in a "starved oxygen" state, meaning there is no flame (the detector bead glows). Air flow is about one third to one half that of the basic FID. One common design places a nonvolatile, electronically heated rubidium bead above the plasma jet. The thermal energy produced by the bead preferentially ionizes compounds containing nitrogen and phosphorous over those containing only hydrocarbons. A selectivity of 50 to 1 for nitrogen and 5000 to 1 for phosphorous over hydrocarbons is reported (McNair and Bonelli, 1969; Willard et al., 1981). The NPD selectivity for nitrogen and phosphorous has made it a desirable detector for the analysis of organophosphorus compounds, triazines, carbamothioates, and carbamates (Marsden, 1989). It has also found applications in the analysis of the nitrogen derivatives, formaldehyde, and acrolein in trapped air samples using adsorbent tubes (Supelco, Inc., 1988).

Another very sensitive and selective ionization type of detector is the electron capture detector (ECD), used for compounds containing electronegative substances such as halogens (McNair and Bonelli, 1969; Marsden, 1989). This detector is unique in that it measures the loss of electrical current rather than its production. The detector uses radioactive isotopes, the most common being ^{63}Ni. During the decay process, this isotope releases beta particles which collide with incoming carrier gas molecules, producing many secondary, low energy electrons. With the placement of electrodes in the detector cavity, these electrons can be captured. They produce a "standing current" (background current) measured by an electrometer. With the introduction of a sample containing electron absorbing molecules, the secondary electrons are captured by these molecules, producing negatively charged ions. The absorption of the secondary electrons by electron absorbing molecules reduces the amount of electrons which can be captured; therefore, the standing current is reduced and produces a negative peak. This peak is inverted during amplification by the electrometer to give a positive response on the integrator or the computer. The ECD, more sensitive than the FID to the compounds discussed above, has found application in the measurement of chlorinated pesticides, polychlorinated biphenyls, TCDD (dioxin isomers), halogenated VOCs, and phthalates (Supelco, Inc., 1986; Rhoades et al., 1988; Donnelly and Sovocool, 1989; Kirshen and Almasi, 1989; Marsden, 1989).

A popular GC detector for the measurement of VOCs used with ECD is the photoionization detector (PID). The PID operates on the principle of generating ions of molecules which have absorbed predetermined emission energy from an UV light source. Some common lamp energies are 9.5, 10.0, 10.2, 10.9, and 11.7 eV. A chamber next to the UV source contains a pair of electrodes. The positive potential of the accelerating electrode creates a field which forces the ions to the collecting electrode. The resulting current is measured. Selecting the proper lamp-ionizing energy is important since the response of compounds depends on their ionizing potentials being lower than that of the UV lamp. Molecules will respond with ionizing potentials up to 0.3 eV higher than the lamp, but with lower efficiencies. Organic compounds which respond to PIDs are aliphatics, aromatics, ketones, aldehydes, heterocyclics, amines, organic sulfur compounds, and some organometallics. Some inorganic compounds that respond are O_2, NH_3, H_2S, HI, ICl, Cl_2, I_2, and PH_3. For maximum selectivity, the energy output of the lamp must be just capable of photoionizing the desired compound. When a 10.2 eV lamp is used, commonly used solvents do not respond (Willard et al., 1981). The lamp of choice for monitoring VOCs has an energy of 10.2 eV (Hinshaw, 1989).

Another detector using light properties in its detector design is the flame photometric detector (FPD). It measures the emission of light produced in a hydrogen/oxygen/air flame from the consumption of column effluent. The FPD is selective for sulfur- or phosphorous-containing compounds, depending upon which gas flows and filter selections are made. The FPD is applicable in the measurement of organophosphorus compounds, some triazines, and carbamothioates (Marsden, 1989).

The most sophisticated ionizing detector for GC, as well as HPLC, is the mass spectrometer (MS). MS is so sensitive that only a few nanograms (and sometimes picograms) of analyte are needed to obtain characteristic information about its structure and molecular weight. Since newer, more easily operated, and cheaper versions have been marketed, the MS is vital in the analysis of environmental contaminants. Two types of MS commonly used today in environmental laboratories are the quadrupole MS and the ion trap MS. Both instruments use either electron impact (EI) or chemical ionization (CI) to fragment compounds to obtain a mass spectrum. In the electron impact mode (hard ionization), the compounds entering the source are bombarded by electrons at a constant electron energy (usually 70 eV). Bonds are cleaved depending upon their bond energies, which results in the formation of positive and negative ion fragments. Fragmentation of compounds is reproducible if conditions such as electron energy, source temperature, and vacuum pressure are kept constant, thus producing a mass spectrum unique to that compound. The mass spectrum is sometimes referred to as the chemical "fingerprint" of a compound. Chemical ionization is a softer means of ionization. A gas such as methane, ammonia, or butane is introduced into the source where it is ionized by electrons. Gaseous ions, which contain much less energy than electrons, bombard the compound to produce relatively large fragments. This type of ionization is more useful in molecular weight determinations than electron impact.

After the ions are formed, they can be analyzed several ways. In a quadrupole analyzer (Figure 19.1), the ions are accelerated into the quadrupole region which consists of four cylindrical rods arranged symmetrically. The ions are filtered through the quadrupole by varying the radio frequency (Rf) and direct current (dc) voltage applied to the four rods in such a way as to allow one atomic mass unit (amu) at a time to pass through the quadrupole. The quadrupole is capable of scanning the mass spectrum, usually 1 to 1000 amu, in a few milliseconds. For trace analysis, monitoring only two or three characteristic ions of the analyte may increase the sensitivity of the method by 100-fold. This technique is called selective ion monitoring (SIM). An electron multiplier (EM) is commonly used to detect ions as they are emitted from the quadrupole rods of the analyzer.

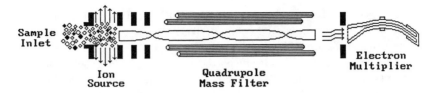

Figure 19.1. Quadrupole analyzer.

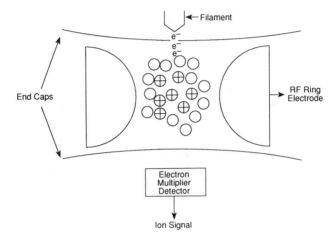

Figure 19.2. Ion trap analyzer.

The use of the ion trap (Figure 19.2), a Rf analyzer, has become a relatively inexpensive universal detector for a GC. Electrons are pulsed from the top of the trap to produce ions in the central trap volume (Figure 19.2). The trap achieves mass selection by making unstable field conditions that expel ions of a given mass-to-charge (m/z) value toward a detector. At constant Rf and zero dc potential, a mass spectrum is produced by clamping the Rf potential on the central ring, expelling ions of progressive m/z value through the lower plate toward the EM detector.

Since MSs operate under vacuum, the interfacing of the MS to other instruments such as the GC and HPLC has been a problem; both the GC and the HPLC operate at atmospheric pressure. The use of capillary columns in GC's have allowed the columns to be inserted directly into the MS source. The reduced flow rate of the columns, usually 1 to 2 ml/min, allows for the MS to maintain operable vacuum conditions. However, many laboratories still use packed GC columns in their MSs. To maintain a vacuum in the system, an interface is necessary in these instruments to reduce the amount of carrier gas entering the MS. Unfortunately, some samples can be lost along with the carrier gas with this type of instrument. Most EPA methods for GC/MS were developed for use with packed columns. Today, many of these methods are being updated for use with capillary columns.

GC/MS has been used to identify polycyclic aromatic hydrocarbons in human tissues (Jongeneelen et al., 1986). Also, GC/MS has been used to screen blood, urine, and tissue samples for volatile organic compounds (Anderson, 1986).

The analysis of biological materials using GC/MS has been of particular importance in the study of the toxicology of dioxins such as as 2,3,7,8-tetrachlorodibenzo-*p*-dioxin (TCDD). TCDD and its isomers must be analyzed in concentrations of parts per trillion. This concentration is attainable by the use of capillary columns, selective ion monitoring, and electron impact modes of operation (Patterson et al., 1987).

The development of a suitable interface between the HPLC and the MS has hindered the use of HPLC/MS in environmental testing. Today, there are at least two versions of interfaces on the market and future environmental applications for HPLC/MS should be forthcoming. As more selective and sensitive detectors are developed for use with HPLC, it will surely become one of the most powerful tools for the environmental analyst. One advantage of HPLC over GC is that some pollutants, such as chlorophenols, can be analyzed without derivatization (Wright et al., 1981).

2. HPLC Detectors

One principal HPLC detector is the UV/visible (vis) detector. It measures changes in light absorbance in column effluent as the effluent passes through a flow cell. This detector is selective toward compounds that absorb UV/vis radiation, especially compounds that have multiple bonds between carbon and oxygen, nitrogen, or sulfur. There are several types of UV/vis detectors. A fixed wavelength detector uses a filter of a specific wavelength and a line lamp; a variable wavelength detector uses a grating monochromator for selection of wavelength and a continuum lamp. Deuterium and xenon lamps offer better sensitivity in the UV range than does the tungsten lamp, although the tungsten lamp can also be used in the visible range. A photomultiplier or photocell is used as the detector (Lindsay, 1987).

Fluorescence detectors measure fluorescence changes in column effluent when exposed to selected wavelengths of light. They are selective toward compounds that instantly absorb UV radiation and then emit radiation at a longer fluorescent wavelength (Lindsay, 1987). In order for a compound to be suitable for fluorescence detection, the ratio of absorbed energy to re-emitted energy should be 0.1 to 1.0. The photomultiplier tube is used as the photodetector. Filter fluorometers use filters to select excitation and emission wavelengths. Spectrofluorometers use diffraction gratings for the selection of appropriate wavelengths.

The fluorescent detector can be used for the detection of nonfluorescing compounds if fluorescent derivatives of those compounds are prepared. Normally, the mobile phase used with the fluorescent detector does not have appreciable fluorescent qualities, unlike UV mobile phases. This reduces background noise, thus increasing the sensitivity of fluorescent detectors over UV detectors (Lindsay, 1987).

The fluorescence detector is useful in trace analysis since excitation and emission wavelengths can be varied, making the detector very selective toward certain compounds (Lindsay, 1987), for example, polynuclear aromatic hydrocarbons (PAHs) which fluoresce naturally (Frost, 1989; Joyce and Schein, 1989). PAHs are barely detectable using an UV/vis absorption detector, but are easily monitored using fluorescence.

In recent years the electrochemical detector (ECD) has become very popular because it is one of the most sensitive and selective of the HPLC detectors. The function of the ECD is the measurement of the current in the flow cell at the column outlet.

The ECD consists of working (WE), auxiliary (AE), and reference (RE) electrodes. The electrochemical reaction (oxidation or reduction) takes place at the WE. The potential across the WE and AE is kept constant using a potentiostat. The RE provides a stable and reproducible voltage to which the potential of the working electrode can be referenced. The detector output tracks the adjustments made by the potentiometer. Electroactive compounds will either be oxidized or reduced at certain potentials; therefore, the selectivity of the ECD can be varied by changing the potential of the WE. Reduction of an analyte is difficult to accomplish since it requires an oxygen free environment. Most applications use oxidations of the analyte (Lindsay, 1987).

There are two types of electrochemical cells: (1) the thin-film cell (amperometric detector) in which the effluent flows over the WE, allowing only 1 to 5% of the electroactive compounds to react, and (2) the flow through cell (coulometric detector) in which the effluent flows through the cell allowing 100% of the electroactive chemicals to react.

An ECD measures either the conductance of the eluent or the current associated with the oxidation or reduction of solutes. In the measurement of conductance, inorganic or organic ions are detected after ion exchange chromatography. Unfortunately, ion exchange chromatography requires an ionic mobile phase. The response of ionic analytes to the ECD depends upon the conductivity of the analytes being greater than the conductivity of the eluent.

A technique called chemical suppression eliminates the solvent ions in the mobile phase, thus reducing background noise. Chemical suppression is an ion exchange mediated acid/base neutralization reaction of eluent ions. For example, in the separation of anions using sodium hydroxide (NaOH), the eluent from the HPLC column passes through a cation exchange membrane where

Na^+ is removed and replaced with H^+, which combines with OH^- to form water. The resulting water plus analyte (anion) then flows into the conductivity detector and the analyte is measured (Joyce and Schein, 1989).

For HPLC, derivatization of a sample is normally used to improve detection of the sample, while in GC it is used to enhance chromatography. Two popular techniques used for derivatization are precolumn off-line or postcolumn on-line (Lindsay, 1987). An example of post-column derivatization is the hydrolyzation of insecticides to primary amines. These primary amines are allowed to react with orthophthaldehyde (OPA) to form strongly fluorescing derivatives which can be measured by a fluorescence detector (Frost, 1989). Another example is in the monitoring of Cr(VI) using post column addition of diphenylcarbizide to produce a color complex which can be measured with a photometric detector (Joyce and Schein, 1989).

For certain analyses it may be necessary to have two detectors in series to measure all of the analytes of interest. For example, two such analytes, iodate and fluoride coelute on a column; however, iodate can be measured with a UV detector and fluoride with a conductivity detector when they are placed in series (Joyce and Schein, 1989).

I. OTHER ANALYTICAL INSTRUMENTS

Very little sample preparation is required for certain environmental methods which often introduce the sample directly into the instrument for analysis.

1. Atomic Absorption (AA) Spectrometry

AA is the measurement of free atoms of inorganic elements which can be produced by subjecting the analytes to flame. The free atoms are then detected and their quantity determined by the amount of absorbance measured.

Two techniques for atomization of a sample are the direct aspiration technique and the furnace technique. With the direct aspiration technique, the sample is aspirated and atomized directly in the flame. With the furnace technique, the sample is placed in a graphite tube in the furnace, evaporated to dryness, charred, and atomized. Caution should be taken when using the latter technique since possible chemical reactions may occur at elevated temperatures resulting in the suppression or enhancement of the element. Specialized atomization procedures can be used if greater sensitivity is needed. These methods include the gaseous hydride method for arsenic and selenium, the cold vapor technique for mercury, and the chelation extraction procedure for other selected metals.

Sample preparation for analysis by AA is minimal. Suspended material in waste water or soil samples must first be solubilized by an acid digest to break down organic materials.

The operation of AA proceeds as follows: a light beam from a hollow cathode lamp (the cathode contains the element to be analyzed) is directed through the flame into a monochromator, then onto a detector that measures the amount of light absorbed. Absorption depends on the presence of free unexcited ground state atoms in the flame. Since the wavelength of the light beam is characteristic of the metal being determined, the light energy absorbed by the flame is a measure of the concentration of the metal in the sample.

2. Atomic Emission (AE) Spectroscopy

AE spectroscopy, also called flame photometry, is similar to AA. The difference between the two instruments is that AE does not use a hollow cathode lamp. Using AE, the analyte of interest is atomized, which raises it to an excited electron state. Upon returning to ground electronic state, the analyte emits characteristic radiation which is measured as it passes through a monochromator or suitable filters (Willard et al., 1981). The AE has found use in many clinical and environmental laboratories to analyze specimens for some elements, that is, lithium and potassium. Also, sodium and strontium are commonly analyzed by AE using a nitrous oxide-acetylene flame (Matusiewicz, 1982).

Zinc, cadmium, lead, copper, and arsenic can be determined in blood using electrochemical detection. This method is at least as sensitive as atomic absorption and is much superior in its capability of providing reproducible results.

3. Gamma Spectroscopy

The principal of this detector is the use of sodium iodide (NaI) to produce gamma rays which excite the atoms of the analyte in the detector. Upon deexcitation, the analyte emits photons of visible light. These photons strike the photocathode of a photomultiplier tube and emit electrons (e^-). The electrons are multiplied by a photomultiplier, producing an electrical pulse which has an amplitude proportional to gamma ray energy. One use for this instrument is the analysis of radon (EG and G Ortec, 1989).

4. Energy-Dispersive X-Ray Fluorescence (EDXRF)

The basis of this instrument is excitation of the analyte by a source of X-radiation. There is an ionization of the atoms in the molecule causing emission of fluorescent X-rays when the ions return to the ground state (secondary X-rays). The energies of these secondary X-rays are characteristic of the emitting element.

Some applications for this instrument are the determination of metal pollutants in waste water, metals in hydrocarbons, and 22 metals on the EPA hazardous substances list (Harding, 1987). This instrument is available as a mobile unit for on-site soil screening (Watson et al., 1989).

5.   Monier-Williams Coulometer

One method for determinating sulfites in food, beverages, water, and soil is by electrochemical analysis using an analytical coulometer. In a modified Monier-Williams procedure, the sample is acidified to release the free sulfite as sulfur dioxide (SO_2). The SO_2 is driven into the M-W apparatus by $N_2(g)$ and measured. Bound sulfite is analyzed by making the acidic sample basic. At a pH of 9, the residual reversible sulfite decomposes to the free form, and then acidifies to release $SO_2(g)$. As the evolved SO_2 enters the first chamber of the coulometer cell containing free iodine, water, and an iodine salt, the SO_2 reacts with the I_2 and H_2O to give sulfuric acid (H_2SO_4), $2I^-$, and $2H^+$. A chain reaction of chemistry is initiated. The generating electrode (first chamber) regenerates I_2 from $2I^-$ and $2e^-$. Hydrogen gas (H_2) and hydroxide ion ($2OH^-$) are generated by the counter generating electrode (second chamber) from H_2O and $2e^-$. To maintain electrical neutrality between the two chambers, OH^- ions diffuse to the first chamber where they are allowed to react with the H^+ ion, yielding water. The instrument contains the electronics to monitor the electrochemical reactions and present the results to the user in readable sulfur units. This type of apparatus is sensitive to quantities of sulfur as low as 10 ng and as high as 100 mg. This modification of the Monier-Williams method is over 100-fold more sensitive than the original acid/base titration for the determination of sulfites (Greyson and Zeller, 1987).

6.   Selective Ion Electrode (SIE)

Fluoride has been determined by electrode in various samples, including water, urine, plasma, and blood. McAnnaley et al. (1979) reported the determination of inorganic sulfide and cyanide in blood using specific ion electrodes in combination with microdiffusion. Selective ion electrodes have principal advantages in that they afford ease of sample preparation for analysis, are inexpensive, and have excellent detection limits. One disadvantage, however, is that selective ion electrodes can be subject to interference. Selective ion electrodes are available for several other elements of environmental interest, that is, calcium and oxygen.

7.   Spectrophotometry

Spectrophotometry is an analytical technique which involves the selective absorption of light by molecules. The energy used in spectrophotometry ranges from the visible wavelength region to the ultraviolet region.

A classic example of the use of this technique is the dithizone method in which a chelating reagent (dithizone) is chelated with metals such as lead (Lerner, 1975), zinc (Dutkiewicz et al., 1980), mercury (Stewart et al., 1977), cadmium

(Iwao et al., 1980), and copper (Beal and Bryan, 1978) in a biological specimen. A colored complex is formed which can be analyzed by spectrophotometry.

A similar technique using silver diethyldithiocarbamate to chelate arsenic has been used in the determination of arsenic in biological specimens such as blood, hair (Curry, 1976), and urine (Corridan, 1974). Detection limits for arsenic, by this method, range from 0.5 to 1 mcg (Pinto et al., 1976).

J. CONCLUSION

The isolation and detection of environmental chemicals are essential to the control of pollution and consequently the future of life on this planet. The environmental chemist must be an expert in both chemistry and the operation of instrumentation and computers.

The tasks of an environmental chemist are complex. These include a multitudinous array of matrices in which any number of the thousands of chemicals are contained. Methods of isolation, detection, and quantitation must be developed to assure analyses of great precision and accuracy. In addition, new chemicals are being added to our society and ultimately to our environment each year, requiring the development of methods of analysis for each chemical.

Considerable emphasis has been placed on the various aspects of the sampling process for environmental analyses. No method of analysis is adequate without strict adherence to the rules of proper sampling protocol. An attempt has been made to include the latest developments in this rapidly changing area of analysis. Therefore, references to supplier literature has been included.

The authors thank Johnette Gothard, James Mozingo, and William Poole for their help and contributions in reviewing and revising this chapter.

REFERENCES

ACGIH. *Threshold Limit Values and Biological Exposure Indices for 1990–1991*. The American Conference of Governmental Industrial Hygienists, Cincinnati, OH, 1990, 51–66.

Anderson, B.A. The determination of carbamate and urea pesticides by thermospray liquid chromatography-mass spectroscopy, *Am. Environ. Lab.* 1(2):41–45, 1989.

Anderson, R.A. Fire gases, in *Analytical Methods in Human Toxicology*, Part 2. A.S. Curry, Ed. Verlag Chemie, Weinheim, Deerfield Beach, FL, 1986, 289–317.

Baker, J. T. *Baker-10 SPE Applications Guide*, Vol. 1. Separations Sciences. J.T. Baker Chemical Company, Phillipsburg, NJ, 1982, 2–10.

Bardodej, Z. and Bardodejova, E. Biotransformation of ethylbenzene, styrene, and alphamethylstyrene, *Am. Ind. Hyg. Assoc. J.* 31:206–209, 1970.

Beal, D.D. and Bryan, G.T. Quantitative spectrophotometric determination of human and rat urinary dimethylamine, *Bioch. Med.* 19:374–382, 1978.

Bryden, G.W. and Smith, L.R. Sampling for environmental analysis. Part 1: planning and preparation, *Am. Lab.* 21(7):30–39, 1989a.

Bryden, G.W. and Smith, L.R. Sampling for environmental analysis. Part 2: sampling methodology, *Am. Lab.* 21(9):19–24, 1989b.

Corridan, J.P. Head hair samples as indicators of environmental pollution, *Environ. Res.* 8:12–16, 1974.

Cram, S.P. Challenges and opportunities of environmental-analytical measurements, *Am. Environ. Lab* 1(1):19–26, 1989.

Curry, A. *Poison Detection in Human Organs*, 3rd ed., Charles C. Thomas, Springfield, IL, 1976, 124–127.

Donnelly, J.R. and Sovocool, G.W. Bromochloro-dioxins and dibenzofurans, *Environ. Lab* 1(5):26–28, 1989.

Draviam, E.J., Kerkay, J., and Pearson, K.H. Separation and quantitation of urinary phthalates by HPLC, *Anal. Let.* 13(b13):1137–1155, 1980.

Dutkiewicz, B., Konczalik, J., and Karwacki, W. Skin absorption and porous administration of methanol in men, *Int. Arch. Environ. Health* 47:81–88, 1980.

Elkins, H.B. Excretory and biological threshold limits, *Am. Ind. Hyg. Assoc. J.* 28:305–314, 1967.

Finnigan, R.E. and Poppiti, J. A look to the future analytical instrumentation, *Environ. Lab.* 1(4):28–32, 1989.

Finnigan, R., Chu, T.Z., Smith, M., and Stevenson, J. GC/MS for analysis of volatiles and extractables, *Environ. Lab.* 1(4):15, 1989.

Fletcher, C.E.,and Barthel, W.F. The electrocapture gas chromatography of paradichlorobenzene metabolites as a measure of exposure, *Bull. Env. Cont. Tox.* 5:354–361.

Friedman, D. Testing methodology in environmental monitoring, *Environ. Sci. Tech.* 24(6):796–798, 1990.

Frost, B.A. Pesticide analysis — selecting techniques for analyzing pesticides in fruits and vegetables, *Environ. Lab.* 1(4):18–23, 1989.

Gautier, M.A. and Gladney, E.S. A quality assurance program for health and environmental chemistry, *Am. Lab.* 19(7):17–22, 1987.

Gottschling, L.M., Beaulieu, H.J., and Melvin, W.W. Monitoring of formic acid in urine of humans exposed to low levels of formaldehyde, *Am. Ind. Hyg. Assoc. J.* 45(1):19–23, 1984.

Greyson, J. and Zeller, S. Analytical coulometry in Monier-Williams sulfite-in-food determinations, *Am. Lab.* 19(7):44–51, 1987.

Groeseneken, D., Veulemans, H., and Masschelein, R. Respiratory uptake and elimination of ethylene glycol monoethyl ether after experimental human exposure, *Br. J. Ind. Med.* 43(8):544–549, 1986.

Grupp, D.J., Everitt, D.A., Bath, R.J., and Spear, R. Use of a transportable XRF spectrometer for on-site analysis of Hg in soils, *Am. Environ. Lab.* 1(2):32–40, 1989.

Harding, A.R. On-stream elemental analysis by EDXRF, *Am. Lab.* 19(12):24–31, 1987.

Hinshaw, J.V. GC troubleshooting — purge-and-trap sampling systems, *LC-GC* 7(11):904–907, 1989.

Hughes, H., Hagen, L., and Sutton, R.A.L. Determination of urinary oxalate by high-performance liquid chromatography, *Anal. Biochem.* 119:1–3, 1982.

Hunter, C.G. and Blair, D. Benzene: pharmacokinetic studies in man, *Ann. Occup. Hyg.* 15:193–199, 1972.

Iwao, S., Tsuchiya, K., and Sakurai, H. Serum and urinary beta-2-microglobulin among cadmium exposed workers, *J. Occ. Med.* 22(6):399–402, 1980.

Jongeneelen, F.J., Bos, R.P., Anzion, R.B.M., Theuws, J.L.G., and Henderson, P.T. Biological monitoring of polycyclic aromatic hydrocarbons, *Scand. J. Work Environ. Health* 12:137–143, 1986.

Joyce, R.J. and Schein, A. Ion chromatography: a powerful analytical technique for environmental laboratories, *Am. Environ. Lab.* 1(2):46–54, 1989.

Keith, L.H., Ed. *Principles of Environmental Sampling.* American Chemical Society, Washington, DC, 1988, 1–458.

Kirshen, N. and Almasi, E. Automated determination of volatile organic contaminants in ambient air and/or soil gas by GC with selective detectors, *Am. Environ. Lab.* 1(1):48–58, 1989.

Lerner, S. Blood lead analysis — precision and stability, *J. Occ. Med.* 17(3):153, 1975.

Lindsay, S. *High Performance Liquid Chromatography, Analytical Chemistry by Open Learning.* John Wiley & Sons, New York, 1987, 1, 53–70, 77–81.

Marsden, P. Pesticide analysis increasing demands — environmental labs will see a rising need for the analysis of non-method 608 pesticides, *Environ. Lab.* 1(4):24–27, 1989.

Matusiewicz, H.K. Determination of natural levels of lithium and strontium in human blood serum by discrete injection and atomic emission spectrometry with nitrous oxide-acetylene flame, *Anal. Chim. Acta.* 136:215–223, 1982.

McAnnaley, B.H., Lowry, W.T., Oliver, R.D., and Garriott, J.C. Determination of inorganic sulfide and cyanide in blood using specific ion electrodes: application to the investigation of hydrogen sulfide and cyanide poisoning, *J. Anal. Toxicol.* 3:111–114, 1979.

McNair, H.M. and Bonelli, E.J. *Basic Gas Chromatography*, 5th ed. Varian Instrument Offices, Palo Alto, CA, 1969, 1, 16–17, 101–108, 189–204.

Obana, H., Hori, S., Kashimoto, T., and Kunita, N. Polycyclic aromatic hydrocarbons in human fat and liver, *Bull. Environ. Contam. Toxicol.* 27:23–27, 1981.

Ortec, E.G. and Ortec, G. Radon measurement by gamma spectroscopy, *Environ. Lab.* 1(5):22–25, 1989.

Patterson, D.G., Jr., Hampton, L., Lapeza, C.R., Jr., Belser, W.T., Green, V., Alexander, L., and Needham, L.L. High resolution gas chromatographic/high-resolution mass spectrometric analysis of human serum on a whole-weight and lipid basis for 2,3,7,8-tetrachlorodibenzo-*p*-dioxin. *Anal. Chem.* 59, 2000–2005, 1987.

Perbettini, L. and Brugone, F. Identification of the metabolites of n-hexane, cyclohexane, and their isomers in men's urine, *Toxicol. Appl. Pharmacol.* 52(2):220–229, 1980.

Pinto, S.S., Varner, M.O., Nelson, K.W., Labbe, A.L., and White, L.D. Arsenic trioxide absorption and excretion in industry, *J. Occ. Med.* 18(10):677–680, 1976.

Rhoades, J.W., Thomas, R.E., Johnson, D.E., and Tillery, J.B. *Determination of Phthalates in Industrial and Municipal Wastewaters.* EPA Contract no. 68-03-2606. National Technical Information Service, 5285 Port Royal Rd, Springfield, VA, 1988, 1–68.

Ryan, J.F. Environmental overview; a review of the U. S. Environmental Protection Agency, *Am. Environ. Lab.* 1(1):28–39, 1989.

Shugar, G.J., Shugar, R.A., and Bauman, L. *Chemical Technicians' Ready Handbook.* McGraw-Hill, Inc., New York, 1973, 113.

Skoog, D.A. and West, D.M. *Fundamentals of Analytical Chemistry*, 3rd ed., Holt, Rheinhart and Winston, New York, 1975, 45.

Spartz, M.L., Witkowski, M.R., Fateley, J.H., Jarvis, J.M., White, J.S., Paukstelis, J.V., Hammaker, R.M., Fateley, W.G., Carter, R.E., Thomas, M., Lane, D.D., Morotz, G.A., Fairless, B.J., Holloway, T., Hudson, J.L., and Gurka, D.F. Evaluation of a mobile FT-IR system for rapid volatile organic compound determination, Part 1: preliminary qualitative and quantitative calibration results, *Am. Environ. Lab.* 1(2):15–30, 1989.

Stewart, W.K., Guirgis, H.A., Sanderson, J., and Taylor, W. Urinary mercury excretion and proteinuria in pathology lab staff, *Br. J. Ind. Med.* 34:26–31, 1977.

Supelco, Inc. Save time by analyzing and confirming U.S. EPA Method 608 pesticides simultaneously, *The Supelco Reporter* 8(5):1–3, 1989.

Supelco, Inc. *Adsorbent Tube Traps Formaldehyde and Acrolein in Large Air Samples.* Supelco GC Bulletin 794C. Bellefonte, PA, 1988, 1–2.

Supelco, Inc. *EPA Procedures for Water Pollution Analyses.* Supelco GC Bulletin 775C. Bellefonte, PA, 1988, 1–10.

Supelco, Inc. *Capillary GC Columns Specifically Tested for Chlorinated Pesticides in US EPA Methodology.* Supelco GC Bulletin 841A. Bellefonte, PA, 1986, 1–4.

Supelco, Inc. *Monitoring Airborne Contaminants in Industrial Atmospheres.* Supelco GC/HPLC Bulletin 769D. Bellefonte, PA, 1985, 1–17.

Supelco, Inc. *Analyzing Trace Amounts of Solvents in Water.* Supelco GC Bulletin 816A. Bellefonte, PA, 1984, 1–5.

U.S. Environmental Protection Agency. Methods of analysis for environmental monitoring. Office of Research and Development, Environmental Monitoring and Support Laboratory, U.S. EPA, Cincinnati, OH. 1989. (Copies may be obtained from National Technical Information Service (NTIS), 5285 Port Royal Road, Springfield, VA 22161).

Waritz, R.S. Biological indicators of chemical dosage and burden, in *Patty's Industrial Hygiene and Toxicology*, Vol. 3, John Wiley & Sons, New York, 1979, 257–318.

Watson, W., Walsh, J. P., and Glynn, B. On-site X-Ray fluorescence spectrometry mapping of metal contaminants in soils at superfund sites, *Am. Lab.* 21(7):60–68, 1989.

Willard, H.H., Merritt, L.L., Dean, J.A., and Settle, F.A. *Instrumental Methods of Analysis*, 6th ed. Wadsworth Publishing Company, Belmont, CA, 1981, 476.

Wright, J. Southwest lab: built on people as well as equipment, *Environ. Lab* 1(5):30–31, 40–41, 1989.

Wright, L.H., Edgerton, T.R., Arbes, S.J., Jr., and Lores, E.M. The determination of underivatized chlorophenols in human urine by combined high performance liquid chromatography mass spectrometry and selected ion monitoring, *Biomed. Mass. Spec.* 8(10):475–479, 1981.

Detection — Bioassay

David J. Schaeffer

A. INTRODUCTION

The classical definition of an ecosystem couples interacting living organisms and nonliving components of the environment to form one physical system. Detection and quantification of the effects of chemicals on an ecosystem is a complex undertaking. Most environmental studies have focused on chemical residue analysis and environmental fate of chemicals, but emerging sciences are focusing on the "health" of ecosystems. Ecoepidemiology is the "study of ecological effects which are prevalent in certain areas (localities) or among certain *population groups, communities, and ecosystems*. Such studies involve not only the *ecotoxicological* description of ecosystem health change but also analyses of their possible causes" (Bro-Rasmussen and Lokke 1984). In the framework of this chapter, the science of "ecosystem health" parallels the sciences of human and veterinary medicine (Gasto, 1980; Rapport et al., 1985; Schaeffer et al., 1988; Slobodkin, 1988). This science analogizes ecological system physicians ("ecophysicians") to health practitioners as shown in Table 20.1 (Schaeffer et al., 1988). Thus, the "health" of an ecosystem after exposure to toxic chemicals must be determined from studies which concentrate on the description of effects, identification of causes, and determination of pathways between these. This chapter addresses the identification and use of diagnostic tools, termed "test systems", which provide both toxicological and ecological information. We begin with some definitions.

- Ecological system — a smaller unit within an ecosystem which is limited by definition, sampling procedures, or a range of other factors, yet realistically

TABLE 20.1. The Tools/Concerns of Human/Veterinary and Ecological System Physicians

Human/Veterinary Physician	Ecological System Physician
Compendium of diseases is available	Diseases of ecological systems largely undefined
	No terminology to describe "ecosystem health status" and "diseases"
Wide body of reference data available for "standard man"	Body of data for ecological systems has little information on normal patterns of change and variability in ecosystems
	Virtually no data from long term, whole ecosystem monitoring
	A "standard ecological system" cannot be defined at this time
Many types of diagnostic tools available	Virtually no proven diagnostic tools
The ranges of use of diagnostic tools and the interpretation of data are well defined	Interpretation of data in many cases is not practicable or possible
Concerns of toxicologists	Concerns of ecosystem toxicologists
(1) Effects of poisons on organisms considered as individuals	(1) Effects of poisons on structure and function of ecosystems
(2) Time to signs, survival time, proportion of surviving individuals, and relations among these and dose of toxin are used for effect assessment	(2) Identification of alternative futures of stressed ecosystems and factors affecting probability that a particular future will be realized
(3) Design of therapy for affected individuals	(3) Design of therapy for affected ecosystems

Source: Adapted from Schaeffer et al., 1988.

deals with the interaction between living and nonliving components of the environment. It permits better classification and a more direct application of scalar issues in both testing and any associated environmental assessment.

- Assay — a discrete measurement or analysis unit employed to examine or assess. *Test systems* range from simple biochemical assays to manipulations of ecological systems. A *test system trait* is a specific measurement with an identified metric made on a test system.

- Control — a nondosed test subjected to the same controlled environmental conditions as are the dosed tests. A control is usually suited to the laboratory but not to the field. Factors such as ecosystem complexity, homeostatic processes, natural variability, and temporal and spatial scales make it impossible to provide strict experimental control in ecosystems. Consequently, experimentation with, or manipulation of, ecological systems uses *ecological references* rather than controls. An ecological reference is a nondosed but spatially defined area subjected to similar, but uncontrolled,

environmental conditions as a manipulated or experimentally dosed area or
ecological system.

- Battery — assays having certain properties such as similar cost or simplic-
ity which measure similar traits or provide complementary information
about the health/diagnosis of an ecosystem may be grouped into batteries.
For example, the acute toxicity of a bactericide can be tested against a
battery of several gram positive bacterial assays and also against a second
battery of gram negative bacterial assays.
- Tier — a group of batteries. In the above example, the two batteries of
bacterial assays comprise a "gram bacteria" tier. Compounds showing
promise in this battery might be tested in a tier of mammalian cell culture
assays. The *screening tier* employs simple and cost effective batteries to
make preliminary determinations of hazard/risk. The *predictive tier* employs more complex, longer term testing to predict environmental/ecologi-
cal consequence. The *confirmatory tier* uses environmental/ecological
analysis to confirm suspect effects. The *surveillance tier* uses long term
monitoring to identify unanticipated long term or low level effects.

When ecosystem analysis or ecological system toxicology is the tier
testing objective, three levels of analysis are possible for each tier (Herricks
and Schaeffer 1987):

- Definitive level — the highest level requires quantifying a minimum set
of properties which exactly define the state or condition of an ecological
system using one or more indicators (although it is possible to quantify
properties and processes in ecological systems, the exact definition of state
or condition has eluded ecologists and proper indicators are still being
defined)
- Classification level — the next level of analysis requires quantifying
properties which could be used to define the state or condition of an
ecological system (measurements of properties or processes made through
time which provide a basis for comparison, or test systems used with
proper references, can operate at the classification level)
- Predictive level — the lowest level and most common analysis requires
quantifying those properties which are predictive of either the classification level or the definitive level (predictive level testing usually uses
laboratory and field-oriented test systems which have uncertain relationships to ecosystem indicators or are not used to develop relationships
between trophic levels)

Classical toxicology is usually defined in an exposure protocol which does
not consider the magnitude and duration of the response. Thus, acute toxicity
is the adverse effects occurring within a short time of administration of a single
dose of a substance or multiple doses given within 24 h. Subacute and subchronic

toxicity are the adverse effects resulting from daily doses over 14 or 90 d, respectively. Because of the spatial and temporal components of ecosystems (see below), medical definitions, in relation to the response, are more appropriate. To the physician, "acute" means "having severe signs and a short course of 12 to 24 h." "Chronic" means "persisting for a long time; the period is undefined and varies with the circumstances." It also has the sense of the disease showing little change or very slow progression over a long period. Differences and similarities between classical and ecotoxicology have been discussed (Schaeffer, 1991).

B. ECOSYSTEM HEALTH

Although human "health" is difficult to define, it is possible to set ranges of parameters which are common for "healthy" people and to develop lists of parameters which define a "healthy" condition or allow identification of "ill" characteristics. Because the lists that characterize whether a human/nonhuman animal as healthy or ill are not known completely, the diagnosis is formalized with progressively more complex testing following initial screening. A similar situation exists in assessing ecosystem health. The ecophysician normally follows a tier testing approach in defining the likely consequence of a potential environmental insult (Herricks and Schaeffer, 1987). While the human diagnostician deals with a limited set of parameters, the ecophysician is confronted by extensive lists which vary with level in the ecological hierarchy and parameters which might be measured. In practice, the number of commonly used ecosystem parameters is small, and comparative measurements are poorly developed. Thus, a "healthy" ecosystem may be defined only by reference to a few parameters (e.g., diversity, productivity), and absence of disease is based on comparison to one or more poorly quantified "ideal" ecosystems.

The way the patient is defined also constrains a health assessment. The physician's patient is an individual, and abundant population statistics allow identification of diseases and assessment of health. Although ecophysicians may be interested in individual species, their work is seldom focused on an individual organism. Ecophysicians replace individual organism analysis with assessments of population effects, but even that falls short of assessing true ecosystem health because of a need for the assessment of interactions between populations in the ecosystem. A more common health assessment procedure is for ecophysicians to observe a disease (often described by a structural or functional anomaly in the ecosystem) and trace its cause. This is similar to human epidemiologic studies.

The physician's ability to discriminate between a healthy and unhealthy (non)human physiology is, like the environmental biologist's, based on extensive study of presumed healthy subjects. Although no tables of standard physiological values exist for ecosystems, individual ecophysicians have developed an understanding of specific ecosystems which allow diagnosis (and perhaps prescription). It is noteworthy that the physician ordinarily does not attempt to identify a disease state from a single value, for experience has shown that disease is multidimensional and a diagnosis depends upon knowledge from several tests. However, the number of proven ecosystem diagnostic tests known is small and the tests are often unspecific, so ecophysicians often rely on single measures (e.g., diversity, productivity) to identify a health state or condition. Existing tests are population specific and often unrelated, but new approaches like hierarchy theory (O'Neill et al., 1986) and interlevel analysis (Allen et al., 1984) connect ecosystem components.

Formally, "disease" is a pathological condition of any part of an organ of the body or of the mind. The pathologies of (non)human animals and plants are well known and systematized by scientists. These comprise a set of functional criteria of health of the individual. The specific "pathologies" of an ecosystem are yet to be defined by ecoepidemiologists. In the broadest sense, some of these pathologies are well known: "...diversity is reduced, horizontal transportation [of nutrients] is increased and the ratio of production/biomass is increased..." (Margalef, 1975). Since the ecosystem may be fundamentally altered by the time scientists attempt to describe and quantify pathology, the etiology of the disease is lost and analysis is primarily forensic or retrospective diagnosis. A prospective diagnosis for ecosystems is required.

Although it is unreasonable to expect that disease states of an ecosystem could be identified from one or a few tests, such a shallow procedure drives virtually all environmental management strategies. For example, single species or limited multispecies testing is the basis for water quality criteria. What is needed is the classification of ecosystem states as "diseased" or "nondiseased" (healthy or unhealthy) using objective criteria for a range of parameters which are appropriate to each ecosystem. Additional criteria are needed to distinguish between acceptable and unacceptable degrees of a diseased state (illness) because recovery from disease is possible and often leads to improved system resistance to future disease insults. To develop these criteria, states of ecosystem health and the properties or characteristics of each state must be systematically compiled and supported by experimental or descriptive diagnostic procedures.

Development of techniques for prospective diagnosis and criteria for interpretation should consider the following three major elements. The first is a functional definition of an ecological disease state with a corresponding assessment of factors which identify a healthy ecosystem. The second is identification of parameters, including allowable ranges and optima which assist in diagnosing ecosystem health. The third is procedures for analysis and interpretation of diagnostic information in the management or protection of ecosystems.

C. SPATIAL AND TEMPORAL SCALES OF TOXICITY TESTING

The driving force for biological testing is assessment of the biological effects of the exposure. Unlike human effects assessment, tests to identify environmental effects must protect many species not just one. Furthermore, the targets of analysis are the processes or characteristics of the environment which may be impaired or damaged and these are not limited to changes in a single species. Testing must identify effects on individual species and assess changes in the interactions between species. Based on spatial and temporal scales, the ecological system is the primary unit of analysis.

A test system selected for broad environmental analysis must produce results with ecological relevance to all ecosystems which may receive a specific insult (Novak et al., 1985). One definition of ecological relevance is "a test result which allows assessment of the effect expected on any process or characteristic of an ecosystem." The search for ecological relevance begins with evaluation of standard toxicity test results. Analysis of lethality does not necessarily provide information on ecological relevance although ecologically relevant information can be developed from a lethality test. For example, if observations of behavior during a test can be related to an impairment of organism function which, in the environment, can lead to lower fecundity or decreased survival of individuals, it may be possible to predict a significant ecological consequence. However, a compound which is toxic to fish in laboratory tests may produce large effects in a population yet still undetectable. For example, due to natural annual variability in the population size, a fisheries population would have to decrease 50% to determine a statistically significant reduction in the catch (Vaughan and Van Winkle, 1982). (See Barnthouse et al., 1990 for methodology for combining laboratory acute toxicity test data with life history data.)

When "effect" is considered throughout the organization and structure of an ecosystem, the spatial scale is expanded beyond the site specific, and time scales are also expanded. The time scale used for analysis varies with the focus of the analysis and can be defined by the half-life of a chemical, the generation time of a species, or the successful maintenance of a community of organisms. The complexity and scale of temporal analysis increases as spatial scales expand. Several scales appropriate for classifying ecosystem type have been proposed (Warren, 1979; Lotspeich and Platts, 1982; Bailey, 1983). These schemes build in assumptions of spatial and temporal scale which may be useful in establishing limitations to testing based on scale and in organizing testing approaches which are defined by scalar issues. An ecologically relevant testing program uses batteries which have internal consistency and are linked to spatial and temporal scales of the environment.

We assume it is possible to identify, define, and even classify ecosystems with enough accuracy so we can work with a "known" unit. In this context we must consider spatial and temporal scales or dimensions and the units of analysis required to deal with them. Failure to adequately define units of analysis is often tied to critical information about ecosystems that is seldom available when needed. Spatial and temporal scales of ecosystems are unidentified, poorly defined, and typically unclassified. Spatial scales operative in ecosystems may vary from microscopic to global. Ecosystems are affected by transient phenomenon and species. Local conditions are linked to widely separated areas through processes which are poorly defined and often not measurable.

Temporal scales also confound ecosystem analysis and information development. Within any ecosystem, species, processes, and relationships between species and processes operate on vastly different time scales. Important time scales for some processes may be measured in parts of a second while species life history may involve a few hours to hundreds of years. Relationships between species or communities reflect variability in process and life history spanning time scales from microseconds to millennia. Temporal effects are usually measured using normal clock time, although the organism's internal (biological) clock time (heart beat, basal metabolic rate) may be more relevant when analyzing some ecosystem components.

There seems to be general agreement that temporal considerations are important for ecosystem structural parameters. Of equal importance are temporal factors in the assessment of ecosystem function. For example, since production in a population determines its capacity for economic exploitation, recovery following stress, and long term maintenance in an ecosystem, the ratio of total annual production (P) to average biomass (B) is an important measure of population differences. The P/B ratio "is equivalent to the average specific production rates of all members in the population weighted for their temporal duration or to the sum of all productivity terms associated with the life history again biased according to the number and duration of the population's members at each ontogenetic stage" (Peters, 1983). For populations of terrestrial and aquatic invertebrates, the P/B ratio declines as $W^{-0.37}$, where W is body mass at maturity. "Apparently, populations of small invertebrates are dominated by highly productive juveniles and those of larger animals by less productive adults and subadults" (Peters 1983). This implies that functional measures will have temporal specificity and that measurement of ecosystem function will be constrained by temporal factors.

1. Spatial and Temporal Scales in Development of Ecological Realism

When spatial and temporal scales of analysis are used to define units of analysis and classify the context of tests, it is possible to address other issues associated with developing ecological realism. A critical question in developing biological testing procedures with ecosystem relevance is how to best develop extrapolation algorithms. Algorithms must form the basis for equations which specify expected consequences of environmental stress. Algorithms defining the step functions relating different levels of biological organization (reflecting both spatial and temporal scales or dimensions) are not known. For example, there are no algorithms that translate laboratory tests such as single loci mutation or LC_{50} data for a species to a meaningful criterion for an ecosystem. If the path is not laboratory to field but field to laboratory, another approach to criteria and standard development is necessary. Development of algorithms from field data, whether experimental or descriptive, requires sifting through data sets with high variability, which often integrate responses from differing spatial and temporal scales. Variability arises, for example, from the number of responses a single species, and even a single individual, can exhibit to the same stimulus.

Time, a primary factor which alters the perspective of the investigator and modifies independence/dependence relationships between variables under study, must be viewed from several perspectives. A *short term, nonsteady state* time increment is part of an organism's life span. An acute, nonrecurring stress such as a chemical spill is typical of this time frame. A *quasi-steady state* increment may be defined in population maintenance or in "generations" of individuals. A low level chronic stress will have less noticeable impact than an acute stress, but often causes subtle changes in the population through generations. With low level chronic stresses, cause and effect become confused because it is difficult to even define or quantify the variables. Furthermore, continuous application of a stress can cause rapid adaptation which may alter the basic relationship of a species with its environment. A widely cited example of rapid adaptation, "industrial melanism" in tree trunk moths in industrial areas of England, occurred because mutant dark moths survived better than the natural population of light moths in polluted (dark) woods (Kettlewell, 1956).

Based on time perspectives, the importance of casual factors and the interpretations of "independence" of a selected variable to individuals and species will change. To some degree it is the changes in dependence and independence which limit the extrapolation of laboratory tests to ecosystems. For example, it is virtually impossible to use short term observations to evaluate thresholds of effect in ecosystems (Woodwell, 1975) since these require analyses based on long term consequences of an action.

Use of time and space perspectives is important in specifying ecosystem relationships or organization. Hierarchies can be defined by space, time, or trophic status, for example. Hierarchy theory is a new approach to the view of ecosystems (Levins, 1981; Patten, 1983; Allen et al., 1984). An example of the use of a test system in a spatial, hierarchical setting is given by Murphy and Wilcox (1986). Their study considered three geographical scales: insular montane ranges, within range riparian habitat, and riparian 1-ha plots. Their predictive tier "test systems" were the numbers and species of mammals and birds in these habitat areas. The confirmatory tier was butterfly diversity. They concluded that the "relatively small effects of habitat area and isolation per se on butterfly diversity suggest that vertebrates provide an adequate protective umbrella for invertebrates at most levels." The limitations of these test systems were also recognized since "the existence of special habitat requirements for butterflies indicates that consideration must be given to the protection and management of such habitat to provide for overall biological diversity."

Temporal variability is also an important factor in toxicity testing. Most organisms are subject, directly or indirectly, to seasonal and daily variations in their environment which affect light, temperature, food, noise, and social contact. It is not surprising that organisms show marked cyclic variation of function. Some of the better documented physiological parameters which show rhythmic daily variation are activity/sleep, body temperature, pain threshold, adrenocortical function, skin histamine sensitivity, liver function, renal function, eosinophil rates, and mitotic rates. This has bearing on toxicological testing. If one of the major control systems of the body is strongly circadian dependent, and if key detoxifying organs like the liver and kidney and cellular systems like mitotic rates, show marked circadian variation, then it is possible for drug action and toxic effects to show circadian variation. In fact, this has been found for many drugs and toxic chemicals (Gall, 1977).

D. RATIONAL DESIGN OF TEST BATTERIES

Of N compounds tested in a screening tier, some are positive and the remainder are negative. Of the N compounds, a known proportion p are positive and the remainder $q = 1 - p$ are negative in the predictive or confirmatory tiers. The *quantities* a, b, c, and d are associated with the four outcomes defined in Table 20.2. The *prevalence* $p = (a + b)/N$ is the proportion of all compounds producing positive results in advanced tier testing. The *sensitivity* $p_1 = a/(a + b)$ is the proportion of all true positive results found in the screening test. The *specificity* $q_2 = d/(c + d)$ is the proportion of all nontoxic compounds which are negative in the screening test. The correlation between the tiers is $\tau = (ad - bc)/(n_1 n_2 n_3 n_4)^{1/2}$, where τ^2 has a chi-square distribution with 1 and N degrees of freedom. τ^2 is a measure of the information content of the relationship. The magnitude and the sign of τ are useful in selecting appropriate assays. The magnitude of τ is related to the number of predictive assays and the number of positive responses used as a criterion. The correlation relationship is nonlinear; it rises to a maximum and then falls off nonsymmetrically. Tiers can be correlated even when the assays within a tier are uncorrelated with each other. An *optimal decision rule* for evaluating the results of a multiple assay battery is a function of test battery size, sensitivity, and specificity, but not sample size (Schaeffer and Janardan, 1987).

TABLE 20.2. Cell Designations for Comparing Confirmatory and Screening Study Results

Confirmatory Tier Results	Screening Tier Results		
	+	−	Total
+	a	b	n_3
−	c	d	n_4
Total	n_1	n_2	N

Rational design of test batteries requires understanding of the mathematical theory and accurate assessment of whether testing will meet the data requirements for test system selection. Of importance to decision making throughout the tier testing process (including interpretation of results from different levels of analysis within any tier) is use of a standardized, comprehensive screening process. Results from a comprehensive screening study include an estimate of the hazard of the contaminant(s), identification of the requirements for further testing, and establishment of a basis for determining the likelihood that a test (or battery) in predictive (or even confirmatory) tiers can economically provide unequivocal data to support decision making. Rational design of ecologically relevant test batteries assumes that inter- and intratier criteria have been used to reduce the number of tests selected for investigation for inclusion in the screening tier. The intertier criterion places the priority on tests providing the maximum information about the potential responses at advanced tier levels and assays which integrate within an organism or across organs/tissues, or which identify the character of the response. As an example, if male infertility is to be assessed by confirmatory testing, a screening test based on blocking of the vas deferens appears preferable to one based on binding of chemicals to gonadal DNA.

1. Improving the Interpretability and Relevance of Bioassays

Tests for initial investigation are chosen from reliable assays which can identify the class and magnitude of effects of concern. Test system selection includes developing dose response curves for selected reference compounds and for mixtures of these. Assume that the response for a given substance is linear with the logarithm of the concentration C_k and that the response of a mixture R_{mix} is the sum of responses for the individual compounds proportioned by their concentrations in the mixture. Specifically, if ϕ_k is the toxicity of the kth component relative to the toxicity of a reference compound, then $R_{mix} = a + b\Sigma^k \phi_k C_k$. Finney (1971) terms this "simple similar action".

An assay included in a test battery should give linear responses with mixtures of known composition when the chemicals in a mixture have a common (but not necessarily known) mechanism of toxic action. Differences in the mechanisms of toxic action of mixture components produce nonlinearity. Both the information returned from, and interpretation of, bioassay data of complex mixtures are improved by adding known amounts of a selected toxic compound to the mixture. The shapes of the response curves for the added toxicant in the mixture and the displacement of the curves from each other are compared statistically (Schaeffer, 1987) using analysis of covariance (ANCOVA).

Consider an assay which uses phenol as the added toxicant. Response curves are obtained for phenol alone and phenol added at the same concentrations to the sample. If none of the compounds in the sample are active in this assay, the two phenol response curves will be superimposable. However, if at least one compound in the sample acts by the same toxic mechanism as phenol, the curve for phenol added to the sample will be offset and parallel to the phenol-alone curve, that is, the joint response will be simple similar action. The amount of this offset (determined by regression analysis) is an estimate of the potency of the sample in units of phenol. If the slopes of the two curves are not parallel, then the interaction of the sample components and phenol are more than or less than additive. For example, addition of phenol to effluent samples inactive in the Microtox test (see below) showed that many samples contributed additional toxicity equivalent to ≥ 1 mg/l of phenol (Schaeffer et al., 1991b).

2. Calibrating Test Systems — Whole Ecosystem Exposure Response Studies

A key element in formulating a science of ecosystem health is developing quantitative ecosystem exposure response curves. High quality data from long term experiments of chemical exposures on ecosystems is limited. Consequently, we suggest that a calibrated scale might be developed by equating ecosystem effects from chemical exposures to radiation effects at the ecosystem level. Radiation effects are directly correlated with chromosome volume, and the nominal and delivered doses of radiation (unlike chemical exposures) are the same. Thus, such a scale could facilitate calibration of laboratory hazard assessment bioassays with bioassessment studies using individual, population, and community measures in the exposed ecosystem, for example, a difference in species response to radiation in the laboratory and field estimates of the community effect on that species.

As summarized by Odum (1971), the effects of chronic gamma radiation exposures on whole communities and ecosystems have been studied in fields and forests at Brookhaven National Laboratory on Long Island, in a tropical rain forest of Puerto Rico, and in a desert in Nevada. The combined effects of chronic exposures to gamma radiation plus neutrons on fields and forests have been studied in Georgia and in Tennessee. Gamma radiation has also been used to study short term effects on a wide variety of communities at the Savannah River Ecology Laboratory in South Carolina. Long term studies have also been carried out of the effects of low level chronic radiation to a lake bed aquatic community. These studies provide data for the "reference ecosystem" where this term is used in the sense of physicians' "standard man".

The relative sensitivities of individual species of, for example, higher plants, can be determined from information on chromosome volume. However, these ecosystem level radiation studies showed that there are community attributes such as biomass, diversity, growth form, or species interactions which may greatly modify responses of species in intact communities. For example, many species in herbaceous communities and early stages of succession have small nuclei making them more resistant than species in mature forests (which have larger nuclei). In addition to differences in chromosome volume, the former communities are also less sensitive because "there is much less 'unshielded' biomass aboveground, and the small herbs can recover more quickly by sprouting from seeds or from protected underground parts" (Odum, 1971).

Long term experiments that evaluate the effects of controlled chemical exposures on aquatic ecological systems and ecosystems will provide data needed to link whole ecosystem radiation and chemical exposure data with each other and with laboratory test data. For example, Perry et al. (1987) used chlorine and ammonia to dose outdoor streams. These ecological systems were constructed as parallel, roughly equivalent channels composed of alternating coarse gravel bottomed riffles and mud bottomed pools, each about 30 m long and supplied with water from the Mississippi River. They also describe a whole lake acidification experiment which simulates impacts at the ecosystem level. Little Rock Lake (Wisconsin), an 18 ha softwater lake with no surface inlets or outlets, consists of two primary basins connected by a relatively narrow isthmus. The lake was divided into two basins during 1984. The long term design includes a preacidification phase (1983–1985), acidification phase of the North basin at a rate of 0.5 pH units every 2 years (1985–1991, ambient pH 6.0), and a postacidification/recovery phase (1991–1993).

E. STUDY DESIGN ASSURANCE — A METHOD FOR SYSTEMATIC MONITORING

Assessing the health of an ecosystem cannot be primarily accomplished using laboratory single species toxicity tests. Rather, assessment must be based on the results of field tests and of tests on samples returned to the laboratory for chemical or physical analysis and bioassay. In the same way that a physician must carry out diagnostic testing in a logical order, the ecosystem physican must also establish a systematic approach to assessment. One such systematic approach, termed study design assurance (SDA), is now presented.

Study design assurance applies principles and practices of statistical auditing to the problem of ecological system health assessment (Schaeffer et al., 1985). The approach has been expanded to include the quality assurance and dynamic feedback elements (Herricks et al., 1988) of the assessment process. The SDA process requires explicit identification of program, management and audit objectives, and tasks. A *program objective* is a goal to be accomplished by the assessment in service to published overall mission objectives of a funding agency. A *management objective* is a goal which requires deciding which actions are to be taken in the environment as a result of information developed from quantifiable audit objectives. An *audit objective* is a quantifiable, nonjudgmental goal which relies on the collection and interpretation of data. Audit objectives are supported by *tasks,* which are the data collection, analysis, and interpretation activities whose successful completion allows accomplishment of audit objectives. Detailed examples and discussions have been published (Herricks and Schaeffer, 1985; Perry et al., 1985; Schaeffer and Novak, 1988; Schaeffer et al., 1988; Schaeffer and Beasley, 1989; Novak and Schaeffer, 1990; Schaeffer et al., 1993).

This example is drawn from our experiences assessing the ecological system health effects of physical and chemical use on military training lands (Novak and Schaeffer, 1990; Schaeffer et al., 1990). A primary concern of the military is long term maintenance of several million acres both for training and natural resource productivity/ecosystem conservation. A program objective to accomplish this is to develop a systematic approach for monitoring and managing changes in the condition of the Army land to maintain ecosystem stability. One management objective was specified as to apply those restoration activities which will enable damaged land to be brought back into use for training while maintaining ecosystem stability. Three related audit objectives are (1) Indicate what changes are occurring on training lands as a result of restoration activities. (2) Of the changes occurring as a result of restoration activities, identify those changes which are desired. (3) Indicate the rate at which the desired changes on the property are occurring. Associated tasks include identification of changes in the types of species present, their abundances, and their spatial distributions.

F. APPROACHES TO ECOSYSTEM HEALTH ASSESSMENT

Ecosystem health assessments usually proceed either along a hazard assessment or a bioassessment path (Herricks and Schaeffer, 1987). Most research has emphasized hazard assessment but this is appropriate only when the exposure is from a single fairly toxic substance originating at one or more identifiable source(s) at relatively high concentrations, and the effect is expected to be acute. Bioassessment must be used when exposure is to multiple agents or those of low, nonspecific toxicity; multiple, often nonspecific, sources are involved, and effects are likely to be low level and chronic. Hazard assessment emphasizes chemical composition, environmental fate, and laboratory toxicology data. Bioassessment emphasizes characterization of the ecosystem, identification of stressors and their relative contributions, and the use of measurements and models to estimate chronic effects (Schmidt-Bleek et al., 1988). Both types of assessments are developed using SDA procedures. Hazard assessment depends on single species laboratory tests and microcosm studies. Bioassessment uses enclosures and other types of microcosms and mesocosms to carry out controlled exposure studies in an ecosystem (Gearing, 1988; Gillett, 1988).

1. Bioassessment Approach

Bioassessment of ecosystem health uses the results of a series of tests which provide quantitative information on "critical" ecosystem components. In the same fashion that respiratory rate and basal metabolism provide system information to the human or nonhuman animal physician, photosynthesis and nutrient cycling provide system information to the ecological physician. Before identifying test systems to measure chemical, physical, and biological effects in ecosystems, specific properties of ecosystems which are affected by chemicals must be specified. These characteristics are discussed in a report of the National Academy of Sciences (1981), in Herricks and Schaeffer (1987), and Schaeffer and Herricks (1992). The latter authors identify 44 measures appropriate to individuals, populations, ecosystem biological components, and major abiotic elements (Schaeffer et al., 1988). The eight critical characteristics, not of equal importance in every ecosystem, are (1) habitat for desired diversity and reproduction of organisms, (2) phenotypic and genotypic diversity among the organisms, (3) a robust food chain supporting the desired biota, (4) an adequate nutrient pool for desired organisms, (5) adequate nutrient cycling to perpetuate the ecosystem, (6) adequate energy flux for maintaining the trophic structure, (7) feedback mechanisms for damping undesirable oscillations, (8) the capacity to temper toxic effects, including the capacity to decompose, transfer, chelate, or bind anthropogenic inputs to a degree that they are no longer toxic within the system.

A requirement for identifying a test system which can be used alone to determine the specific effects of an insult to the ecosystem is that there exist among the 44 measures one unique relationship between a measure and a critical characteristic. Such a unique relationship does not often exist. For example, the chemistry of hydrocarbon aerosols suggests that exposure could alter nutrient cycling, one measure of which is total N_2 production. Since each measure can have more than one test system associated with it, the absence of unique relationships prevents the identification of a unique test system. The complex interactions among the 44 measures also make it unlikely that definitive evaluations of an ecosystem can be accomplished using a "standard" set of a small number of test systems.

Systematic selection of the right test system for a study requires that test systems be classified logically. The scientific evaluation criteria were identified by Novak et al. (1985) and include media (air, water, soil), exposure duration (continuous, continual, episodic) and intensity (mg/l, mg/m³), the organisms used in the test system (bacteria, fish, mammals, plants), exposure route (ingestion, inhalation, dermal contact), likely effects on test system species (genetic, toxic, bioaccumulation), and measurement methods. Additional practical evaluation criteria include test system variability, accuracy, precision, sensitivity, convenience and cost, technical requirements for personnel and laboratory facilities, and the status of the test (consensus, standard, tentative, developing).

Ecologists know that the relative importance of the eight critical characteristics in a given ecosystem depends on the geographic location of the ecosystem (i.e., the ecoregion) and whether the system is aquatic or terrestrial (Odum, 1971). Nonetheless, identification of specific test systems requires that these "generally known" differences be made explicit in the context of the 44 measures. To accomplish this, Herricks and Schaeffer (1987) applied the principles for hierarchial analysis (Allen et al., 1984) to the eight critical characteristics. Table 20.3 gives the results as a matrix to rank selection of critical ecosystem characteristics. Table 20.4 is a matrix of the availability of test systems for ecosystem measurements also presented by Schaeffer and Herricks (1993).

Each critical ecosystem characteristic in Table 20.3 (rows) is associated with some of the 44 measures. For example, nutrient cycling is associated with measures such as acclimation (individual level), adaption (population level), and guild composition (system level). Since nutrient cycling, as measured by annual production of dry matter in kilograms per hectare per year, is directly related to millimeters of rainfall (Odum, 1971); nutrient cycling will be more important in determining the structure and dynamics in a prairie terrestrial ecosystem than in a desert terrestrial ecosystem. These types of considerations produced priority scores (Table 20.3) of 1 and 2 for this characteristic in these ecosystems. Carrying this example further, test systems to evaluate the effects of chemicals on a terrestrial desert ecosystem would emphasize food chains but not genetic diversity. However, test systems to evaluate food chains and genetic diversity would be useful in a prairie ecosystem.

Each critical characteristic in Table 20.3 is associated with one or more measures and each measure is represented by one or more taxonomic groups

TABLE 20.3. Matrix to Prioritize Selection of Critical Ecosystem Characteristics for Test System Selection

Critical Ecosystem Characteristics	Subtropical A[a]	Subtropical T	Prairie A	Prairie T	Mediterranian A	Mediterranian T	Steppe A	Steppe T	Desert A	Desert T
Habitat	2[b]	1	3	1	3	1	2	1	3	2
Genetic diversity	2	1	3	1	3	2	3	2	3	3
Food chain	2	1	3	1	3	2	3	2	3	1
Nutrients	1	1	2	1	2	1	3	1	3	2
Nutrient cycling	1	1	2	1	2	1	2	1	3	2
Energy flux	1	1	3	1	3	2	3	2	3	2
Desirable oscillations	2	2	2	1	3	1	3	1	3	1
Toxics assimilation	1	1	3	2	3	2	3	2	3	3

[a] A — aquatic ecosystem, T — terrestrial ecosystem.
[b] Measurement of critical item is 1 — important, 2 — likely to provide useful information but utility may depend on seasonal conditions, 3 — likely to provide limited information.

in Table 20.4. An entry in Table 20.4 does not imply that a test system using a specified taxonomic group is available for every type of ecosystem in Table 20.3. These tables thus identify the theoretical and implementation bounds on test system selection. Selection of specific test systems is constrained by the type of ecosystem, critical characteristics and their measures, trophic level(s) amenable to measurement or experimentation, and test systems available at a given trophic level. The tables also show where new test systems could be developed. When laboratory testing is relevant to ecosystem analysis, Tables 20.3 and 20.4 guide selection of tests. The research must then systematically show the relevance of the laboratory testing to the ecosystem assessment.

2. Some Representative Bioassessment Test Systems

Tables 20.3 and 20.4 show that specific design of a bioassessment study depends on the type of ecosystem. Consequently, we exemplify the general principles by considering a specific ecosystem. This example is based on a preliminary field study of the effects of chronic exposures of aerosolized hydrocarbons on a midwest prairie ecosystem (Schaeffer et al., 1990). Equivalent test systems are available for freshwater (Urban and Cook, 1986; Perry et al., 1987) and marine (U.S. EPA, 1989) aquatic systems. We emphasize that the protocols discussed here will require extensive modification depending on the type of ecological sysem, spatial factors, purposes of the study, and other factors identified through SDA analysis.

Vegetation Studies

Disturbance is an important and pervasive influence in natural plant communities, and its effects on various aspects of community structure and dynamics

TABLE 20.4. Availability of Test Systems for Ecosystem Measurements

Measurement	Bacteria	Fungi	Bryophytes and Algae	Vascular Plants	Protozoa	Molluscs
Ecosystem level[a]						
Biomass production	X	X	X	X	X	X
Nitrogen fixation	X	X	U	U	NS	NS
Soil retention	NS	NS	NS	NS	NS	NS
Community level[b]						
Production/decomposition	X	X	X	X	X	X
Guild structure	X	X	X	X	X	X
Predator/prey	NS	NS	NS	NS	X	X
Vegetative analysis	NS	NS	X	X	NS	NS
Diversity	X	X	X	X	X	X
Genetic alteration	U	U	U	X	U	U
Bioaccumulation	X	X	X	X	X	X
Individual/population level						
Reproduction	X	X	X	X	X	X
Life table/production	NS	NS	X	X	X	X
Disease/mutation	X	X	X	X	X	X
Behavior	NS	NS	NS	NS	NS	U

Taxonomic Category

TABLE 20.4. (Continued)

Measurement	Taxonomic Category						
	Arthropods	Earthworms	Nematoda	Amphibians	Reptiles	Birds	Mammals
Ecosystem level[a]							
Biomass production	X	X	X	X	X	X	X
Nitrogen fixation	NS	X	U	NS	NS	NS	NS
Soil retention	NS	NS	NS	NS	NS	NS	NS
Community level[b]							
Production/ decomposition	X	X	X	X	X	X	X
Guild structure	X	X	X	X	X	X	X
Predator/prey	X	X	X	X	X	X	X
Vegetative analysis	NS	NS	NS	NS	NS	NS	NS
Diversity	X	X	X	X	X	X	X
Genetic alteration	U	U	U	U	U	U	X
Bioaccumulation	X	X	X	X	X	X	X
Individual/population level							
Reproduction	X	X	X	X	X	X	X
Life table/production	X	X	X	X	X	X	X
Disease/mutation	X	X	X	X	X	X	X
Behavior	X	U	X	X	X	X	X

a Assessments for biomass production, nitrogen fixation, and predators were developed by an expert group (Glennon, 1982). NS indicates lack of direct ecological relevance for specified measurement, X indicates a testing or analysis method exists for one or more species in toxonomic group, U indicates uncertain availability of testing or analysis method.

b Recovery is also appropriate at this level — it is possible to measure colonization following stress; measurement of recovery is problematic because an adequate definition for recovery does not exist.

have been well studied (White, 1979; Runkle, 1981). Disturbance is defined as a sudden change in the resource base of a landscape unit that produces readily detectable population responses. A hierarchical series of responses to these changes can be predicted. Whole plant physiological responses may influence the demography of both plants and plant parts. Many plant population characteristics such as density, dispersion, growth rates, survivorship, age structure, genetic variation, species interactions, and life history patterns may be altered by disturbance. Differential responses among co-occurring species may in turn alter community level characteristics such as species composition and richness, relative abundance distributions, and successional changes. At the ecosystem level, these responses result in changes in gross and net primary productivity, decomposition, mineralization, and other processes. This response hierarchy varies with the type, frequency, intensity, duration, timing, and heterogeneity of the disturbance events.

Vegetation studies are useful in evaluating the effects of acute and chronic exposures. Chronic or acute exposures may directly affect plant growth and reproduction. Differential responses among herbaceous species can produce changes in species composition, relative abundances, and successional patterns of the herbaceous plant community. It may be possible in some situations to evaluate the effects that differences in the intensities and combinations of exposures have on the patterns in plant community structure and dynamics. Comparisons of plant communities on exposed sites and on unexposed and manipulated reference sites with comparable soil types, topography, and fire histories can be used to evaluate responses to disturbances. Furthermore, it may be possible to correlate results from laboratory tests such as root and shoot elongation and algal growth made on soils and soil elutriates with field vegetation data (Thomas et al., 1986).

Plant Community Responses

Ten sampling points are located and permanently marked with numbered galvanized steel stakes at 10-m intervals along two 100-m transects 50-m apart. These sampling points form the centers of 10-m^2 circular plots. Plant species composition and abundances are recorded for each of the 20 circular plots in each site in late April or early May, late July, and September. Canopy coverage (Daubenmire, 1959) is computed by averaging the midpoints of the ratings for each species for the 20 plots at each sampling period. For species encountered and rated on more than one date per year, the maximum coverage rating is used. The frequency of occurrence of each species is calculated as the percent of the 20 plots in which the species occurred. Species diversity and relative abundance distributions are calculated for each site based on the canopy cover data, then compared among sites by calculating community and site similarity indices (Hellawell, 1978).

The aboveground biomass per unit area is assessed concurrently with the species cover sampling in each of the sample sites on each sampling date. For destructive sampling, a 100-m transect is established 4 m from, and parallel to, each transect. At each transect, all of the aboveground biomass is clipped in 20.2 m × 0.5 m quadrats. Plant material is separated into grasses/sedges and forbs/woody components, and into living vs. dead material within each of these categories. The plant material is then oven dried to a constant weight at 90°C and weighed.

Species Level Responses

To evaluate potential effects of acute and chronic exposures at the species level, growth and reproduction responses of the dominant grass *Andropogon gerardii* are measured simultaneously in the control and treatment plots before and after experimental exposures. At 3-week intervals throughout the growing season, 15 individual tillers of *A. gerardii* are clipped at ground level. Leaves and stems are separated and total and individual leaf areas are measured. Leaves and stems are then oven dried to constant weight at 90°C. Specific leaf mass, (SLM), and shoot, leaf, and stem biomass are determined for each tiller. Growth analysis methods are used to interpret whole plant performance in treated and reference stands (Hunt, 1978). Replicate values for whole plant dry weight (W) and total leaf area per plant (La) are calculated. These primary data are used to obtain growth analytical quantities including relative growth rate, $Rw = (1/W)(dW/dT)$, relative leaf area growth rate, $Rl = (1/La)(dLa/dT)$, leaf area ratio, $F = La/W$, and unit leaf rate, $E = (1/La)(dW/dT)$. Instantaneous values of these are calculated using the methods of Hunt and Parsons (1974). In addition, flowering stem densities and heights are measured in 10.2 m × 0.5 m quadrats randomly placed within *A. gerardii* stands in both the exposure and reference plots.

Soil Chemistry

Soil fertility indices and indicators of mineralization potential of organic nitrogen (N) and phosphorus (P) are studied during the growing seasons in each year. The fertility indices (exchangeable cations, pH, inorganic N, total N, Bray P, labile inorganic and organic P) may not be sensitive enough to detect the subtle changes induced by exposure to weakly toxic organics. Therefore, these indices are measured only twice yearly. However, the mineralization potential for N and P are fairly sensitive to disturbances and are measured every 3 to 4 weeks. Fertility indices are determined using standard procedures. Mineralization of organic N is determined by a modification of the Stanford and Smith (1972) method which involves heating the samples with 2 mol/l KCl at 90°C followed by shaking, filtration, and determination of inorganic N. Mineralization of P is measured by recording resin-extractable P before and after a low temperature incubation and by a phosphatase enzyme assay.

Plant Chemistry

Plant chemical analyses must be coordinated with vegetation studies. The dominant plants are analyzed for major cations N, P, Fe, and Zn every month during the growing season. Vegetation obtained for aboveground productivity measurements and rhizomes and roots obtained for belowground studies are similarly analyzed. This sampling regime should detect any exposure-induced changes in the nutrient status of the vegetation.

Decomposition and Mineralization of Vegetation

Decay patterns in litter decomposition and mineralization studies exhibit the characteristic temperature-modified exponential functions. Patterns in temperate ecological systems have been described. In prairie ecosystems, nitrogen and phosphorus are immobilized by microbes on decaying litter in a pattern inverse to that for decay. These decomposition and mineralization patterns of decaying prairie foliage provide an integrated measurement of the activities of heterotrophic microbes and arthropod detritivores with the aerosols. Deviations from the expected patterns indicate treatment-induced changes that affect ecological system characteristics. The procedures for litterbag decomposition and mineralization studies are well known (Wieder and Lang, 1982).

Aboveground Insects

Risser et al. (1981) used the Osage IBP site in Oklahoma to conduct the most complete survey of aboveground herbivorous insects of the Tallgrass Prairie. Biomass of aboveground insects (estimated by drop trap) varied between 15 and 140 mg/m^2 from April to November over a 2-year period. The abundant insect groups represent a great diversity of feeding patterns, and potentially these insects may affect plants in many different ways. Satisfactory estimates of how much primary productivity is consumed and/or destroyed by these herbivores are not yet available; in old fields and salt marshes, aboveground insects have been estimated to ingest 10% or less of net primary production (Wiegert and Evans, 1967). However, such estimates of consumption reflect only some of the impacts that herbivores have on plants. Homopteran populations are sensitive to subtle changes in the chemistry and species composition of the plant community and are therefore believed to be excellent biomonitors for disturbance studies.

Soil Invertebrates

Earthworms, arthropods, and nematodes are relatively sensitive to the quantity and quality of plant roots and plant detritus and are believed to both affect and be affected by microbial populations and activities (Seastedt, 1984a,b). This group of consumers is traditionally recognized as potential regulators of nutrient cycling processes (Swift et al., 1979). The larger organisms (macroarthropods and earthworms) within this group are probably more important for comminution-related processes (Anderson and Ineson, 1983), while the smaller organisms (microarthropods, nematodes) are more important for microbial interactions. Since earthworms are a standard laboratory bioassay system (Glennon, 1982), the potential for developing algorithms to link laboratory test data with ecosystem effect is improved. Herbivores, including root-feeding nematodes and white grubs *(Scarabaeidae),* effect net primary productivity and are sensitive to changes in plant chemistry.

The numbers and trophic composition of soil invertebrates are measured together with plant root and rhizome biomass and necromass. Greater densities of these organisms per unit of plant biomass are hypothesized to occur in stressed sites due to plant chemistry changes (i.e., reduced secondary plant substances in stressed vegetation, higher nitrogen content). The root and rhizome standing crop data can also be used as an index of site fertility and/or site stress (Chapin, 1980). Root to shoot ratios provide a measurement of plant carbon allocation, patterns that reflect whether the system is limited by photosynthetic or nutrient processes. The patterns of living and dead plant mass and nutrients, in relation to consumer numbers and composition, provides considerable insights into the functioning of the system. Rhizomes, earthworms, and larger arthropods are sampled in the spring and autumn by excavating 0.1 m^2 by 30-cm deep soil cores. Microarthropods and nematodes are sampled by taking smaller soil cores and mechanically extracting the fauna.

Measurements for Acute Exposure

A subset of plant, soil, and invertebrate measurements can be used to study the short term response to exposures, particularly when exposures are delivered at random intervals. Soil (top 5 cm) and vegetation (roots, leaves, stems) samples are collected especially for this purpose. Plant physiological measurements have been described (Knapp, 1985). The relative health and vigor of both the plant foliage and microbes can be measured by analyzing the organic and inorganic nitrogen content of rainwater that passes through the canopy (throughfall) (Seastedt, 1985). Invertebrate measurements (bumper traps, sticky traps, and pitfall traps) provide a relative index of insect activity within these plots, while buried bag incubations using exposed and unexposed surface soils allow assessment of short term soil microbial responses. Exclosures and inclosures are the best way to conduct these experiments.

3. Hazard Assessment Approach

An ecosystem may be acutely exposed to relatively high concentrations of a pure substance or a defined mixture. The most common events of this type are spills and volatile emissions such as those which occur from fertilizer misuse, transportation accidents, and ruptures and explosions during chemical manufacture and storage. Chemical analyses of contaminated media (e.g., soil,

TABLE 20.5. Test System Selection Criteria

The test organism:

1. Can be maintained in good condition in the laboratory during acclimation and testing, with relative ease and safety
2. Is available in sufficient numbers and at reasonable costs during most of the year
3. Is representative of species likely to be exposed (habitat and mode of exposure)
4. Is amenable to culture in order to provide life stages and be used in life cycle testing
5. Is methodologically sound
6. Is important in ecological processes
7. Is commercially and recreationally important
8. Is widely distributed in its natural environment
9. Has an available data base
10. Is weakly correlated with other assays in the same battery
11. Is strongly correlated with assays in the definitive tier or with ecological system responses

air, water) provide concentration data but no direct information about toxicity. Initial scoping studies should use easily implementable, reproducible, and inexpensive bioassays. It is convenient to classify biological test systems as genetic effects on individuals (bacterial, plant, insect, and cell mutations and sister chromatid exchange), somatic effects on individuals (acute/chronic stress and lethality), and chronic effects on ecosystems (model ecosystems, field monitors, surveys, macrocosms). We have found that an initial battery consisting of Microtox™ and planarians (used as a sentinel species; NRC, 1991) can be used for both aquatic and terrestrial ecosystems if testing is modified to use a reference toxicant (Schaeffer et al., 1991b). A second terrestrial battery can be devised using white rats, bobwhite quail, houseflies, wild oats, and earthworms, and an aquatic systems battery can use daphnia, fathead minnow, rainbow trout, mysid shrimp, alga, and duckweed. Since a given species may be assayed for several traits (e.g., mutagenicity, acute toxicity) these sequences provide a starting point for customizing batteries to assess specific contamination problems.

Criteria to guide the selection of individual species for use in ecotoxicity testing are given in Table 20.5 (modified from Glennon, 1982). Criteria based on biological groups (Table 20.6) are used to establish the quality of impacted areas in the International Great Lakes (IJC, 1987). The Microtox™ test and a planarian behavioral test are used to illustrate these types of tests. Whether used with single substances or mixtures, the ecological relevance and interpretability of tests are increased by modifying the standard assay as discussed above.

Planarian Behavioral Test

Five mature asexual *Dugesia dorotocephala* are placed in a beaker containing a standard medium (50 ml) away from direct sunlight, at 19 to 20°C. The medium is gently withdrawn using a pipet and sample is gently added

TABLE 20.6. Characteristics of Various Biological Groups for Use in Aquatic Monitoring Programs

	Bacteria	Phytoplankton	Zooplankton	Periphyton	Macrophytes	Benthos	Fish
Presence in areas of concern habitats	+	+	+	+	+	+	+
Ease of quantitative sampling	–	+	–	+	–	+	–
Temporal heterogeneity	–	–	–	+	+	+	+
Spatial heterogeneity	0	+	–	+	–	+	–
Mobility (site representativeness)	0	0	0	+	+	+	+
Sample preservation	0	+	+	+	+	+	+
Taxonomy	–	–	+	–	+	+	+
Known responses to pollutants	–	+	+	+	–	+	+

Note: + = suitable; – = not suitable; 0 = marginal.

(time 0). Observations are made on the organisms starting at 1 min and continuing each minute for 5 min, and then every 10 min until 60 min. Usually, a dilution series (e.g., 100%, 50%, 10%) is used. An acute, nonlethal characteristic behavioral response of toxic exposure is reduction of ventral body surface contact with the tube. The sample can be tested alone and with the addition of a reference toxicant such as phenol. Subchronic effects are determined by continuing exposure for 3 to 90 d. With short exposures (1 to 30 d), effects can include readsorption of the head, lesions, and immobility. With longer exposures, effects include neoplasms, severe lesions, and malformations such as a second head.

Microtox™

The Microtox™ test is an instrumented bioassay which uses the freeze dried marine bioluminescent bacteria *Photobacterium phosphoreum* as the bioassay organism. Measurements are made of the reduction in light emission when the bacteria are exposed to a toxicant. The result is expressed as the median effective concentration (EC_{50}), that is, the concentration causing a 50% reduction in light emission. The test has been used with soil samples and semisolid and solid wastes. Advantages of the test are its low cost ($25 to $50/ sample), short analysis time (about 60 min for a complete response curve), repeatability, use of standardized organisms, portability permitting field use, and applicability to a wide variety of media and toxic chemicals.

Estimated toxicity values from this test correlate highly with toxicity measured in many standard bioassay systems such as 96-h LC_{50} in the brackish water harpacticoid *Nitocra spinipes* (Tarkpea et al., 1986) and 24 to 96 h LC_{50} in fish (Bulich, 1982) and mammals (Burton et al., 1986). One study compared Microtox™ EC_{50}s for effluents with 96-h fish LC_{50}s using a "25%" criterion: any sample with a LC_{50} or an EC_{50} of <25% on dilution was scored as "toxic" while values of ≥25% were scored as "nontoxic". A total of 257 complex effluent samples were tested; 235 were assayed simultaneously using Microtox™ and the fathead minnow test while 155 were tested using Microtox™ and the *Daphnia* acute bioassay. The agreements were fish and Microtox™ 209/235 (89%) and *Daphnia* and Microtox™ 132/155 (85%). Similarly, using a criterion of 750 mg/l (EC_{50} or 96-h LC_{50}) or less as "toxic" (+), the agreement between Microtox™ and a 96-h fathead minnow test for 31 hazardous waste samples was Microtox™ (–) Fish (–) (51%), Microtox™ (+) Fish (+) 7 (23%), Microtox™ (+) Fish (–) (6.5%), Microtox™ (–) Fish (+) (9.5%) (Microbics Inc., Bulletin M105). Arranged as an ordered 2 × 2 contingency table, the exact Jonckheere-Terpstra test for these data showed a significant dependence between Microtox™ EC_{50} and fish LC_{50} values.

The standard Microtox™ test is easily modified to improve the interpretation of chemically uncharacterized complex samples. One series of tests using phenol as the added toxicant found that 12 of 27 effluents which were not directly toxic enhanced the toxicity of the added phenol (Schaeffer et al., 1991b).

G. PUTTING IT TOGETHER — AN ECOSYSTEM DIAGNOSIS EXAMPLE

Few studies have examined the long term effects of chemical exposures on ecosystems. Young et al. (1987) reviewed the final results of a long term field study of ecosystems contaminated with 2,3,7,8-tetrachlorodibenzo-*p*-dioxin (TCDD). This work nicely illustrates both the hazard assessment and bioassessment paths. It also shows the difficulties imposed by spatial and temporal scales in extrapolating short term data to long term ecological system effects.

The polychlorinated dibenzo-*p*-dioxins (PCDDs) consist of 75 isomers that differ in the number and position of attached chlorine atoms. The isomer considered here, 2,3,7,8-TCDD, occurs as a contaminant of products made from trichlorophenol. It is the most toxic synthetic compound ever tested under laboratory conditions. In certain animal models TCDD is a potent teratogen, hepatocarcinogen, and liver tumor promoter. It induces several hepatic enzymes such as microsomal BPH and affects hepatic plasma membrane proteins such as the EGF receptor. "There is general agreement that 2,3,7,8-TCDD is exceedingly stable, readily incorporated into aquatic and terrestrial ecosystems, extraordinarily persistent, and virtually impossible to destroy" (Eisler, 1986).

From 1962 to 1970, four ecological systems in Test Area C-52A at Eglin Air Force Base (Florida) received massive quantities of the herbicides 2,4,5-trichlorophenoxyacetic acid (2,4,5-T) and 2,4-dichlorophenoxyacetic acid (2,4-D). Test Area C-52A (8 km²) is a grassy plain surrounded by a forest stand dominated by longleaf pine *(Pinus palustris),* sand pine *(Pinus clausa),* and turkey oak *(Quercus laevis).* The exposed portion was a cleared, leveled area of about 3 km² occupied mainly by broomsedge *(Andropogon virginicus),* switchgrass *(Panicum virgatum),* wooly panicum *(Panicum lanuginosum),* and low growing grasses and herbs. The total exposed area of about 3 km² actually consisted of four separate testing grids and received a total of about 73,000 kg 2,4,5-T and 77,000 kg 2,4-D. Based on chemical analysis of archived samples of the 2,4,5-T and 2,4-D, an estimated 2.8 kg of TCDD had been applied to the test sites.

Following the hazard assessment path, residues were determined in soil profiles. Further, the significant contribution of grooming to the total body burden of small mammals was shown in laboratory studies which compared levels in beachmouse *(Peromyscus polionotus)* exposed to 0 (controls) or 2.5 ppb TCDD in an alumina gel dusted on the fur on the central thoracic and abdominal regions, side, back, and tail.

Ecological surveys of plants, mammals, fish, and other taxonomic levels and other types of studies illustrate the bioassessment approach. For example, studies using "tagged" beachmice obtained from a reference at Eglin and beachmice indigenous to the test sites illustrate field to lab and lab to field extrapolation (Schaeffer and Beasley, 1989). Liver and pelt concentrations of the TCDD between the tagged and native beachmice were the same after 3 months, which suggests that body burden levels of TCDD were obtained at this site within at least 3 months.

What is the final ecological system health assessment? During 1969 to 1984, 341 species associated with the test area were observed and identified. TCDD residues were found in 32 different species of mammals, birds, insects, reptiles, amphibians, and fish. "Examination of the ecological niches of the species positive for TCDD residue suggested that the commonality was a close relationship to contaminated soil. Studies spanning more than 50 generations of the beachmouse *Peromyscus polionotus* concluded that exposure to soil concentrations of TCDD in the range of 0.1 to 1.5 ppb have had minimal effect upon the health and reproduction of this species" (Young et al., 1987).

H. CONCLUSIONS

The science of ecosystem health is in the process of being defined, and independent development of similar ideas is occurring. Unfortunately, some excellent work which conceptualizes and develops ecosystem health methodologies is obscure. In this regard, Gasto (1980) provides a fitting summary and conclusion to this chapter.

The study of sick ecosystems requires a methodology that can evaluate these ecosystems completely and analytically. The result of the analysis and synthesis of the ecosystem should identify the causes of poor functioning and identify the measurements needed to correct the ecosystem. The clinical methodology must involve a general review of the patient, a general clinical exam or anamnesis (Bunyan and Stanley, 1982), diagnosis, and finally treatment. The clinical exam involves the study, observation, and measure of signs. Not all the signs are of value in the prediction of ecosystem function, while others allow interpretation of the general function of the ecosystem and set up the real symptoms of the sickness. One of the first tasks of ecophysicans in coming years must be the identification and description of these signs and their measures. The general review of the patient begins with recording information about the history of the ecosystem and the qualifications of the ecophysican(s) carrying out the examination. The information file should include the geographic location of the patient, the type of ecological unit in the study, and the administrator/owner of the unit. An outline of the patient's characteristics should be prepared, which includes climate, external aspects of vegetation and fauna, the use(s) of the ecosystem, and general symptoms. Signs of poor function or structure can be used as a positive proof of sickness. Diagnosis is of fundamental importance, and the implementation of a diagnostic study is through assessment as considered in this chapter.

REFERENCES

Allen, T.F.H., O'Neill, R.V., and Hoekstra, T.W. Interlevel relations in ecological research and management: some working principles from hierarchy theory. General Technical Report RM-110, U.S. Department of Agriculture, Rocky Mountain Forest and Range Experiment Station, Fort Collins, CO, 11, 1984.

Anderson J.M. and Ineson, P. Interactions between soil arthropods and microorganisms in carbon, nitrogen and mineral nutrient fluxes from decomposing leaf litter, in *Nitrogen as an Ecological Factor*, J.A. Lee, S. McNeill, and F. Raison, Eds. Oxford University Press, Boston, l983, 413–431.

Bailey, R.G. Delineation of ecosystem ecoregions, *Environ. Manage.* 7:365–373, 1983.

Barnthouse, L.W., Suter, G.W., II, and Rosen, A.E. Risks of toxic contaminants to exploited fish populations: influence of life history, data uncertainty and exploitation intensity, *Environ. Toxicol. Chem.* 9:297–311, 1990.

Bro-Rasmussen, F. and Lokke, H. Ecoepidemiology — a casuistic discipline describing ecological disturbances and damages in relation to their specific causes: exemplified by chlorinated phenols and chlorophenoxy acids, *Regulat. Toxicol. Pharm.* 4:391–399, 1984.

Bulich, A. A practical and reliable method for monitoring the toxicity of aquatic samples, *Process Biochem.* March/April:45–47, 1982.

Bunyan, P.J. and Stanley, P.I. Toxic mechanisms in wildlife, *Regulat. Toxicol. Pharmacol.* 2:106–145, 1982.

Burton, S.A., Petersen, R.V., Dickman, S.N., and Nelson, J.R. Comparison of *in vitro* bacterial bioluminescence and tissue culture bioassays and *in vivo* tests for evaluating acute toxicity of biomaterials, *J. Biomed. Mater. Res.* 20:827–838, 1986.

Chapin F.S. The mineral nutrition of wild plants, *Ann. Rev. Ecol. Syst.* 11:233–260, 1980.

Daubenmire, R. Measurement of species diversity using canopy coverage classes, *Northwest Sci.* 33:43–66, 1959.

Eisler, R. Dioxin hazards to fish, wildlife, and invertebrates: a synoptic review, *U.S. Fish Wildl. Serv. Biol. Rep.* 85 (1.8):37, 1986

Evans, G.C. *The Quantitative Analysis of Plant Growth.* Blackwell Scientific Publications, Oxford, England, 1972, 734.

Finney, D.J. *Probit Analysis,* 3rd ed. Cambridge University Press, Cambridge, England, 1971, 230–268.

Gall, D. Temporal variations in toxicity, in *Current Approaches to Toxicology.* Year Book Medical Publishers Inc., Chicago, IL, 1977, 12–21.

Gasto, J. *Ecologia: El Hombre y la Transformación de la Naturaleza.* Editorial Universitaria, Santiago, Chili, 1980, 573.

Gearing, J.N. The role of aquatic microcosms in ecotoxicologic research as illustrated by large marine systems, in *Ecotoxicology: Problems and Approaches,* S.A. Levin, M.A. Harwell, J.R. Kelly, and K.D. Kimball, Eds. Springer-Verlag, New York, 1988, 411–470.

Gillett, J.W. The role of terrestrial microcosms and mesocosms in ecotoxicologic research, in *Ecotoxicology: Problems and Approaches,* S.A. Levin, M.A. Harwell, J.R. Kelly, and K.D. Kimball, Eds. Springer-Verlag, New York, 1988, 367–410.

Glennon, J.P. Surrogate species Workshop. Project 1247. Life Systems, Inc., Cleveland, OH, 1982.

Hellawell, J.M. *Biological Surveillance of Rivers.* Water Research Centre, Stevenage Laboratory, Elder Way, Stevenage, England, 1978, 33.

Herricks, E.E. and Schaeffer, D.J. Can we optimize biomonitoring?, *Environ. Manage.* 9:487–492, 1985.

Herricks, E.E. and Schaeffer, D.J. Selection of test systems for ecological analysis, *Water Sci. Technol.* 19:47–54, 1987.

Herricks, E.E., Schaeffer, D.J., and Perry, J.A. Biomonitoring: closing the loop in the environmental sciences, in *Ecotoxicology: Problems and Approaches,* S.A. Levin, M.A. Harwell, J.R. Kelly, and K.D. Kimball, Eds. Springer-Verlag, New York, 1988, 351–366.

Hunt, R. *Plant Growth Analysis.* Edward Arnold, London, 1978, 67.

Hunt, R. and Parsons, I.T. A computer program for deriving growth-functions in plant growth analysis, *J. Appl. Ecol.* 11:297–307, 1974.

International Joint Commission. Guidance on characterization of toxic substances. Problems in Areas of Concern in the Great Lakes Basin. International Joint Commission, Winsor, Ontario, Canada, 177, 1987.

Kettlewell, H.B.D. Further selection experiments on industrial melanism in lepidoptera, *Heredity* 10:287–301, 1956.

Knapp, A.K. Effect of fire and drought on the ecophysiology of *Andropogon gerardii* and *Panicum virgatum* in Tallgrass Prairie, *Ecology* 66:1309–1320, 1985.

Levins, R. Ecosystem properties relevant to ecotoxicology, Appendix B, in *Testing for Effects of Chemicals on Ecosystems.* National Academy Press, Washington, DC, 1981, 103.

Lotspeich, F.B. and Platts, W.S. An integrated land-aquatic classification system, *North Am. Fish. Manage.* 2:138–149, 1982.

Margalef, R. Human impact on transportation and diversity in ecosytems. How far is extrapolation valid?, in *Proceedings of the First International Congress of Ecology: Structure, Functioning and Management of Ecosystems.* Center for Agricultural Publishing and Documentation, Wageningen, Netherlands, 1975, 233–241.

Murphy, D.D. and Wilcox, B.A. Butterfly diversity in natural habitat fragments: a test of the validity of vertebrate-based management, in *Wildlife 2000: Modeling Habitat Relationships of Terrestrial Vertebrates,* J. Verner, M.L. Morrison, and C.J. Ralph, Eds. University of Wisconsin Press, Madison, WI, 1986, 287–292.

National Academy of Sciences. *Testing for Effects of Chemicals in Ecosystems.* National Academy Press, Washington, DC, 1981,103.

National Research Council. *Animals as Sentinels of Environmental Health Hazards.* National Research Council, National Academy Press, Washington, DC, 1991, 158.

Novak, E.W. and Schaeffer, D.J. Developing comprehensive field studies to identify subchronic and chronic effects of chemicals on terrestrial ecosystems, in *In Biological Hazards of Environmental Pollutants,* S.S. Sandhu, W.R. Lower, F.J. DeSerres, W.A. Suk, and R.R. Tice, Eds. Plenum Press, New York, 1990, 109–118.

Novak, E.W., Porcella, D.B., Johnson, K.M., Herricks, E.E., and Schaeffer, D.J. Selection of test methods to assess ecological effects of mixed aerosols, *Ecotox. Environ. Saf.* 10:361–381, 1985.

Odum, E.P. *Fundamentals of Ecology.* W.B. Saunders, Philadelphia, 1971, 574.

O'Neill, R.V., DeAngelis, D.L., Waide, J.B., and Allen, T.F.H. *A Hierarchical Concept of Ecosystems.* Princeton University Press, Princeton, NJ, 1986.

Patten, B.C. On the quantitative dominance of indirect effects in ecosystems, in *Analysis of Ecological Systems: State-of-the-Art in Ecological Modeling,* W.K. Lauenroth, G.V. Skogerbee, and M. Flug, Eds. Elsevier, New York, 1983, 27–37.

Perry, J.A., Schaeffer, D.J, Kerster, H.W., and Herricks, E.E. The environmental audit. II. Application to Idaho's water quality monitoring program, *Environ. Manage.* 9:199–208, 1985.

Perry, J.A., Troelstrup, N.H., Jr., Newsom, M.H., and Shelley, B. Whole ecosystem manipulation experiments: the search for generality, *Water Sci. Technol.* 19:55–72, 1987.

Peters, R.H. *The Ecological Implications of Body Size.* Cambridge University Press, Cambridge, England, 1983, 329.

Rapport, D.J., Riegier, H.A., and Hutchinson, T.C. Ecosystem behavior under stress, *Am. Nat.* 125:617–640, 1985.

Risser, P.G., Birney, E.C., Blocker, H.D., May, S.W., Parton, W.J., and Wiens, J.A. *The True Prairie Ecosystem.* Hutchinson Ross, Stroudburg, PA, 1981, 557.

Schaeffer, D.J. A new approach for using short-term tests to screen complex mixtures, *Regulat. Toxicol. Pharmacol.* 7:417–421, 1987.

Schaeffer, D.J. A toxicological perspective on ecosystem characteristics to track sustainable development. VII. Ecosystem health, *Ecotox. Environ. Saf.* 22:225–239, 1991.

Schaeffer, D.J. and Beasley, V.R. Ecosystem health. II. Quantifying and predicting ecosystem effects of toxic chemicals: can mammalian testing be used for lab-to-field and field-to-lab extrapolations?, *Regulat. Toxicol. Pharmacol.* 9:296–311, 1989.

Schaeffer, D.J. and Herricks, E.E. Biological monitors of pollution, in *Handbook of Hazardous Materials,* M. Corn, Ed. Academic Press, San Diego, CA, 1993, in press.

Schaeffer, D.J. and Janardan, K.G. Designing batteries of short-term tests with largest intertier correlation, *Ecotox. Environ. Saf.* 13:316–323, 1987.

Schaeffer, D.J. and Novak, E.W. Integrating epidemiology and epizootiology information in ecotoxicology studies. III. Ecosystem health, *Regulat. Toxicol. Pharmacol.* 16:232–241, 1988.

Schaeffer, D.J., Perry, J.A., Kerster, H.W., and Cox, D.K. The environmental audit. I. Concepts, *Environ. Manage.* 9:191–198, 1985.

Schaeffer, D.J., Herricks, E.E., and Kerster, H.W. Ecosystem health. I. Measuring ecosystem health, *Environ. Manage.* 12:445–455, 1988.

Schaeffer, D.J., Seastedt, T.R., Gibson, D.J., Hartnett, D.C., Hetrick, B.A.D., James, S.W., Kaufman, D.W., Schwab, A.P., Herricks, E.E., and Novak, E.W. Ecosystem health. V. Use of field bioassessments to select test systems for relevant impact assessments or hazard evaluations. training lands in Tallgrass Prairie, *Environ. Manage.* 14:81–93, 1990.

Schaeffer, D.J., Tehseen, W.M., Johnson, L.R., McLaughlin, G.L., Hassan, A.S., Reynolds, H.A., and Hansen, L.G. Cocarcinogenesis between cadmium and Aroclor 1254 in planarians is enhanced by inhibition of glutathione synthesis, *Qual. Assur. Good Prac. Regul. Law* 1:1–11, 1991a.

Schaeffer, D.J., Goehner, M., Grebe, E., Hansen, L.G., Hankenson, K., Herricks, E.E., Matheus, G., Miz, A., Reddy, R., and Trommater, K. Evaluation of the "reference toxicant" addition procedure for testing the toxicity of environmental samples, *Bull. Environ. Contam. Toxicol.* 47(4):540–546, 1991b.

Schaeffer, D.J., Cox, D.K., Herricks, E.E., and Keevin, T.M. The design of NEPA studies: application of the study design assurance process to the Upper Mississippi River navigation impact studies, *Environ. Professional* 15:88–94, 1993.

Schmidt-Bleek, F., Peichl, L., Reiml, D., Behling, G., and Müller, K.W. A concept for detecting unsuspected changes in the environment early, *Regulat. Toxicol. Pharmacol.* 8:308–327, 1988.

Seastedt, T.R. Microarthropod of burned and unburned Tallgrass Prairie, *J. Kansas Entomol. Soc.* 57:468–476, 1984a.

Seastedt, T.R. Belowground macroarthropods of annually burned and unburned Tallgrass Prairie, *Am. Midl. Nat.* 111:405–408, 1984b.

Seastedt, T.R. Canopy Interception of nitrogen in bulk precipitation by annually burned and unburned Tallgrass Prairie, *Oecologia* 66:88–92, 1985.

Slobodkin, L.B. Intellectual problems of applied ecology, *BioScience* 38:337–342, 1988.

Stanford, G. and Smith, S.J. Nitrogen mineralization potential of soils, *Soil Sci. Soc. Am. Proc.* 36:465–472, 1972.

Swift, M.J., Heal, O.W., and Anderson, J.M. *Decomposition in Terrestrial Ecosystems.* Blackwell Scientific Publishers, Oxford, England, 1979, 372.

Tarkpea, M., Hansson, M., and Samuelsson, B. Comparison of the Microtox test with the 96-hr LC50 test for the harpacticoid *Nitocra spinipes, Ecotox. Environ. Saf.* 11:127–143, 1986.

Thomas, J.M., Skalski, J.R., Cline, J.F., McShane, M.C., Simpson, J.C., Miller, W.E., Peterson, S.A., Callahan, C.A., and Green, J.C. Characterization of chemical waste site contamination and determination of its extent using bioassays, *Environ. Toxicol. Chem.* 5:487–501, 1986.

Urban, D.J. and Cook, N.J. Hazard evaluation division standard evaluation procedure ecological risk assessment. EPA 540/9-85-001. U.S. Environmental Protection Agency, Washington, DC, 96, 1986.

U.S. Environmental Protection Agency. Biomonitoring for control of toxic effluent discharges to the marine environment. Report 625/8-89/015. CERI, Technology Transfer, U.S. Environmental Protection Agency, Cincinnati, OH, 58, 1989.

Vaughan, D.S. and Van Winkle, W. Corrected analysis of the ability to detect reductions in year-class strength of the Hudson River white perch *(Morone americana)* population, *Can. J. Fish. Aquat. Sci.* 39:782–785, 1982.

Warren, C.E. Toward classification and rationale for watershed management and stream protection. EPA-600/3-790. U.S. Environmental Protection Agency, Washington, DC, 143, 1979.

White, P.S. Pattern, process and natural disturbance in vegetation, *Bot. Rev.* 45:229–299, 1979.

Wieder, R.K. and Lang, G.E. A critique of analytical methods used in examining decomposition data from litterbags, *Ecology* 63:1636–1642, 1982.

Wiegert, R.G. and Evans, F.C. Investigations of secondary productivity in grasslands, in *Secondary Productivity of Terrestrial Ecosystems,* K. Petruseqicz, Ed. Polish Academy of Science, Krackow, Poland, 1967, 499–515.

Woodwell, G.M. Threshold problems in ecosystems, in *Ecosystem Analysis and Prediction,* S.A. Levin, Ed. SIAM, Philadelphia, 1975, 9–21.

Young, A.L., Cockerham, L.G., and Thalken, C.E. A long-term study of ecosystem contamination with 2,3,7,8-tetrachlorodibenzo-*p*-dioxin, *Chemosphere* 16:1791–1815, 1987.

Human Health Risk Assessment

Norbert P. Page

A. INTRODUCTION

This past decade has seen a dramatic shift in the nature and complexity of risk assessments. Standardized approaches have been developed along with research to improve methodology and reliability. For decades, risk assessors have essentially considered the effects on humans from exposures to individual substances. With the growing concern for the risks of environmental pollution with multiple pathways and media, the risks of effects on ecology as well as humans, and from the risks of exposure to multiple chemicals (especially at hazardous waste sites), risk assessments have reached a complexity not previously known.

Risk assessment and risk management are complex processes which are normally fraught with major uncertainties from the beginning. Nevertheless, they are integral components of our national effort to assure a healthy environment. Risk assessments now involve not only human health but also the effects of environmental pollutants on other environmental organisms, for example, wildlife and aquatic species. This chapter will emphasize human risk assessments and Chapter 22 will discuss ecological effects and assessment. However, the basic concepts are similar for human and ecological assessments and risk management.

The increase in the complexity of risk assessments and the numerous organizations performing them resulted in the use of risk assessment terms in quite differing ways with confusion in terminology and assessment concepts. Two main concepts, *risk* and *hazard* were used in different ways resulting in

0-8493-8851-1/94/$0.00+$.50

confusion to the regulators and public. To bring consistency in the risk assessment process, the National Academy of Sciences published recommended definitions and the main principles of risk assessment and risk management (NRC, 1983). The NAS defines *risk assessment* as "the determination of the probability that an adverse effect will result from a defined exposure" while *risk management* is defined as "the process of weighing policy alternatives and selecting the most appropriate regulatory action based on the results of risk assessment and social, economic, and political concerns."

The risk assessment process was organized into four primary activities:

- *Hazard identification* — characterizes the innate toxic effects of agents
- *Dose response assessment* — characterizes the relationships between doses and incidences of adverse effects in exposed populations
- *Exposure assessment* — measures or estimates the intensity, frequency, and duration of human exposures to agents
- *Risk characterization* — estimates the incidence of health effects under the various conditions of human exposure

The most complex risk assessments are those performed for hazardous waste sites due to the potential for chemical and biological interactions with the simultaneous exposure to many substances. To complicate matters, multiple exposure pathways and media and differing susceptibilities of the exposed populations may be involved. However, any procedure discussed for multiple pollutants could be applied to the assessment of a single pollutant as well, except for the integration of the multiple chemical risks.

The EPA and even state environmental agencies have developed standardized procedures for the routine conduct of risk assessments for environmental pollution, including hazardous waste sites. The EPA procedures serve as the core or guidance for most state procedures. For this reason, the procedures used by the EPA will be discussed here.

The EPA baseline risk assessment is a key component of the Remedial Investigation and Feasibility Study (RI/RS) of a hazardous waste site. It is conducted after the collection, analysis, and validation of waste site monitoring data. Essentially, the risk assessment determines whether contaminants at a site are a current or a future risk to human health or other environmental organisms under a "no action" alternative (i.e., without remediation of the site). The baseline risk assessment thus provides the basis for determining whether remedial actions are necessary. The EPA has provided detailed guidance for the conduct of waste site risk assessments (U.S. EPA 1988a,b, 1989, 1991). The reader is referred to the review by Page and Donahue (1993) for details about the procedures used in the risk assessment of health effects related to toxic wastes and hazardous waste sites.

The EPA, while adopting the NAS process and definitions in principal, has organized the baseline risk assessment for superfund sites into a slightly different operational concept. It also consists of four basic steps:

- Data collection and evaluation
- Exposure assessment
- Toxicity assessment
- Risk characterization

The EPA superfund toxicity assessment incorporates both hazard identification and dose response assessments (U.S. EPA 1989).

B. DATA COLLECTION/DATA EVALUATION

It is in this initial phase that specific data to be used in the baseline risk assessment are collected and analyzed. Since hundreds of substances may be present in a site, often only a subset of those chemicals is normally selected for evaluation in the risk assessment. These chemicals are known as the "chemicals of potential concern". Also, only chemicals that are present as a result of waste disposal activities are evaluated in the baseline risk assessment. To accomplish this, a comparison is made of chemical concentrations detected at the site and the normal background concentrations to determine whether the chemicals that were measured at the site actually resulted from anthropogenic contamination. This is especially true for inorganic chemicals normally present in the soil and some organic chemicals such as polycyclic aromatic hydrocarbons (PAHs). For example, background levels of arsenic, lead, manganese, or PAHs may represent a greater risk than any additional contamination of the site by dumping activities. In such a case, it would not be appropriate to allocate great resources to remediate a hazardous waste site if the primary toxins are natural background toxins. Indeed, remediation may be of questionable long term benefit if the natural sources remain. For each chemical in the Chemicals of Potential Concern list, the percent contribution of risk, known as the risk factor, is estimated using specific concentration data and toxicity criteria provided in the EPA guidelines.

C. EXPOSURE ASSESSMENT

The exposure assessment is an obvious key phase in the risk assessment process since without an exposure toxic chemicals are not hazardous. All potential exposure pathways are carefully considered with an analysis of contaminant releases, their transport and fate, and the population exposed.

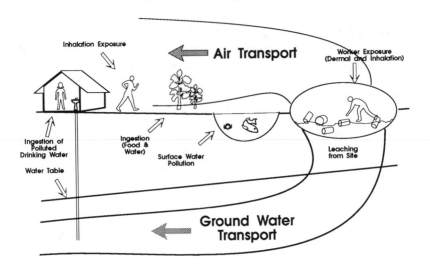

Figure 21.1. Exposure pathways for chemicals in hazardous waste sites (modified from U.S. EPA 1988b).

There are three primary steps:

- Characterization of the exposure setting
- Identification of exposure pathways
- Quantification of the exposure

The populations of concern may be humans, animals, or plants residing on or near the site, workers performing the cleanup, and the public and organisms in the environment some distance from the site that may be exposed to the toxic chemicals following transport away from the site. The exposures may be brief, intermittent, or protracted. Pathways and media of concern for humans include ingestion (soil, water, and contaminated food including fish, wildlife, and plants), inhalation of contaminated air, and dermal exposure (from swimming and soil). All possible exposure scenarios are considered to assess the toxicity and risk that might occur with these differing types of exposure (Figure 21.1). In some situations, such as in the use of a pharmaceutical, the risk assessment may pertain to only one chemical and one route of exposure.

The physical environment and the potentially exposed populations are first determined. The physical environment includes climate, meteorology, geologic setting, vegetation, soil type, ground water hydrology, and location and description of surface water. The assessor determines the location of current populations relative to the site, current and future land uses, and subpopulations of potential concern living on, nearby, or some distance from the site who may be exposed to chemicals that may migrate from the site.

It is assumed that the future land use will be residential. Subpopulations having increased sensitivity or potential for high exposure, such as infants, the elderly, pregnant women, and those with chronic illness, may be sensitive populations and often considered separately. Sources, exposure points, routes of the toxins, and the release and receiving media, along with the fate and movement of substances in the environment, are determined. The exposure pathway describes the movement from the contaminating source until it reaches the exposed populations.

Pollutants may be transported away from the source, be physically, chemically, or biologically transformed, and accumulate in various media. Assessment of the chemical fate requires knowledge of the organic carbon and water partitioning coefficient at equilibrium (K_{oc}), chemical partitioning between soil or sediment and water (K_d), partitioning between air and water (Henry's Law Constant), solubility constants, vapor pressures, octanol/water partition coefficient (K_{ow}), and bioconcentration factors. These factors are integrated with the data on sources, releases, and routes to determine the exposure pathways of importance.

Exposure pathways may include

- Contaminated groundwater — ingestion (drinking water), dermal contact (bathing), and inhalation of volatile organic compounds (showering)
- Surface water and sediments — incidental ingestion and dermal absorption of contaminants, e.g., by children playing in streams or residents swimming in lakes
- Contaminated food — ingestion of contaminated fish tissue by fishermen or the public, vegetables and fruit grown in contaminated soils or covered with contaminated dust, meat, and dairy products
- Surface soils — ingestion and dermal absorption of contaminants by children trespassing at a site
- Fugitive dust and VOC emissions — inhalation by nearby residents or on-site workers
- Subsurface soil and air-borne contaminants — future land-use conditions during construction activities
- Contaminated breast-milk — nursing infants whose mothers were exposed to highly toxic lipophilic contaminants released from a site

The final step in the EPA exposure assessment is quantifying exposure for the identified exposure pathways for the reasonable maximum exposure (RME) case. The RMEs are conservative estimates and, although they are possible, they are normally well above the average expected exposure.

Point concentrations and the average concentrations over the duration of the exposure period are estimated for each chemical of potential concern. The EPA recommends applying a 95th UCL on the arithmetic mean concentration because of the uncertainty associated with available monitoring data. Often this exceeds the maximum detected concentration at the site, which also results in a conservative estimate of exposure and thus risk.

Field data often are not available to estimate the exposure point concentrations, and exposure models may be needed. For example, predicting the air concentrations to downwind residents may require chemical emission and air dispersion models. Residential wells downgradient of a site may show no signs of contamination but may become contaminated in the future by a migrating plume in the groundwater. For those situations, groundwater solute transport models are used to estimate if and when the plume and each chemical of potential concern would reach these wells. The EPA has recommended specific fate and transport models for all media which can be used to estimate exposure point concentrations.

As many exposure pathways are quantified for a 30- or 70-year exposure period, the environmental fate of specific chemicals must be evaluated. For example, since trichloroethylene (TCE), a common pollutant at superfund sites, often undergoes biodegradation in groundwater, it is unrealistic to assume that an individual will be exposed to currently measured groundwater concentrations of TCE over a 70-year period. On the other hand, some TCE degrades to vinyl chloride, a known human carcinogen, and vinyl chloride may actually increase in concentration over time. Thus, assuming exposure to present day levels of vinyl chloride might result in an underestimation of risk.

Finally, exposure point concentrations, expressed as milligrams of contaminant per kilogram of body weight per day (mg/kg/d), are combined using a simple intake equation. Chronic daily intakes (CDIs) are estimated for most pathways in a baseline risk assessment.

The generic equation for calculating chemical intakes is

$$I = C \times \frac{CR \times EF \times ED}{BW} \times \frac{1}{AT}$$

where I is intake (mg/kg/d), C is chemical concentration (mg/l water), CR is contact rate (l/d), EF is exposure frequency (d/year), ED is exposure duration (year), BW is body weight (kg), and AT is averaging time (d).

Variations of this generic formula are used to determine dose of ingested chemicals from a number of sources including contaminated drinking water, swimming water, fish and shellfish, fruits and vegetables, meat, eggs and dairy products, and soil. Similar equations quantify dermal contact with chemicals in water, soil, and sediments, and from the inhalation of airborne chemicals.

Usually site-specific exposure parameter values are not available, in which case default EPA standardized values are used (Table 21.1). Upper bound estimates are used to be consistent with the RME philosophy. For example, the 90th percentile of water ingested per day by a resident (2 l/d) is used for

TABLE 21.1. EPA Reasonable Maximum Exposure Assumptions

Factor	Assumption
Body weights	Adults — 70 kg
	Children (1–6 years) — 16 kg
Water intake	Adults — 2 l/d
	Children — 1 l/d
Inhalation rates (air intake)	Adults — 20 m³/d
	Children — 5 m³/d
Soil ingestion	Children (1–6 years) — 200 mg/d
	Children (>6 years) — 100 mg/d
Food consumption	Fin fish — 54 g/d
	Beef — 0.112 kg/meal
	Eggs — 0.064 kg/meal
Exposure times	Showering — 12 min/d
	Swimming — 2.6 h/d
Exposure frequency	Pathway-specific (typically 350 d/year for ingestion of contaminated food and water and inhalation)
Exposure duration	Pathway-specific (typically 30 years for residential exposure pathways for adults)
Average time of exposure	Carcinogenic chemicals — 70 years (365 d/year)
	Noncarcinogenic chemicals — number of days in the exposure duration

Source: U.S. EPA 1988a, 1989.

estimating exposure by drinking contaminated groundwater from a well 350 d/year for 30 years. An exception to the conservative approach is that median body weights are employed when estimating exposure. The *averaging time* equals the numbers of days in the exposure period when evaluating noncarcinogenic effects and the number of days in a 70-year lifetime when evaluating carcinogenic effects.

Use of upper bound exposure values and maximum detected chemical concentrations may result in unrealistic estimates of exposure, so alternative methods that yield more accurate estimates can be employed. One method is computer modeling, such as Monte Carlo simulation, that considers the variability in chemical data and exposure parameters. A goal is to derive the most realistic upper bound estimates of exposure.

More accurate methods for estimating exposure may also involve pharmacokinetic models which model the behavior of chemicals in the body. One example is the integrated uptake/biokinetic model (IU/BK), a computerized pharmacokinetic model which estimates the blood lead concentrations of children under age seven. In another example of a pharmacokinetic model, exposure of nursing infants to chemicals is estimated.

D. TOXICITY ASSESSMENT

Obviously the most desirable data for human health evaluations are human data. Unfortunately, the few epidemiology studies that are available usually have technical problems with exposure histories being the most uncertain. Nevertheless, these studies are given the highest priority when possible in the dose response assessment. The usual situation is that human studies can provide qualitative evidence but not data adequate for quantitative assessment of environmental exposures and the presence of adverse effects and dose response relationships. In practice, animal data are often used for the quantitative evaluation. This is a valid and accepted procedure since a general premise of toxicology is that effects in laboratory animals are similar to those seen in humans at comparable dose levels. There are some exceptions to this assumption which are attributable to differences in the pharmacokinetics and metabolism of the xenobiotics.

Supporting data such as pharmacokinetic and mechanistic data may allow a comparison of the toxic effect in animals and humans. If defined differences are noted they should be taken into account in the toxicity assessment. Cell studies for cytotoxicity, mutations, and DNA damage can predict a chemical's potential for toxic and biologic activity. Without specific toxicity data, structure/activity studies may help predict toxicity of chemicals based on knowledge of the toxicity of similar chemicals. Structure activity relationships are only generalizations and specific exceptions often occur.

Toxicokinetic analysis can assess the absorption of chemicals into the body, the body's ability to metabolize the chemical, and its storage and elimination. For inhalable materials, particle size is of the utmost importance in assessing potential toxicity. Particles that are 1 to 3 μm in size have the highest probability of being deposited in the lung.

The nature of toxic effects will depend on the intensity and duration of the exposure. Acute effects result from a single or short term exposure, but subchronic effects are observed after exposures of weeks to a few months. Chronic effects are of two types: those that develop after long term, continuous exposure and those that develop with time, such as cancer and nephropathy that have a long latency period. Testing for chronic toxicity in rodents and dogs entails exposure for at least 1 year. For carcinogenicity, the long latency period requires that exposure and observation include the major portion of an animal's lifespan. Thus, animals with a short lifespan, such as rats and mice, are used with the exposure period normally for 2 years or more. Adverse reproductive effects and birth defects are also major concerns for hazardous waste chemicals. The usual test species for developmental and reproductive toxicity testing are rabbits, rats, and mice. Details of toxicity test methods are presented in Chapter 20.

E. DOSE RESPONSE ASSESSMENT

The procedures used by the EPA to quantitatively evaluate risk are quite different for noncarcinogenic and carcinogenic effects. The basis for the different approaches is that noncarcinogenic effects occur after exposure to threshold dose but cancer initiation is not normally considered to have a threshold to elicit a response.

1. Assessment of Noncarcinogenic Effects

Historically, the allowable daily intake (ADI) procedure has been used to calculate permissible chronic exposure levels. The basic concept is that an ADI is determined by applying safety factors (to account for the uncertainty in the quality of the data) to the highest dose in human or animal studies that has been demonstrated not to cause toxicity. The ADI approach has been used extensively by the Food and Drug Administration and World Health Organization and has recently been promoted by the Consumer Product Safety Commission for the evaluation of chronic exposure to products such as art materials that might be hazardous. The EPA has slightly modified the ADI approach for their purposes. For chronic noncarcinogenic effects, the EPA acceptable safety level is known as the *reference dose* (RfD). The RfD is defined as "an estimate of a daily exposure level for the human population, including sensitive subpopulations, that is likely to be without an appreciable risk of deleterious effects during a lifetime."

The critical study used in the calculation of an RfD is one having the lowest exposure level at which a statistical or biological increase in an adverse effect has been demonstrated. The EPA default position is that humans are as equally sensitive as the most sensitive animal species unless scientific evidence is available to decide otherwise. In determining the RfD, the highest dose level which did not produce an effect or the dose which did produce an effect is divided by uncertainty factors to provide a margin of safety for allowable human exposure. The no observed adverse effect level (NOAEL) is the highest dose tested that did not produce an effect. Effects may vary from lethality to minor functional decrements. When a NOAEL is not available, a lowest observed adverse effect level (LOAEL) can be used to calculate the RfD. A dose response for noncarcinogenic effects is illustrated in Figure 21.2 which also identifies the threshold, NOAEL, and LOAEL. Any toxic effect might be used for the NOAEL or LOAEL so long as it is the most sensitive toxic effect and considered likely to occur in humans.

Uncertainty factors of 10 are used to account for the confidence level of the data as indicated in Table 21.2. If a LOAEL must be used, an additional safety factor is included. A modifying factor (0.1 to 10) is also included to allow the risk

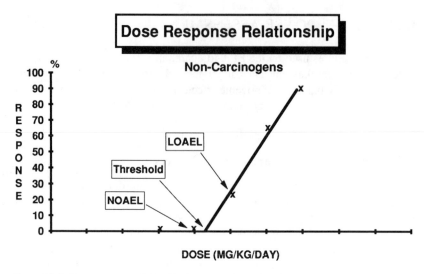

Figure 21.2. Dose response relationship for noncarcinogenic effects.

TABLE 21.2. Safety Factors for Risk Assessment of Noncancer Effects

10	Human variability
10	Extrapolation from animals to humans
10	Use of less than chronic data
10	Use of LOAEL instead of LOAEL
10	Incomplete data base
MF	Modifying factor (>0–10)

assessors to use scientific judgment in upgrading or downgrading the summation safety factor as appropriate. For example, if a particularly well-designed and well-conducted study is the basis for the risk assessment, a modifying factor of less than one may be incorporated. When a poor study must be used in the analysis, an additional factor of greater than one can be incorporated to compensate for the uncertainty associated with the poor quality of the critical study.

The general formula for deriving the RfD is

$$RfD = \frac{NOAEL/LOAEL}{UF_1 \times UF_2 \ldots \times MF}$$

As the data become more uncertain or unreliable, higher safety factors are applied. Two examples of RfD calculations are provided below. In the first, data from a high quality epidemiologic study are available and thus a simple factor of only 10 is used to account for intrahuman variability. For the second calculation, a LOAEL from a less than chronic study was used. The uncertainty factors used in this case are 10 for human variability, 10 for the use of an animal study, 10 for duration of a study that was less than chronic exposure, and 10 for the use of a LOAEL instead of a NOAEL.

1. Critical study is an epidemiology study with a NOAEL of 50 mg/kg/d:

$$RfD = \frac{50 \text{ mg/kg/d}}{10} = 5 \text{ mg/kg/d}$$

2. Critical study is a 90-d study in rats with a LOAEL of 50 mg/kg/d:

$$RfD = \frac{50 \text{ mg/kg/d}}{10 \times 10 \times 10 \times 10} = 0.005 \text{ mg/kg/d}$$

RfDs can be derived for several types of toxic effects, including chronic oral, chronic inhalation, and developmental toxicity. The basic procedures are essentially the same. Data from all toxicity studies are reviewed and a series of dose response relationships that exist for the various toxic effects are used to calculate the RfD. Studies having a long term exposure period are obviously the most reliable to predict chronic effects that might result from extended or lifetime exposure of humans. Without long term data, data from subchronic or acute exposures may also be used. However, as one diverges from the chronic exposure, additional safety factors are incorporated to accommodate the related uncertainties in the data.

Health advisories (HA) may also be used to assess the effects of acute or short term exposures to chemicals in drinking water. HAs are used for guidance of allowable exposures of 1 and 10 d and are calculated like RfDs but use data from studies with durations appropriate to the HA (1 and 10 to 14 d exposures). It is assumed that a 10-kg child drinks 1 l of water per day and a 70-kg adult consumes 2 l of water per day. For calculating a 10-d HA, a 10 to 14-d exposure study is desired.

2. Assessment of Cancer Risks

For cancer risk assessment and to extrapolate from high doses (normally used in animal tests or from occupational exposures) to the lower dose levels to which the general population are usually exposed, mathematical models are used. The carcinogenic evaluation is a two step process: (1) initially a qualitative evaluation of the data (the "weight of evidence" approach), then classification of the evaluated substance and (2) a calculated risk estimate for those substances that are classified as known or probable human carcinogens from the initial step.

TABLE 21.3. EPA Cancer Assessment Categories

Group A — human carcinogen	Sufficient human evidence for causal association between exposure and cancer
Group B1 — probably human carcinogen	Limited evidence in humans
Group B2 — probably human carcinogen	Inadequate evidence in humans, sufficient evidence in animals
Group C — possible human carcinogen	Limited evidence in animals
Group D — not classifiable as to human carcinogenicity	Inadequate evidence in animals
Group E — no evidence of carcinogenicity in humans	At least two adequate animal tests or both epidemiology and animal studies which are negative

In conducting the weight of evidence analysis, all available data are critically reviewed to determine the likelihood that the evaluated agent is a "human carcinogen". It is then assigned a carcinogen classification ranging from A to E as listed in Table 21.3 based on the extent that the substance has been proven to be carcinogenic in animals or humans (or both). Evidence is sufficient in humans on the basis of an epidemiology study that clearly demonstrates a causal relationship between exposure to the substance and cancer in humans. Where there may be alternative explanations for the observed effect, the data are determined to be limited in humans. Sufficient animal evidence consists of an increase in cancer in more than one species or strain of laboratory animals or in more than one experiment. Data from a single experiment can also be considered sufficient animal evidence if there was a high incidence or unusual type of tumor induced. Normally, however, a carcinogenic response in only one species, strain, or study, is considered as only limited animal data.

Agents assigned a classification of category A or B, human or probably human carcinogens, are then subjected to a quantitative risk assessment. For those assigned to category C, possible human carcinogens, the scientific evaluator can decide on a case-by-case basis whether a quantitative assessment is appropriate.

The key risk assessment parameter derived from the carcinogen risk assessment process as used by the EPA is the "slope factor", a toxicity value that quantitatively defines the relationship between dose and response. The slope factor is a plausible upper bound estimate of the probability that an individual will develop cancer if exposure is to a chemical for a lifetime of 70 years. The slope factor is expressed as $mg/kg/d^{-1}$. Figure 21.3 illustrates a dose response relationship for a carcinogen and then demonstrates how a mathematical model extrapolates from observed data to predict risk at low doses. As

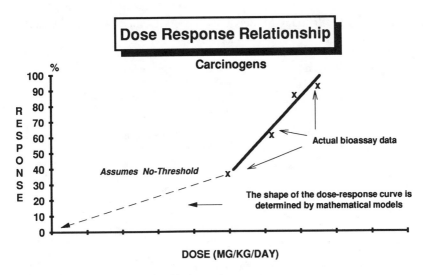

Figure 21.3. Dose response relationship for carcinogens.

illustrated, the low dose risk extrapolation assumes linearity with no threshold for the effect.

Several mathematical models have been used to extrapolate from carcinogenic results observed at high doses to predict risk at low doses. The model chosen is based on an understanding of the mechanism for the carcinogenic response. The EPA default model is the linearized multistage model (LMS). The LMS model yields the slope factor, also known as the q^{1*} (pronounced Q-one-star). Other models that have been used are listed in Table 21.5. Physiologically based pharmacokinetic models (PB-PK) are relatively new and are being used when biological data are available. The PB-PK model quantitates the absorption of a foreign substance and its distribution within the organism and tissue compartments. Some compartments store the chemical (bone and adipose tissue) while others biotransform or eliminate the substance (liver or kidney, respectively). The absorption, distribution, metabolism, and elimination are quantified with mass balance equations. Gargus (1991) has recently reviewed the PB-PK models and PB-PK risk assessments for volatile substances.

F. CONVERSION OF ANIMAL DOSES TO EQUIVALENT HUMAN DOSES

Either the comparative ratio of body weights or differences in surface area can be used to extrapolate effective doses between animals and humans. Comparing dose effectiveness on a relative body weight basis was used at one time (mg chemical/kg body weight). From studies with pharmaceuticals and industrial

TABLE 21.5. Primary Models Used for Assessment of Nonthreshold Effects

Linearized multistage model	Assumes that there are multiple stages for cancer
	Fits curve to the experimental data
	Linear from upper confidence level to zero
One hit model	Assumes there is a single stage for cancer and that one molecular or radiation interaction induces malignant change
	Very conservative
Multihit model	Assumes several interactions needed before cell becomes transformed
	Least conservative model
Probit model	Assumes probit (log-normal) distribution for tolerances of exposed population
	Appropriate for acute toxicity; questionable for cancer
Physiologically based pharmacokinetic models	Incorporates pharmacokinetic and mechanistic data into the extrapolation
	Data rich requirements and, while promising, are currently of limited availability

chemicals it was noted that adjustments based on differences in body weight lacked precision. In fact, the studies by Pinkel (1958) and Freireich et al. (1966) with chemotherapeutic agents suggested that surface area was the most satisfactory means for dose scaling among species, including humans. The human dose equivalent is based on the assumption that different species are equally sensitive to the effects of the substance per unit of body surface area. Since surface area is approximately proportional to the two thirds power of body weight, the equivalent human dose (in mg/d) can be calculated by multiplying the animal dose by the ratio of human to animal body weights (a parameter which can be conveniently measured) raised to the two thirds power. Use of surface area correlations for dose conversion has been under reevaluation since there has not been agreement between the EPA and FDA on the issue. The FDA uses simple body weight conversion parameters rather than surface area correction. Travis and White (1988) evaluated a large group of chemicals and claim that the use of body weight raised to the three fourths power yields a much better fit for dose conversion than the two thirds power of body weight.

Risk can be expressed as concentration of the substance in the environmental medium where human contact can occur. For example, an air unit risk is the risk per mg/m^3 while a water unit risk is risk per mg/l drinking water. These media risk estimates are calculated by dividing the slope factors by 70 kg (average weight of man) and multiplying by 20 m^3/d (average inhalation rate of an adult) or 2 l/d (average water consumption rate of an adult).

Baseline risk assessments of hazardous waste sites use carcinogen slope factors and noncarcinogen reference doses which have been developed by the EPA when they are available. They can be obtained from the Integrated Risk Information System (IRIS), an EPA data base. Without data in IRIS, other EPA or ATSDR documents (Health Effects Assessment Summary Tables [HEAST], criteria documents, and toxicology profiles) can be used.

G. RISK CHARACTERIZATION

The final phase in the risk assessment process is risk characterization. In the risk characterization phase, exposure and toxicity assessments are integrated to yield probabilities of effects occurring in humans exposed under specific exposure conditions. Quantitative risks are calculated for each population exposed for the appropriate media and pathways. The results can then be added to determine estimates of carcinogenic and noncarcinogenic health risks. The risk characterization step quantifies the risks and bridges the gap between risk assessment and risk management.

The quantification of risk is initially conducted for individual chemicals and then summated for the mixtures of chemicals in each pathway. When needed, the risks can be combined across exposure pathways to determine an overall risk to the exposed populations. The methods used to combine chemicals differ for carcinogens and noncarcinogens. Before quantifying the cancer and noncancer risks, the data often must be adjusted for exposure duration, routes of exposure, and absorption.

- Exposure duration — In the estimation of cancer risks, an average lifetime exposure is used since less than lifetime exposures are converted to equivalent lifetime values. For noncarcinogens, short term exposures are not converted to equivalent lifetime values. Without short term toxicity values, the chronic RfD can be used as an initial screening value by comparing the ratio of the short term exposure value to the chronic RfD. If the ratio is less than one, concern for adverse health effects is considered low.
- Exposure routes — Ideally, toxicity values used for assessing each exposure pathway should be derived from studies that employ the same route of exposure, for example, oral toxicity test for assessing human oral exposure. When effects are systemic in nature and when the absorption is comparable for the exposure routes, the extrapolation between exposure routes is acceptable. If toxic effects are localized at the site of exposure, extrapolation from one route to another is not considered scientifically acceptable. For example, when toxicity observed in an inhalation study is strictly of the respiratory system, such data are of questionable value in the assessment of the effects of human oral exposure to the substance.

- Absorption adjustments — To determine the absorbed dose it may be necessary to multiply the exposure dose (dermal or inhalation intakes) by absorption values (percent penetration or absorbed from the lung). Those adjustments may be needed to match the exposure estimate with the toxicity value if one is based on an absorbed dose and the other is based on the ambient concentration. Adjustments may also be necessary to account for different vehicles or media of exposure such as water, food, or soil. As an example, the estimate of risk may be associated with incidental ingestion of contaminated soil. Most RfDs and some slope factors are expressed as the amount of material administered per unit time and unit body weight (exposure dose), whereas exposure estimates for the dermal route of exposure are eventually expressed as absorbed doses. An adjustment may be required in this case. For example, if the RfD based on the exposure dose is found to be 20 mg/kg-d^{-1} and the absorption efficiency is 10%, then the adjusted RfD would be 2 mg/kg-d^{-1} (20 mg/kg-d^{-1} × 0.10).

1. Risks for Individual Substances

Potential carcinogenic risks are expressed as increased probability of developing cancer during a person's lifetime. For example, a 10^6 increased cancer risk represents an increased lifetime risk of 1 in 1,000,000 of developing cancer. The multistage model assumes that the slope factor is linear in the low dose portion of the dose response curve with carcinogenic risk, thus directly proportional to chemical intake in that range. For carcinogenicity, the probability of an individual developing cancer over a lifetime is estimated by multiplying the slope factor (mg/kg-d^{-1}) for the substance by the chronic (70-year average) daily intake (mg/kg/d). This risk is considered a conservative estimate since the upper bound estimate for the slope factor is used, with the "true risk" likely being less. When the calculated carcinogenic risk exceeds 10^{-2}, the following equation may be used to estimate carcinogenic risk: cancer risk = $1 - e^{(-CDI \times SF)}$.

For noncancer effects, the exposure level is compared with a RfD derived for similar exposure periods. The comparison provides a ratio of exposure to toxicity (E/RfD) which is referred to as the "non-cancer hazard quotient". The noncancer hazard quotient assumes that a level of exposure (RfD) exists, below which it is unlikely that sensitive subpopulations will encounter adverse health effects. Three exposure durations are considered: chronic, subchronic, and shorter term. For humans, chronic exposures range from 7 years to a lifetime while subchronic human exposures cover the range of 2 weeks to 7 years. Chronic exposure is normally the greatest concern for exposures associated with hazardous waste sites.

2. Risks from Multiple Substances

People are normally exposed to more than one substance at a time for many industrial and environmental exposures. This is even more likely for hazardous waste sites where many chemicals are dumped into the same site. Procedures for combining the risks of the individual substances to obtain an overall risk from the exposure to several substances simultaneously are used. As a default position, the EPA assumes dose additivity. However, the approaches differ for carcinogenic vs. noncarcinogenic effects. For carcinogens, the risks of the individual substances are assumed to be independent of action and thus the risk estimates are simply added without regard to species (including humans), classification (B or C), tumor type, or mechanism.

In assessing risk for noncarcinogens, a hazard index (HI) is derived. In this case it is assumed that simultaneous subthreshold exposures to several chemicals can result in an adverse effect. It is assumed that the magnitude of the adverse effects will be proportional to the sum of the ratios of the subthreshold exposures to that of acceptable exposures. The HI is simply calculated by the summation of the hazard quotients as previously described, that is, $HI = E1/RfD1 + E2/RfD2 + \ldots + Ei/RfDi$. HIs are calculated for the three usual exposure durations, chronic, subchronic, and shorter term. When the HI exceeds one, there is reason to be concerned for potential health effects.

This is a simplistic approach and the severe limitations and uncertainties are acknowledged by the EPA. RfDs vary in precision and may not be based on the same toxic effect. Thus, combining the hazard quotients to produce the HI is likely based on different effects which have different toxicological significance. The assumption of dose additivity is most acceptable when substances induce the same toxic effect by the same mechanism. The HI approach is simply used as a screening approach. Segregation of hazard indices by toxic effect and mechanism of action may be necessary, especially when many substances are involved and where the HI exceeds one. In those situations, the hazard quotients of the various substances are segregated by toxic effect and affected organ and then added. When information on mechanisms of action and chemical interactions is available it can be used in deriving a more scientific HI.

3. Risks from Multiple Pathways

Individuals are often exposed to substances by more than one exposure pathway, for example, drinking of contaminated water and inhalation of contaminated dust. The total exposure to various chemicals via various routes will equal the sum of the exposures by all pathways. The HI cannot be simply added for all pathways. Initially, the assessor identifies reasonable exposure pathway combinations and then determines whether individuals would consistently face the reasonable maximum exposure by more than one pathway.

For carcinogens, the cancer risks for the same subpopulation can be added for each exposure pathway contributing to exposure, assuming additivity and if necessary correcting for exposure periods. In practice, the noncarcinogen HIs are calculated for the three exposure durations separately. If any of the HIs exceed unity, the assessor may segregate the contributions of the different chemicals according to the major toxic effects.

H. INTEGRATION OF RISK ASSESSMENT AND RISK MANAGEMENT

Risk characterization presents the estimates of potential carcinogen risk and noncarcinogen hazards posed by a hazardous waste site under future land use conditions as well as current uses. The baseline risk assessment does not determine whether the site should be remediated; that is a risk management decision. In many cases, unacceptable risks associated with current land use conditions have been remediated before the risk assessment is performed to reduce the potential for significant exposure. Often, it is the predicted future land use that determines the remediation requirements for superfund sites. In many cases, potential residential use of a site may yield higher risks than current land uses. To encourage consistent risk management decisions, the National Contingency Plan (U.S. EPA, 1990) has issued guidelines that, for known or suspected carcinogens, acceptable exposure levels are concentration levels that represent an excess upper bound lifetime cancer risk of 10^{-4} to 10^{-6}, with a 10^{-6} increased cancer risk as the point of departure. Cancer risks below 10^{-6} usually do not require a remedial response, but risks greater than 10^{-4} may call for remediation or institutional controls to restrict access to the site.

When a site is to be remediated, the risk assessment is used to develop health-based cleanup criteria or remediation goals for the chemicals of concern. Typically, the remediation goals are considered along with various remedial alternatives. The remediation goals are derived using the same assumptions as discussed previously. Toxicity criteria data and exposure parameters are used to calculate the concentration of the chemicals of concern that correspond to a target risk level (e.g., the concentration of chemical A in groundwater that corresponds to 10^{-6} cancer risk from exposure via ingestion, dermal absorption, and inhalation while showering). Areas at the site which require remediation can then be identified by comparing the current chemical concentrations with the derived remediation goals.

I. ECOTOXICITY ASSESSMENT

Ecotoxicity assessments are considerably more complex than the human health assessments due to the need to consider the effects of the hazardous materials on a community of many different species of organisms rather than on only one as is the case for the human assessment. The ecotoxic effects may affect various species differently so there may be a reduction in population size for certain species and changes in the community structure of the ecosystem. Ideally, data from studies of the hazardous substances present in the hazardous waste site on the ecosystem in or around the hazardous waste site would be ideal for the ecotoxicity evaluation. Normally, however, ecotoxicologists study the effects of a single substance on specific populations of one species. In addition, it may be necessary to use data from one species to predict the toxic effects for several species. The data used for the ecotoxicity evaluation should be based on species that are representative or good indicators for the indigenous species.

Many standardized ecotoxicity tests have been developed by the EPA, Organization for Economic Cooperation and Development (OECD), American Society for Testing and Materials (ASTM), and the American Public Health Association (APHA). The test organisms have been chosen on the basis of being representative of a particular genus or class of organism, widely available, amenable to routine maintenance in laboratories, and having a good background of information on the species. When there is a wide array of environmental organisms (the usual situation), it is desirable to have data from tests with several species and from different taxonomic groups to obtain an indication of the natural variability in response. The OECD test guidelines include studies with algae, *Daphnia* (water flea), fish, avian species, earthworms, and terrestrial plants. Several species of fish are used in the aquatic ecotoxicity tests including the guppy, bluegill sunfish, rainbow trout, carp, fathead minnow, sheepshead minnow, red killifish, channel catfish, and zebra fish. The avian species studied include bobwhite quail, ring-necked pheasants, and mallard ducks. Other marine and estuarine species may be studied such as blue crabs, mussels, crayfish, midge larvae, and mysid shrimp.

As can be observed from this discussion, risk assessment and risk management represent complex processes. This is especially so for environmental pollution that may involve many substances and several environmental pathways and routes of exposure. The evaluation and possible remediation of hazardous waste sites represent one of the nation's most intense applications of risk assessment and risk management, and they have had a major impact on the methods used in general for the assessment of toxic substances.

REFERENCES

Freireich, E., Gehan, E., Rall, D., Schmidt, L., and Skipper, H. Quantitative comparison of toxicity of anticancer agents in mouse, rat, hamster, monkey, and man, *Cancer Chemotherapy Rep.* 50(4):219–244, 1966.

Gargas, M. Chemical-specific constants for physiologically-based pharmacokinetic models, *CIIT Activities* 11(3):1–9, 1991.

National Research Council. Risk assessment in the federal government: managing the process. National Academy Press. Washington, DC, 1983.

Page, N. and Donohue, S. Health risk assessment of toxic wastes, in *Effective and Safe Waste Management: Interfacing Sciences and Engineering with Monitoring and Risk Analysis.* R. Jolley and R. Wang, Eds. Lewis Publishers, Inc., Chelsea, MI, 1993, 293–313.

Pinkel, D. The use of body surface area as a criterion of drug dosage in cancer chemotherapy, *Cancer Res.* 18(1):853–856, 1958.

Travis, C. and White, R. Interspecific scaling of toxicity data, *Risk Anal.* 8(1):119–125, 1988.

U.S. Environmental Protection Agency. Guidance for conducting remedial investigations and feasibility studies under CERCLA. Interim Final. OSWER Directive 9355.3-01. Office of Emergency and Remedial Response. U.S. Environmental Protection Agency. Washington, DC, 1988a.

U.S. Environmental Protection Agency. Superfund exposure assessment manual. EPA/540/1-88/001. Office of Remedial Response. U.S. Environmental Protection Agency. Washington, DC, 1988b.

U.S. Environmental Protection Agency. Risk assessment guidance for superfund. Vol. 1. Human health evaluation manual (Part A). Interim Final. EPA/540/1-89-002. Office of Emergency and Remedial Response. U.S. Environmental Protection Agency. Washington, DC, 1989.

U.S. Environmental Protection Agency. National contingency plan, *Federal Register* 55:8666, 1990.

U.S. Environmental Protection Agency. Human health evaluation manual, supplemental guidance: standard default exposure factors. OSWER Directive 9285.6-03. U.S. Environmental Protection Agency. Washington, DC, 1991.

Chapter

22

Ecological Risk Assessment

Donald J. Rodier and Maurice G. Zeeman

A. INTRODUCTION

In this chapter, the basic principles of ecological risk assessment will be discussed as well as the context within which these assessments are used. Ecological risk assessment can be thought of as a logical process which estimates the likelihood that undesirable ecological effects may occur or are occurring as a result of human activities. These activities include the accidental or deliberate release of chemicals into the environment, physical disruptions such as construction, or the alteration of land for agricultural purposes. Although the process is not limited to chemicals, most risk assessments (human and ecological) within many state and federal agencies such as the U.S. Environmental Protection Agency have been oriented towards such substances. Since most of the experience lies with assessing the risks of man-made chemicals such as pesticides and a wide array of industrial chemicals, our discussion will focus on the concepts and procedures used to assess the ecological risks of man-made chemicals, or as they are commonly called xenobiotics. However, the basic concepts of ecological risk assessment are also applicable to nonchemical stressors such as physical habitat alterations and changes in environmental conditions such as water temperature and pH.

Although ecological risk assessments, especially those oriented towards xenobiotics, have been in common use for at least two decades, most state and federal agencies have focused their attention towards human health issues. Recently, a panel of outside experts, the Science Advisory Board (SAB), urged the EPA to give equal emphasis to ecological issues as it had to human health problems (U.S. EPA, 1990). As a result, the EPA is now reestablishing its priorities to address ecological problems (Habicht, 1990; Fisher, 1991). In

0-8493-8851-1/93/$0.00+$.50
© 1993 by CRC Press, Inc.

addition, the EPA is now developing agency-wide guidelines for ecological risk assessment (U.S. EPA, 1992). The National Academy of Science (NAS) is also seeking ways to improve ecological risk assessments. In 1991, the NAS Committee on Risk Assessment Methodology convened a workshop in Airlie, Virginia to discuss ecological risk assessment. Several case studies were presented and discussed (Barnthouse, 1992; Huggett et al., 1992; Kendall, 1992).

We must note that ecological issues and human health issues are not mutually exclusive. Indeed, a healthy environment is essential for mankind. In turn, a healthy environment depends on a society which is cognizant of the influence of man's activities on the environment. In their recommendations to the EPA, the SAB strongly emphasized the need for risk-based priorities which reflect a proper balance between ecological health, human health, and welfare concerns. In making this recommendation, the SAB noted the vital link between ecosystems and human health (U.S. EPA, 1990).

The relationship between the environment and human health is illustrated by the tragedy which occurred to residents who lived in the region of the Minamata Bay in Japan. In the 1930s, a chemical plant began discharging mercury into the Minamata Bay. In the 1950s, cats and pigs in the area began showing abnormal behavior and the condition was described as the disease of the dancing cats. Fish kills in the bay were also reported. Villagers also described how sea birds could be readily captured at sea because the birds were disoriented and often fell into the water. In the 1960s, villagers began showing symptoms such as sensory impairment, hearing loss, ataxia, and speech impairment. Approximately 6% of the infants born near Minamata had cerebral palsy (Eisler, 1987). By 1982, 1800 out of the 200,000 population were diagnosed as having methylmercury poisoning. Over 50 people died and 700 were left permanently paralyzed (Kudo and Miyahara, 1984). The cause of what is now known as Minamata disease is due to methylmercury. It is estimated that over 600 tons of mercury were discharged from the chemical plant between 1932 and 1968 (Doi et al., 1984). The mercury had been converted to methylmercury by aquatic microorganisms (Zakrzewski, 1991). The methylmercury accumulated in fish and then in birds. Fish was a staple resource to the villagers of Minamata Bay. Thus, the ingestion of mercury-contaminated fish eventually led to Minamata disease. This example demonstrates how man's activities altered the environment with detrimental consequences to himself and how organisms in the environment can serve as early monitors of adverse effects.

B. APPLICATIONS OF ECOLOGICAL RISK ASSESSMENTS

Good risk assessments are complex analytical exercises. Risk assessments consist of a variety of analyses based upon real data (when it is available) and numerous assumptions, many of which cannot be verified. The assessment must be completely candid in distinguishing fact from assumptions, and all the uncertainties in the overall assessment must be discussed.

Although ecological risk assessments use standardized scientific methods and procedures, the final product, the risk assessment itself, functions as a practical tool in the regulatory decision making process.

One of the emerging uses of both human health and ecological risk assessments is to set priorities for research and budget needs. As discussed by Travis and Blaylock (1992), environmental problems such as global warming, ozone depletion, endangered species and wetlands, toxic air pollutants, and carcinogens are competing for dwindling financial resources. Risk assessment offers a credible and defensible way of establishing priorities for environmental problems. For example, risk assessments may play a role in deciding which hazardous waste sites should receive the highest priorities for cleanup. Likewise, risk assessments may be used to rank pollutants being released from an industrial plant to determine which pollutants should receive the highest priority and what studies or other regulatory actions are needed. In addition to setting priorities, human health and ecological risk assessments are used on a regular basis to assess the risk of man-made chemicals. For example, ecological risk assessments may be used to determine if a new pesticide is likely to cause adverse effects of such magnitude as to warrant a restriction or ban on its manufacture and use (Urban and Cook, 1986).

Although risk assessment plays a role in the regulatory process, the results of risk assessment may be only one consideration. Many statutes such as the Federal Insecticide Fungicide and Rodenticide Act (FIFRA) and the Toxic Substances Control Act (TSCA) require that considerations of the benefits of a particular chemical to society be considered as well as its risk to humans, the environment, or both. This weighing of risks vs. benefits belongs in the domain of *risk management* (Figure 22.1). Although this chapter deals with only the basic principles of ecological risk assessment, the reader should be aware that the risk assessment itself is only one of the inputs to the regulatory process. Because a high quality risk assessment process is one of the essential scientific components in rational regulatory decisions, the risk assessment process must be conducted in an open manner and subjected to the highest peer review standards, as well as be relatively isolated from outside influences such as politics, economic considerations, etc. Such influences can occur in both directions, since risk assessors often develop a viewpoint on what they feel the appropriate management should be. Scientific objectivity is key in both risk assessment and risk management analyses. For this reason, the National Research Council (NRC) of the National Academy of Sciences recommends that the activities of risk assessment be kept separate from risk management (NRC, 1983). There is a definite need for communication between the risk assessor and the risk manager before the initiation of the risk assessment and upon its completion. This is a crucial issue which will be discussed later. Since the risk assessment plays a significant role in protecting environmental resources, the

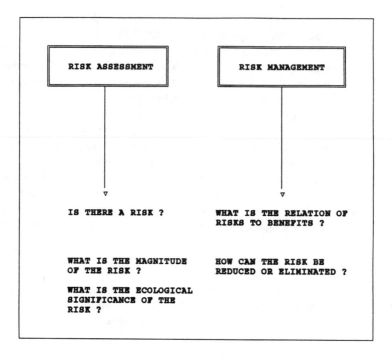

Figure 22.1. The roles of risk assessment and risk management.

resources to be protected must be addressed adequately in the assessment and the results of the risk assessment in turn must be related to the resource of concern. Without this critical linkage, the risk assessment will not serve a useful purpose in the regulatory process.

C. THE SCIENTIFIC DISCIPLINES OF ECOLOGICAL RISK ASSESSMENT

In order for a chemical to pose a risk in the environment, two conditions must be met. First, the chemical has to have an inherent ability to cause some undesirable or hazardous ecological effect. Second, the chemical has to be present in the environment (and exposure must occur) at sufficient levels to cause adverse effects in the organisms of concern.

1. Ecotoxicology

The science that addresses the adverse ecological effects of xenobiotics is known as ecotoxicology. Text books and short reviews of this subject are presented by Connell and Milton (1984), Connell (1986), Rand and Petrocelli

(1985), Ramade (1987), Cairns (1988), Moriarity (1988a,b), Cairns and Mount (1990), and Harris et al. (1990). Unlike human toxicology which assesses effects on one species — man — ecotoxicology is concerned with the effects of chemicals on a variety of organisms found in natural populations, communities, and ecosystems. Because of the natural complex interactions between populations which form communities and the interactions of communities and habitats to form ecosystems, one can appreciate the potential complex effects that a chemical can induce once it is released into the environment. Although the effects of a chemical in a particular environment may be complex, the method of assessing such effects needs to be made in an orderly and logical fashion. Without such an approach, the public or industry may spend a tremendous amount of time and money without achieving a significant insight into the significance of the chemical on ecological components. To help understand the potential complexity and need for an organized approach to assessing the effects of a chemical, we will examine the possible ways a chemical can affect natural populations.

To conceptualize how a chemical might affect a particular population of organisms, a fault tree approach (Barnthouse et al., 1986) is employed and is shown in Figure 22.2. Figure 22.2 shows that toxic chemicals can exert undesirable effects on natural populations by adversely affecting mortality, reproduction, and growth and development. Obviously, if there is high mortality (above the natural death rate) a population may be at risk of declining. If individuals within a population fail to grow and develop, the population may be also be in jeopardy. In addition, if the population is a commercially valuable one, the utility of that population is likely to be reduced or lost. Of course, if individuals within a population fail to reproduce, the population is likely to become extinct.

Accepting that growth and development, mortality, and reproduction are the vital factors for sustaining natural populations, how do toxic chemicals exert their effects on these factors? The fault tree shown in Figure 22.2 subdivides effects on growth and development, mortality, and reproduction into two categories: direct effects and indirect effects. Each category is then subdivided into natural causes and toxic chemicals. The toxic chemicals category is further subdivided into short term effects and long term effects.

The term "direct effects" means there is a direct influence of either toxic chemicals or natural causes on growth and development, mortality, and reproduction. In evaluating the hazard of a chemical, the direct toxic action or direct effects of a particular chemical is evaluated by testing either the species of concern or more commonly by testing substitute or surrogate species.

The term "indirect effects" is not as easy to define, interpret, or measure as are direct effects. Indirect effects are direct effects to other trophic levels upon which a particular population is dependent. Indirect effects include disruptions in the forage/prey and competition/predation relationships of a

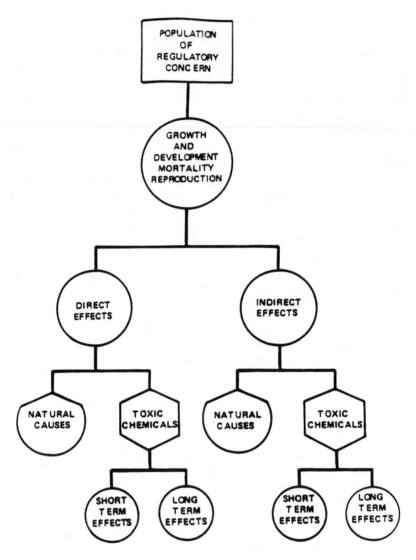

Figure 22.2. Population fault tree.

particular population with other species. Indirect effects may be thought of as food web or ecosystem perturbations. As an example, consider a chemical that is highly toxic to algae (phytoplankton) but less toxic to aquatic invertebrates (zooplankton). If such a chemical entered a lake in sufficient quantities, the phytoplankton populations are likely to be reduced. The zooplankton which graze upon the algae for food are then likely to decrease. Then, the fish which depend upon the zooplankton for food are also less likely to grow and develop.

Mortality in these fish is likely to increase not only because of starvation through a decrease in the food supply but also because they are likely to become more susceptible to predation and disease. Reproductive success in these fish populations is certain to decrease because of reduced food intake.

Both direct and indirect effects are commonly elicited by toxic chemicals. In addition, we know that populations are subject to varying natural cycles. This natural variation is influenced by the environment in which a particular population resides. The term "natural causes" includes the natural variation in population maintenance influenced solely by environmental factors such as climatic variability, availability of food, parasitism, and diseases.

Direct and indirect effects caused by toxic chemicals can be brought about through short or long term exposure to a chemical or direct and indirect exposure (e.g., via food intake) or a combination of these exposure routes. In a full evaluation of the ecotoxicity of a particular chemical, both the short term effects (such as mortality) and long term effects (such as effects on mortality, growth and development, and reproduction) are usually measured.

2. Exposure Assessment

The second prerequisite for ecological risk to occur is that the chemical is (or is likely to be) present at sufficient levels or concentrations to elicit specific adverse effects. Physical/chemical properties such as the volatility and solubility of a chemical will largely determine in which environmental compartments (air, land, water, or biota) a particular chemical will eventually reside. Factors such as molecular weight, solubility, and vapor pressure must therefore be considered when assessing the potential exposure of a chemical to the environment. However, the interactions of chemicals with the environment is much more complex than the relationships between the three factors just mentioned. For instance, the chemical sulfur dioxide is formed during the combustion of coal and readily escapes to the atmosphere, resulting in the formation of smog. One would expect little exposure to the aquatic environment based on simple physical/chemical properties of sulfur dioxide. However, some of the most deleterious effects of acid precipitation are due to sulfur dioxide combining with water to form sulfuric acid which then enters the aquatic environment during periods of rainfall.

In a particular environmental medium a chemical may not be distributed uniformally. In the aquatic environment, the water solubility of the chemical will largely determine whether the compound will remain in the water column, adsorb to suspended solids, or enter sediments. Thus, a wide array of different biota may be exposed due to a single factor such as solubility. In addition, water solubility and molecular weight are important factors which may determine if a chemical is likely to bioconcentrate in organisms and serve as a route of exposure to other organisms which feed or prey upon the contaminated organisms.

In addition to physical and chemical properties which determine where a chemical may partition into the environment, a chemical may be transformed into lesser or more toxic compounds than the parent compound. This can be brought about by chemical reactions such as hydrolysis, oxidation, reduction, or reactivity with other ambient chemicals. Microorganisms can also play a significant role in the transformation of compounds. Many compounds are assimilated and thus their exposure potential is reduced. On the other hand, microbial transformations can result in the exposure of organisms to compounds more toxic than the parent compound. In the earlier discussion of Minamata disease, the metal mercury was shown to be transformed into the highly toxic and bioavailable methylmercury due to microbial action (Kudo and Miyahara, 1984; Zakrzewski, 1991).

All aspects of the potential exposure pathways and processes that a chemical may undergo will not be discussed in this chapter. However, while exposure assessment issues can be complex, they nevertheless have to be addressed if a risk assessment is to be accurate and credible. A concise discussion of the exposure aspects of ecotoxicology is presented by Connell (1986).

D. THE BASIC ELEMENTS OF ECOLOGICAL RISK ASSESSMENT

The elements of risk assessment were first formalized by the National Academy of Sciences National Research Council (NRC, 1983). In addressing risk to human health, the NRC described the risk assessment process as consisting of four steps: (1) hazard identification, (2) dose response assessment, (3) exposure assessment, and (4) risk characterization (Figure 22.3).

Hazard identification entails determining if a particular chemical can elicit adverse human health effects (the chief emphasis at the time was on cancer). Typically, animal assays are used to evaluate the oncogenic potential as well as other human health effects. Dose response assessment (Figure 22.4) presents the relationship between the animal assay data and effects to humans. Because seldom do we experiment with humans, the dose response assessment typically must extrapolate from high dose to low doses and then from animals to humans (NRC, 1983). Exposure assessment evaluates the exposure of a specific chemical to humans. Considerations are given to the significant routes of exposure (inhalation, dermal absorption, dietary intake, or a combination of routes), where the exposure occurred (workplace, home, or both), and the ways in which exposure could be lowered or removed. Risk characterization integrates the dose response assessment and exposure assessment of a particular chemical to describe the nature and magnitude of the human health risk and any uncertainties in the estimates.

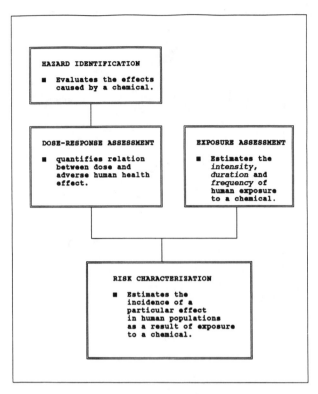

Figure 22.3. The NRC paradigm for human health risk assessment, (NRC, 1983).

The ecological risk assessment process, as now practiced, is conceptually similar to the NRC (1983) paradigm for human health in that both the potential effects and exposure of a chemical are assessed and the results integrated through risk characterization. However, the terminology used in ecological risk assessment has not always been uniform. For instance, a comparison of ecological effects and exposure has been referred to as "hazard assessment" (SETAC, 1987). The term hazard assessment also refers to the characterization of the adverse ecological effects elicited by a chemical (U.S. EPA, 1979). Suter (1990a) distinguishes ecological risk assessment from ecological hazard assessment in that the former yields probabilistic estimates of risk, considers risk/benefit decisions, has explicit endpoints, and can be performed with available data. This separation is not always clearly demarcated, and many assessments dubbed as ecological hazard assessments may in fact be considered risk assessments. Because of this confusion between ecological hazard and risk assessment, the EPA is now developing a framework to help standardize terminology and provide basic principles for ecological risk assessment (U.S. EPA, 1992).

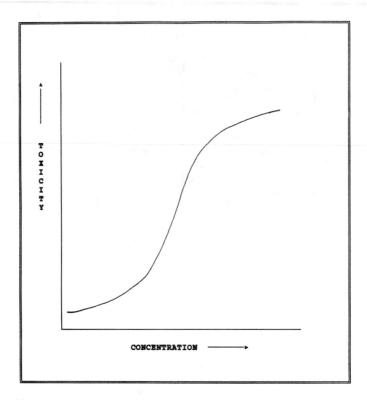

Figure 22.4. Representative dose response curve.

In this chapter, the term hazard assessment will be restricted to the characterization of adverse effects caused by a particular chemical, the term exposure assessment will refer to the process of evaluating the potential exposure of a given chemical in the environment, and risk characterization will refer to the integration of hazard and exposure to yield an estimation of the likelihood of adverse effects. We must note that most ecological risk assessments conducted for chemicals do not yield true probabilistic estimates of risk as is commonly done with human health risk assessments for cancer. Although probabilistic estimates are often lacking, the fact that considerations are given both to potential effects (hazards) and exposure qualifies them to be considered risk assessments. The objective is to reach a level of development for these assessments so probabilistic estimates can be employed. However, regardless of how probabilistic a risk assessment may be, the need for professional judgment is paramount (U.S. EPA, 1992). Thus, ecological risk assessments should be carried out by teams of qualified scientists whose expertise includes biology, toxicology, and the chemical sciences (including physical, organic, inorganic, and biochemistry). Many of the aspects of exposure assessment will require the services of engineers as well. The basic steps of an ecological risk assessment are

1. Define the problem.
2. Obtain the necessary information/data.
3. Assess the hazard potential.
4. Assess the exposure potential.
5. Integrate the hazard and exposure assessment (risk characterization).
6. Summarize and present the results.

1. Defining the Problem

Ecological risk assessments, unlike human health risk assessments, may consider several levels of diverse effects to individuals, populations, communities, and ecosystems. In addition, some of the endpoints used in ecological risk assessments may not be as readily understood to risk managers as the endpoints (e.g., cancer) used in human health risk assessments. As discussed earlier, the ecological risk assessment must yield information that can be used by the risk manager. One of the critical points in the risk assessment process is defining what is to be protected. This is often not an easy task, but if it is not undertaken the assessment can go awry.

Because risk assessments involve identifying what is to be protected as well as using specific measurements in the hazard and exposure assessments, the term "endpoint" by itself can often be ambiguous. Because of this ambiguity, Suter (1990b) proposed that the endpoints for ecological risk assessment be divided into two groups: (1) assessment endpoints and (2) measurement endpoints. Assessment endpoints are the environmental values which are to be protected and are the goals or focuses of an ecological risk assessment. Examples of assessment endpoints include the protection of an endangered species, protection of an economic resource (viable fisheries), or protection of water quality. In selecting assessment endpoints Suter (1990b) recommends the following criteria: (1) social relevance, (2) biological relevance, (3) unambiguous operational definitions, (4) accessibility to prediction and measurements, and (5) susceptibility to the hazard.

The measurement endpoints are those endpoints actually used during the hazard assessment. For instance, toxicity tests on several aquatic organisms may constitute the basic measurement endpoints used for establishing water quality criteria for aquatic life (assessment endpoint). In some instances the assessment and measurement endpoints may be the same. For example, if the assessment endpoint was to maintain a harvestable crab population in the Chesapeake Bay, the measurement endpoints may be effects seen on the blue claw crab itself. If the assessment endpoint was an endangered species, testing of the species would be impractical, and perhaps a closely related or similar species would serve as a surrogate and be the measurement endpoint. In considering the selection of measurement endpoints, Suter (1990b) recommends the following criteria: (1) corresponds to or is predictive of an assessment endpoint, (2) is readily measured, (3) is appropriate to the scale of the disturbance, (4) is appropriate to the route of exposure, (5) contains the appropriate temporal dynamics, (6) has low natural variability, (7) is diagnostic of the effect being measured, (8) can be broadly applicable, (9) is a standard measurement, and (10) contains an existing data set.

2. Gathering the Information/Data

Once the assessment endpoint has been identified, an investigator must then obtain the necessary information and data to conduct the risk assessment. The two basic groups of data needed for an ecological risk assessment are (1) data on hazard and (2) data on exposure.

For assessing the intrinsic hazard of chemicals, laboratory tests are typically used. Because not all species can possibly be tested, a commonly used approach is to use a limited number of surrogate species. This means that test species are selected that are representative of naturally occurring organisms. There are many criteria for selecting surrogate species, but practical considerations such as ease of culture, sensitivity, availability, and having an existing data base dominate (U.S. EPA, 1982). The usual hazard assessment typically uses measurement endpoints such as the adverse effects of a chemical on mortality, growth and development, and reproduction.

In assessing hazard, a tiered testing approach (Bascietto et al., 1990; Maki and Slimak, 1990; Walker, 1990; Zeeman and Gilford, 1993) is usually used. That is, tests which measure effects on mortality are conducted first. Effects of a chemical on mortality are measured chiefly through acute toxicity tests which are typically of short exposure duration. For aquatic toxicity tests they are typically 4 d or less, and for terrestrial toxicity they are typically 5 to 14 d (U.S. EPA, 1985). If the chemical shows high toxicity, additional longer term tests are conducted to determine sublethal effects such as effects on growth and reproduction. Effects on growth and development are measured through subchronic or full chronic tests. For aquatic and terrestrial organisms these measurements can be made with either a partial or full life cycle test. Effects on reproduction are typically measured through chronic tests. For aquatic organisms this could be a full life cycle test. For terrestrial organisms, appropriate reproductive tests such as an avian reproductive test are used.

Although commonly used, single species assays have been criticized for not enabling one to evaluate ecosystem level effects (Cairns, 1986, 1988). However, if appropriate surrogate species are chosen, an investigator may be able to identify potential direct and indirect effects which can be evaluated further. Thus, if an investigator was evaluating the toxic effects of a chemical on aquatic organisms, testing surrogate species which represented trophic levels such as primary producers (phytoplankton), primary consumers (zooplankton), and predators (fish) would yield more complete information on toxicity than simply testing one or two trophic levels. Results could be evaluated further using microcosms or mesocosms to confirm effects which might occur in the natural environment.

In gathering information about exposure, the investigator is chiefly concerned with magnitude, timing, and duration. For chemicals not yet released into the environment, predictive methods (chiefly models) are used. In instances where a contaminant has been identified in the environment, actual monitoring of the chemical might be undertaken. The ideal method would be a combination of modeling and monitoring.

In addition to the magnitude, timing, and duration of the chemical exposure, an investigator also needs to characterize the biota which might be present. The extent to which the biota are identified depends upon the type of risk assessment being conducted. A risk assessment concerned with water quality in a river may not need to know exactly what species are present. However, if specific species were the subject of the assessment, then a knowledge of the geographical distribution and natural history of that species and perhaps other species would be important.

3. Assessing the Hazard

The chief measurement endpoints used in assessing the ecological hazard (and risk) of chemicals are mortality, growth and development, and reproduction.

Effects on mortality are chiefly expressed as the LC_{50} or LD_{50} (the concentration or dose that is calculated to kill 50% of the test animals). In addition to the LC_{50} or LD_{50}, the dose response of the organism to the chemical is extremely important. Depending upon the toxicity and mode of action of a chemical or the susceptibility of a species, a steep or shallow dose response curve may be seen (Figure 22.5). Knowing the slope of a dose response curve is often important in a risk assessment since incremental increases in exposure may have significant or inconsequential effects, depending upon the slope of the curve.

Effects other than mortality can be expressed as an EC_{50}. This is basically an effective concentration for which some effect was calculated to affect 50% of the test organisms. Commonly encountered EC_{50}s are an EC_{50} for growth and an EC_{50} for immobilization (i.e., the organisms appear paralyzed). For all practical purposes, immobilization can be considered identical to mortality since immobilized zooplankton are likely to be readily preyed upon by predators. For sublethal effects, the results are often expressed as a maximum acceptable toxicant concentration (MATC). The number is ascertained by statistically determining the highest no observable effect concentration (NOEC) and the lowest observable effect concentration (LOEC). The MATC is usually calculated as the (geometric) mean of the NOEC and the LOEC.

In many instances, the measurement endpoint used (effects on a surrogate test species) may not correspond to the assessment endpoint of interest. Often,

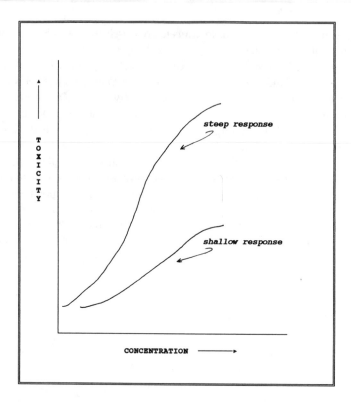

Figure 22.5. Steep and shallow dose response curves.

the investigator must extrapolate effects upon one species to another or must estimate chronic effect data from acute data. Data bases and methods for conducting such extrapolation have been described by Barnthouse et al. (1986, 1990) and Suter et al. (1983). Unfortunately, such data bases have been developed for only a few aquatic organisms including fish and invertebrates.

Sometimes, the investigator may not have any data at all upon which to evaluate the hazard of a chemical. This situation is commonplace in what is now known as the Office of Pollution Prevention and Toxics (OPPT) and was formerly known as the Office of Toxic Substances (OTS), U.S. EPA. Under section 5 of TSCA, manufacturers of new industrial chemicals do not have to initiate ecotoxicity tests for new chemicals (Auer et al., 1990; Nabholz, 1991; Zeeman and Gilford, 1993). Thus, the EPA evaluates the ecotoxicity of thousands of new chemicals using only structure activity relationships (SAR), unless there are previous data on a given chemical and the manufacturer knows that the data exists and submits it to the EPA. SAR entails the use of analogs (chemicals for which there are data) and quantitative structure activity relationships (QSAR — quantitative relationships between toxicity and physical/chemical properties such as molecular weight and the octanol/water partition coefficient). Discussions of this approach are presented by Veith et al. (1984), Lipnick (1989), Auer et al. (1990), and Nabholz (1991). Chemical classes for which such QSARs have been developed were compiled by Clements (1988).

It is obvious from the foregoing discussion that the quantity and quality of data available to assess the hazard of a particular chemical will vary. Because of the problems encountered with the quality and quantity of data, uncertainty factors are often employed to accommodate data gaps. For example, the OPPT commonly uses uncertainty factors which range from 10 to 1000 (U.S. EPA, 1984; Bascietto et al., 1990; Nabholz, 1991). The factors correspond inversely to the type and quantity of data. For example, if an acute LC_{50} was derived from a QSAR and that was the only data available, an uncertainty factor of 1000 would be applied to the LC_{50} to arrive at a concentration of concern which would be used in a risk assessment. If the QSAR estimated acute LC_{50} was 10 mg/l, that figure could be divided by 1000 to yield an effect concentration of 0.01 mg/l. That figure would be used in the risk assessment and would serve as an approximation of the measurement endpoint to determine if there would be any adverse effects as a result of exposure to that chemical. When there are QSAR acute toxicity estimates for three or more suitable species for the same chemical, or if such actual acute toxicity data exists for the chemical, an uncertainty factor of 100 can be used. When adequate chronic data are available (such as adverse effects on growth and reproduction), an uncertainty factor of 10 is employed. The uncertainty factors described are based upon differences between acute and chronic effects, differences in species sensitivity, and differences between effects observed under laboratory conditions and actual field conditions. Discussions on the use of these uncertainty factors are presented by Auer et al. (1990), Nabholz (1991), and Zeeman and Gilford (1993).

4. Assessing Exposure

The exposure assessment consists of determining into which environmental compartment (e.g., air, water, soil, sediment) a chemical may be introduced or occur, estimating or measuring the amounts or concentrations of the chemical in these receiving environment(s) and identifying which biota may be potentially exposed to the chemical.

For assessing chemicals, a knowledge of the life cycle of the chemical is important. That is, a knowledge of the manufacturing, processing, use, and disposal of a particular chemical is needed. Environmental exposure could occur during all phases of the life cycle or be limited to one or two phases such as use or disposal. For example, specialized chemicals such as pesticides are deliberately introduced into the environment to control or mitigate damage caused by agricultural or ornamental pests. Thus, the use pattern of the pesticide will enable one to evaluate where in the environment the pesticide is likely to occur and in what quantities or concentrations. Other chemicals such as surfactants may be released into the environment through disposal once the intended task of the surfactant is completed.

In assessing exposure one needs to evaluate the initial loading rates and environmental fate (including biological and chemical transformations, persistence, solubility, volatility, and photodegradation). The duration and intensity of the exposure are very important. Many chemicals are often discharged daily but others (such as dyes) may be used infrequently. Even infrequently discharged chemicals can be devastating if the exposure coincides with critical life stages of organisms (as in spawning or nesting).

The final product of an exposure assessment is a summary of the estimated environmental concentrations expressed as concentrations (mg/l), or amounts (g/m² or hectare), or dose (mg/kg body weight). In addition, the magnitude and duration of the exposure is also summarized. Finally, the exposure assessment identifies which biota may be exposed to the chemicals. For some chemicals such as pesticides which are applied in specific geographical areas, a precise identification of which biota may be exposed, including endangered species, is possible. For other chemicals which are used in a widespread manner, the identification of biota may be general. For example, surfactants may be discharged to many rivers throughout the U.S. The exposure assessment in this case would be limited to a generic description of the biota which may be affected (fish, aquatic invertebrates, phytoplankton).

5. Risk Characterization

Risk characterization is the process of integrating the ecological hazard and exposure assessment to estimate the likelihood that some adverse effect may occur as a result of exposure to a particular chemical.

The most commonly used method of risk characterization for assessing the ecological effects of chemicals is the quotient or ratio method (Barnthouse et al., 1986, Urban and Cook, 1986; Nabholz, 1991; Rodier and Mauriello, 1993). The quotient method is essentially a "common sense" method whereby a concentration for a particular toxicological endpoint is compared to an estimated environmental concentration for the same chemical:

$$\frac{\text{estimated environmental concentration}}{\text{toxicological endpoint concentration}} = \text{quotient}$$

As the quotient approaches or exceeds one or more, a risk of the toxicological endpoint manifesting itself in the environment is inferred. Naturally, if the quotient is less than 1, less of a risk is inferred.

The following is a simple example of the quotient method:
A particular chemical was tested by conducting a 96-h fish acute test using the fathead minnow *(Pimephales promelas)*. The 96-h LC_{50} was determined to be

0.025 mg/l. The estimated environmental concentration was 0.05 mg/l. The quotient method would be used as follows:

$$\frac{\text{estimated environmental concentration}}{\text{96-h LC}_{50}} = \frac{0.05 \text{ mg} / 1}{0.025 \text{ mg} / 1} = 2$$

In this particular case the quotient was 2, which indicates a very high risk of a 50% mortality to fish. A quotient of 1 would have also indicated a risk of 50% mortality to fish. Had the quotient been <1, less risk of a 50% mortality would be inferred.

In the above example, the LC_{50} was directly compared to the exposure concentration. There was no provision or allowance for the uncertainty regarding that acute value and how it would relate to "real world" conditions. The hypothetical example below demonstrates how a risk assessment might be conducted on a chemical for which there was some toxicity data.

We will assume that the chemical is a component used to manufacture lubricants. The following aquatic ecotoxicity test data was available for the compound:

- 96-h rainbow trout *(Oncorhynchus mykiss)* LC_{50} = 0.3 mg/l
- 48-h aquatic invertebrate *(Daphnia magna)* EC_{50} = 0.1 mg/l
- 96-h algal *(Selenastrum capricornutum)* EC_{50} = 0.5 mg/l

Information about the relative water solubility of the compound was supplied also. This information consisted of the logarithm of the octanol/water partition (log K_{ow} or log P as it is commonly called). The octanol/water partition coefficient is a measurement of the propensity of a particular chemical to partition into either *n*-octanol or water when the chemical is introduced into a mixture of the two (Walker, 1990). Thus, highly water soluble compounds will partition into the water while chemicals with low water solubility will partition into the *n*-octanol. The significance of this partitioning is that, in the environment, compounds with low water solubility are likely to accumulate in adipose (lipid) tissue of living organisms. The compound may accumulate until toxic levels are reached and kill the organism, or the compound could accumulate in the lipid tissue of a particular organism and not be lethal to that organism. However, other organisms which may feed upon that particular organism will then be exposed to that compound. Although the log K_{ow} can actually be measured in the laboratory, it is frequently estimated (Veith et al., 1979). In addition to estimating the bioconcentration potential of industrial chemicals, the log K_{ow} is also useful for estimating the toxicity of many compounds (Veith et al., 1984; Clements, 1988; Auer et al., 1990).

In this hypothetical example, the log K_{ow} was determined to be 2.8 (the antilog of the ratio of the solubility of the chemical in octanol compared to

TABLE 22.1. Estimated Environmental Concentrations Using the Simple Dilution Model for Mean and Low Flow Conditions

	Concentration (μg/l)	
Percentile	Mean Streamflow	Low Streamflow
10	2	300
20	0.6	100.0
30	0.2	33.0
40	0.13	4.0
50	0.055	1.0
60	0.01	0.3
70	0.003	0.06
80	0.001	0.0015
90	0.0004	0.0006

water is 630), indicating that the compound is not likely to bioconcentrate significantly. Typically, compounds with a log K_{ow} of 3.5 (the antilog of ratio of solubility in octanol compared to water is 3162) are considered to have a high bioconcentration potential (U.S. EPA, 1983).

In evaluating the environmental exposure of the chemical, it was determined that approximately 200,000 kg of the chemical would be processed annually at various lubricant manufacturing sites throughout the U.S. During the processing, about 2 kg/d of the chemical were estimated to be discharged directly to receiving streams. A simple stream flow dilution model was used to estimate the exposure concentrations of the chemical in the receiving streams. The dilution model simply divides the amount of chemical released (in kg/d) by the stream flow (in liters of water/day) for a particular river. Stream flow data were obtained for the various rivers where lubricant plants were located. The final concentrations are typically low and the units are expressed as mg/l (ppm) or μg/l (ppb). The chemical is assumed to mix immediately with the water and no losses to occur due to volatilization or partitioning. Thus, this represents a conservative or "worst case" exposure scenario. Calculations are normally performed for an estimated mean stream flow and a low stream flow (the latter is also referred to as a 7Q10 which is the lowest 7-d stream flow average expected during a 10-year period). The results of the calculations, shown in Table 22.1, are expressed as percentiles of known flow rates of specific stream reaches in the U.S. The percentiles identify the estimated concentrations based on such flow rates. Thus, streams with the lower flow rates are expected to have estimated higher concentrations than those with higher flow rates. This example was chosen to demonstrate one method of estimating concentrations of a chemical in streams. In real life, chemicals are not usually directly discharged to receiving bodies of water. Normally, chemicals are processed through publicly owned (waste water) treatment works (POTWs). In estimating the concentrations of a chemical before and after passage through a POTW, considerations are given to the volatility of the chemical, its adsorptive properties (through use of the K_{ow}), and biological half-life.

Because the hazard data for this chemical was limited, an uncertainty factor of 100 was applied to the most sensitive species, *Daphnia magna* (EC_{50} = 0.1 mg/l or 100 ppb) to arrive at a concern level of 0.001 mg/l (0.001 ppm) or 1 µg/l (1 ppb) (0.1 mg/l ÷ 100). The uncertainty factor of 100 accounted for uncertainty due to acute-to-chronic effects and differences between laboratory and field effects.

The concern level for the lubricant additive would thus be 1 µg/l. Using the quotient method, it is obvious that under mean flow conditions in the 10th percentile, the quotient of the exposure concentration divided by the concern level is 2. Under low stream flow conditions in the 10th to 50th percentiles, the quotient of the exposure concentration divided by the concern level ranges from 1 to 300 and the concern level is clearly exceeded. The relationship of the risk of exceeding the concern level to the assessment endpoint (risk to aquatic life) is discussed below.

6. Summarizing and Presenting the Results

The foregoing example was simplified to demonstrate the basic principles of an ecological risk assessment. Most investigators would agree that more information would have been desirable to evaluate more fully the hazards and risks of the chemical. However, with the available information, how would an investigator clearly communicate the results of the ecological risk assessment to a risk manager?

As discussed under Section D1 (Defining the Problem), there has to be some goal or environmental value to be protected. Without this, the risk assessment would not have been performed in the first place. In the hypothetical example of the lubricant additive, the chemical could have been evaluated under TSCA, which strives to protect man and the environment from unreasonable risks. The lubricant additive was being discharged to rivers from various processing plants. The assessment endpoint can therefore be thought of as protection of the aquatic life in those rivers from unreasonable risks. The three surrogate species upon which the hazard assessment was based and the effects that were measured can be thought of as the measurement endpoints. The three species, the rainbow trout, the aquatic invertebrate, and the alga, therefore served as surrogates for the aquatic life in rivers.

In presenting the results of the risk assessment, the assessor discusses the high toxicity of the compound (compounds with a LC_{50} or EC_{50} below 1 mg/l [1 ppm] are typically considered as highly toxic). The investigator would also point out that the compound is toxic to a wide array of aquatic life: primary producers (as represented by *S. capricornutum),* primary consumers (as represented by *D. magna),* and top carnivores (as represented by *O. mykiss).* The no observed effect concentration (NOEC) for this compound was not determined by this testing. Thus, there is some uncertainty since the only test data consisted of short term assays with mortality as the primary effect. Since only an EC_{50} was available, we do not know the no effect concentration for any of the surrogate species. In addition, experience has shown that compounds as toxic as this one are likely to adversely affect sublethal parameters such as growth and development and reproduction at even lower concentrations.

The investigator is likely to conclude that the lubricant additive presents a distinct risk to the aquatic life in certain rivers (the assessment endpoint). The assumptions used in this case would be (1) the aquatic organisms present in the rivers would be as sensitive as the surrogate species, (2) the lubricant additive would be present long enough and at sufficient concentrations to elicit the measured adverse effects, and (3) the aquatic organisms would come in contact with the additive. Of course there are uncertainties in the overall risk assessment: (1) the organisms in the rivers may be more *or* less sensitive than the surrogate species and (2) the duration of exposure may not be sufficient enough to elicit any adverse effects.

Because the investigator is not sure how toxic the chemical actually is, a recommendation might be to obtain additional test data; specifically a fish early life stage test and a daphnid reproduction test. The fish early life stage would permit one to evaluate the effects of the chemical on the larval stages of fish (typically the most sensitive life stage) with regard to growth and development, and the daphnid reproduction test would yield an estimate of the effects of the chemical on the reproductive capacity of aquatic invertebrates.

Keep in mind that the above risk assessment identified a risk to aquatic life. It did not identify an "unreasonable" risk. Such a decision would be made by the risk manager who would weigh the economic or social benefits of the chemical along with its risk. Control measures such as requiring disposal through a waste water treatment plant or otherwise limiting or even prohibiting discharge to water until additional testing is obtained could also be considered by the risk manager.

The important point of this example is that the results of the risk assessment have to be clearly explained in terms of the assessment endpoint. The severity of the risk also has to be identified. In this case we identified the severity in a semiquantitative fashion by classifying the chemical as highly toxic. The uncertainties in the assessment also have to be listed. In this instance, a discussion regarding the sensitivity of surrogate species compared to naturally occurring species and the duration of exposure was included. The use of uncertainty factors to account for acute-to-chronic effects and differences in laboratory to field sensitivity was also included.

If the risk manager decides to obtain additional information, the risk assessment process begins anew. New data results in a new risk assessment and the risks are then reevaluated.

E. DISCUSSION AND CONCLUSIONS

The foregoing example of the lubricant additive was necessarily short and simple. In reality, additional data and methodologies are considered and the reader is referred to Auer et al. (1990) and Nabholz (1991) for a more detailed discussion. However, the example risk assessment does represent a real and commonly encountered situation where there is minimum initial information upon which to base the hazards and ecological risks of new chemicals. The important points to remember are

1. For a risk to occur, a chemical has to elicit certain toxic effects and either be present in the environment or estimated to be present in the environment in sufficient concentrations to cause the adverse effect. Thus, risk is a function of hazard and exposure.
2. To be meaningful, a risk assessment must have an environmental value that is to be protected, an assessment endpoint, and the methods used and the measurement endpoints must be relatable to that assessment endpoint.
3. A risk assessment can be effective even if semiquantitative techniques such as the quotient method are used.
4. The results of the risk assessment must be communicated clearly to the risk manager.
5. Ecological risk assessments are often iterative. Regardless of the initial information used, when additional information is obtained, the process is repeated.

F. SUMMARY

Environmental problems are receiving increasing attention in state and federal agencies. Ecological risk assessments can play an important role in establishing overall environmental priorities as well as evaluating individual problems. Risk is basically a function of hazard and exposure. To be effective in the regulatory process an ecological risk assessment must establish a clear relationship between the measurements which were used to evaluate risk and the environmental value that is to be protected.

DISCLAIMER

This document has been reviewed by the Office of Pollution Prevention and Toxics, U.S. EPA and approved for publication. Approval does not signify that the contents necessarily reflect the views and policies of the agency nor does mention of trade names or commercial products constitute endorsement or recommendation for use.

ADDENDUM

After this manuscript had been submitted to the editors, the EPA published a *Framework for Ecological Risk Assessment*. This report (EPA/630/R-92/001), prepared under the direction of the Risk Assessment Forum, outlines a basic and flexible process for conducting and evaluating ecological risk assessments within the EPA. The term "hazard assessment" as used in this chapter has now been replaced with the term "characterization of ecological effects". The term "exposure assessment" has been replaced with "characterization of exposure".

REFERENCES

Auer, C.M., Nabholz, J.V., and Baetcke, K.P. Mode of action and the assessment of chemical hazards in the presence of limited data: use of structure activity relationships (SAR) under TSCA Section 5, *Environ. Hlth. Perspect.* 87:183–197, 1990.

Barnthouse, L.W., Suter, G.W., II, Bartell, S.M., Beauchamp, J.J., Gardner R.H., Linder, E., O'Neill, R.V., and Rosen, A.E. User's manual for ecological risk assessment. ORNL Publication No. 2679. Oak Ridge National Laboratory (ORNL), Oak Ridge, TN, 1986.

Barnthouse, L.W., Suter, G.W., and Rosen, A.E. Risks of toxic contaminants to exploited fish populations: influence of life history, data uncertainty and exploitation intensity, *Environ. Toxicol. Chem.* 9:297–311, 1990.

Barnthouse, L.W. Case studies in ecological risk assessment, *Environ. Sci. Technol.* 26(2):230–231, 1992.

Bascietto, J., Hinckley, D., Plafkin, J., and Slimak, M. Ecotoxicity and ecological risk assessment: regulatory applications at EPA, *Environ. Sci. Technol.* 24(1):10–14, 1990.

Cairns, J. The myth of the most sensitive species, *BioScience* 36(10):670–672, 1986.

Cairns, J. Putting the eco in ecotoxicology, *Reg. Toxicol. Pharmacol.* 8:226–238, 1988.

Cairns, J.C. and Mount, D.I. Aquatic toxicology, *Environ. Sci. Technol.* 24(2):154–161, 1990.

Clements, R.G. (Ed.) Estimating toxicity of industrial chemicals to aquatic organisms using structure activity relationships. Office of Toxic Substances, U.S. Environmental Protection Agency, Washington, DC, 1988.

Connell, D.W. and Milton, G.J. *Chemistry and Ecotoxicology of Pollution.* John Wiley & Sons, New York. 1984, 444.

Connell, D. Ecotoxicology — a new approach to understanding hazardous chemicals in the environment, *Search* 17:27–31, 1986.

Doi, R., Ohno, H., and Harada, M. Mercury in feathers of wild birds from the mercury polluted area along the shore of the Shiranui Sea, Japan, *Sci. Total Environ.* 40:155–167, 1984.

Eisler, R. Mercury hazards to fish, wildlife, and invertebrates: a synoptic review. Biological Report 85(1.10). Fish and Wildlife Service, U.S. Department of Interior, Laurel, MD, 1987.

Fisher, L.J. Course setting at the Environmental Protection Agency, *Environ. Carcin. Rev. (J. Environ. Sci. Hlth.) C*(8)2:205–213, 1991.

Habicht, F.H. Strategies for meeting our goals, *EPA J.* 16(5):8–11, 1990.

Harris, H.J., Sager, P.E., Regier, H.A., and Francis, G.R. Ecotoxicology and ecosystem integrity: the Great Lakes examined, *Environ. Sci. Technol.* 24(5):598–603, 1990.

Hoffman, D.J., Rattner, B.A., and Hall, R.J. Wildlife toxicology, *Environ. Sci. Technol.* 24(3):276–283, 1990.

Huggett, R.J., Unger, M.A., Seligman, P.F., and Valkirs, A.O. The marine biocide tributylin, *Environ. Sci. Technol.* 26(2):233–237, 1992.

Kendall, R.J. Farming with agrochemicals, the response of wildlife, *Environ. Sci. Technol.* 26(2):239–245, 1992.

Kudo, A. and Miyahara, S. Mercury dispersion from Minamata Bay to the Yatsushiro Sea during 1975–1980, *Ecotoxicol. Environ. Safety* 8:507–510, 1984.

Lipnick, R.L. Narcosis, electrophile, and proelectrophile toxicity mechanisms: application of SAR and QSAR, *Environ. Toxicol. Chem.* 8:1–12, 1989.

Maki, A.W. and Slimak, M.W. The role of ecological risk assessment in decision making, in *Ecological Risks — Perspectives from Poland and the U.S.* W. Gradzinski, E.B. Cowling, and A.I. Breymeyer, Eds. National Academy of Science, Washington, DC, 1990, 77–87.

Moriarity, F. Ecotoxicology, *Human Toxicol.* 7:437–441, 1988a.

Moriarity, F. *Ecotoxicology — The Study of Pollutants in Ecosystems*, 2nd ed. Academic Press, New York. 1988b, 289.

Nabholz, J.V. Environmental hazard and risk assessment under the United States Toxic Substances Control Act, *Sci. Total Environ.* 109/110:649–665, 1991.

National Research Council. *Risk Assessment in the Federal Government: Managing the Process.* National Academy Press, Washington, DC. 1983, 189.

Ramade, F. *Ecotoxicology.* John Wiley & Sons, New York. 1987, 262.

Rand, G.M. and Petrocelli, S.R. *Fundamentals of Aquatic Toxicology.* Hemisphere Publishing Corporation. Washington, DC, 1985, 666.

Rodier, D.J. and Mauriello, D. A. The quotient method of ecological risk assessment under TSCA: a review, in *Environmental Toxicology and Risk Assessment*, Vol. 1. ASTM STP 1179. J. Hughes, W. Landis, and M. Lewis, Eds. American Society for Testing and Materials, Philadelphia, 1993, 80–91.

SETAC. Research priorities in environmental risk assessment. Report of a workshop held in Breckenridge, Colorado. Society of Environmental Toxicology and Chemistry, Rockville, MD. 1987, 103.

Suter, G.W., II. Environmental risk assessment/environmental hazard assessment: similarities and differences, in *Aquatic Toxicology and Risk Assessment,* Vol. 13. ASTM STP 1096. W.G. Landis and W.H. van der Schalie, Eds. American Society for Testing and Materials, Philadelphia, 1990a, 5–15.

Suter, G.W., II. Endpoints for regional ecological risk assessments, *Environ. Manage.* 14:9–23, 1990b.

Suter, G.W., II, Vaughan, D.S., and Gardner, R.H. Risk assessment by analysis of extrapolation error, a demonstration for effects of pollutants on fish, *Environ. Toxicol. Chem.* 2:369–378, 1983.

Travis, C.C. and Blaylock, B.P. Setting priorities for environmental policy, *Environ. Sci. Technol.* 26(2):215, 1992.

Urban D.J. and Cook, N. Ecological risk assessment. EPA 540/9-85-001. Office of Pesticide Programs, U.S. Environmental Protection Agency, Washington, DC, 1986.

U.S. Environmental Protection Agency. Toxic substances control act. Discussion of premanufacture test policies and technical issues; request for comment, *Fed. Reg.* 44:16240–16292, 1979.

U.S. Environmental Protection Agency. Surrogate species workshop. Office of Toxic Substances, Washington, DC, 1982.

U.S. Environmental Protection Agency. Testing for environmental effects under the Toxic Substances Control Act. Office of Toxic Substances, Washington, DC, 1983.

U.S. Environmental Protection Agency. Estimating "concern levels" for concentrations of chemicals in the environment. Office of Toxic Substances, Washington, DC, 1984.

U.S. Environmental Protection Agency. Toxic Substances Control Act test guidelines, final rules, *Fed. Reg.* 50(188):39252–39516, 1985.

U.S. Environmental Protection Agency. Reducing risk: setting priorities and strategies for environmental protection. SAB-EC-90-021. Science Advisory Board, U.S. Environmental Protection Agency, Washington, DC, 1990.

U.S. Environmental Protection Agency. Peer review workshop report on a framework for ecological risk assessment. EPA/625/3-91/022. Risk Assessment Forum, U.S. Environmental Protection Agency, Washington, DC, 1992.

Veith, G.D., De Foe, D.L., and Bergstedt, B.V. Measuring and estimating the bioconcentration factor of chemicals in fish, *J. Fish. Res. Board Can.* 36:1040–1048, 1979.

Veith, G.D., De Foe, D.L., and Knuth, M. Structure-activity relationships for screening organic chemicals for potential ecotoxicity effects. *Drug Metab. Rev.* 15(7):1295–1303, 1984.

Walker, J.D. Chemical fate, bioconcentration, and environmental effects testing and decision criteria, *Tox. Assess.* 5:103–134, 1990.

Zakrzewski, S.F. *Principles of Environmental Toxicology.* American Chemical Society, Washington, DC, 1991, 270.

Zeeman, M. and Gilford, J. Ecological hazard evaluation and risk assessment under EPA's Toxic Substances Control Act: an introduction, in *Environmental Toxicology and Risk Assessment*, Vol. 1, ASTM STP 1179, J. Hughes, W. Landis, and M. Lewis, Eds.. American Society for Testing and Materials, Philadelphia, 1993, 7–21.

Index

A

AA, *see* atomic absorption spectrometry
abiotic characteristics, of water, 360
abiotic factors, altering of, 8
abnormalities, observations of, 393
above ground biomass, 548
Abrus precatorius, 272
absorption, 49
acetate, 67
acetic acid, 172
acetone, 171
2-acetylaminofluorene, 460
acetylation, 372
acetylcholinesterase, 16
acetyl transferases, 93
ACGIH, *see* American Conference of
        Governmental Industrial Hygienists
acid rain, 9, 361
acids, 337
aconitase, 15
acrylonitrile, 460
actinium, 236
active transport, 22, 24
acute delayed neurotoxicity test, 474
acute exposure, 18, 550
acute radiation sickness, 254
acute radiation syndrome (ARS), 254
acute response, 532
acute toxicity, 4, 189, 471, 531, 551
acylation, 64
additive effects, production of, 412
additive response, 16
S-adenosyl methionine (SAM), 67–68
adenylate energy charge, 396
ADI, *see* allowable daily intake
adrenal, 222, 225
adsorption, of gases to solids, 30
AE, *see* atomic emission spectroscopy
aflatoxins, 278, 280
Agkistrodon
    contortrix, 264
    mokasen, 265

piscivorus, 265
agricultural runoff, 387
AHH, *see* aryl hydrocarbon hydroxylase
Ah receptor, 218, 220, 221, 226
air pollution, toxicity of, 287–319
    air pollution regulations and standards,
        293–297
        national ambient air quality standards,
            293–294
        national emissions standards for
            hazardous air pollutants, 294–295
        new Clean Air Act amendments, 295–
            297
            acid rain, 296
            air toxics, 295–296
            chlorofluorocarbons, 296–297
    biological effects of air pollution, 297–
        310
        air toxics, 308–310
        criteria pollutants, 298–308
            carbon monoxide, 302–304
            lead, 307–308
            nitrogen dioxide, 300–302
            oxidants, 298–300
            particulate matter, 304–306
            sulfur dioxide, 306–307
    difficulties in study of, 291–293
    historical perspective, 287–288
    primary emissions and secondary
        pollutants, 288–291
    principles of risk assessment for inhaled
        chemicals, 310–312
        dose response assessment, 311
        hazard identification, 310–311
        human exposure evaluation, 311
        risk characterization, 312
ALA, *see* aminolevulinic acid
ALAD, *see* aminolevulinic acid dehydroge-
        nase
ALAS, *see* aminolevulinic acid synthetase
alcohol dehydrogenase, 52, 59, 66, 172
aldrin epoxidase, 83, 84
aldrin hydroxylase, 82

E

O